BIOLOGICAL AND MEDICAL PHYSICS, BIOMEDICAL ENGINEERING

BIOLOGICAL AND MEDICAL PHYSICS, BIOMEDICAL ENGINEERING

The fields of biological and medical physics and biomedical engineering are broad, multidisciplinary and dynamic. They lie at the crossroads of frontier research in physics, biology, chemistry, and medicine. The Biological and Medical Physics, Biomedical Engineering Series is intended to be comprehensive, covering a broad range of topics important to the study of the physical, chemical and biological sciences. Its goal is to provide scientists and engineers with textbooks, monographs, and reference works to address the growing need for information.

Books in the series emphasize established and emergent areas of science includingmolecular,membrane, and mathematical biophysics; photosynthetic energy harvesting and conversion; information processing; physical principles of genetics; sensory communications; automata networks, neural networks, and cellular automata. Equally important will be coverage of applied aspects of biological and medical physics and biomedical engineering such as molecular electronic components and devices, biosensors, medicine, imaging, physical principles of renewable energy production, advanced prostheses, and environmental control and engineering.

For other titles published in this series, go to www.springer.com/series/3740

Mark C. Williams · L. James Maher, III
Editors

Biophysics of DNA-Protein Interactions

From Single Molecules to Biological Systems

Springer

Editors
Mark C. Williams
Department of Physics
Northeastern University
Boston, MA
USA
mark@neu.edu

L. James Maher, III
Department of Biochemistry
and Molecular Biology
Mayo Clinic College of Medicine
Rochester, MN
USA
maher@mayo.edu

ISBN 978-0-387-92807-4 e-ISBN 978-0-387-92808-1
DOI 10.1007/978-0-387-92808-1
Springer New York Dordrecht Heidelberg London

Library of Congress Control Number: 2010935301

Printed on acid-free paper

Springer is part of Springer Science+Business Media (www.springer.com)

Contents

Part II Complex Protein-DNA Interactions

Contributors

Claire A. Adams
Department of Microbiology and Immunology,
Center for Structural Biology, University of Kentucky, Lexington,
KY 40536, USA

Sankar Adhya
Laboratory of Molecular Biology, NCI, NIH, 37 Convent Drive, Bethesda,
MD 20892-4264, USA

Anjum Ansari
Physics Department, M/C 273, University of Illinois at Chicago,
845 W. Taylor St., Chicago, IL, 60607 USA

R.H. Austin
Department of Physics, Princeton University, Princeton, NJ, 08544

Daniel Barsky
Physical and Life Sciences Directorate, Lawrence Livermore National Laboratory,
7000 East Avenue, Livermore, CA 94550, USA

Penny J. Beuning
Department of Chemistry & Chemical Biology, Center for Interdisciplinary,
Research on Complex Systems, Northeastern University,
360 Huntington Ave 102 Hurtig Hall, Boston, MA 02115, USA

Bram van den Broek
Physics of Life Processes, Leiden Institute of Physics, Leiden University,
Niels Bohrweg 2, 2333 CA, Leiden, The Netherlands

David Dunlap
Cell Biology Department, Emory University, 615 Michael St.,
Atlanta, GA 30322, USA

Laura Finzi
Department of Physics, Emory University, 400 Dowman Dr.,
615 Michael St., Atlanta, GA 30322, USA

Michael G. Fried
Center for Structural Biology, Department of Molecular and Cellular Biochemistry, University of Kentucky, 741 South Limestone, Lexington, KY 40536-0509, USA
michael.fried@uky.edu

Gourisankar Ghosh
Department of Chemistry and Biochemistry, University of California, 9500 Gilman Drive, La Jolla, San Diego, CA 92093-0375, USA

De-Bin Huang
Department of Chemistry & Biochemistry, University of California, 9500 Gilman Drive La Jolla, San Diego, CA 92093-0375, USA

Tom Huxford
Department of Chemistry & Biochemistry, San Diego State University, 5500 Campanile Drive, San Diego, CA, 92182-1030, USA

Serguei V. Kuznetsov
Physics Department, M/C 273, University of Illinois at Chicago, 845 W. Taylor St., Chicago, IL 60607, USA

Ted A. Laurence
Physical and Life Sciences Directorate, Lawrence Livermore National Laboratory, 7000 East Avenue, Livermore, CA 94550, USA

Sanford H. Leuba
Departments of Cell Biology and Physiology and Bioengineering, Hillman Cancer Center, Petersen Institute of NanoScience and Engineering and University of Pittsburgh Cancer Institute, University of Pittsburgh School of Medicine and Swanson School of Engineering, 5117 Centre Avenue, 2.26a, Pittsburgh, PA 15213, USA

Dale Lewis
Laboratory of Molecular Biology, NCI, NIH, 37 Convent Drive, Bethesda, MD 20892-4264, USA

De Pei Liu
National Laboratory of Medical Molecular Biology, Institute of Basic Medical Sciences, Chinese Academy of Medical Sciences & Peking Union Medical College, Beijing100005, People's Republic of China

Jie Liu
Section of Microbiology, University of California-Davis, Davis, CA 95616, USA

Michael A. Lomholt
Department of Physics and Chemistry, MEMPHYS – Center for Biomembrane Physics, University of Southern Denmark, Campusvej 55, 5230, Odense M, Denmark

L. James Maher, III
Department of Biochemistry and Molecular Biology,
Mayo Clinic College of Medicine, Rochester, MA, USA

Carlo Manzo
Institute for Bioengineering of Catalonia, Barcelona, Spain

Manana Melikishvili
Department of Molecular and Cellular Biochemistry, Center for Structural Biology, University of Kentucky, Lexington, KY 40536, USA

Ralf Metzler
Physics Department T30g, Technical University of Munich, 85747 Garching, Germany

Scott W. Morrical
Department of Biochemistry, University of Vermont College of Medicine, Burlington, VT 05405, USA

Jaylene N. Ollivierre
Department of Chemistry & Chemical Biology, Northeastern University, Boston, MA 02115, USA

Ioulia Rouzina
Department of Biochemistry, Molecular Biology, and Biophysics, University of Minnesota, Minneapolis, MN, USA

Jana Sefcikova
Department of Chemistry & Chemical Biology, Northeastern University, Boston, MA 02115, USA

Michelle C. Silva
Department of Chemistry & Chemical Biology, Northeastern University, Boston, MA 02115, USA

Richard A. Steinman
Department of Medicine, Hillman Cancer Center, University of Pittsburgh School of Medicine, University of Pittsburgh Cancer Institute, 5117 Centre Avenue, Pittsburgh, PA 15213, USA

Česlovas Venclovas
Institute of Biotechnology, Graičiūno 8, LT-02241 Vilnius, Lithuania

Haowei Wang
Physics Department, Georgia Institute of Technology,
400 Dowman Dr, Atlanta, GA 30322, USA

Y.M. Wang
Department of Physics, Washington University in St. Louis,
Saint Louis, MO, 63130

Mark C. Williams
Department of Physics, Northeastern University, Boston, MA, USA

Gijs J.L. Wuite
Department of Physics and Astronomy, Faculty of Sciences, VU University, De Boelelaan 10811081 HV, Amsterdam, The Netherlands

Li Xin
251 Bayview Boulevard, Suite 100, Baltimore, MD 21224-2815, USA

Guo Ling Zhou
Department of Genetics, Harvard Medical School, Center for Computational and Integrative Biology, Massachusetts General Hospital, 185 Cambridge Street, Boston, MA 02114-2790, USA

Chiara Zurla
Department of Bioengineering, Georgia Institute of Technology, Atlanta, GA, USA

Part I
Binding DNA and Changing its Shape

Chapter 1
Introduction to Biophysics of DNA–Protein Interactions: From Single Molecules to Biological Systems

Mark C. Williams and L. James Maher, III

The field of biophysics grows as new methods are applied to increasingly complex biological problems. For example, the field of single molecule biophysics has made new contributions to many areas of biology, from molecular to cellular levels. These insights enhance our quantitative understanding of biology. Recent studies that combine in vitro bulk biochemical methods, single molecule experiments, and in vivo approaches to molecular interactions have advanced many areas of biology. Particularly relevant illustrative examples are provided by the field of DNA–protein interactions. This book focus on some of these examples, involving application of in vitro bulk solution methods to probe interactions at the molecular level, single molecule methods, and in vivo measurements in the same systems. The book presents an overview of modern quantitative methods used for studying DNA–protein interactions, and displays the various levels of complexity involved in studying these interactions in biological systems. These chapters should therefore be of interest to all workers in the field of DNA–protein interactions.

Section 1 covers DNA–protein interactions that involve only one or two proteins, illustrating approaches to deduce mechanistic insights into the biophysical principles at play in these "simple" interactions. One important question concerns the mechanism by which binding proteins search for and identify specific binding sites. This mechanism is important for many proteins that recognize specific sequences and bind to them to regulate cellular processes. The most commonly studied example is that of a transcription factor, which recognizes specific sequences that regulate transcription of a gene. It is now appreciated that many DNA-binding proteins reach their specific sites on DNA using mechanisms more complex than

M.C. Williams (✉)
Department of Physics, Northeastern University, Boston, MA, USA
e-mail: mark@neu.edu

L.J. Maher, III
Department of Biochemistry and Molecular Biology,
Mayo Clinic College of Medicine, Rochester, MN, USA
e-mail: maher@mayo.edu

M.C. Williams and L.J. Maher, III (eds.), *Biophysics of DNA-Protein Interactions*,
Biological and Medical Physics, Biomedical Engineering,
DOI 10.1007/978-0-387-92808-1_1,

three-dimensional random diffusion from solution. Alternative mechanisms involve a reduction of the volume to be searched. This volume reduction likely comes from a sequence-nonspecific DNA-binding mode allowing diffusion in one dimension along the DNA. Thus, *Wang and Austin* in Chap. 2 describe experiments to measure how the model Lac repressor protein searches in one dimension to find a specific binding site. In their study, they directly image a fluorescent Lac repressor molecule as it searches a single DNA molecule, remaining close to the DNA for long periods of time. They also address the question of how the protein molecule slides along the DNA – Does the protein remain in DNA contact and truly slide, or does it exhibit small micro-hops in order to sample DNA-binding states? While this question has not yet been conclusively answered, it may be possible to test predictions of different hypothetical mechanisms using further developments in fluorescence imaging methodology or other single molecule methods.

The mechanism of DNA sliding is further addressed in Chap. 3 by *Barsky et al.*, which discusses how it is physically possible to slide along DNA and what biophysical aspects of DNA–protein interactions may affect sliding rates. This chapter also provides a concise overview of different systems in which DNA sliding may be important. These systems extend well beyond the canonical transcription factor search problem. For example, the processivity clamp is a critical component of the DNA replication fork, and this protein encircles DNA and must slide at high rates within the replication complex.

Beyond DNA sliding mechanisms, Chap. 4 by *Metzler et al.* discusses the advantages of sliding compared to searching the fluid volume. While it is clear that some enhancement of the DNA search rate is obtained by searching only the DNA itself, the one-dimensional search process proceeds slowly, as diffusion along DNA is slower than diffusion in solution. Measuring one-dimensional diffusion in single molecule experiments typically involves stretched DNA. However, it is clear that DNA in solution and in vivo is always compacted by naturally entropic elasticity and packaging proteins. How does DNA compaction affect the DNA search time? It has been shown that the theoretical optimum search time must involve both one-dimensional diffusion along DNA as well as some diffusion in solution. This chapter further describes single molecule experiments that directly demonstrate compaction effects by measuring protein search times while the DNA tension is varied in an optical tweezers instrument. When the DNA is under less tension, proteins hop quickly from one part of the molecule to another, either through solution or by direct transfer from one DNA segment to another part. To measure diffusion, the authors determine the time required for restriction endonuclease cleavage of the target DNA molecule. The authors demonstrate that the rate of restriction endonuclease digestion is increased on more compact DNA molecules under little tension than when DNA is stretched. This suggests that the localization of a specific sequence by a restriction endonuclease occurs through a combination of sliding along DNA and other diffusion mechanisms.

Once a protein arrives at a specific DNA site, recognition of that site is a complex problem that is not yet well understood from a biophysical standpoint. In Chap. 5, *Ghosh et al.* describe exemplary X-ray crystallography experiments that probe in detail how members of the dimeric NF-κB transcription factor family recognize related DNA-binding site sequences. These authors have compared structures in atomic detail for an extensive array of different complexes. This provides a nearly unique depth of understanding for the NF-κB family of proteins. Emphasized by this work is the fact that the binding site sequence, in itself, may be insufficient to guide homing to unique sites. Additional specification is likely provided by cooperative contacts with additional bound proteins. The authors touch on the interesting observation that transcription factors such as NF-κB can recognize folded RNA molecules with sequence and structure specificity.

It is increasingly understood that DNA site recognition may involve mutual structural reorganization by the protein and DNA partner. This reorganization can dramatically alter the dynamic structure and trajectory of DNA. Two chapters are devoted to these issues. In Chap. 6, *Ansari and Kuznetsov* describe extremely fast kinetic experiments measuring the dynamics of sequence-specific DNA-bending proteins during their interaction with DNA. At issue is the mechanistic question of whether site binding and DNA bending are concerted or separable events. Laser-induced temperature jump relaxation kinetics with fluorescence resonance energy transfer detection has been applied to this fascinating problem.

While the mechanism by which sequence-specific binding proteins locate and recognize their DNA sites is well studied and of great interest, many proteins that bind DNA do not recognize specific double-stranded DNA (dsDNA) sequences. For example, a class of DNA-bending proteins known as high mobility group type B (HMGB) proteins bind dsDNA at any sequence and bend the DNA at that location. The biophysics of the bending process is of particular interest because protein binding changes the character of DNA, endowing DNA with a collection of rapidly interconverting new shapes. In effect, the DNA appears to be more flexible in the presence of HMGB proteins. In Chap. 7, *Maher* reviews the biophysics of HMGB proteins, including in vivo data from bacterial experiments. This work indicates that so-called architectural DNA-binding proteins (e.g., the bacterial HU protein, or HMGB proteins introduced experimentally into bacteria from higher cells) play important roles in regulating genetic switches that require strong DNA bending and looping.

Rather than recognizing a specific sequence of dsDNA, single-stranded DNA (ssDNA)-binding proteins (SSBs) recognize DNA in its denatured form. Because ssDNA is so different structurally from dsDNA, the ability of these proteins to distinguish ssDNA is not difficult to imagine. These proteins typically contain oligonucleotide/oligosaccharide binding folds (OB-folds) that facilitate simultaneous stacking of aromatic residues with ssDNA bases as well as cationic binding to the ssDNA backbone. However, the proteins must locate and bind ssDNA that occurs transiently during DNA replication or repair. For example, ssDNA is transiently exposed by helicases and other motor proteins at the replication fork. In the case of bacteriophage replication, the fork moves extremely rapidly, so bacteriophage SSBs must search DNA optimally in order to facilitate rapid DNA replication. In Chap. 8 *Williams and Rouzina* describe the biophysical mechanisms that govern the interactions of SSBs from bacteriophage T4 (T4 gene 32 protein) and T7 (T7 gene 2.5 protein), as determined through single DNA molecule stretching experiments, combined with the results of other biochemical measurements. They describe how these

SSBs search dsDNA primarily via one-dimensional diffusion along the dsDNA molecule. Because high concentrations of these proteins are present in the infected cell, many proteins simultaneously search dsDNA, resulting in a rapid process that requires little contribution from solution diffusion. Thus, SSBs bind initially to dsDNA, sliding to locate transient sites of ssDNA as DNA replication forks propagate along the molecule. These authors also discuss the mechanism by which both T4 gene 32 protein and T7 gene 2.5 protein binding to DNA is regulated by salt-dependent conformational changes involving their C-termini, a common protein-binding regulatory mechanism. The authors go on to compare traditional replication SSBs from bacteriophages to retroviral nucleocapsid (NC) proteins, which also bind preferentially to ssDNA. However, retroviral NC proteins differ significantly from SSBs in that their overall preferential binding to ssDNA relative to dsDNA is weak, and they therefore only weakly destabilize dsDNA. Termed "nucleic acid chaperones," these proteins facilitate single-stranded nucleic acid folding and secondary structure rearrangement during the retroviral replication process. Single molecule studies are ideal for examining the mechanism by which these proteins facilitate nucleic acid melting and reannealing. The described experiments illustrate the capability of single molecule force spectroscopy to measure DNA binding thermodynamics and kinetics, which quantitatively elucidate the biophysics of DNA–protein interactions.

Thus, the first 8 chapters describe biophysical mechanisms by which individual proteins interact with DNA in simplified experimental system. Increasingly relevant (but less tractable) systems attempt to mimic multi-component complexes as found in living cells. Studies of single proteins interacting with DNA reveal the mechanisms of individual proteins, but it is desirable to understand these results in the context of more complex biological systems. The second half of the book describes examples of these complex protein–DNA interactions in the context of important biophysical processes.

Chapter 9 by *Finzi et al.* reports a detailed biophysical analysis of the molecular mechanism of the bacteriophage lambda epigenetic switch, which regulates the choice between the quiescent lysogenic state and the violently reproductive lytic state in the host cell. This switch mechanism is a model for other more complex regulatory processes. It involves the interaction of multiple protein clusters positioned at a distance along DNA. Using single molecule tethered particle methods, Finzi et al. characterize the functional interactions between a single DNA molecule and the lambda repressor protein (CI). The authors characterize the evidence for protein–protein interactions tethered by a long segment of looped viral DNA. Estimation of kinetic parameters permits characterization of some of the biophysical features of this complex system.

In Chap. 10, *Liu and Morrical* describe some of the complex DNA–protein interactions involved in bacteriophage T4 recombination processes. During late stages of bacteriophage infection, replication occurs rapidly through recombination, allowing replication of several phage genomes to occur in parallel. This recombination-dependent replication involves multiple protein–DNA and protein–protein interactions, including both DNA-binding proteins, such as T4 UvsY and T4 gene 32 protein, as well as DNA motors such as the helicase UvsX.

These proteins work together to convert SSB-coated homologous ssDNA strands into a presynaptic filament competent for the recombination events that are required for additional replication forks to be created. The chapter describes how characteristics of the DNA–protein interactions, such as preferentially binding to wrapped or stretched ssDNA or dsDNA, facilitate transitions during the recombination-dependent replication process.

While recombination-dependent replication in bacteriophage T4 involves the reorganization of several DNA–protein interactions, the *Escherichia coli* replication fork presents a dauntingly complex system of interactions between proteins and DNA. Bacterial DNA replication is much more complex and regulated than bacteriophage replication, and the required proteins are in constant exchange as replication proceeds. In Chap. 11, *Ollivierre et al.* describe interactions between several of the *E. coli* replication proteins that form the holoenzyme, the core protein complex that replicates the bacterial genome. This chapter describes examples of the molecular details of protein–protein and protein–DNA interactions involved in replication. The focus is placed on the replication of damaged DNA. Here regulated DNA polymerase switching is observed. In this process, specialized DNA polymerases are substituted for the replicative polymerase in order to copy across the damaged DNA template and allow replication to continue past a lesion. The authors briefly describe much more complex DNA replication processes in eukaryotic cells. The results illustrate how a wide array of biochemical and biophysical methods, as well as complementary in vivo measurements, are needed to shed light on the extremely complex biological mechanisms required for DNA replication in living systems.

The concept of DNA damage recognition and repair in eukaryotes is again emphasized by *Adams et al.* in Chap. 12. The authors discuss their pioneering work on the entirely sequence-nonspecific repair protein O^6-alkylguanine-DNA-alkyltransferase (AGT). This protein reverses DNA damage through an irreversible transfer of the DNA alkyl lesion to a protein side chain. Using clever biophysical methods including protein–protein cross-linking and analysis of changes in DNA topology, the authors build a compelling model for a highly cooperative protein–DNA filament. As with all good models, the work raises many new questions about the mechanism by which this sequence-nonspecific protein functions.

In eukaryotic cells, processes such as replication and transcription, replete with protein–protein and protein–DNA interactions as discussed above, must also occur in the context of highly packaged genomic DNA. Eukaryotic DNA is packaged into chromatin, a highly organized structure with multiple levels of compaction. The exact structure of biological chromatin is not known in detail, but many aspects are beginning to be understood. It is known, for example, that the first level of DNA compaction involves histone octamers around which DNA is tightly wrapped. The forces required to disrupt nucleosome structures have been quantified, and monomeric nucleosome core particles are thermodynamically stable in vitro. Higher order chromatin structure is not very well understood, and in vivo chromatin dynamics remain mysterious. In Chap. 13, *Leuba and Steinman* introduce single molecule experiments on artificial chromatin structures, which may shed light on their dynamics.

Finally, *Zhou et al.* describe in Chap. 14 the application of low-resolution but extremely powerful genome-wide methods to explore the overall organization of eukaryotic DNA within the nuclei of cells. Practitioners of the Chromatin Conformation Capture (3C) method, these authors summarize how this approach, and its new relatives, can be applied to determine which sequences are held close together in three-dimensions within the nuclear compartment. These methods provide semi-quantitative measurements, ultimately promising a series of distance constraints that might do for cell biology what the Nuclear Overhauser Effect has done for nuclear magnetic resonance spectroscopy. The present chapter surveys the concept of long-range DNA regulatory loops, reviews classic examples, identifies key implicated proteins, and offers a look at where the future may lead in solving this ultimate problem in protein–DNA biophysics.

Chapter 2
Single-Molecule Imaging of LacI Diffusing Along Nonspecific DNA

Y.M. Wang and R.H. Austin

2.1 Introduction

Transcription factors, restriction enzymes, and RNA polymerases are proteins that function by binding to their specific target sites on DNA [1, 2]. The DNA targets for these proteins are typically a few tens of base pairs long, while the chromosomes contain over a million base pairs of DNA (*E. coli*, for example, has 4.6 million base pairs); therefore, before reaching their targets, it is inevitable that DNA-binding proteins encounter nonspecific DNA first. In this process, protein–nonspecific-DNA binding does occur (although with weaker affinity than DNA target binding [3]) and this interaction affects the specific-DNA targeting rate of the protein. In order to regulate the targeting rate of DNA-binding proteins, which is an important step for gene expression regulations, the mechanisms of protein interaction with nonspecific DNA must be elucidated.

The notion that nonspecific DNA influences the targeting rate of DNA-binding proteins gained renewed credence when faster-than-diffusion target binding of LacI (Lactose repressor protein) was observed in 1970. LacI binds to its lactose operator target (*lacO*) 100 times faster than allowed by the 3D diffusion limit [4]. This faster-than-diffusion binding was explained by the facilitated-diffusion model, in which a DNA-binding protein interacts with nonspecific DNA before reaching its target [5–18].

A facilitated-diffusion process is composed of three main types of protein's motion around DNA: sliding, where the protein translocates (or slides) along nonspecific DNA base pairs without loosing contact (Fig. 2.1, green arrows); hopping, where the protein dissociates from DNA briefly, performing free 3D diffusion, and lands back on DNA at a location that is shorter than the DNA's persistence length away from the dissociation site (Fig. 2.1, black arrows); and jumping, where the protein's DNA landing location is not correlated to the dissociation site [15].

Y.M. Wang (✉)
Department of Physics, Washington University in St. Louis, Saint Louis, MO, 63130

M.C. Williams and L.J. Maher, III (eds.), *Biophysics of DNA-Protein Interactions*, Biological and Medical Physics, Biomedical Engineering,
DOI 10.1007/978-0-387-92808-1_2,

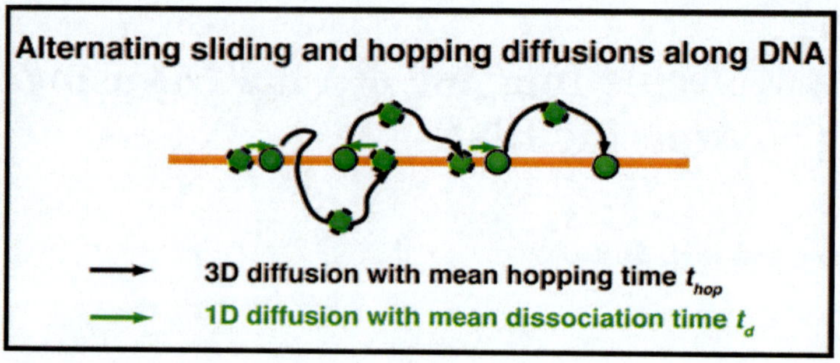

Fig. 2.1 Sliding and hopping diffusions of a protein (*green*) on an elongated DNA molecule (*orange*). Typical experimentally observed seconds-long protein trajectories on DNA should have many sliding and hopping alternations convolved. The mean 1D sliding time (or nonspecific DNA dissociation time) and 3D hopping time are t_d and t_{hop}, respectively

Figure 2.1 shows that for the typical elongated DNA geometry used in single-molecule fluorescence imaging experiments, a protein performs a series of alternating sliding and hoping motions along DNA. In a cell, the targeting process can be further complicated by road blocks of other bound proteins on DNA and different configurations of the chromosomal DNA due to confinement [19–21]. In this chapter, we focus on sliding and hopping diffusions of proteins on DNA since most single-molecule fluorescence imaging experiments use low concentrations of elongated DNA molecules for which jumping and in vivo effects are absent.

In order to calculate the effects of facilitated diffusion on the targeting rate of proteins, all three components of the facilitated-diffusion process as well as the switching kinetics among the different components should be quantified. Hopping and jumping motions are 3D Brownian motions of proteins in water, and the dynamics of 3D Brownian motions are relatively well understood, in contrast to sliding. For the sliding motion, two major issues await investigation due to limitations of current single-particle-tracking techniques: (1) The translocation (or sliding) mechanism of a protein on DNA sequences is not fully understood. There are two main translocation models: one is the DNA-sequence-independent model, where proteins are insensitive to DNA sequences and perform Brownian diffusion on DNA; the other is the DNA-sequence-dependent model, where the sliding diffusion characteristics are time dependent – anomalous subdiffusion in short millisecond timescales and Brownian diffusion in long second timescales [22]. (2) The mean sliding time and hopping time, or the protein-nonspecific-DNA dissociation rate constant (or nonspecific-DNA dissociation time, or sliding time) t_d and 3D hopping time t_{hop}, respectively, are not precisely known. Uncertainties in these parameters will decrease the accuracy of protein targeting rate calculations.

This chapter emphasizes studies of the sliding component of facilitated diffusion (and along with it the sliding–hopping alternation kinetics) using single-molecule fluorescence imaging methods. LacI is used as a model protein, and DNA molecules are elongated along fused-silica surfaces. The elongated DNA configuration offers a simple platform for revealing the dynamics of protein translocation on DNA base pairs during sliding, which is essential for understanding facilitated diffusion for any DNA configurations in cells. Why use the single molecule method? Conventional bulk measurements have been very powerful in validating and characterizing facilitated diffusion [4, 8, 10, 13, 16, 18, 23–25]; however, in order to characterize the sliding mechanisms of proteins on DNA, the motion of proteins on DNA has to be tracked. Thus, single-molecule imaging is an ideal candidate for these studies.

This chapter is divided into three parts: First, we present our initial study of LacI diffusion on nonspecific DNA in the timescale of seconds performed in 2006. Prior to this study, two articles had reported on the direct imaging of proteins diffusion on DNA. One was by Nobuo Shimamoto [12] and the other was by Yoshie Harada et al. [26]. In the first article, it is not clear whether the proteins interact with a single DNA molecule, or an array of DNA molecules; and in the second article the diffusion characteristics are not analyzed. Our study addresses both issues. Next, we discuss the limitations of current instrumentation and the single-molecule point spread function (PSF) centroid tracking method for technically demanding studies of molecule sliding, which require millisecond temporal resolution and nanometer spatial resolution. We introduce our new single-molecule image deconvolution (SMID) method, which meets these technical demands and we present the results of our preliminary application of SMID to LacI sliding studies. Using the SMID method, sliding and sliding–hopping alternation characteristics can be extracted from single protein-diffusing-on-DNA images in millisecond and sub-millisecond exposure times; and thus the SMID method increases the temporal resolution of single-particle tracking by at least 100-fold from $\approx$ 300 ms to milliseconds and sub-milliseconds (Sect. 2.4.1). Lastly, we outline additional studies necessary for single-molecule investigations of protein sliding on DNA using SMID.

2.2 Early LacI Diffusion Experiments in the Timescale of Seconds

2.2.1 Facilitated Target Association Rate Calculation Assuming Brownian Sliding

In this section, we discuss our single-molecule imaging studies of one-dimensional diffusion of LacI repressor proteins along elongated DNA in the timescale of seconds. Our analysis of the LacI transcription factors' diffusion yielded four main results: (1) LacI diffuses along nonspecific DNA in the form of 1D Brownian motion, (2) the observed 1D diffusion coefficients D_1 vary over an unexpectedly large range, from 2.3×10^{-12} cm²/s to 1.3×10^{-9} cm²/s, (3) the lengths of DNA covered by these 1D diffusions vary from 120 nm to 2920 nm, and (4) assuming that the LacI sliding is Brownian for all timescales, the mean values of D_1 and the diffusional lengths indeed predict a LacI target binding rate 90 times faster than the 3D diffusion limit.

The expected association rate $k_{a(3D)}$ by which DNA-binding proteins find their specific target sequences on double-stranded DNA in a random 3D search is $4\pi D_3 l_{seq}$ per unit protein concentration, where l_{seq} is the effective DNA target length, and $D_3 = k_B T / 3\pi\eta a = 9\times10^{-7}$ cm²/s is the 3D diffusion coefficient of the protein in solution [4, 15, 27], where k_B is the Boltzmann constant, T is the temperature, η is the viscosity of the solvent, and $a \approx 5$ nm is the typical diameter of the protein. With $l_{seq} \approx 3$ bp (or 1 nm), the protein–DNA association rate $k_{a(3D)}$ should be 10^8/M/s. The original in vitro study on LacI-*lacO* binding by Riggs et al. was with 45.5 kbp DNA of 15.5 μm in length, and the *lacO* association rate $k_{a(Exp)}$ was measured to be 10^{10}/M/s, 100 times higher than the diffusion limit of $k_{a(3D)} \approx 10^8$/M/s [4] (the 10^{10}/M/s binding rate was also reported in [13, 15, 28]).

It has been proposed that such high rates can be achieved if the protein undergoes a facilitated-diffusion process in which the protein performs a combination of 1D diffusion along the DNA and 3D diffusion in solution. In this model, the key to faster targeting lies with the nonspecific DNA sequences that flank the target site. By 3D diffusion, a protein most likely will run into a segment of nonspecific DNA first. After nonspecific binding, the protein will diffuse along the DNA for a certain time and eventually dissociate. By doing so, the effective concentration of protein near the DNA increases, and thus the targeting rate should change. This facilitated-diffusion modified protein-target association rate k_a per protein concentration has been derived by Halford and Marko [15]:

$$k_a = \left(\frac{1}{D_3 l_d} + \frac{L l_d c}{D_1}\right)^{-1} = k_{a,3D}\left(\frac{l_{seq}}{l_d} + \frac{D_3}{D_1} l_{seq} L l_d c\right)^{-1}, \tag{2.1}$$

where D_1 is the 1D diffusion coefficient of the nonspecifically bound protein along the DNA, L is the total length of the DNA molecule, l_d is the maximum DNA contour distance $x_{max} - x_{min}$ covered by the protein before dissociation, c is the concentration of the target, and $\left(\frac{l_{seq}}{l_d} + \frac{D_3}{D_1} l_{seq} L l_d c\right)^{-1}$ is the acceleration factor to $k_{a,3D} = D_3 l_{seq}$. In order to evaluate the facilitated diffusion model directly, it is necessary to know D_1 and l_d, which can only be obtained by imaging protein-DNA binding dynamics using single-molecule measurements. In fact, if these values do not fall within a certain range, "facilitated" diffusion can actually slow the search times.

Note that in this facilitated-targeting rate calculation (and other numerous calculations), sliding was assumed to be the DNA sequence-insensitive Brownian diffusion for all timescales [15, 21, 29, 30]. In the Brownian sliding model, the 1D mean square displacement of proteins on DNA is $\langle \Delta x^2 \rangle = 2D_1 t$, where D_1 is the 1D sliding diffusion coefficient and t is the observation time. Should protein translocation follow a different model from Brownian motion (i.e., if D_1 is time dependent as $D_1(t)$, or if the time dependence $\langle \Delta x^2 \rangle$ of differ from the power of one), the protein search speed through DNA sequences will change. As a consequence, the effect of facilitated diffusion on the protein's target association rate will vary.

2.2.2 Experiments

We used a LacI fusion-protein consisting of a green fluorescent protein (GFP)–GFP13 (S65T):*lacI-I12* fusion (GFP-LacI), and stained the DNA with the dimeric cyanine dye BOBO-3. DNA constructs of Lambda Zap vector with 256 tandem copies of *lacO* (*lacO*$_{256}$) were used. *LacO*$_{256}$-DNA was 42.06 kbp long with a contour length of 14.3 μm, and the 9.22 kbp *lacO*$_{256}$ insertion started at 24.02 kbp. The nonspecific sequences of the DNA construct are identical to that of λ DNA. The synthesis methods for the fusion protein and the *lacO*$_{256}$-DNA, and the sample preparation method are described in [31]. There were *lacO*$_{256}$-DNA dimers as well as monomers in the solution; the dimers were formed by the sticky-end-hybridization of two *lacO*$_{256}$-DNA monomers (Fig. 2.2a). After the LacI-DNA and BOBO-3 incubation, the GFP-LacI concentration was 50 nM and the *lacO*$_{256}$-DNA concentration was 11 pM (0. 3μg/ml). The DNA intercalating cyanine dyes are known to stretch DNA by 30% in length at 1 dye/5 bp [32], so at our concentration of 1 dye/10 bp, the DNA molecules were stretched by 15% to 16 μm. Since BOBO-3 produced no obvious effect on the DNA-configuration-dependent LacI-DNA

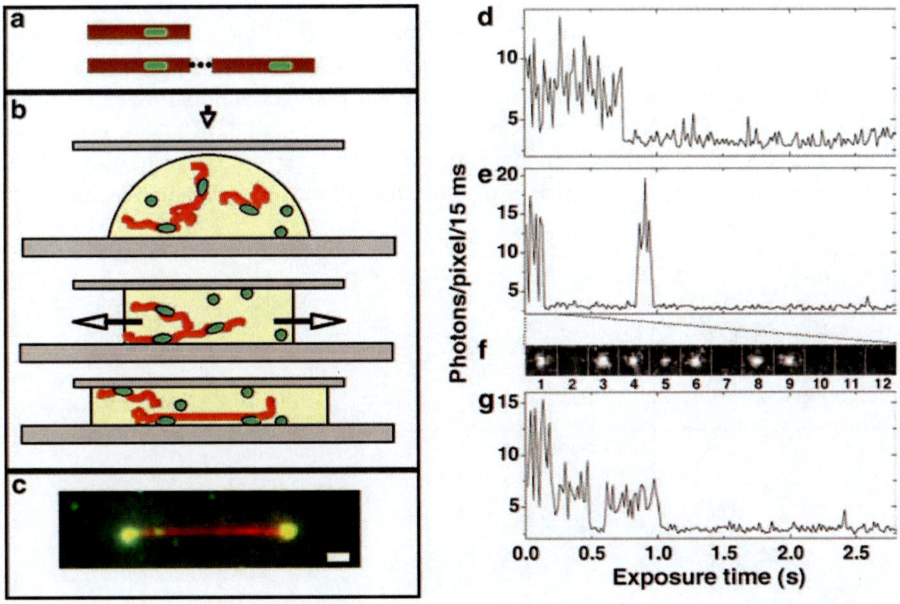

Fig. 2.2 (**a**) Schematics of a GFP-LacI (*green*) bound *lacO*$_{256}$-DNA monomer and dimer (*red*). (**b**) Elongation of the DNA. (**c**) Frame-averaged superposed image of GFP-LacI bound to an elongated *lacO*$_{256}$-DNA dimer. The scale bar is 1 μm. (**d**) A GFP-LacI monomer of frequent blinking and unitary bleaching. (**e**) A GFP-LacI monomer that blinked, recovered the first bleaching in 3 s, and finally irreversibly bleached. (**f**) The GFP-LacI dots for the first 12 frames of (**e**), showing blinking at frames 2 and 7, and bleaching at frame 10. (**g**) A GFP-LacI dimer with two bleaching events

specific binding [31], we expect that its effect on the LacI DNA-specific binding (which is less DNA-configuration dependent) will be negligible [33]. A catalytic oxygen scavenging solution was used to maximize dye lifetimes [31]. 1 μl of the DNA+LacI solution and 4 μl of the oxygen scavenging solution were deposited onto a fused-silica chip.

A glass cover slip was used to flatten the solvent, and the edges of the cover slip were then sealed with nail polish. As the cover slip flattened the droplet, hydrodynamic flow elongated the DNA dimers, and the two LacI-$lacO_{256}$ sites stuck to the surface, creating an anchored elongated DNA molecule (Figs. 2.2b, c, and 2.3c) stretched up to 90% of its native contour length. The tension on DNA was a few pico-Newtons [34]. DNA was not observed to stick to fused-silica surfaces at our pH of 8.0 and BSA concentrations, and the elongated DNA molecules were effectively suspended from the surface, as evidenced by the DNA's transverse motion of ± 50 nm (data not shown). Thus, unbound GFP-LacI molecules interacted only with free unattached and nonspecific DNA. Note that the sticking of GFP-LacI to fused-silica surfaces occurred only at the deposition step as the air–water interface moved over the chip surface. After the cover slip was sealed, the free GFP-LacI molecules (≈2 nM) diffused in the solution freely and did not stick to the surface, as evidenced by observation of the freely diffusing GFP-LacI near the surface (data not shown).

The single-molecule experiments were performed using a prism-type Total Internal Reflection Fluorescence Microscopy (TIRFM) method (Fig. 2.3a and b). The laser excitation was synchronized to the 3.4 Hz data acquisition rate of the I-CCD camera. The emitted photons from BOBO-3 and GFP were collected using a 100X TIRF oil-immersion objective (N.A. = 1.45), went through a custom-designed dichroic mirror and emission filter set (Chroma Technology

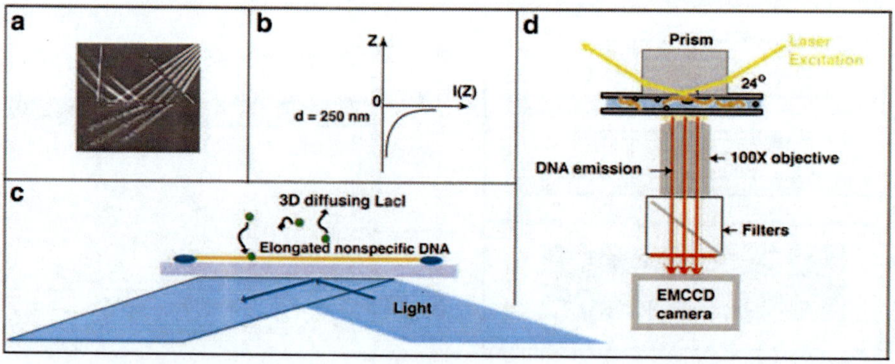

Fig. 2.3 Schematics of our experimental setup. (**a**) Light goes from air into a prism. At the prism–water interface, TIRF occurs at an inclination above the critical angle. (**b**) The decay of the evanescent light intensity $I(z)$ in z direction. The penetration depth d is ≈ 250 nm. (**c**) Schematic of our protein–DNA interaction system. A single DNA molecule (*orange*) is elongated along the fused-silica surface (*grey*) for TIRF imaging (*blue light*), and the 3D diffusing LacI molecules (*green*) run into the DNA molecule and bind to it. (**d**) Schematics of our imaging system. DNA and proteins are excited by yellow and blue laser light, respectively, and the emission photons go through an 100X objective and are collected by a single-photon sensitive CCD camera

Corp, Rockingham, VT), and were recorded by an I-CCD camera (I-PentaMAX:HQ Gen III, Princeton Instruments, Trenton, NJ; Fig. 2.3d). The PSF width of the diffraction-limited optical system was measured to be 280 nm, and the imaging pixel size was 117 nm. The pixel count of the camera was converted to a photon count using known conversion factors [31]. The mean 488 nm illumination intensity over the illumination areas of 30 μm × 50 μm was 1000 W/cm^2. The centroid location of a GFP-LacI dot was determined by fitting its 1D fluorescence intensity profile to a Gaussian. The number of detected photons per PSF per frame (between 50 and 300 photons) limited the position measurement accuracy to be between 10 and 50 nm [35].

Knowledge of the fluorescence characteristics of single free GFP-LacI monomers and dimers attached to fused-silica surfaces is essential in determining the single-molecule nature of a bound protein. GFP-LacI monomers blink frequently (short fluorescence dips to near instrumental noise level), and bleach with no recovery (Fig. 2.2d). At our excitation intensity of 1000 W/cm^2, mean exposure time of 10 ms, and synchronized imaging frequency of 3.4 Hz, the mean net observation time of each GFP-LacI molecule was 5 s before it bleached (giving a total laser exposure time of 0.15 s). The mean number of photons emitted by the bound GFP-LacI molecules before bleaching was $\approx 4 \times 10^4$ photons. This 5 s observation time gave the *instrumental* limit to the maximum mean distance we observed GFP-LacI motion on DNA in this experiment.

2.2.3 Results and Analysis

An image sequence of a single GFP-LacI molecule diffusing along DNA is shown in Fig. 2.4b. This is 1 out of 70 walks that were observed and chosen for its large net displacement. Figure 2.4a shows the frame-averaged superposed image of the anchored DNA and the diffusing GFP-LacI on DNA. Time-lapse images of the diffusing protein show clear relative displacements (Fig. 2.4b), with one immobile anchoring site used as a reference point. We know that we were observing a single GFP-LacI dimer from the fluorescence time trace in Fig. 2.4d, which clearly shows two bleaching steps. Both GFP-LacI monomers (80%) and dimers (20%) have been observed to diffuse on DNA. As is evident in Fig. 2.4d, fluorescence time traces of bound GFP-LacI molecules were identical to that of single immobile GFP-LacI (Fig. 2.2d-g), with the same blinking rate and characteristic bleaching time of $\approx$ 0. 15 s (5 s net observation time). The DNA locations of the diffusing protein at different frames are correlated and localized, thus at $D_3 = 10^8$ nm^2/s and our protein concentration of a few proteins/μm^3, the chance for two different proteins landing consecutively on the same location of DNA is 1 in 1000. Figure 2.4d plots the distribution of all relative displacements $\langle D_1 \rangle =$ of the walk. This is a Gaussian of SD = 130 nm centered near zero, which is typical for Brownian motion with limited data points.

Now, we discuss our analysis showing that individual protein diffusion trajectories, which consist of multiple measurements x_i until the protein disassociates, are

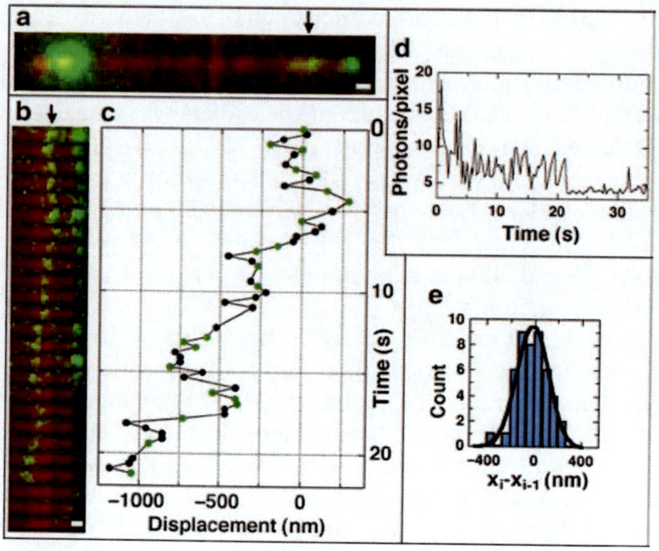

Fig. 2.4 (**a**) Frame-averaged, superposed image of a GFP-LacI molecule diffusing along DNA. The two *large dots* at the DNA ends are LacI-$lacO_{256}$ sites, and the *green segment* on the nonspecific DNA (*arrow*) is the trace of the diffusing GFP-LacI. (**b**) Image series of the diffusing protein (*arrow*) of selected clear relative displacements corresponding to *green dots* in (**c**), the displacement versus *t* curve of the diffusing protein. (**d**) Fluorescence time trace of the diffusing GFP-LacI. It is a dimer. (**e**) Gaussian distribution of $x_i - x_{i-1}$. The scale bar is 0.5 μm

Brownian in nature, and we obtain 1D diffusion constant D_1 for these trajectories. Qian et al. have derived an expression to tell (1) whether a single diffusion trajectory is Brownian and if so (2) obtain the diffusion constant of the trajectory [36]. This method calculates the mean square displacement $\text{MSD}_{(n,N)}$ for all available time intervals of a single diffusion trajectory

$$MSD_{(n,N)} = \frac{\sum_{i=1}^{N-n} (x_{i+n} - x_i)^2}{N-n} = 2D_1 n\Delta t + 2\sigma_s^2, \quad (2.2)$$

where, N is the total number of positions measured, n is the measurements index going from 1 to N, Δt is the time interval between two consecutive position measurements, and σ_s is the measurement accuracy associated with each x_i. We can obtain D_1 of a single diffusion trajectory from its $\text{MSD}_{(n,N)}$ to high precision by weighted linear-fitting $\text{MSD}_{(n,N)}$ to n, taking $\text{MSD}_{(n,N)}$'s variances at different n into consideration. As n increases, the number of available measurement points for $\text{MSD}_{(n,N)}$ averaging decreases, and the variance in $\text{MSD}_{(n,N)}$ increases as

$$\sigma_{n,N}^2 = (2D_1 n\Delta t)^2 (2n^2 + 1) / [3n(N - n + 1)]. \quad (2.3)$$

If a single trajectory is Brownian, then its $\text{MSD}_{(n,N)}$ at n below a cutoff n_c will be a linear function of t, with n_c determined by a set fractional $\text{MSD}_{(n,N)}$ uncertainty in Eq. 2.3. We chose n_c to be where $\sigma_{n,N}/(2D_1 n\Delta t)$ is 50%. We plot $\text{MSD}_{(n,N)}$ versus n

for trajectories with N > 10, where there are at least three $MSD_{(n,N)}$ values whose fractional variances are <50% Eq. 2.3. We also used only trajectories with less than five contiguous GFP blinks. Since $\sigma_s < 50$ nm, $MSD_{(n,N)}$, which is the square of the difference of two position measurements, has an offset of 2500 nm$^2 < 2\sigma_s^2 < 5000$ nm^2. These photon noise offsets were subtracted in the $MSD_{(n,N)}$ versus n curves.

Figure 2.5a plots displacement x versus time for 70 trajectories. The 15 trajectories in color are the walks for which we have obtained D_1, and the center black line is a stationary GFP-LacI stuck to the fused-silica surface (not DNA). Figure 2.5c plots $MSD_{(n,N)}$ versus n for these 15 trajectories in linear scale and Fig. 2.5d in log–log scale at low n values, respectively. The log–log plots are all straight lines with the slope of 1 at low n, clearly indicating that the 1D trajectories are Brownian motions. The dashed line in Fig. 2.5d is a fit of 2 with weighted error 3 to all n points below n_c of the topmost trajectory. The intercepts at $n = 1$ are $2D_1\Delta t$ for each particular walk, as can be seen by inspection of 2 (Fig. 2.5d). Thus, while all the walks are Brownian in nature, the different intercepts at $n = 1$ indicate that there is a large distribution in diffusion coefficients and there is not a unique, single value for D_1. We also plotted the distributions of nondegenerate relative displacements $x_i - x_{i-n}$ for the first 15 positions of all 70 trajectories for $n = 1$, 2, and 3 in Fig. 2.5b; the displacements are all Gaussians centered at zero with SD increasing with n. This result further demonstrates that LacI's diffusion trajectories are truly Brownian in nature, regardless of the variations in individual diffusion coefficients. Two other papers on single-molecule imaging studies of

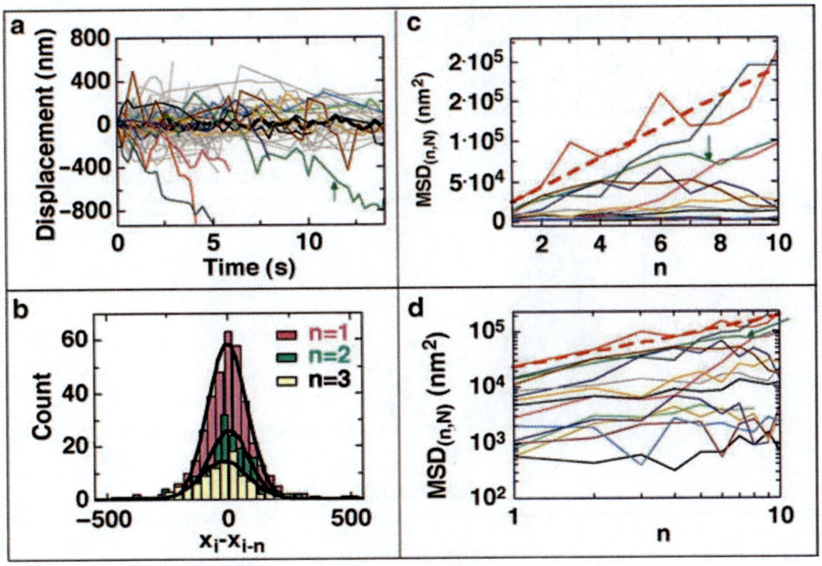

Fig. 2.5 (**a**) x versus t for 70 trajectories. (**b**) Nondegenerate $x_i - x_{i-n}$ distributions. (**c**) $MSD_{(n,N)}$ versus n for the 15 colored walks in linear scale and (**d**) in log–log scale. The *arrows* in (**a**), (**c**), and (**d**) denote the walk in Fig. 4. The *dashed lines* in (**c**) and (**d**) are the fit of 2 to the top trajectory

protein diffusion on DNA were published in the same year (2006) [37, 38], with [38] reporting a similar large distribution in 1D diffusion coefficients for Rad51 on aligned DNA molecules, so our result here may be of some generality.

Figure 2.6a shows the distribution of the 15 D_1 values (corrected to DNA contour length), which span a large range from 2.3×10^2 nm²/s to 1.3×10^5 nm²/s. Figure 2.6b shows that the different D_1 values are distributed randomly along the DNA, showing a lack of correlation between the D_1 and the regions on the λ DNA on which the protein has diffused. Figure 2.6c shows the distribution of the l_d in DNA contour length. Because λ DNA has large sequence variance in the nonspecific region with ± 30% difference in local AT and CG concentrations, it is possible that the diffusion constants are a function of local sequence. It is also possible that the large distribution in D_1 is caused by conformational distributions in the protein [39]. Further experiments are needed to answer these questions.

Finally, we use our data to examine the question of the extent to which facilitated diffusion can enhance the LacI target binding rate. Just as there is a distribution in the 1D diffusion coefficient D_1, there is also a distribution in the diffusion lengths l_d, whose value is further compromised by the mean observation time to bleaching of the GFP of 5 s. Since the final target binding is the result of many diffusion events

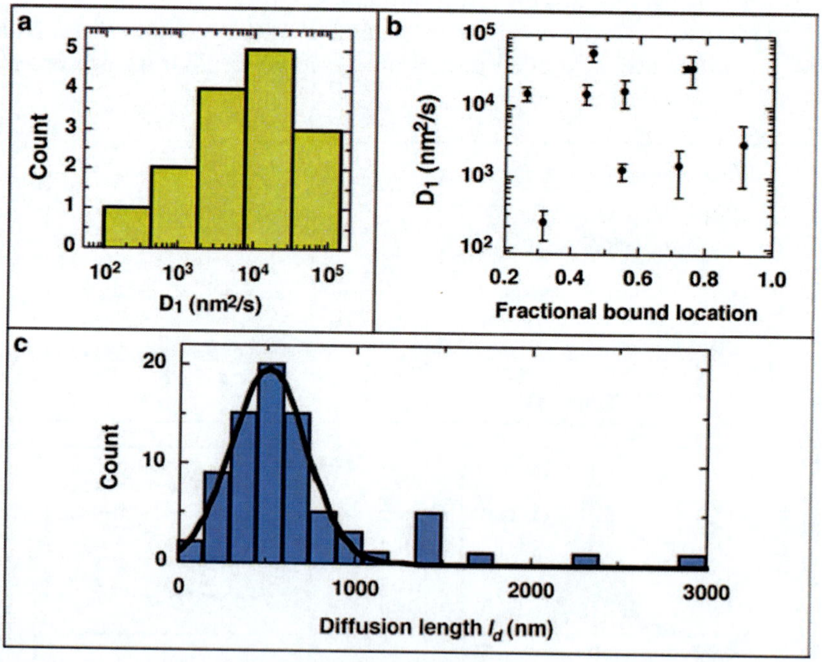

Fig. 2.6 (**a**) D_1 distribution of the 15 trajectories. (**b**) D_1 versus fractional bound location on the nonspecific segment of the $LacO_{256}$-DNA dimer. The error bars were obtained from the fit of $MSD_{(n,N)}$ to n with weighted errors at each n given by 3. (**c**) Histogram of $x_{max} - x_{min} = l_d$ for the 70 trajectories. The *solid line* is a Gaussian fit with a mean of 500 ± 220 nm (mean ± SD). Values in (**a**) and (**c**) have been adjusted to DNA contour length

on nonspecific DNA, we use the mean $< l_d >$ (probably a lower bound due to bleaching) of 500 nm, and the mean diffusion coefficient of $< D_1 >= 2.1\times10^{-10}$ cm²/s in 1. Also using Riggs' concentration of 1 *lacO*/1670 μm^3, $D_3 \approx 4 \times 10^{-7} cm^2/s$ for LacI tetramers (a ≈ 10 nm), and $L = 15.5$ μm, the accelerating factor in 1 is 93 ± 20, which thus resolves the 100-fold discrepancy between the theory and the experimental data. We conclude from these measurements that facilitated diffusion increases the LacI-*lacO* binding rate well over the apparent diffusion limit. This result demonstrates that facilitated diffusion in the form of 1D Brownian motion is the mechanism responsible for the faster-than-diffusion binding of LacI to *lacO*, and quite possibly, the reason also for the observed faster-than-diffusion binding in other protein–DNA interactions.

2.2.4 *Concern for the Interpretation of the "Sliding Length" $< l_d >$ and D_1*

There are two concerns for the above interpretation of experimental results: (1) The observed mean "sliding length" $< l_d >$, which is the maximum DNA contour distance $x_{max} - x_{min}$ covered by the protein before dissociation, quite likely is not the pure sliding length defined in Eq. 2.1; rather it is the combined distance of many sliding and hopping cycles before the permanent dissociation of proteins from DNA. This statement stems from the estimation that LacI's nonspecific DNA dissociation rate constant is on the order of milliseconds (≈ 0.6 ms to 5 ms, or milliseconds to seconds) [3, 19, 21, 30]. (2) For the same reason, D_1 is the "effective" 1D diffusion coefficient of the combined sliding and hopping trajectory in the timescale of seconds, rather than the pure sliding diffusion coefficient.

With the three questionable parameters in Eq. 2.1 – (1) Brownian sliding (Sect. 2.2.1), (2) the sliding length (Sect. 2.2.3), and (3) the 1D diffusion coefficient D_1 (Sect. 2.2.3) – the LacI target association rate calculation due to facilitated diffusion should be reevaluated. In order to correctly calculate facilitated target association rate using Eq. 2.1, the true sliding displacement versus time relation, sliding length, and sliding dissociation rate constant should be obtained.

2.3 New Challenge: Millisecond Timescale Single-Molecular Tracking

To reiterate key limitations of the above LacI diffusion experiments in the timescale of seconds for sliding mechanism studies of LacI on DNA: (1) We know that the LacI proteins stay around DNA for the observation time of seconds; what we do not know is whether the proteins slide for the whole time or many hopping cycles are convolved. (2) If the protein diffusion pathway on DNA is a combination of sliding and hopping, the sliding–hopping-alternation kinetics is not known (i.e., the mean sliding time t_d and hopping time t_{hop} are not known). (3) In the sliding motion, how the protein displacement changes with time is not known. These three factors must be addressed for correct calculation of the target association rate of proteins.

Faster single-molecule tracking than the current centroid tracking method in seconds timescales appears to be the answer to the above three questions. Then, how much faster is fast enough? For (1) and (2), millisecond tracking resolution may be sufficient according to Sect. 2.2.4. For (3), the sliding motion characteristics studies, is milliseconds tracking also sufficient? Below, we list predicted displacement versus time characteristics of different sliding models in 10^{-7} s to 1 s timescales.

Other than the DNA-sequence-independent Brownian sliding model, the alternative sliding models are the DNA-sequence-dependent sliding models. There are four simple DNA-sequence-dependent models describing four different protein translocation energy landscapes along a stretch of DNA sequences (Fig. 2.7a I–IV): (1) The energy at each binding site n is independent of others. The translocation energy barrier from site n to site $n' = n \pm 1$ is the difference between the protein binding energies of the two sites, if positive and zero, if negative [22, 40]. (2) In order to move from site n to site n', the protein needs to completely dissociate from the DNA first over a threshold level EM = Max[E(n)]. The translocation energy barrier is EM - E(n) [22]. (3) The threshold energy Et is lower than EM, and the translocation energy barrier is the maximum of the energy differences and zero – Max[Et - E[n], E[n′] - E[n],0] [22, 41, 42]. (4) Two-state model in which Et separates the reading regime ($E[n] < Et$), where the translocation is the same as in model III from a sliding regime ($E[n] > Et$), where the sliding motion is on a flat energy landscape [22, 43, 44].

In contrast to the DNA-sequence-independent sliding model, where the translocation mechanism yields Brownian diffusion for all timescales as $\langle n^2 \rangle = 2D_1 t$ the four DNA-sequence-dependent protein translocation mechanisms predict distinct diffusion patterns at different timescales. Figure 2.7b shows the mean square displacement versus time ($\langle n^2 \rangle$ versus t) log–log plots of the four sequence-dependent diffusion models from $t = 5\times10^{-7}$ s to 1 s. All diffusions are Brownian with the $\langle n^2 \rangle$ versus t slope of one in the log–log plot at $\approx t > 30$ ms (left vertical dashed line). At $t < 30$ ms, all models exhibit subdiffusion behavior, in which $\langle n^2 \rangle$ is proportional to $A(t)t^{b(t)}$, with $b < 1$ (b[t] being the slope of the lines at time t and $log(A[t])$ being the vertical offset of the line, according to the expression $log\langle n^2 \rangle = logA[t] + b[t]logt$).

It is clear that diffusion studies of sliding below 30 ms are necessary to differentiate Brownian sliding from subdiffusive sliding, provided that sliding lasts longer than 30 ms. This 30-ms tracking resolution cannot be afforded by the current centroid method, which is limited to temporal resolution of 300 ms (see Sect. 2.4.1), let alone 1 ms sliding time. While the diffusion of proteins on DNA in the timescales of seconds has been reported in many recent single-molecule protein–DNA interaction studies [20, 37, 38, 45–48, 48–56], no millisecond timescales studies have been reported. Alternative higher temporal resolution single-molecule

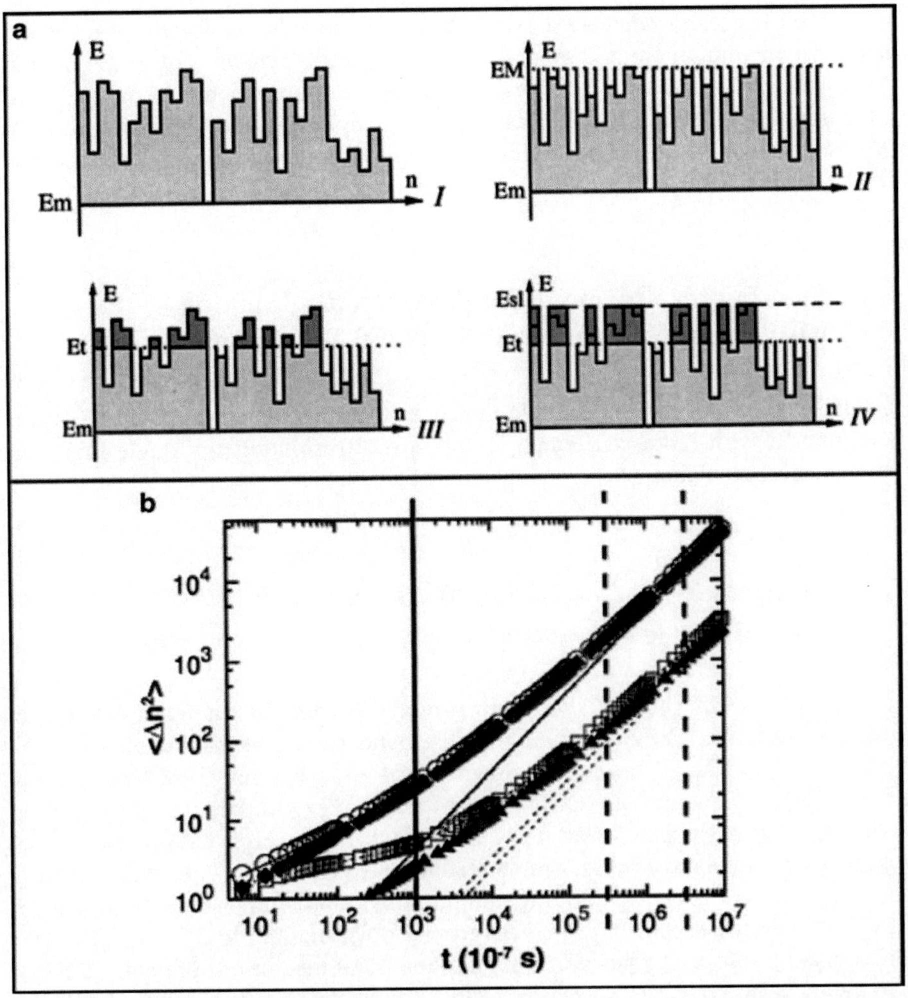

Fig. 2.7 Protein translocation models and the corresponding diffusions, adapted from [22]. In the calculations of total bound energy between a protein and DNA bases, the hydrogen bond energy between each base pair and the protein used is ε. (**a**) Schematics of the four energy landscapes for translocation of RNA-polymerase along 30 base pairs T7 DNA. *Em* is the minimum protein interaction energy and *EM* is the maximum interaction energy of all base pairs. The threshold level *Et* (*dotted line*) is set to *EM* for model II and to an intermediate value for models III and IV. For model IV, all energy levels above *Et* are redefined to a common sliding energy *Esl* (*dashed line*). (**b**) Mean square site displacement $\langle n^2 \rangle$ versus *t* for the four different models, with $\varepsilon = k_B T$ and $Et = 0$. Model I, *circles*; model II, *triangles*; model III, *diamonds*; model IV, *squares*. The *slanted straight lines* with slope of one, indicating the slope of Brownian motions, are fits to the models from 0.6 to 1 s. The *solid vertical line* marks 0.1 ms, the lowest timescale for our studies using single-molecule image deconvolution. The *left vertical dashed line* at 30 ms, where the subdiffusive curves begin to deviate from the Brownian diffusion line, marks the fastest camera frame rate for centroid measurements in the high camera-reading-noise regime (without pixel binning and partial screen imaging). The *right vertical dashed line* at 300 ms marks the fastest camera frame rate in the low camera-reading-noise regime

millisecond timescale studies are necessary. Below, we describe our new single-molecule image deconvolution method to study the subexposure dynamics of single mobile fluorescent molecules with millisecond temporal resolution and nanometer spatial resolution. In Sect. 2.4, we use this method to analyze images of diffusing LacI on DNA with exposure times approaching the predicted nonspecific DNA dissociation time $t_d \approx 1$ ms.

2.4 Pushing the Envelop: Single-Molecule Imaging with Higher Temporal and Spatial Resolution

In this section, we introduce single-molecule image deconvolution (SMID) as a new method with increased temporal and spatial resolution for single-molecule tracking measurements.

2.4.1 Rationale for Developing the Single-Molecule Image Deconvolution Method

Ideally, to discern subdiffusive sliding models from the standard Brownian sliding model, centroid versus time measurements should be used to obtain $\langle n^2 \rangle$ versus t relation at all timescales down to below $t_d \approx 1$ ms. However, current single-molecule imaging EMCCD cameras (Photomax, Andor, etc.) do not offer faster than 10 MHz pixel reading speed, which in the high background noise regime corresponds to a 30-Hz frame reading rate (or 100 Hz with pixel binning and partial screen imaging, although binning is not recommended due to loss of PSF information) and in the low background noise regime to 3 Hz (or 10 Hz). These maximum frame rates correspond to temporal resolutions of centroid measurements between ≈ 30 and 300 ms. Does this mean that one can study 30 ms diffusion characteristics with centroid measurements? The answer is no. Tracking more than 10 centroids of the molecule at different times is necessary for studying single-particle diffusion characteristics [36, 45], and this requirement lifts the centroid measurement temporal resolution to 0.3 s in the high camera-reading-noise regime (right vertical dashed line in Fig. 2.7b) and to seconds in the low camera-reading-noise regime.

Thus, in order to discriminate different displacement versus time characteristics at millisecond timescales, centroid measurements with the best temporal resolution of 100 ms (high background noise regime and loss of spatial resolution due to pixel binning; not suggested for single-molecule tracking studies) are not sufficient. The single-molecule image deconvolution method introduced below will have millisecond temporal resolution and nanometer spatial resolution for single-molecule tracking.

2.4.2 Single-Molecule Image Deconvolution (SMID)

A Gaussian fit to the PSF of a single fluorophore contains two fitting parameters: centroid and standard deviation (SD, denoted by s in this chapter). The centroid is considered the center of the PSF, while the SD is the width of the PSF (Fig. 2.8d). In contrast to centroid measurements, which have been extensively used for 1D and 2D localization studies, PSF SD measurements have played little role in single-molecule tracking studies. However, the SD of a point light source can carry additional localization and dynamic information about the particle that is inaccessible by centroid measurements: (1) for a stationary molecule, the axial location (distance away from the focal plane) of the particle can be obtained only from SD measurements [57–59] and (2) for a moving particle, regardless of exposure time, the captured image will contain additional blurring. Figure 2.8 shows that the moving molecules in (b) and (c) are obviously blurred compared to the stationary molecule in (a). In one recent study, SD measurements of a moving GFP-LacI in *E. coli* have been reported [37]; however, for both immobile and mobile single molecules, correct interpretation of the SD values requires further theoretical and experimental studies. Here, we introduce an algorithm to deconvolve the blurred image of a moving molecule to obtain its trajectory during exposure by measuring the SD of the image.

Our single-molecule image deconvolution method uses a mathematical algorithm that deconvolves the blurred snapshots of a moving molecule with subexposure

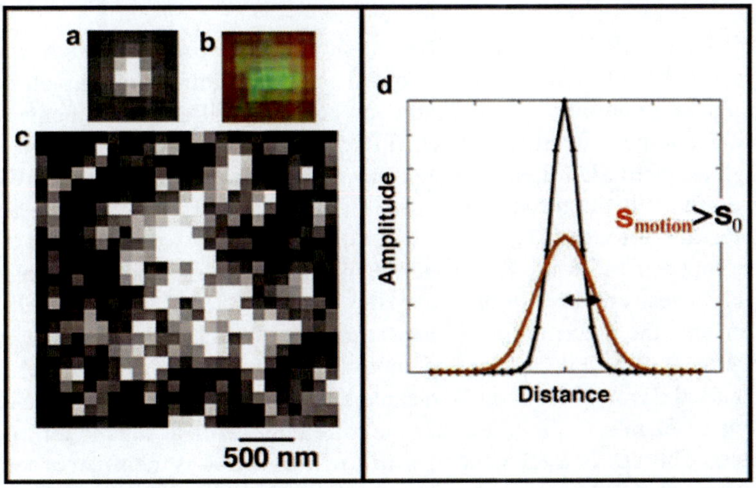

Fig. 2.8 (**a**) An immobile GFP-LacI molecule on surface, (**b**) an 1D diffusing GFP-LacI molecule on DNA, (**c**) a 3D diffusing GFP-LacI molecule in solution. The SDs of the PSF are larger for the diffusing LacI molecules. (**d**) Schematic Gaussian PSF of immobile (*black*) and diffusing proteins (*red*); SD of the diffusing protein s_{motion} is > than that of the immobile one s_0

temporal resolution. This method is based on the principle that a fluorescent particle emits photons along its pathway throughout the whole exposure time, and the number of photons emitted at a specific position x depends on the net time the molecule spends at that position during the exposure. We introduce the pathway distribution function $g(x)$ to describe the net time distribution that a single fluorescent molecule spends at location x during the exposure. The final fluorescence intensity profile $I(x)$ of the moving molecule is the convolution of the molecule's PSF $f(x) \propto exp(-x^2 / 2s_0^2)$ and its pathway distribution function $g(x)$,

$$I(x) \propto f(x) * g(x), \tag{2.4}$$

where s_0 is the SD of the PSF of the immobile molecule at focus. By fitting the final convolved snapshots of a moving single molecule to Eq. 2.4, we can extract the appropriate $g(x)$. We call this procedure for obtaining the pathway distribution $g(x)$ for a convolved image of a moving single fluorophore the *Single-Molecule Image Deconvolution (SMID) method* [60].

There could be more than one suitable pathway that yields the same convolved image. It is fortunate, however, that there are usually only a few theoretical models from which to choose. By combining SD measurements with the additional centroid-versus-time information at different timescales in the analysis, the $g(x)$ selection can be narrowed to fewer possibilities.

2.4.3 Precision Analysis Associated with SD Measurements

In order to use SMID to discern different subexposure time pathways, which in our case reflect different sliding mechanisms and sliding–hopping-alternation kinetics, it is important to obtain the precision of SD measurements. If the expected SD values of various models differ by more than the error of the experimentally measured SD for each model, the models can be differentiated. We have performed the full SD measurement error analysis for single immobile fluorescent molecules at focus [61].

Error analysis for centroid measurements of immobile molecules has been thoroughly investigated and applied to many systems [35, 62]. As with the centroid measurements, the precision of SD measurements is affected by the same experimental settings of a finite number of photons per PSF N, the camera background noise standard deviation σ_b, and the camera's finite pixel size a. We have derived an analytical expression for PSF SD measurement error as a function of these parameters. The SD measurement error in x or y direction is the square root of the mean square error $\langle(\Delta S_i)^2\rangle$

$$\langle(\Delta s_i)^2\rangle = \frac{s_0^2 + \frac{a^2}{12}}{N} + \frac{16\pi\left(s_{0x}^2 + \frac{a^2}{12}\right)^{3/2}\left(s_{0y}^2 + \frac{a^2}{12}\right)^{1/2}(\sigma\_b^2 + \langle b\rangle)}{3a^2N^2}, \tag{2.5}$$

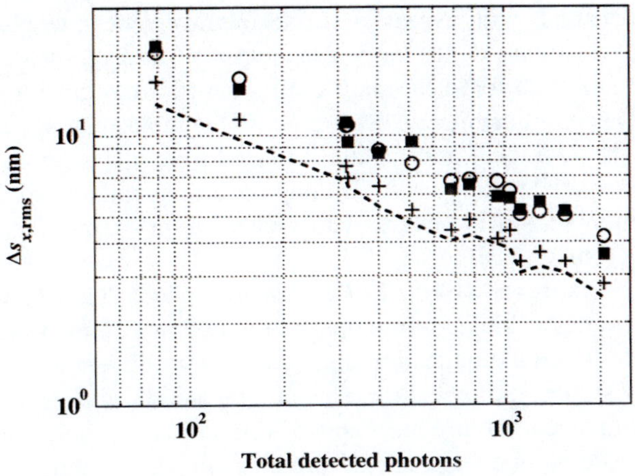

Fig. 2.9 SD measurement error, $\Delta s_{x,\,rms}$, versus the number of detected photons N studied by using four different methods: experimental measurements (*solid squares*), simulations (*circles*), numerical integrations (*crosses*), and analytical calculations (*dashed line*). Each experimental $\Delta s_{x,\,rms}$ data point is the SD of a Gaussian fit to the s_x distribution of a single streptavidin-Cy3 monomer adsorbed to surfaces

where, $\langle b \rangle$ is the mean background photon count. Figure 2.9 shows our SD measurement error results using four methods of study: experimental measurements, simulations, numerical integrations, and analytical calculations. At low PSF photon count of $N \approx 100$, the SD error is ≈ 15 nm; at high PSF photon count of $N \approx 1,000$, the SD error is ≈ 4 nm. This SD error expression is approximately 1.25 times higher than the precision of centroid measurements under comparable experimental conditions. The experimental results in Fig. 2.9 were obtained using a different camera (Andor EMCCD), a different excitation wavelength of 532 nm, and a different pixel size of $a = 79$ nm from the measurements in Sect. 2.2. In addition, 2D Gaussian fitting to PSF was used instead of the 1D fitting in Sect. 2.2.

2.4.4 SMID Enables Single-Image Molecular Dynamics Studies

In the above section, we introduce a new method, SMID, for single-molecule localization and tracking studies with higher temporal and spatial resolution than currently afforded by centroid measurements. The higher temporal resolution is achieved through the following mechanism: while many centroid measurements at different times are required for tracking a single molecule, SD measurement of only one image can offer insight into single-molecule dynamics at the shorter subexposure timescales with higher precision. Literally, one millisecond exposure image can be used to differentiate mechanisms that differ in SD by nanometers, enabling single-image studies of molecular dynamics with at least 100-fold improvement in temporal resolution. For the spatial resolution, while centroid analysis is based on measuring the mean location of the molecule for each image during the exposure time, measurement of the PSF SD provides a quantitative description of the molecule's spatial distribution. Consequently, the additional information extracted from this distribution constitutes improvement in spatial resolution. These improved temporal and spatial resolutions enable description of SMID studies with another term – single-image molecular analysis (SIMA) studies [60].

Our analytical expression of the PSF SD error provides the level of precision for SMID measurements of single immobile fluorescent molecules. When this expression is extended to axial localization and subexposure time studies, the dynamics of various very similar biological systems investigated in vitro and in vivo can be elucidated and their underlying mechanisms differentiated, such as our studies on differentiating sliding mechanisms of proteins on DNA. When the difference between the SDs measured for molecules displaying different characteristic motions is small, our error analysis will serve as a means for proper discrimination. SMID with full error analysis should be applied to reanalyze existing single-molecule tracking studies, as well as to all future fluorescence particle tracking experiments for a more thorough description of a particle's dynamics.

2.5 Application of SMID to LacI Diffusion on DNA Studies

Now, we discuss preliminary results and considerations about the application of SMID to LacI diffusion on DNA studies. Issues relevant to the interpretation of experimental results for the correct diffusion model selection are discussed.

2.5.1 *SD Measurements of Diffusing LacI on DNA*

Figure 2.10 shows 1D SD measurements for diffusing GFP-LacI on DNA using exposure times of 5, 10, and 15 ms (using the same set of data in Sect. 2.2). To illustrate that the SDs of moving proteins differ from that of stationary proteins, s_x distributions of diffusing proteins are compared to s_0 distribution of stationary proteins. The mean of the s_0 distribution, $\langle s_0 \rangle = 132\text{nm}$ ndicates that our imaging system is diffraction limited (Fig. 2.10a), and the SD of the s_0 distribution is the error associated with each PSF SD measurement using our imaging system. In Fig. 2.10b, the mean and spread of the diffusing proteins' s_x distribution at 5 ms exposure are 165 ± 32 nm (mean ± SD), and they are apparently larger than those of the stationary proteins of 132 ± 23 nm. Figure 2.10c shows that as exposure times increase to 10 ms and 15 ms, $\langle s_x \rangle$ values increase accordingly, to 176 ± 50 nm and 183 ± 62 nm, respectively.

Although Fig. 2.10 validates the applicability of SMID for protein diffusion on DNA studies, the data may not be directly useful for sliding dynamics analysis. For

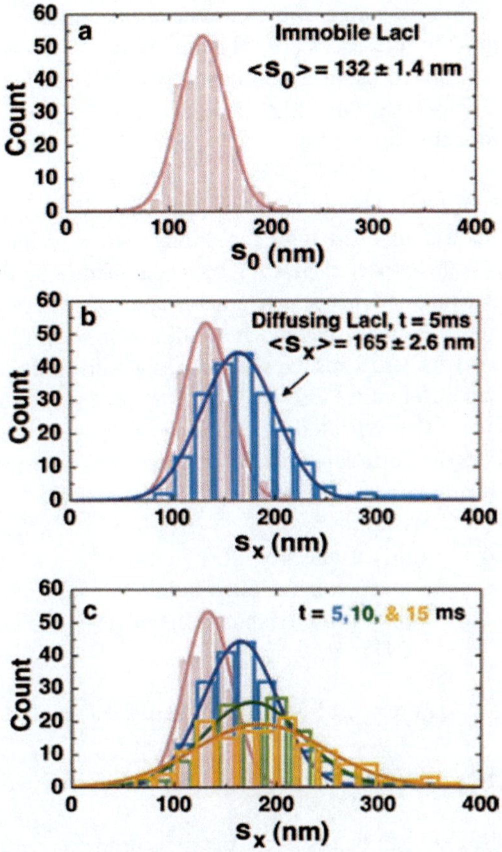

Fig. 2.10 **(a)** s_0 distribution of immobile GFP-LacI (*pink*) with $\langle s_{0x} \rangle = 132 \pm 1.4$ nm (mean ± error to the mean). **(b)** s_x for diffusing GFP-LacI on DNA at 5 ms exposure time (*blue*) with $\langle s_x \rangle = 165 \pm 2.6$ nm. **(c)** s_x distributions for 5, 10, and 15 ms exposure times, and the $\langle s_x \rangle$ values are 165 ± 3.5 nm, 176 ± 4.5 nm, and 183 ± 5.2 nm, respectively

exposure times of 5–15 ms, whether hopping motion is convolved in the images, and if so, the extent to which it is convolved are unknown. These unknowns render the data insufficient for studying the sliding movements of proteins without a full analysis of their hopping motions as well.

2.5.2 *Exposure Times of 0.1–2 ms Should Be Used for LacI Sliding Studies*

For LacI sliding studies, shorter exposure times of 0.1–2 ms should be used. We chose this range of exposure times for three reasons: (1) 0.1 ms is the shortest exposure time at our peak laser power to detect PSF with a signal-to-noise ratio of more than 3;

(2) 0.1–2 ms falls within the range of the current estimated dissociation time of LacI from nonspecific DNA. In references [19, 21], the dissociation time of LacI from nonspecific DNA was calculated using experimental results to be between 0.3 and 5 ms in vivo in *E. coli*; alternatively, when the theoretical limit of association was used for calculation, the dissociation time is between 1 ms and 1 s [3, 30]. Although there have not been experimental measurements of the LacI's nonspecific DNA dissociation rate constant, our exposure times should satisfy a majority of the estimated dissociation times to guarantee that we will be imaging the sliding motion with minimal perturbation from hopping; and (3) the expected SDs for Brownian sliding at the 0.1 ms and 2 ms exposures differ by 1.6 nm, and this is an attainable level of precision for SMID according to Sect. 2.4.3. For the sequence-dependent subdiffusion sliding models, the expected SD difference at the 0.1 ms and 2 ms exposure times would be less than 1.6 nm. Using this difference, we can differentiate Brownian sliding from subdiffusive sliding.

Let us now calculate the expected SD difference between the 0.1 ms and the 2 ms images for the sequence-independent Brownian sliding model. For Brownian diffusions, D_1 is constant in all timescales and the pathway density distribution function $g(x)_{Sliding}$ is a Gaussian as $g(x)_{Sliding} \sim \exp\left(-x^2/4D_1 t\right)$, with $2D_1 t$ being a reasonable estimate for the variance of $g(x)_{Sliding}$. Since the convolution of two Gaussians is another Gaussian with variance being the sum of the two variances, the imaged Brownian sliding protein should have a Gaussian intensity profile $I(x)_{Sliding}$, with $s^2_{Sliding} = s_0^2 + 2D_1 t$ as

$$I(x)_{Sliding} \propto \exp\left(-x^2/2(s_0^2 + 2D_1 t)\right). \tag{2.6}$$

Using $\langle D_1 \rangle \approx 10^5\,\mathrm{nm}^2/\mathrm{s}$ [45], the expected SD difference between $t = 0.\ 1$ ms and $t = 2$ ms is *1.6 nm.*

For the 1.6 nm difference to be resolved, the measurement error of each mean SD value at a specific exposure time should be less than one-half of the difference, or 0.8 nm. Assuming each image contains a conservative number of 200 photons (which corresponds to a SD measurement error of 10 nm according to Fig. 2.9), the number of images required for the mean SD error to be less than 0.8 nm is $N_{frames} \approx 160$ images, according to the relation for error to the mean of $10\ \mathrm{nm}/\sqrt{N_{frames}} = 0.8$ nm.

From the above calculations, if the mean measured SDs increases with t according to $\mathrm{SD}(t) = \sqrt{130^2 nm^2 + 2D_1 t}$, and differ at 0.1 ms and 2 ms exposure times by ≈ 1. 6 nm, then the sliding is Brownian in millisecond timescales. If the mean SDs increase from 0.1 ms to 2 ms slower than $\sqrt{130^2 nm^2 + 2D_1 t}$, or differ at 0.1 ms and 2 ms exposure times by < 1. 6 nm, then the sliding is subdiffusive.

2.5.3 *Will Hopping Be an Issue for the 0.1–2 ms Exposure Times?*

During facilitated diffusion, after the first sliding motion, the protein dissociates from DNA and performs a hopping motion. Are the exposure times of 0.1 ms and 2 ms too long such that the protein images may have hopping motions convolved? The

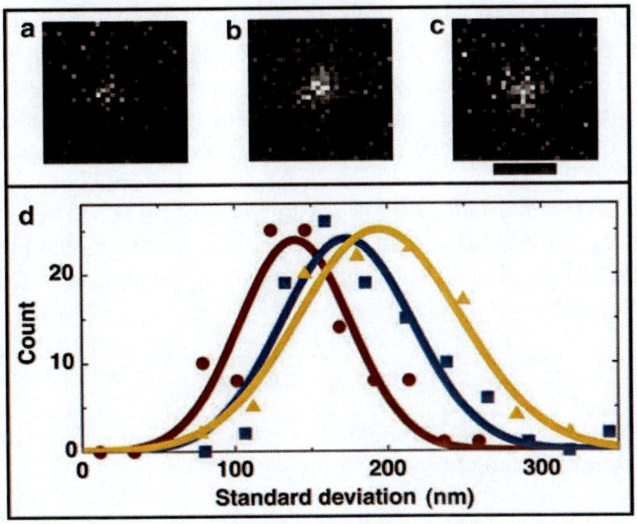

Fig. 2.11 Images of representative single 3D diffusing molecules with exposure times of (**a**) 0.3 ms, (**b**) 0.7 ms, and (**c**) 1 ms. The spread of the molecules increases with exposure time and the SD values are 135 nm (**a**), 180 nm (**b**), and 204 nm (**c**), respectively. (**d**) SD distributions for exposure times of 0.3 ms (*red*), 0.7 ms (*blue*), and 1 ms (*yellow*). The SD values are 139.5 ± 3.6 nm (mean ± standard error of the mean), 173.3 nm ± 4.2 nm, and 194.5 ± 5.2 nm, respectively. The mean of the SD distribution increases with exposure time, indicating that the observed molecules are indeed 3D diffusing molecules. Scale bar, 1 μm

answer is yes; it is possible that occasionally hopping motion may be convolved in the images. Statistically, if the mean sliding dissociation time is 1 ms, some short slides will inevitably be followed by hopping during exposure and thus some hopping will be convolved in the images. Whether the hopping-convolved images can be differentiated from the purely sliding images will depend on the fraction of time the protein spends hopping during exposure. If the percentage of time for hopping is too high to be ignored, for example, if 1 ms out of a 2 ms exposure is hopping, then the hopping-convolved image will differ dramatically from that of a protein that only moves by sliding. Figure 2.11 shows 3D diffusing molecules with 0.3 ms, 0.7 ms, and 1 ms exposure times and the respective mean SD values of 140 nm, 173 nm, and 195 nm. It is evident that the "blur" of 3D hopping proteins is significantly larger than $s_{sliding} \approx 130$ nm of the sliding proteins and should be easily discerned if convolved. If the hopping time is short, for example, 0.1 ms out of 2 ms, then the percentage of emitted photons from hopping will be too low to affect the overall SD of the image significantly.

2.5.4 Effect of Defocusing

It is known that the width of a defocused PSF increases [58]. Will defocusing of the sliding proteins affect the expected SD values significantly? Here, we will estimate the effect using DNA transverse fluctuation analysis. In the transverse and axial directions, a bound-sliding protein should fluctuate the same as the DNA. An elongated λ DNA molecule that extends to the typical 60–90% of its contour length fluctuates transversely with maximum excursion of 100 nm and frequency of 10 Hz (100 ms each cycle, unpublished data). This transverse motion can be approximated to be fluctuation in the axial direction. In the 0.1–2 ms range of exposure times, DNA should have barely moved (or at most defocused by 10 nm) and the defocusing effect on the protein SD should be negligible according to the following relation: the standard deviation $s(\Delta z)$ of a defocused PSF changes with the defocusing distance Δz as [63]

$$s(\Delta z) = s_0 \sqrt{1 + (\Delta z / D)^2}, \tag{2.7}$$

where $D \approx 400$ nm is the depth of field of our imaging system. If we use $z_{Max} = 10$ nm, there is essentially no change to s_0.

2.5.5 *Effect of DNA Longitudinal Fluctuation*

Since a protein slides longitudinally along DNA, the longitudinal fluctuation of DNA can introduce blur to the PSF. Here, we estimate the maximum effect DNA longitudinal fluctuation can have on $s_{sliding} = s_x$. According to a simple calculation, the longitudinal displacement of a segment of DNA under tension can be calculated from its measured transverse displacement by assuming that the DNA functions like a spring under tension. For the measured transverse displacement of ± 100 nm at the center of a λ DNA molecule stretched to 10 μm, the corresponding longitudinal displacement at the center segment is $\sqrt{(100\ nm)^2 + (5\ \mu m)^2} - 5\ \mu m$ = 1 nm. Putting this 1 nm into 6 (in the place of $2D_1t$), the change to the observed SD should be minimal.

This theoretical estimation awaits experimental validation described in Sect. 2.6.1.2, where the DNA longitudinal fluctuations will be investigated through SD measurements of single specifically bound proteins.

2.6 Prospects for LacI Sliding Studies Using SMID

More experiments and analysis are necessary to unravel the sliding mechanisms of proteins on DNA using SMID. Here, we lay out necessary investigations in detail. We first describe three different, but complementary approaches that should be used in the study: analytical calculations, simulations, and experimental measurements. The intensity profiles of a sliding protein on DNA with exposure times from 0.1 ms to 2 ms should be obtained using each approach and the results should be compared in order to determine the most appropriate mechanism of LacI protein translocation along nonspecific DNA sequences, or the most appropriate sliding mechanism. If experimental results disagree with the existing translocation models, alternative sliding models should be proposed.

2.6.1 *Analytical SD Calculation of Sliding Proteins*

In order to obtain analytical expressions for the fluorescence intensity profile and SD of a sliding protein according to the different translocation models, the following studies should be carried out: (1) develop pathway distribution functions g(x) for single trajectories, (2) quantify the effects of defocusing and DNA fluctuation on SD measurements, (3) assess SD measurement error associated with different sliding models, and (4) finally consolidate all analyses to arrive at an expression for expected SD of sliding proteins on DNA for different translocation models.

2.6.1.1 Obtain the Explicit Pathway Distribution Function *g(x)* for Individual Trajectories

When we calculate the sliding PSF SD value $s_{Sliding}$ in Sect. 2.5.2 using the deconvolution method, the pathway distribution function used was the location distribution at time t for an ensemble of independent trajectories, which is a Gaussian distribution as $g(x) = exp(-x^2 / 4D_1t)$, where the mean square displacement of many different trajectories at time t is $\langle x^2 \rangle = 2D_1t$. Although it is a good approximation, this $g(x)$ is not exactly appropriate for studying individual trajectories, such as the single sliding proteins we are analyzing. To our knowledge, the pathway distribution function for single Brownian trajectories has not been formulated, in contrast to the well known $P(x) = exp(-x^2 / 4D_1t)$ relation for an ensemble of independent trajectories.

We have performed preliminary simulations of single Brownian trajectories and observed that the SD of the pathway distribution for single trajectories $s_{SingleTrajectory}$ is at least two times smaller than that of an ensemble of independent trajectories. Additional simulations and analytical calculations are necessary to formulate $s_{SingleTrajectory}$ numerically and analytically.

2.6.1.2 Quantify the Defocusing and DNA Fluctuation Effects Using Single Specifically Bound GFP-LacI Molecules

Although we have estimated in Sects. 2.5.4 and 2.5.5 that the defocusing and DNA fluctuation should have minimal effects on protein sliding measurements, for mechanisms discrimination studies at nanometer precision, the precise effect should be experimentally quantified as an offset to the measured SD.

While the transverse fluctuation of DNA molecules has been extensively investigated [64, 65], the longitudinal fluctuations of elongated DNA molecules have not been studied. Since λ DNA has two *lacO* sites in its sequence, a single LacI can be bound specifically to an elongated λ DNA and the exact $s_{Fluctuation}$ of the protein due to defocusing and DNA fluctuation can be obtained. The DNA molecule will be anchored at both ends onto a fused-silica surface, and the *lacO* sequences between the two anchors can be identified by the specifically bound protein. This study should provide the control measurement for all external effects that may increase the expected SD values of sliding proteins. The difference in SDs between the single specifically bound LacI on DNA and immobile proteins on surfaces will be the SD offset, Δs_{Offset}, resulted from defocusing and DNA longitudinal fluctuation for the experimental setting.

2.6.1.3 Calculate SD Error $\Delta s_{\mathrm{Sliding}}$ for Sliding Protein Images

In Sect. 2.4.3, we have developed the SD measurement error expression for immobile single molecules. For mobile molecules, such as sliding proteins on DNA, the SD measurement error should be developed for correct assessment of the measurement precision. Using the new single trajectory pathway distribution functions $g(x)$ for Brownian sliding and subdiffusion slidings to be developed in Sect. 2.6.1.1, the expected intensity profiles of sliding proteins will differ from that of immobile proteins. By using the same method for SD error derivation for immobile single molecules and replacing the intensity profile of the immobile molecules with that of the sliding proteins, SD measurement error for sliding proteins can be obtained [61].

2.6.1.4 SD of the Final Sliding Protein Images

When we combine all factors that may affect the observed $s_{Sliding}$ values, the final theoretical $s_{Sliding}$ should include SD due to the sliding motion $s_{SingleTrajectory}$, the defocusing and DNA fluctuation offset Δs_{Offset}, and SD measurement error $\Delta s_{Sliding}$ as

$$s_{Sliding} = \sqrt{s^2_{SingleTrajectory} + (\Delta s_{Offset})^2} \pm \Delta s_{Sliding}. \tag{2.8}$$

2.6.2 *SD Simulations of Sliding Proteins*

In order to simulate the fluorescence intensity profiles of sliding proteins, proteins' trajectories need to convolve with the PSF. All other experimental effects should also be included in the simulation. These effects are DNA fluctuation, defocusing, evanescent light intensity decay, and statistical error originated from fluctuating numbers of photons per PSF, finite camera pixel size, and background noise.

The simulation of sliding LacI images can be carried out in the following order: (1) Simulate 1D Brownian diffusion trajectories along DNA with the known $\langle D_1 \rangle = 10^5\,\mathrm{nm}^2/\mathrm{s}$ and subdiffusion trajectories with a starting b value of 0.5. How do we choose the step size (or time interval for simulation)? This depends on how many photons will be there in an image. If there are 100 photons in a 0.1 ms image, then the step size cannot be smaller than the interval during which one photon is detected, or 0.1 ms/100 = 1 μs. (2) Incorporate the transverse and axial direction fluctuations into the 1D trajectories according to DNA's transverse fluctuations, thus forming a 3D trajectory. (3) At each step, insert a PSF. The PSF should only have one photon, and the location of the photon will be randomly distributed within the PSF according to the PSF Gaussian photon distribution. The PSF SD will change according to the axial location 7. (4) Correct each photon count by a weighting factor, since the decaying evanescent excitation (Fig. 2.3b) does not excite the fluorophore at all depths equally. (5) Bin the simulated weighted photons for the whole image into pixels to include the camera pixelation effect. (6) Add the background noise to the binned image. (7) Fit the final intensity profile to a 2D Gaussian to obtain the SD value for the image. (8) Obtain the SD distribution for many simulated images for a specific exposure time. (9) Add the DNA fluctuation and defocusing offset s_{Offset} to the mean SD. (10) Plot and compare the mean SD values at different exposure times with experimental data.

2.6.3 *Experimental SD Measurements of Sliding LacI on λ DNA*

In Sect. 2.5.1, SD distributions of diffusion LacI on DNA were measured for three exposure times of 5 ms, 10 ms, and 15 ms. Although changes in SD with exposure time are apparent, these exposure times might be too long for hopping to be negligible. In order to ensure detection of genuine sliding motion, exposure times of 0.1–2 ms should be used.

The mean SD values from SD distributions at various exposure times should be plotted. If SDs increase with t as SD(t) = $\sqrt{s_0^2 + CD_1 t}$, where C is the coefficient for single trajectories that will be calculated in Sect. 2.6.1.1, the sliding is Brownian. If the SD increases are less, then the sliding is subdiffusive.

2.6.4 *Resolution Limits, Alternative Translocation Models, and Extensions*

During experiments, the following items might be of concern and may introduce further complications to the studies.

2.6.4.1 Resolution Limitation

It is possible that even with nanometer resolution of the SD measurements, different sliding models may not be differentiated if the SD difference between the models is less than the SD measurement error, which in our case is ≈ 0.8 nm (Sect. 2.5.2). One example is the DNA-sequence-dependent sliding Models I and III (Fig. 2.7), where their $\langle n^2 \rangle$ values at 2 ms are almost identical. If this occurs, then this study can at least rule out subdiffusion models with a *b* value significantly lower than 0.5, or models with a very low *A* or D_1 value, or the combination of the three.

2.6.4.2 Obtain the Realistic Sliding–Hopping Alternation Kinetics by Simulation

In spite of recent experimental and theoretical investigations of protein hopping on DNA [30, 66], the real hopping interval distribution and nonspecific DNA dissociation time distribution remain unclear. Simulation might offer a solution to the problem. The whole sliding and hopping process to seconds long can be simulated using Brownian dynamics to search for a combination of sliding characteristics and hopping time distribution that yield the known LacI 1D Brownian diffusion with $\langle D_1 \rangle = 10^5\,\text{nm}^2/\text{s}$ in the timescale of seconds [45]. The results should provide validation for the interpretation of experimental results.

2.6.4.3 Uncertain Sliding Dissociation Time and Possible Alternative Sliding Model

If the unknown sliding time is more than 5 ms, the lowest exposure time used in experiments in Sect. 2.5.1, then reinterpretation of data in Fig. 2.10 for the 5-ms exposure time suggests an alternative mechanism of translocation: one that diffuses fast on millisecond timescales and slow on second timescales.

Let us calculate the expected SD ($s_{Sliding}$) for the Brownian sliding model assuming that the mean sliding time is longer than 5 ms. Since the sliding is Brownian for all timescales, D_1 is constant at all timescales. Putting the experimental values of t = 5 ms and $\langle D_1 \rangle = 10^5$ nm²/s [45] into 6, $s_{Sliding}$ = 133.7 nm. This calculated $s_{Sliding}$ is significantly lower than the measured mean value of $\langle S_x \rangle = 165 \pm 2.6$ nm (Fig. 2.10b). This result suggests faster diffusion in millisecond timescales than in second timescales, contradicting the existing Brownian sliding and subdiffusion sliding models.

2.6.5 *Relating Physical and Biological Significance*

In some respects, the *rate* at which transcription factors find their specific sites is not a very interesting biological problem. While you can worry about the speed needed to find transcription factor sites in rapidly growing cells such as *E. coli*, for some cells there would seem to be plenty of time for the proteins to find their sites given the known range of on and off nonspecific dissociation dynamics and the probable size of the specific site at which the transcription factor binds tightly. It is possible that a far more important biological problem is as follows: what is the physics of how tight association to a specific sequence of basepairs occurs as opposed to the secondary issue of how the site is found.

The common thread linking the specific and nonspecific binding is the DNA sequence, just as in the protein problem the common thread linking folding rates and activity is the amino acid sequence. In this paper, we discuss a new way to analyze the dynamics of transcription factor dynamics on a DNA molecule in terms of pathway distribution function g(x), the position-(and time) dependent dynamics of the transcription factor moving along the DNA molecule. It is somewhat comparable to the distribution of conformation states g(E) that the Frauenfelder group developed some years ago which characterized the probability distribution of a protein on a free energy landscape and the subsequent dynamics [39, 67]. We can ask if, perhaps, the g(x) we introduce here could play a similar role for the movement of a transcription factor along the DNA as it moves in a, perhaps, sequence-directed manner toward deep specific binding sites. This approach could elucidate the basic biologically relevant mechanisms.

2.7 Summary

In this chapter, we present single-molecule fluorescence imaging studies of LacI protein diffusing along elongated DNA molecules and detail the studies necessary to discriminate different sliding mechanisms of proteins on DNA using theoretical calculations, simulations, and experimental measurements. A new method, SMID, that can improve the temporal resolution of single-molecule tracking experiments by 100-fold is developed and applied to LacI diffusion studies. The experimental and analytical methods presented here should advance the single-molecule imaging field by providing a new method that improves the temporal and spatial resolutions of single-molecule tracking. The sliding mechanism research should provide fundamental insights into how proteins interact with nonspecific DNA sequences and help to quantify the effect of facilitated diffusion on gene regulation.

References

1. Ptashne M (1992) A genetic switch: phage lambda and higher organisms, 2nd edn. Blackwell, Cambridge, MA
2. Alberts B, Johnson A, Lewis J, Raff M, Roberts K, Walter P (2002) Molecular biology of the cell, fourth edn. Garland Science, New York

3. Revzin A (1990) The biology of nonspecific DNA protein interactions. CRC Press, London
4. Riggs AD, Bougeois S, Cohn M (1970) The *lac* repressor-operator interaction. 3. Kinetic studies. J Mol Biol 53:401–417
5. Adam G, Delbruck M (1968) Reduction of dimensionality in biological diffusion process. In: Rich A, Davidson N (eds) Structural chemistry in molecular biology, Freeman, San Francisco, pp. 198–215
6. Berg OG, Blomberg C (1976) Association kinetics with coupled diffusional flow. Special application to the Lac repressor-operator system. Biophys Chem 4:367–381
7. Berg OG (1978) On diffusion-controlled dissociation. Chem Phys 31:47–57
8. Winter RB, von Hippel PH (1981) Diffusion-driven mechanisms of protein translocation on nucleic acids. 2. The Escherichia coli repressor-operator interaction: equilibrium measurements. Biochemistry 20:6948–6960
9. Jack WE, Terry BJ, Modrich P (1982) Involvement of outside DNA sequences in the major kinetic path by which EcoRI endonuclease locates and leaves its recognition sequence. Proc Natl Acad Sci U S A 79:4010–4014
10. Ricchetti M, Metzger W, Heumann H (1988) One-dimensional diffusion of Escherichia coli DNA-dependent RNA polymerase: A mechanism to facilitate promoter location. Proc Natl Acad Sci U S A 85:4610–4614
11. Ruusala T, Crothers DM (1992) Sliding and intermolecular transfer of the Lac repressor - Kinetic perturbation of a reaction intermediate by a distant DNA-sequence. Proc Natl Acad Sci U S A 89:4903–4907
12. Kabata H, Kurosawa O, Arai I, Washizu M, Margarson SA, Glass RE, Shimamoto N (1993) Visualization of single molecules of RNA polymerase sliding along DNA. Science 262:1561–1563
13. Hsien M, Brenowitz M (1997) Comparison of the DNA association kinetics of the Lac repressor tetramer, its dimeric mutant LacI^{adj} and the native dimeric Gal repressor. J Biol Chem 272:22092–22096
14. Shimamoto N (1999) One-dimensional diffusion of proteins along DNA. J Biol Chem 274:15293–15296
15. Halford SE, Marko JF (2004) How do site-specific DNA-binding proteins find their targets? Nucleic Acids Res 32:3040–3052
16. Misteli T (2001) Nuclear structure - Protein dynamics: Implications for nuclear architecture and gene expression. Science 291:843–847
17. Stanford NP, Szczelkun D, Marko JF, Halford SE (2000) One- and three-dimensional pathways for proteins to reach specific DNA sites. EMBO J 23:6546–6557
18. Gowers DM, Halford SE (2003) Protein motion from non-specific to specific DNA by three-dimensional routes aided by supercoiling. Embo J 22:1410–1418
19. Elf J, Li G-W, Xie XS (2007) Probing transcription factor dynamics at the single-molecule level in a living cell. Science 316:1191–1194
20. Biebricher A, Wende W, Escudé C, Pingoud A, Desbiolles P (2009) Tracking of single quantum dot labeled EcoRV sliding along DNA manipulated by double optical tweezers. Biophys J: Biophys Lett 96:L50–L52
21. Li G-W, Berg OG, Elf J (2009) Effects of macromolecular crowding and DNA looping on gene regulation kinetics. Nat Phys 5:294–297
22. Barbi M, Place C, Popkov V, Salerno M (2004) A model of sequence-dependent protein diffusion along DNA. J Biol phys 30:203–226
23. Berg OG, Ehrenberg M (1982) Association kinetics with coupled three- and one-dimensional diffusion : Chain-length dependence of the association rate to specific DNA sites. Biophys Chem 15:41–51
24. Gowers DM, Wilson GG, Halford SE (2005) Measurement of the contributions of 1D and 3D pathways to the translocation of a protein along DNA. Proc Natl Acad Sci U S A 102:15883–15888
25. Porecha RH, Stivers JT (2008) Uracil DNA glycosylase uses DNA hopping ad short-range sliding to trap extrahelical uracils. Proc Natl Acad Sci U S A 105:10791–10796

26. Harada Y, Funatsu T, Murakami K, Nonoyama Y, Ishihama A, Yanagida T (1999) Single-molecule imaging of RNA polymerase-DNA interactions in real time. Biophys J 76:709–715
27. Berg OG, von Hippel PH (1985) Diffusion-controlled macromolecular interactions. Annu Rev Biophys Biophys Chem 14:131–160
28. Barkley MD (1981) Salt dependence of the kinetics of the *lac* repressor-operator interaction: role of nonoperator deoxyribonucleic acid (DNA) in the association reaction. Biochemistry 20:3833–3842
29. Hu L, Grosberg AY, Bruinsma R (2008) Are DNA transcription factor proteins Maxwellian demons? Biophys J 95:1151–1156
30. Wunderlich Z, Mirny LA (2008) Spatial effects on the speed and reliability of protein-DNA search. Nucleic Acids Res 36:3570–3578
31. Wang YM, Tegenfeldt J, Reisner W, Riehn R, Guan X-J, Guo L, Golding I, Cox EC, Sturm J, Austin RH (2005) Single-molecule studies of repressor-DNA interactions show long-range interactions. Proc Natl Acad Sci U S A 102:9796–9801
32. Perkins TT, Smith DE, Larson RG, Chu S (1995) Stretching of a single tethered polymer in a uniform flow. Science 268:83–87
33. Kalodimos CG, Biris N, Bonvin AMJJ, Levandoski MM, Guennuegues M, Boelens R, Kaptein R (2004) Adaptation in nonspecific and specific protein-DNA complexes. Science 305:386–389
34. Smith SB, Finzi L, Bustamante C (1992) Direct mechanical measurements of the elasticity of single DNA molecules by using magnetic beads. Science 258:1122–1126
35. Thompson RE, Larson DR, Webb WW (2002) Precise nanometer localization analysis for individual fluorescent probes. Biophys J 82:2775–2783
36. Qian H, Sheetz MP, Elson EL (1991) Single particle tracking: Analysis of diffusion and flow in two-dimensional systems. Biophys J 60:910–921
37. Blainey PC, van Oijent AM, Banerjee A, Verdine GL, Xie XS (2006) A base-excision DNA-repair protein finds intrahelical lesion bases by fast sliding in contact with DNA. Proc Natl Acad Sci U S A 103:5752–5757
38. Graneli A, Yeykal C, Robertson R, Greene E (2006) Long-distance lateral diffusion of human Rad51 on double-stranded DNA. Proc Natl Acad Sci U S A 103:1221–1226
39. Austin RH, Beeson K, Eisenstein L, Frauenfelder H, Gunsalus I, Marshall V (1974) Activation energy spectrum of a biomolecule: Photodissociation of carbonmonoxy myoglobin at low temperatures. Phys Rev Lett 32:403–405
40. Slutsky M, Mirny LA (2004) Kinetics of protein-DNA interaction: Facilitated target location in sequence-dependent potential. Biophys J 87:4021–4035
41. Barbi M, Place C, Popkov V, Salerno M (2004) Base-sequence-dependent sliding of proteins on DNA. Phys Rev E 70:041901
42. Gerland U, Moroz JD, Hwa T (2002) Physical constraints and functional characteristics of transcription factor–DNA interaction. Proc Natl Acad Sci U S A 99:12015–12020
43. Berg OG, von Hippel PH (1989) Selection of DNA binding sites by regulatory proteins statistical-mechanical theory and application to operators and Promoters. J Mol Biol 193:723–750
44. von Hippel P, Rees WA, Rippe K, Wilson KS (1996) Specificity mechanisms in the control of transcription. Biophys Chem 59:231
45. Wang YM, Austin RH, Cox EC (2006) Single molecule measurements of repressor protein 1D diffusion on DNA. Phys Rev Lett 97:048302
46. Gorman J, Chowdhury A, Surtees JA, Shimada J, Reichman DR, Alani E, Greene EC (2007) Dynamic Basis for one-dimensional DNA scanning by the mismatch repair complex Msh2-Msh6. Cell 28:359–370
47. Kim JH, Larson RG (2007) Single-molecule analysis of 1D diffusion and transcription elongation of T7 RNA polymerase along individual stretched DNA molecules. Nucleic Acids Res 35:3848–3858
48. Tafvizi A, Huang F, Leith JS, Fersht AR, Mirny LA, van Oijen AM (2008) Tumor Suppressor p53 Slides on DNA with low friction and high stability. Biophys J: Biophys Lett 95:L01–L03

49. Bonnet I, Biebricher A, Porté P-L, Loverdo C, Bénichou O, Voituriez R, Escudé C, Wende W, Pingoud A, Desbiolles P (2008) Sliding and jumping of single EcoRV restriction enzymes on non-cognate DNA. Nucleic Acids Res 36:4118–4127
50. Gorman J, Greene EC (2008) Visualizing one-dimensional diffusion of proteins along DNA. Nat Struct Mol Biol 15:768–774
51. van Mameren J, Peterman EJG, Wuite GJL (2008) See me, feel me: methods to concurrently visualize and manipulate single DNA molecules and associated proteins. Nucleic Acids Res 36:4381–4389
52. Komazin-Meredith G, Mirchev R, Golan DE, van Oijen AM, Coen DM (2008) Hopping of a precessivity factor on DNA revealed by single-molecule assays of diffusion. Proc Natl Acad Sci U S A 105:10721–10726
53. Laurence TA, Kwon Y, Johnson A, Hollars CW, O'Donnell M, Camarero JA, Barsky D (2009) Motion of a DNA sliding clamp observed by single molecule fluorescence spectroscopy. J Biol Chem 283:22895–22906
54. Lin Y, Zhao T, Jian X, Farooqui Z, Qu X, He C, Dinner AR, Scherer NF (2009) Using the bias from flow to elucidate single DNA repair protein sliding and interactions with DNA. Biophys J 96:1911–1917
55. Kurita H, Torii K, Yasuda H, Takashima K, Katsura S, Mizuno A (2009) The effect of physical form of DNA on ExonucleaseIII activity revealed by single-molecule observations. J Fluoresc 19:33–40
56. Fan H-F, Li H-W (2009) Studying RecBCD helicase translocation along χ-DNA using tethered partical motion with a stretching force. Biophys J 96:1875–1883
57. van Oijen A, Köhler J, Schmidt J, Müller M, Brakenhoff G (1998) 3-Dimensional super-resolution by spectrally selective imaging. Chem Phys Lett 292:183–187
58. Speidel M, Jonas A, Florin E-L (2003) Three-dimensional tracking of fluorescent nanoparticles with subnanometer precision by use of off-focus imaging. Opt Lett 28:69–71
59. Huang B, Wang W, Bates M, Zhuang X (2008) Three-dimensional super-resolution imaging by stochastic optical reconstruction microscopy. Science 319:810–813
60. DeCenzo S, DeSantis M, Wang YM (2010) Single-image separation measurements of two unresolved fluorophores. Opt Express 18:16628–16639
61. DeSantis M, DeCenzo S, Li JL, Wang Y (submitted) Precision analysis for standard deviation measurements of immobile single fluorescent molecule images. Opt Express 18:6563–6576
62. Yildiz A, Tomishige M, Vale RD, Selvin PR (2004) Kinesin walks hand-over-hand. Science 303:676–678
63. Schütz GJ, Pastushenko VP, Gruber HJ, Knaus H-G, Pragl B, Schindler H (2000) 3D imaging of individual ion channels in live cells at 40 nm resolution. Single Mol 1:25–31
64. Quake SR, Babcock H, Chu S (1997) The dynamics of partially extended single molecules of DNA. Nature 388:151–154
65. Crut A, Lasne D, Allemand JF, Dahan M, Desbiolles P (2003) Transverse fluctuations of single DNA molecules attached at both extremities to a surface. Phys Rev E 67:051910
66. Loverdo C, Bénichou O, Voituriez R (2009) Quantifying hopping and jumping in facilitated diffusion of DNA-binding proteins. Phys Rev Lett 102:188101
67. Austin R, Beeson K, Eisenstein L, Frauenfelder H, Gunsalus I, Marshall V (1975) Dynamics of ligand binding to myoglobin. Biochemistry 14:5355–5373

Chapter 3
How Proteins Slide on DNA

Daniel Barsky, Ted A. Laurence, and Česlovas Venclovas

3.1 Introduction

Protein–DNA interactions are required for all the major functions of DNA: transcription and regulation, replication and repair, even the packaging of DNA into chromosomes. Not only are protein–DNA interactions crucial for all these cellular activities, but they are also, in our view, among the most fascinating macromolecular interactions because of their dynamics. In this chapter, we focus on DNA sliding by proteins, particularly diffusive sliding. Such sliding is typically part of the search for a target on the DNA itself or for another protein bound to the DNA. Of particular interest here are the proteins known as DNA sliding clamps that can remain bound to the DNA while diffusing vast distances along the double helix of DNA. We do not yet know the detailed mechanisms of protein sliding on DNA, but we aim to familiarize the reader with what is known observationally and to provide some discussion of potential mechanisms.

Passive vs. active sliding. Proteins that interact with DNA can be divided into two groups: those that actively move along DNA ("active sliders") and those that do not. The first group contains proteins such as DNA and RNA polymerases and helicases, and because of the "processivity" of their functions (e.g., incorporating one base after another), it is intuitively obvious that these proteins can slide along DNA, that is, remain in contact with DNA while moving along it. Many proteins in the second group also slide or hop along DNA, either through attachment to active sliders or by diffusion ("passive sliders"). Among the members of the second group are proteins such as DNA replication processivity factors that function to promote the retention of an active slider (such as a DNA polymerase) on the DNA. Because they remain bound to DNA independent of binding to other proteins, the processivity factors have been dubbed "DNA sliding clamps". In isolation on DNA, a DNA sliding clamp becomes a passive slider. DNA sliding clamps exist in all life forms. Examples include PCNA (eukaryotic and archaeal), β clamp (bacterial), UL42/44 (viral), and gp45 (of phages). Other passive

D. Barsky (✉)
Physical and Life Sciences Directorate, Lawrence Livermore National Laboratory,
7000 East Avenue, Livermore, CA 94550, USA
e-mail: barsky@cornell.edu

M.C. Williams and L.J. Maher, III (eds.), *Biophysics of DNA-Protein Interactions*,
Biological and Medical Physics, Biomedical Engineering,
DOI 10.1007/978-0-387-92808-1_3,

sliders are not obviously sliders at all. In a sense, the fact that sequence-specific proteins such as restriction endonucleases and transcription factors can slide along DNA may appear surprising. After all, a biomolecule that binds its specific target sequence with nanomolar specificity might be expected to bind and stay put. The very interactions that stabilize specific binding (i.e., the "binding specificity") would seem to ensure that sliding does not occur. Nevertheless, even proteins that bind their DNA targets very tightly slide or hop along DNA to get there.

Brief historical perspective. The concept that proteins passively slide on DNA to locate specific binding sequences is at least four decades old. As with many ideas in science, it is difficult to determine when the idea was first formulated. In the early 1980s, Otto Berg, Peter von Hippel, and colleagues wrote a series of landmark papers. In one of those papers, Berg et al. [1] cited a 1968 book chapter by Adam and Delbrück [2]. Hence, we know the concept is at least four decades old. A defining (and likely independent) moment occurred in 1970 when it was reported that the lacI repressor protein could locate its target site about two orders of magnitude faster than predicted by normal diffusion-collision mechanism (i.e., three-dimensional (3D) diffusion and random collision) [3]. Given the sensitivity of the result to the salt concentration, it was immediately interpreted as evidence that the DNA electrostatically attracts the protein [3]. Later, the search for the specific DNA target was modeled as a two-step process in which the search was "facilitated" by first binding the DNA molecule nonspecifically and then sliding to the target [1]. In the ensuing decades, a combination of biochemical assays and theoretical considerations gave rise to an increasing appreciation of the importance of sliding along DNA for many DNA-binding proteins. The latest chapter in the lacI repressor story is presented by Wang and Austin in this book.

Sliding, hoping, and jumping. In their papers, Otto Berg, Peter von Hippel, and colleagues developed a theory of this facilitated diffusion model and followed it with further experiments on lacI [1, 4, 5]. They laid out four possible modes of DNA target searching by proteins – more generally, modes of moving from one site to another on DNA: (1) *sliding* along DNA via continuous one-dimensional (1D) diffusion without dissociation, (2) *hopping*, where the protein effectively diffuses along a single molecule of DNA but does so via a series of dissociation and rebinding events, (3) *jumping*, which, in contrast to facilitated diffusion, amounts to ordinary 3D diffusion between DNA sites, and (4) *intersegmental transfer*, where the protein swaps sites on the DNA via a looped intermediate (Fig. 3.1). To date with some minor differences all models of proteins moving on DNA have defined these same modes [6, 7]. The first, second, and third modes present a continuum transition from 1D diffusion on DNA to 3D diffusion. The fourth mode, intersegmental transfer, does not fit so neatly into this continuum. This mode requires at least two DNA-binding domains in a special arrangement.

While some experimentation and refinement of this theory took place in the ensuing decades, there has been an explosion of activity in just the last few years, particularly through direct observation of protein sliding on DNA. In our review of that activity, we consider only passive sliding. For investigations of active sliders such as polymerases, there are recent reviews [8, 9]. Many proteins have been observed passively sliding on DNA (Table 3.1). These include proteins responsible for detecting DNA damage (hOgg1, MutM, Msh2–Msh6, Ada, Rad51), proteins

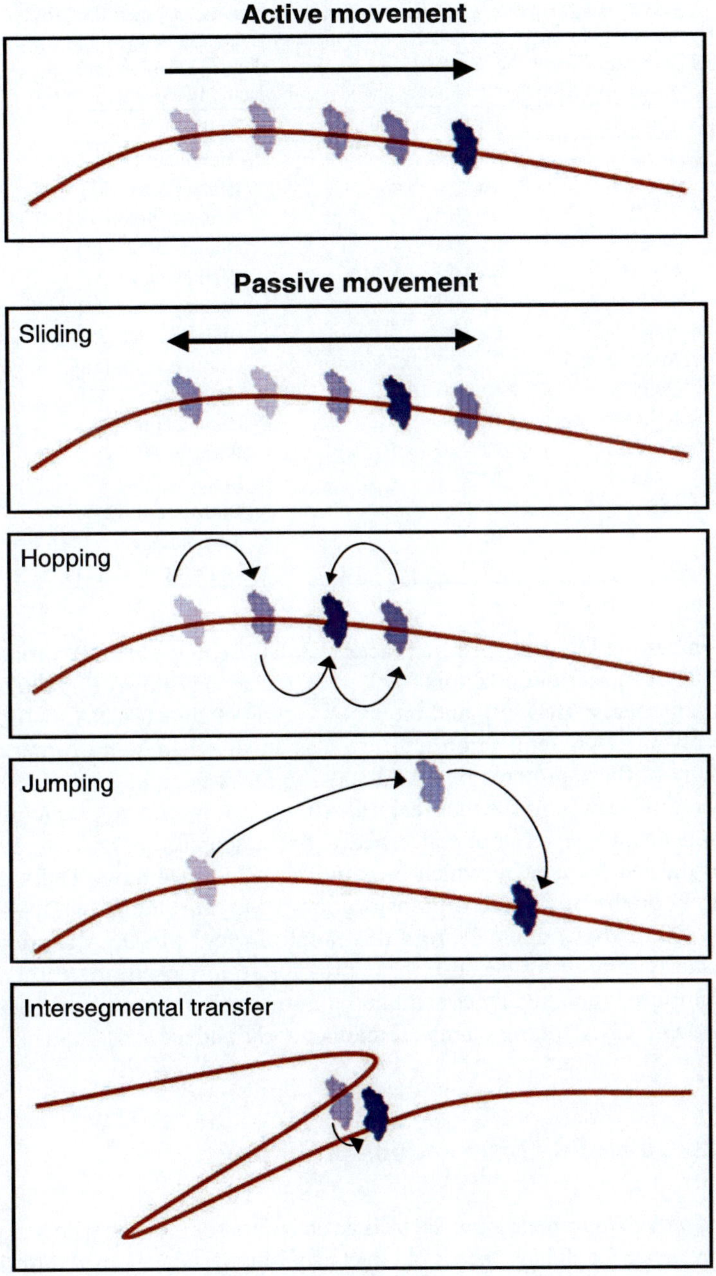

Fig. 3.1 Modes of proteins searching for DNA targets. This chapter focuses on passive sliding

Table 3.1 Proteins and their diffusion constants, as measured by single-molecule methods. All of the measurements use single-molecule tracking, except the *Eco*RI [30] and the β clamp [11] measurements which use a biochemical assay and single-molecule FRET, respectively

Protein	Diffusion constant (m^2/s)	References
*Eco*RV	3×10^{-15}	Biebricher et al. [20]
*Eco*RV	10^{-14}	Bonnet et al. [21]
*Eco*RI	3×10^{-15}	Rau and Sidorova [30]
hOgg1	6×10^{-13}	Blainey et al. [22]
MutM	4×10^{-14}	Blainey et al. [22]
AVP–pVIc	2×10^{-12}	Blainey et al. [31]
*Bam*HI	6×10^{-13}	Blainey et al. [31]
Msh2–Msh6	9×10^{-14} to 2×10^{-16}	Gorman et al. [18]
Rad51	10^{-13} to 10^{-14}	Graneli et al. [32]
RNAP	10^{-14}	Harada et al. [23]
LacI	10^{-13} to 2×10^{-16}	Wang et al. [27]
C-Ada	10^{-12}	Lin et al. [25]
p53	3×10^{-13}	Tafvizi et al. [26]
PCNA	10^{-12}	Kochaniak et al. [24]
β clamp	10^{-14}	Laurence et al. [11]

involved in cutting DNA in specific places (*Eco*RI, *Eco*RV, *Bam*HI), promoter and repressor DNA transcription factors (lacI, p53), the adenoviral AVP–pVIc complex, the RNA polymerase (RNAP), and the DNA sliding clamps (PCNA, β clamp) that aid critically in DNA replication and are also involved in some forms of DNA repair. Many of these proteins bind to a specific DNA feature; they slide on DNA in the course of searching for that feature. Although a fascinating topic in its own right, in this chapter we are not concerned with the search for DNA targets per se, but rather the mechanism by which proteins actually move along DNA. We thus focus mainly on the first mode of translocation, *bona fide* sliding on DNA, and on its nearest alternative, hopping. From this point of view, the DNA sliding clamps are particularly interesting because these proteins form closed rings around DNA and are therefore topologically constrained to remain bound. As we will see below, the diffusion of DNA sliding clamps remains poorly understood.

3.2 Experimental Observations of Sliding

Monitoring protein movement along DNA is an active area of research with many papers reporting evidence for sliding, speeds of target acquisition, and measured diffusion constants. Until recently, biochemical assays commonly involving various pathways and traps were the most frequently used methods, but these have now been largely overtaken by single-molecule assays, especially single-molecule tracking by total internal reflection fluorescence (TIRF) microscopy. Other single-molecule methods include atomic force microscopy (AFM) and fluorescence resonance energy transfer (FRET).

Intuitively, it might be expected that it would be difficult to distinguish between sliding and hopping on DNA, and indeed, this is the case [10]. What is known about translocation in terms of distance and time is dictated by what can be observed experimentally. In this respect, the very short time and length scales involved in translocation over a small number of base pairs can make the distinction between sliding and hopping very difficult. Most experiments observe movement over many seconds and across hundreds to tens of thousands of base pairs, although some recent reports probe distances as short as tens of base pairs with time resolution in submicroseconds [11].

Because active sliding tends to be directional, it can be characterized by a simple rate, with units of base pairs per second (or in SI units, m/s). Passive sliding is a diffusional or Brownian process and therefore is not properly characterized by a rate, but rather by a diffusion constant (D). The diffusion constant is related to the average distance that a particle moves (Δx) in a time (t) by Einstein's simple relation:

$$\left\langle \Delta x^2 \right\rangle = 2nDt \tag{3.1}$$

where n is the number of dimensions involved in diffusion (i.e., 1, 2, or 3). This has two important consequences. First, the time it takes for a protein to cover a distance by random search grows very rapidly, as the square of the distance. Second, Δx is a distance independent of direction, so, for example, after a time interval t a protein sliding on DNA with diffusion constant D is just as likely to be Δx "beyond" or "behind" its original position. Despite this distinction between active and passive sliding, it is common nonetheless to use dynamic adjectives such as "fast" or "slow" to characterize passive sliding with relatively high or low diffusion constants.

3.2.1 Biochemical Assays

To date, a great wealth of data has come from biochemical assays. By measuring the rate of product production, biochemical assays that measure the productivity and processivity of active sliders, such as DNA polymerases, provide the sliding rate for these proteins. For example, we know that DNA polymerase III holoenzyme moves along the template strand, reading it and building the opposite strand at the rate of roughly one base every 2 ms [12], a measurement more recently confirmed by single-molecule methods [13]. For the passive sliders, similar strategies can be used, even in cases where no product is produced (nonenzymatic sliders). Although biochemical assays are not ideal for monitoring motion, certain clever experiments can reveal critical information about sliding. In many cases, the biochemical assays make use of DNA topology such as closed loops or even catenated DNA (where one loop runs inside another, as in links of a chain) and careful placement of targets and starting points [14].

One strategy is to load the protein onto DNA at a known location and then measure the time it takes to reach a target, as has been done for a DNA sliding clamp, the *Escherichia coli* β clamp more than a decade ago in the O'Donnell lab. In a control experiment, the β clamp was loaded onto a nicked, circular DNA plasmid (7,200 bp), and the β clamp remained on the circular plasmid with a half-life of over an hour – a long time relative to the life of a bacterial cell [15]. When the circular DNA plasmid was cut open by a site-specific endonuclease, the β clamp disassociated from the DNA within a few minutes by sliding off one of the free ends of the "linearized" plasmid. Although those experiments were intended as qualitative observations of sliding on DNA and not designed to measure the diffusion constant, we can nevertheless calculate at least a lower limit for the diffusion constant. After 1 min of linearizing (endonuclease) reaction, plasmids were analyzed by approximately 15 min of gel filtration, after which little or none of the sliding clamps were present on the linearized DNA [15]. Using this time (~15 min), and assuming that clamps were at random positions on the 7,200-bp plasmid upon linearization, a "survival probability" analysis [16] yields a relation between the time τ, interval length L, and diffusion constant D, such that

$$\tau = L^2 / (D\pi^2), \tag{3.2}$$

and this gives a limit of $D \geq 10^{-15}$ m²/s.

Rau and Sidorova devised a method for measuring sliding on DNA based on the ratio of the dissociation rate of the *Eco*RI restriction endonuclease from DNA fragments containing one specific binding site vs. the dissociation rate of *Eco*RI from DNA fragments containing two specific binding sites [30]. The measurements enabled the authors to determine the sliding rate (Table 3.1), finding it relatively insensitive to salt concentration and osmotic pressure, and indicating that the "protein moves smoothly along the DNA probably following the helical phosphate-sugar backbone of DNA." They estimate that the sliding rate they measure is 2,000-fold slower than the diffusion of free protein in solution. A factor of 40–50 can be accounted for by rotational drag that would result from following a helical path on the DNA. They suggest that the other factor of 40–50 could reflect the requirement for making and breaking salt bridges between the DNA and the protein, or to the disruption of water structure at the protein–DNA interface as the two surfaces move past each other.

Besides biochemical approaches, there have been a variety of spectroscopic approaches, new and old. Older approaches such as fluorescence recovery after photobleaching (FRAP) are poorly suited to such measurements. Extracting information on the diffusion of sliding clamps along DNA would require synchronized loading of clamps on many aligned and uniformly stretched DNA molecules. Even so, some useful information can be obtained from older optical methods. For example, fluorescence anisotropy allows one to create a nonequilibrium situation very simply by exciting only the fluorophores whose excitation dipoles happen to be aligned with excitation light. Although definitive information on diffusive motion is difficult to obtain by this approach, Austin et al. [17] were able to place an upper bound on the diffusion rate of *E. coli* RNA polymerase on DNA using this technique. The newer, single-molecule biophysical methods are much more effective in revealing diffusion-based sliding.

3.2.2 Single-Molecule Methods

There has been an explosion of single-molecule research for measuring the sliding of proteins on DNA. The diffusion constants of many passive sliders have now been measured primarily by single-molecule methods. Fourteen examples are listed in Table 3.1. Single-molecule measurements are ideal for measuring diffusion of proteins such as sliding clamps on DNA because single-molecule methods directly observe the Brownian motion involved in diffusion. Diffusion constants are then determined by statistical analysis of the Brownian motion. Two single-molecule approaches have been used to observe diffusion of proteins on DNA. First, *single-molecule tracking* measures the position of a protein with an attached fluorophore moving on stretched DNA [18–27]. Second, *single-molecule FRET* monitors distance changes between a donor-labeled protein moving on acceptor-labeled DNA using FRET efficiency changes [11, 28, 29]. The latter methodology does not require stretched DNA or even surface attachment to reveal diffusion.

Single-molecule tracking. This is the most common single-molecule method used to measure protein sliding on DNA. The positions of fluorescently labeled DNA-binding proteins are dynamically monitored as they move on stretched DNA [18–27]. For a recent review, see Gorman and Green [19]. The fluorescent labels for the proteins are small organic fluorophores (such as Alexa 488 or Cy3B) [21–26], fluorescent proteins [27], or quantum dots [18, 20, 24]. These methods directly monitor the position of the fluorescently labeled proteins by measuring their position using a CCD camera (Fig. 3.2a). Excitation is provided by total internal reflection of the fluorescence (TIRF), providing a wide area of excitation that only penetrates ~200 nm into the solution. This is an excellent approach to image wide areas with low background from solution. One exception is provided by Biebricher et al. [20], where the authors take advantage of especially bright quantum dots and use bright field imaging to monitor diffusion away from the glass surface. One current drawback of the TIRF approach is that the CCD cameras used typically have time resolutions in the order of 10–100 ms. This prevents monitoring of faster diffusive motions, which is unfortunate, since the diffusive motion may vary over different length and time scales. For example, there is some evidence of multiple modes of diffusion in the case of PCNA [24]. The time resolution limit is not intrinsic to the method and may be improved with newer technology, although signal-to-noise issues will ultimately limit how fast the position may be monitored. The chapter in this book by Wang and Austin discusses methods to improve the time resolution.

When monitoring the position of the protein directly, stretching of the DNA is necessary (at least to the point of straightening) so that position measured using the camera can be directly linked to position on the DNA. Careful preparation of stretched or elongated DNA is characteristic of this methodology. The mode of stretching is another primary methodological distinction between these studies. The DNA may be attached to the surface at both ends [18, 21, 24], attached at one end and elongated using fluid flow [22, 25, 26], or stretched between beads in optical traps [20, 23]. The proximity of the surface is a possible issue with these studies (except for Biebricher et al. [20] where the DNA is away from the surface),

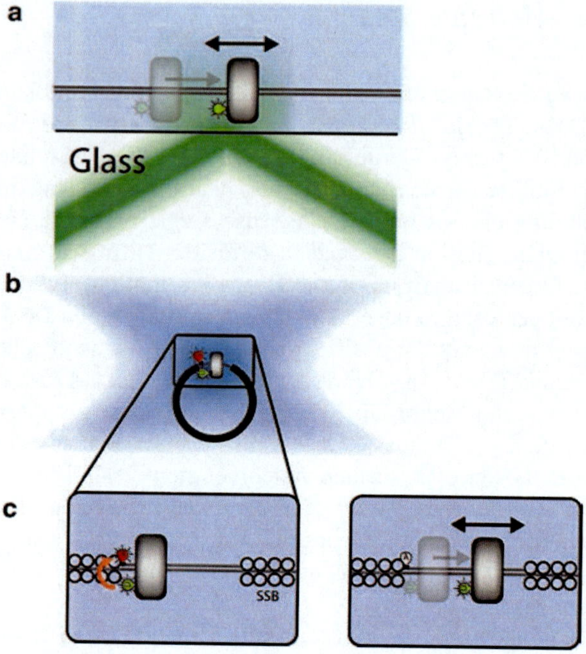

Fig. 3.2 Single-molecule methods used to track proteins sliding on DNA. (**a**) Single-molecule tracking monitors the position of proteins on stretched DNA using total internal reflection excitation in combination with high sensitivity CCD cameras. The position is obtained by fitting the centroid of the observed fluorescence, obtaining accuracy down to ~50 nm. Typical time resolution is about 30–100 ms. (**b**) Fluorescence resonance energy transfer (FRET) between donor-labeled protein and acceptor-labeled DNA may be used to track position of protein on DNA. Using correlations, fluctuations in distance on faster time scales (<1 ms) may be monitored. (**c**) *Left panel*: When the protein labeled with donor (*green bulb*) is near the acceptor label (*red bulb*), FRET occurs (*orange arrow*). *Right panel*: When the protein diffuses away from the acceptor label, FRET decreases or disappears

but these studies have included several controls that confirm that the biochemical systems operate correctly. In the experiments that stretch the DNA with flow, the flow velocity can affect the diffusion of the proteins, especially in cases where "hops" between sites are expected.

Care must also be taken when interpreting data obtained using large labels such as fluorescent protein domains and quantum dots since they are similar in size to the labeled proteins and can clearly affect diffusive behavior. In fact, Kochaniak et al. [24] used a large quantum dot to detect helical sliding by the logic that if the protein is constrained to follow the helical shape of the DNA by revolving once about every 10 bp, the larger hydrodynamic radius caused by the attached quantum dot would slow down the measured diffusion considerably more if the protein was constrained to rotate while sliding. These issues are explored further in Sect. 3.3.

Single-molecule FRET. The second major single-molecule method for monitoring diffusive motion of proteins on DNA is FRET [11, 28, 29]. FRET involves the non-radiative transfer of electronic excitation energy from donor to acceptor fluorescent molecules via a weak dipole–dipole coupling mechanism. FRET requires a resonance between the emission of a donor molecule and the absorption of an acceptor molecule, and depends critically on the relative orientations of the dipole moments of the donor and acceptor molecules, although for fluorescently labeled biomolecules orientational averaging typically occurs. Most importantly, FRET has a strong dependence on the distance between the two fluorophores:

$$E = 1/\left[1 + (R/R_0)^6\right], \tag{3.3}$$

where E is the fraction of donor excitations transferred to the acceptor, R is the distance between donor and acceptor, and R_0 is the distance at which $E=0.5$ (R_0 is between 4 and 7 nm for typical fluorophores). For experiments monitoring proteins sliding on DNA, the relative motion of a donor-labeled protein and acceptor-labeled DNA (donor and acceptor labels can be swapped) is monitored by measuring E as a function of time for each single protein. The FRET methodology is complementary to the tracking methodology described above because the length scales probed are so different between the methods. FRET allows detection of proximities (i.e., distances between labels) in the range of 2–8 nm; whereas, by using CCD cameras to monitor the position of the protein, the tracking methods approach the limits of their sensitivity well above this range at around 50 nm. In addition, movement of the DNA itself will not affect FRET experiments, but will cause additional errors in tracking experiments. At longer time scales (≥1 ms), there is sufficient signal to monitor distance changes by calculating E as a function of time. At shorter time scales (<1 ms), fluorescence correlations reveal FRET intensity fluctuations caused by distance fluctuations.

Using the FRET method and fluorescence correlation spectroscopy, we determined a diffusion constant of $D \sim 10^{-14}$ m²/s for the *E. coli* β clamp sliding on DNA [11, 33]. Our experiments achieved fine time-resolution measurements using a single detection volume (~2 fl) defined by confocal microscopy, detecting photons with avalanche photodiodes (APD), and obtaining time resolutions down to 100 ns [11]. By loading a single protein (β clamp) onto a small section of each large DNA plasmid, we were able to monitor sliding clamps diffusing on DNA via FRET without any immobilization, that is, while both protein and DNA were freely diffusing in solution (Fig. 3.2b and c). Given that a protein with diffusion constant D moves a mean-square distance given by (3.1), $\Delta x^2 = 2D\Delta t$ within time Δt, and the fact that FRET efficiency decreases from nearly 1 to nearly 0 within about 5 nm, we can obtain limits on the diffusion constants measured. The time resolution of 100 ns limits one to measuring diffusion constants $D < 10^{-10}$ m²/s. In our detection methodology, proteins and DNA can diffuse in and out of the detection volume, giving rise to "bursts" of fluorescence. Bursts of fluorescence due to FRET are observed when sliding clamps are loaded on DNA, and then the DNA–clamp complexes diffuse in and out of the confocal detection volume. Diffusion of the DNA–sliding clamp complex out of the confocal detection volume in free solution meant that we could not observe the sliding motion for arbitrarily long periods of time and therefore limited our measurements to $D \geq 10^{-14}$ m²/s. The diffusion constant of the β clamp sliding on DNA was very close to this lower limit.

Single-molecule FRET measurements are commonly performed using TIRF microscopy in combination with a CCD camera to monitor several protein/DNA systems simultaneously [28, 29]. Unfortunately, current CCD cameras sensitive enough for single-molecule spectroscopy do not have sufficient time resolution (typically 30–100 ms, down to 1 ms in some cases), limiting measurement to slow diffusive motions $D < 10^{-14}$ m²/s. For many proteins that slide on DNA, it is important to sacrifice the ability to simultaneously monitor many protein–DNA systems for better time resolution.

To date, our measurement of the *E. coli* β clamp has been the only single-molecule FRET measurement of DNA sliding, and it is interesting to compare it with what was more recently reported by single-molecule tracking for the sliding of human PCNA. The proteins are functionally and structurally similar. By single-molecule tracking, Kochaniak et al. determined an overall diffusion constant for PCNA that, at $D = 3 \times 10^{-12}$ m²/s, is two orders of magnitude higher than that for the β clamp [24]. These authors also obtained evidence that 85–99% of the motion reflected the clamp in a slower diffusing mode that likely tracks the groove of DNA. The remainder of the time, the clamp diffuses much more quickly without tracking DNA groove. It is intriguing that for the β clamp, we were unable to detect any evidence for fluctuations that would be expected for $D = 10^{-12}$ m²/s, precisely where our correlation-based detection is most sensitive [11]. Indeed, the lack of fluctuations at the appropriate time scale for such fast diffusion was one evidentiary fact from which we deduced that $D \sim 10^{-14}$ m²/s. Assuming that the dimeric β clamp behaves quite similarly to the trimeric PCNA in terms of sliding behavior, we believe that the large discrepancy between the two measurements likely involves the two modes of diffusion proposed by Kochaniak et al., based on their experimental results [24]. If the sliding clamp spends 2–15% of its time in the fast mode, fluorescence correlation amplitudes for that motion would be correspondingly small (i.e., 2–15% of the total fluctuations, well within the noise level of our previous experiments). In fact, the lack of significant fluctuations at the appropriate time scale is consistent with the multiple diffusion mode interpretation of Kochaniak et al. In the end, additional, improved experiments that measure the diffusion of the sliding clamps over multiple length and time scales will be necessary to resolve the puzzle produced by these seemingly discordant results.

3.3 Interpreting the Observations

What factors affect the sliding of proteins on DNA? In a Newtonian sense, sliding should be controlled by entities that can exert a force on proteins. Major components that can induce or retard the motion of the DNA sliding protein include other proteins, solvent, and DNA. In fact, the thermal forces that induce sliding through Brownian motion are mediated by the same objects that can also retard the motion. We can avoid confusing the issue by focusing only on the frictional or retarding forces. The "other proteins" could bind only the DNA sliding protein or they too could bind the DNA. These proteins might be actively sliding proteins, such as polymerases, or other passively sliding proteins. Since we are interested in how proteins slide on DNA, we will not discuss explicit protein–protein interactions except for the important case of a passively sliding protein affecting the active sliding of another, such as a DNA sliding clamp attached to a DNA polymerase (see below). That leaves just the solvent and the DNA to affect the sliding motion, i.e., to produce retarding forces.

Three drag terms. The solvent, through its viscosity, will affect the sliding according to the shape of the protein. This "three-dimensional (3D) solvent drag" is the same drag that occurs in ordinary 3D solvent-phase diffusion, and it is characterized by frictional drag term α, which can be calculated by solving the Navier–Stokes equations. For simple shapes, α is given by the product of the viscosity of the medium, the size of the protein, and a geometrical factor (a simple example is given below).

The DNA will affect sliding in two possible ways: (1) frictionless forces that nevertheless create additional solvent drag by compelling the protein to take a longer path (e.g., a helical one) through the solvent and/or to rotate while moving along the DNA, and (2) frictional drag due to a series of energetic barriers that the DNA presents to the sliding protein. For simplicity in terminology, we call (1) "DNA-induced solvent drag" and (2) "DNA surface drag." Together, the 3D solvent drag, the DNA-induced solvent drag, and the DNA surface drag constitute three drag terms that completely describe the retarding forces for proteins sliding on DNA (Fig. 3.3). What can create confusion is that observations of DNA sliding cannot usually detect the actual path (e.g., rotations or revolutions) of the protein and thus assume simple 1D diffusion, sometimes called "effective 1D diffusion". At least one author makes a distinction between 1D and 2D sliding, with 1D/2D indicating that the protein is/is not required to rotate as it slides, respectively [34]. This terminology has not been widely followed; we and others simply describe both of these as "1D" diffusion that either does or does not follow a helical path along the DNA.

In focusing on the passive sliding of a protein, sliding on DNA is most simply characterized by the effective 1D diffusion constant, a quantity that can be measured as previously discussed in Sect. 3.2. To date, over a dozen such diffusion constants have been measured. The most striking feature of the diffusion constants listed in Table 3.1 is that they span a very large range, covering four orders of magnitude, and in a few cases a single protein ranges over three orders of magnitude. So far, the fastest sliding proteins are AVP–pVIc, Ada (C terminus), and PCNA, with diffusion constants of 10^{-12} m²/s to 2×10^{-12} m²/s. At the other extreme are the lacI repressor and the Msh2–Msh6 repair protein, whose sliding is characterized by diffusion constants as low as 2×10^{-16} m²/s. (The sliding of SSB on single-stranded DNA might be even slower [29], but it is not directly comparable to sliding on double-stranded DNA because the structure of single-stranded DNA is so different from double-stranded DNA). To put these numbers in perspective, we compare them in two ways. First, we compare the measured diffusion constants with the solvent-phase 3D diffusion constants of small molecules and proteins. Second, we interpret the diffusion constants in terms of distance and time scales in sliding on DNA. For the first, it may be helpful to point out that a "3D" diffusion constant can be equally valid for characterizing 1D diffusion (recall that the Einstein relation always uses the same diffusion constant, but changes through the factor n according

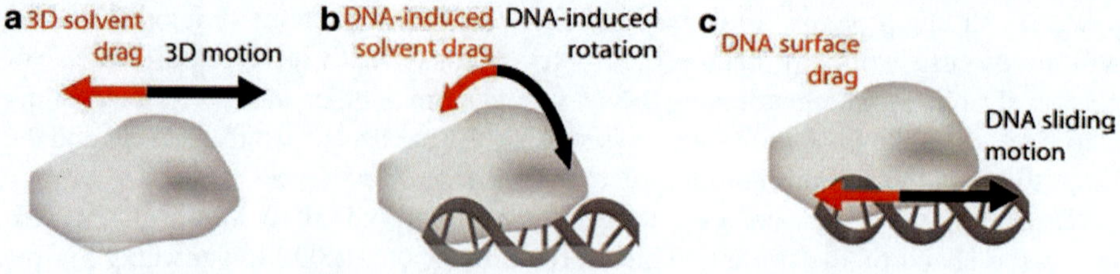

Fig. 3.3 The three drag terms relevant to DNA sliding. From *left to right*: 3D solvent drag, the drag that determines the "ordinary" diffusion constant for a protein in solution; DNA-induced solvent drag, the "extra" solvent drag that a protein experiences while sliding along DNA due to requirements to rotate or revolve around the DNA; DNA surface drag, the drag that a protein experiences due to the DNA directly inhibiting the sliding of the protein

to the dimensionality). The projection of the 3D diffusion of a protein onto a single coordinate axis is an idealized model of diffusion on a frictionless, rod-like DNA, i.e., with neither DNA-induced solvent drag nor DNA surface drag. Therefore, the 3D diffusion constant can be viewed as the baseline 1D diffusion constant for sliding on DNA. The actual diffusion constant for sliding on DNA will be reduced from that by the DNA-induced solvent drag and DNA surface drag terms.

The 3D solvent drag. The diffusion constant for (3D) solvent-phase diffusion decreases as the size of the molecule increases. The diffusion constant of water molecules in pure water is about 3×10^{-9} m^2/s. Proteins typically exhibit diffusion constants that are one or two orders of magnitude smaller. For a protein diffusing in solution the diffusion constant can be estimated by Stokes–Einstein relation:

$$D = k_B T/\alpha \tag{3.4}$$

where k_B is the Boltzmann constant, T is the temperature, and α is the frictional drag term introduced above. For the simplest case of a spherical protein, α is given by:

$$\alpha = 6\pi\eta R, \tag{3.5}$$

where η is the viscosity of water and R is the radius of the sphere. The physicist's infamous spherical approximation is not as bad as it might appear. For example, we can approximate the donut-shaped β clamp as a sphere, using a=4.5 nm, the approximate cylindrical radius of the β clamp; and at temperatures 25–40°C (i.e., 298–313°K) this gives an estimate of D in the range of 5×10^{-11} to 8×10^{-11} m^2/s. To avoid the crude spherical approximation, one can also solve the Navier–Stokes equations numerically for the molecular structure of the β clamp [35], using the program, hydropro [36]. At 27°C this calculation gives D=5.7×10^{-11} m^2/s, which agrees very well with our previous estimate. These solvent-phase diffusion constants characterize 3D diffusion, not necessarily diffusion on DNA; however, as reasoned above, the 1D diffusion rate for the sliding clamp on DNA cannot be larger than its ordinary (3D) diffusion constant in solution. Hence, we can postulate an upper limit for diffusion on DNA of $D\sim10^{-10}$ m^2/s.

The reported diffusion constants for the DNA sliding clamps are at least two orders of magnitude lower than this.

The DNA-induced solvent drag. As described above, the nature of protein–DNA interactions may give rise to an induced rotation and/or revolution of proteins sliding on DNA. In 1979 Schurr wrote a short report in response to early suggestions that low effective diffusion constants for proteins sliding on DNA were due to large energy barriers for proteins sliding on DNA, calculating the extra solvent drag that would occur if the lacI protein were forced to rotate during its translocation along the DNA axis [37]. He introduced the DNA-induced solvent drag term (with other words) to compete with the DNA surface drag term, and estimated for the lacI protein the DNA-induced solvent drag was responsible for a two order of magnitude reduction in the effective diffusion constant. To keep the model simple, and perhaps because of the lack of insights from crystal structures of protein–DNA complexes, Schurr assumed that the protein would rotate on the DNA. He found for the total solvent drag the formula,

$$\alpha = 6\pi\eta R\left[1+(4/3)\left(2\pi R/b\right)^2\right] \tag{3.6}$$

where b is the distance along the DNA traveled in one rotation, about 3.4 nm for the DNA double helix. In (3.6) the first term is simply the usual Stokes drag (Fig. 3.3a) and the second term is due to the rotational drag (Fig. 3.3b). This model was recently updated to include a term for revolution of a protein on the DNA [38], allowing one to calculate the DNA-induced solvent drag for a nonrotationally symmetric protein as well as for a particle attached to the DNA-binding protein, such as a bead or quantum dot. The observation of rotation/revolution during sliding has been mostly indirect. Using the fact that the rotational/revolutionary part of the DNA-induced solvent drag has a $1/R^3$ dependence on the radius of the protein (or similar for the radius of the revolution) while the 3D solvent drag has a $1/R$ dependence on the radius of the protein, Blainey et al. estimated the contribution of the DNA-induced solvent drag for a handful of proteins based on the effective diffusion constant and concluded that all undergo rotation-coupled sliding [31]. Different arguments for the rotation-coupled sliding of PCNA have been made (see below) [24].

While it may be intuitive that some proteins would follow the helical form of the DNA, by tracking a groove or the phosphate backbone, it is less obvious that the DNA sliding clamps would be required to rotate while sliding – that is, sliding on DNA more like a nut on a bolt than like a washer. To illuminate the subject better, instead of using the above formula for α, we can reason as follows. If the β clamp were constrained to follow the groove of the DNA, it would be slowed down by the additional viscous drag as the clamp rotated once around the DNA for every 10 bp or so of linear diffusion – what we termed "DNA-induced solvent drag" (above). We can account for this drag by calculating the rotational diffusion constant of the protein as it rotates on its axis. The hydropro program [36] gives a rotational diffusion constant $D_{rot} = 2.3\times10^6$ to 3×10^6 rad^2/s, depending on the axis of the rotation. Using the average value and the rotational analog to the Einstein relation,

$$\left\langle\theta^2\right\rangle = 2D_\gamma t, \tag{3.7}$$

we find that it takes 7 μs to make a full rotation around the DNA helix. By contrast, using the solvent viscosity-induced diffusion constant $D \sim 10^{-10}$ estimated above, the time it would take to linearly diffuse the distance of one helical turn by simple linear diffusion is just 0.1 μs. Thus, if the clamp were constrained to follow the groove of DNA and therefore rotate as it moved along the DNA, the viscosity of water would slow the clamp down by a factor of about 70 relative to simple linear diffusion. This retardation is nearly as much as the factor of 110, estimated several decades ago by Schurr for the lacI repressor, which turns out to bind to DNA with an altogether different topology [37]. The factor of 70 would give an effective diffusion constant, $D_{eff} \sim 10^{-12}$ m^2/s, that would be measured in our experiments if the diffusion of the sliding clamp were governed only by the viscosity of water and the requirement to follow the groove of DNA (i.e., the combined 3D solvent drag and DNA-induced solvent drag). Because our experiments on the β clamp provided a diffusion constant estimate two orders of magnitude smaller than this value, we have concluded that diffusion of the sliding clamp is retarded well beyond these two drag terms. This implies that there must be significant DNA surface drag (discussed below). It seems logical that the clamp would follow a helical path at least while attached to the polymerase; however, our measurements neither confirmed nor refuted this tendency in the absence of the polymerase.

While it is reasonable to expect essentially the same behavior for the other major DNA sliding clamp, PCNA, a first publication by Kochaniak et al. suggests otherwise [24]. These authors report an apparent diffusion constant of about 10^{-12} m^2/s, which is what we have just predicted for the combined effects of 3D solvent drag and DNA-induced solvent drag. Yet it may be premature to conclude that for PCNA there is almost no DNA surface drag. First, Kochaniak et al. use two types of viscogens that should differentiate rotational and translational motion, thereby finding evidence for rotationally coupled sliding. Nevertheless, when a large quantum dot (12 nm in radius) was attached to the protein, the protein-dot couple appeared to slide without any rotation, prompting Kochaniak et al. to propose a second, much faster sliding mode that was not coupled to rotation. They proposed that the apparent diffusion constant reflects the sum of a fast linear diffusion and a slow rotationally-coupled diffusion:

$$D_{app} = f\,D_{fast} + (1-f)\,D_{slow}, \tag{3.8}$$

where D_{fast} is 10^{-11} m^2/s or larger and f is the fraction that diffuses at this fast rate. This result leaves D_{slow} somewhat undefined since it can be close to 10^{-12} m^2/s for a very small f of ~1–2%, but D_{slow} can be about 10^{-14} m^2/s for $f \sim 10\%$. Thus, resolution of this issue awaits improved resolution in the experiments.

DNA surface drag. In addition to protein–solvent interactions, protein–DNA interactions can play a major role in limiting the rate of sliding. Most proteins that diffuse on DNA must first adhere, but for the adhesion to cause frictional drag, it must vary as the protein moves. The electrostatic attraction between protein and DNA, for example, does not necessarily create drag. When there is a sufficiently large number of attractive interactions between proteins and DNA, a number of slip–stick interactions can occur, converting kinetic energy of the whole protein moving into randomized kinetic energy of vibrational motions (i.e., molecular friction). Thus, it is reasonable to postulate a series of local energy minima separated by energetic barriers, where the local minima are due to attraction and adhesion between protein and DNA, while barriers arise as these attractions vary with the changing details of the intermolecular interactions. Therefore, while the presence of attractive forces between the protein and DNA does not necessarily create drag, in practice intermolecular drag is likely. It is interesting to consider that not all sliding proteins must intrinsically adhere to the DNA. The DNA sliding clamps, owing to their ring-like topology, could theoretically remain on the DNA in the absence of any attractive, sticking forces. We will return to these topics in Sect. 3.4 below.

By modeling a protein sliding on DNA as a particle diffusing in a rough potential energy well, it has been shown by Zwanzig that the roughness of the energy landscape determines the diffusion rate [39]. In particular, the author shows that in a rough energy landscape characterized by an average barrier height of ε, the effective diffusion constant D_{eff} due to the rough energy landscape can be related to the diffusion constant in the absence of the roughness D by the expression,

$$D_{\text{eff}} = D\exp\left[-\left(\varepsilon/k_B T\right)^2\right] \tag{3.9}$$

More precisely, this result is for a Gaussian distribution of barrier heights, where ε is the standard deviation of the distribution. Although, as Zwanzig points out, this 1D model cannot fully account for the multidimensional dynamics of proteins, it is a useful phenomenological model, allowing us to estimate the energetics that give rise to the drag involved in sliding. Other formulations, based on an Arrhenius model, have been employed [22, 24, 40]. The form of Zwanzig's expression means that large differences in the diffusion constant can reflect small differences in the roughness of the energy landscape. For example, we predicted a 3D diffusion constant for the β clamp sliding on DNA of $D \sim 10^{-10}$ m²/s, and measured a diffusion constant of $D \sim 10^{-14}$ m²/s. If the entire difference in these two diffusion constants was due to DNA surface drag, the average barrier height would be ~3 $k_B T$. Slutsky and Mirny suggest that free 1D diffusion requires energy barriers between positions on DNA to be less than about 2 $k_B T$ [6]. This result implies that the β clamp diffusion constant $D \sim 10^{-14}$ m²/s must include DNA-induced solvent drag. Blainey et al. and Gorman et al. estimated energy barriers from measured diffusion constants and found that the estimated energy barriers were within the 2 $k_B T$ limit [18, 22]. However, Gorman et al. found this to be the case only if the Msh2–Msh6 proteins followed a helical path along DNA, revolving around the DNA with sliding (as we just showed for the β clamp). More recently, Blainey et al. used Zwanzig's formulation to estimate energy barriers near 1 $k_B T$ for several DNA-binding proteins (not all included in Table 3.1) and provided evidence that the proteins track the groove of DNA as they slide [31]. Overall, it appears that DNA surface drag is a dominant factor in sliding. Even the fastest AVP–pVIc protein is retarded by an additional factor of 2–5 by DNA surface drag above retardation due to the combination 3D solvent drag and DNA-induced solvent drag [31].

Distance and time scales. Next, we consider what the enormous range of measured diffusion constants (Table 3.1) implies in terms of time and length scales. A natural measure of distance along DNA is the distance from one base pair to the next, ~3.4 Å for B-form DNA, the conformation thought to be most prevalent under biological conditions. Using the Einstein relation we have calculated (Table 3.2) the time it would take a hypothetical protein to diffuse 1, 10 or 100 bp, as well as the mean displacement that occurs in 1 ms. The fastest diffusing proteins ($D \sim 10^{-12}$ m^2/s) travel over 100 bp in a millisecond. The slowest proteins scarcely move on the millisecond time scale.

Sliding vs. hopping. In addition to the body of experimental observations, there has also been a growing body of theoretical arguments concerning the movement of proteins on DNA. In just the last few years several dozen theoretical papers have been published. A recent paper by Wunderlich and Mirny on DNA target searching gives a good summary of the latest theoretical developments for proteins that diffuse only a certain distance along DNA (sliding mode) before dissociating from the DNA (hopping or jumping mode) [10]. The average such distance is defined to be the protein "sliding length," (s) an experimentally determined parameter that is key to the remainder of the analysis. The experimental determination of s can be, however, thwarted by the presence of very small hops, which, the authors point out, would be missed by single-molecule observations. For some proteins, the theory predicts very small hops, with a median distance of about 1 bp, but such small hops could arguably be considered simply part of the sliding mechanism with one important difference: hops are more likely to lead to jumps. Because such small hops do not affect the diffusion constant, the authors conclude "the major contribution of hops is to the duration of sliding rather than to its rate" [10].

Within the observational limitations of the various experiments it has been difficult to separate true sliding from hopping. An approach to remedy this is presented in the chapter of this volume on lacI diffusion by Wang and Austin. They note that with millisecond dissociation times estimated for lacI, there is expected to be a significant amount of hopping. In general, fast dissociation kinetics (short dissociation times) can implicate hopping when compared to mean sliding times. On the other hand, long dissociation times would indicate little hopping. At the extreme are the DNA sliding clamps that not only display very long dissociation

Table 3.2 Effect of diffusion constant on mean time for three log displacement as well as the mean displacement that occurs in 1 ms, based on the Einstein relation for one-dimensional diffusion given by (3.1)

D (m^2/s)	Time (μs) (for 1 bp)	Time (ms) (for 10 bp)	Time (ms) (for 100 bp)	Displacement (bp) (in 1 ms)
10^{-16}	578	57.8	5780	1.3
10^{-15}	57.8	5.78	578	4.2
10^{-14}	5.78	0.578	57.8	13.2
10^{-13}	0.578	57.8 (μs)	5.78	41.6
10^{-12}	57.8 (ns)	5.78 (μs)	0.578	131.5
10^{-11}	5.78 (ns)	0.578 (μs)	0.058	415.9

times but must be actively reloaded onto DNA. Such proteins are not expected to display hopping. Nevertheless, in a sort of gray area in the definition of hopping, it is possible that hop-like diffusion can occur where the clamp protein is not in contact with DNA. This case would pose an important exception to the above theory since the sliding clamp topology (closed ring around DNA) implies that hops never lead to jumps (zero probability), and moreover, the clamp remains bound to the DNA without the *prima facie* requirement for any overall protein–DNA attraction.

In the earliest report of lacI sliding on DNA, it was noted that the protein–DNA interactions are dominated by electrostatic forces, a result gleaned from the sensitivity of the sliding to the ionic strength of the solution [3]. In fact, this has become a standard method for distinguishing hopping from sliding. In order to experimentally increase hopping, one can increase the ionic strength of the solution. As mentioned above, DNA sliding clamps cannot hop in the ordinary sense, but they can potentially undergo hop-like diffusion ("ice skate" [41]). Kochaniak et al. found that the diffusion coefficient of PCNA along DNA changed by only a factor of 2 upon a 13-fold change in ionic strength (from 41 to 541 mM). While this is a very small effect, it may indicate some smoothening of the energy landscape due to interference in transient ionic bonds between protein and DNA.

Other proteins may share at least some of the topological character of the DNA sliding clamps. There is evidence, for example, that Rad51 and Msh2–Msh6 complexes also form rings or "clamp-like" structures on DNA. In the case of Rad51, this inference is made on the basis of (1) previous structural considerations, (2) lack of helical extension of the DNA templates, (3) no observations of proteins moving past one another, (4) implausibility of a helical form sliding so well along DNA [32]. In the case of Msh2–Msh6, the sliding clamp model is consistent with both crystal structures and electron micrographs that show the Msh2–Msh6 complex completely encircling DNA. Gorman et al. suppose that the complex would have to "at least partially unfold to allow dissociation, and this requirement for a large-scale structural reorganization may account for the long time periods that Msh2–Msh6 is able to slide on DNA" [18]. "Hopping [of the complex] is inconsistent with the finding that the diffusion coefficients were insensitive to salt concentration, different proteins bound to the same DNA were unable to bypass one another, and they were resistant to challenge with excess competitor oligonucleotide" [18].

Diffusion constants and biological function. While most of the research in the area of passive sliding is aimed at proteins that slide (and hop, jump, etc.) in search of targets on DNA, DNA sliding clamps must slide to serve their function. For proteins that search for a target, the diffusion constant conveys a key aspect of the speed of the survey search from a given starting point. For proteins that slide as their primary function, the diffusion constant describes the drag imposed on that function. The β clamp again provides an excellent example. The reported value of $D \sim 10^{-14}$ m^2/s implies significant friction (attractive or sticky interactions) between the protein and DNA, since this value is at least two orders of magnitude slower than would be predicted for a protein of this size diffusing through water alone. This may appear surprising since the polymerase III core (α, θ and ε subunits) replicates DNA approximately 50-fold faster when the β clamp is included in the reaction [42]. Nevertheless, we can reason that even the "slow" diffusion constant is large enough to imply almost no friction for the DNA polymerase to work against. In isolation the β clamp moves from one base pair to the next (0.34 nm) in 1/100th the time the polymerase takes to insert a single base (~1 ms). Note that the polymerase, a Brownian motor, presumably also moves forward by random diffusion on a submillisecond time scale, but each freshly inserted DNA base becomes the ratchet tooth that prevents backward motion.

Even though the β clamp easily moves from one base to the next in just 6 μs, longer distances require much more time due to the nature of diffusion (Table 3.2). Since the β clamp spans a distance of about 12 bp on dsDNA (~4 nm), the time to diffuse just one clamp width away is about a millisecond. We have suggested that this relatively slow sliding of the clamp may play a role in preventing the polymerase–clamp complex from excessive drift during the polymerization process. It is noteworthy that recent single-molecule observations of HIV reverse transcriptase (an RNA polymerase that does not attach to a sliding clamp) recorded frequent events involving polymerase sliding away from the polymerization site [28]. The relatively slow sliding of the β clamp on DNA could mean that the attached polymerase need not maintain contact with the DNA to keep its place on the template. Indeed, such a feature may be required for efficient switching between polymerases and other DNA replication and repair factors. It is known, for example, that in the absence of the β clamp, the main bacterial polymerase – the α subunit – loses its grip on the template and rapidly dissociates from the DNA. It is therefore plausible that the polymerase often loses direct contact with the DNA during synthesis, especially during polymerase swapping events [43, 44]. During these transitions, it may be the β clamp that maintains the correct position on the DNA. Thus, by sticking significantly as it passes along the DNA, the β clamp may help prevent rapid polymerase drift.

Ultimately, do all DNA-binding proteins slide on DNA? Possibly not, but it may be difficult to rule out any sliding. One can imagine proteins that cannot slide, for example, because they are too stiff (see below). In some cases there is evidence that sliding does not occur over long distances. One particularly interesting example is RecA which facilitates homologous recombination by matching single-stranded DNA to a complementary region in double-stranded DNA. In complex with single-stranded DNA, RecA was shown to reach its target on double-stranded DNA independent of its position, indicating that it did not reach its target by sliding along DNA [45]. Nevertheless, this work falls short of proving that RecA does not slide on DNA at all.

3.4 Models and Mechanisms of Sliding

3.4.1 Implications of Search

Results from theoretical treatments of DNA target search have implications for the nature of sliding and hopping on DNA. Several reports, based on theoretical considerations, suggest that many proteins interact with DNA in at least two modes.

In one mode the protein slides and/or hops along DNA with high diffusion rate. In another mode the protein binds tightly to a particular site and remains there. Two DNA interaction modes provide an obvious solution to the apparent paradox of protein sliding over great distances while also tightly binding to the specific target sequence. The mechanism for switching between these modes invokes the general idea that protein–DNA interactions are nonspecific during sliding/hopping and specific during tight binding. One diffusion rate for each protein is probably inadequate to describe target acquisition. During fast diffusion there is too little interaction to detect the specific target. Therefore, a slower diffusion mode may be required by the biology (and implicated by kinetics experiments). In the language of proteins searching for a target, the two modes would be a fast diffusion over nonspecific DNA and a slow diffusion close to the target site. This also seems to fit with the crystal structures of proteins bound "specifically" and "nonspecifically" to DNA (as discussed below). It should be mentioned that there has been some controversy about the thermodynamics of such a switch. These considerations place limits on the frequency of conformational changes that control the modes [46].

Kinetics studies over the past decade, such as those performed with *Eco*RI, suggested that while proteins are sliding along DNA, they pause at sites that resemble their recognition sites [47]. This provided evidence for a fast and slow mode of sliding, fitting well with the aforementioned theories of DNA target search. From X-ray crystallography, we have structures for a handful of proteins bound "specifically" to a DNA target or bound "nonspecifically" to generic DNA. These two states are thought to be indicative of target binding vs. target searching, perhaps revealing intermolecular contacts in sliding quickly vs. sliding slowly, and the large differences in the intermolecular conformations have been interpreted in this way. We will now discuss the theoretical arguments for multiple DNA interaction modes and then the experimental evidence supporting this view. Work has focused on analyzing crystal structures of proteins bound to DNA, as well as simplified simulations and calculations. Although mechanisms largely remain unclear, some clues are available from structural analyses and simulations, as discussed below.

3.4.2 Analysis and Implications of Crystal Structures

The ways in which proteins interact with DNA are varied. The protein may wrap around the DNA partly or completely, may grab the DNA deeply with many intermolecular contacts or barely touch the DNA, and may twist, splay, or sharply bend the DNA. The nature of the protein–DNA interactions can strongly affect the structure–function relationship of the complex, especially for enzymatic proteins such as the restriction endonucleases. A decade ago Jones et al. published a summary of protein–DNA interactions, analyzed from 26 protein–DNA complexes and compared with some 36 protein–protein interactions [48]. The main findings were that the protein–DNA interface is largely polar, the protein–DNA interactions are often mediated by water molecules, and most of the protein–DNA contacts are to the sugar-phosphate backbone, not the bases. Positively charged arginine residues were the most likely to be involved in the protein–DNA interface, followed by polar threonine, asparagine, and positively charged lysine residues. Among the least observed residues to contact the DNA were the negatively charged aspartic and glutamic acid residues. Traditionally the emphasis in the crystallography of protein–DNA complexes was to understand how specific sequence recognition occurs. Due to the high stability of the sequence-specific complex, this turns out to be easier than obtaining structures for complexes with nonspecific DNA sequences. At that time the only structure of a protein bound to DNA "nonspecifically" was the structure of *Eco*RV bound to noncognate DNA. This remains an illuminating example when compared with the structure of the protein bound specifically to the cognate DNA (Fig. 3.4). The gist of the comparison is that there are many fewer contacts overall and no contacts to bases in the nonspecific complex.

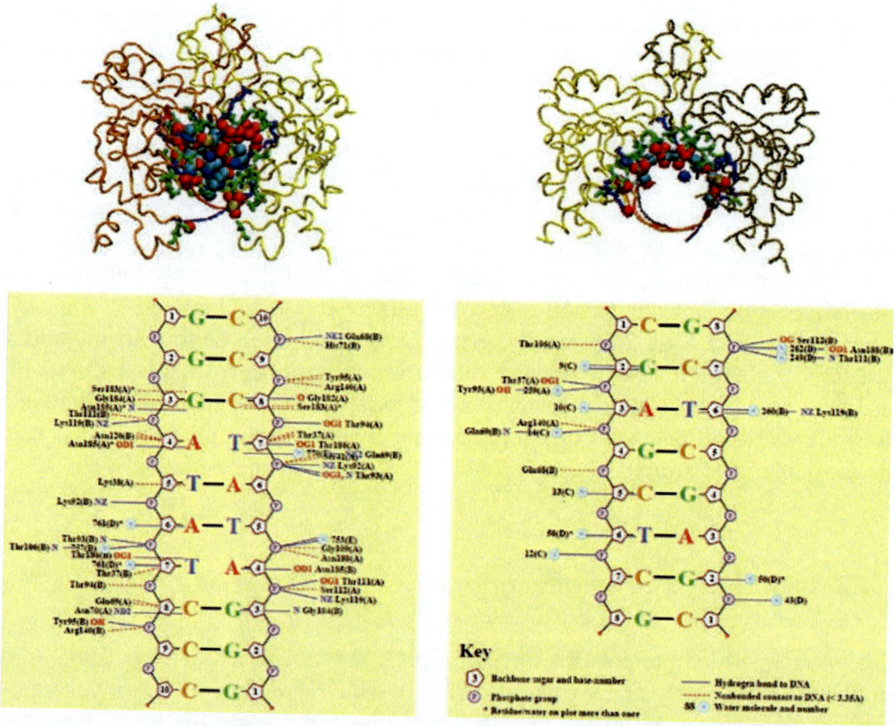

Fig. 3.4 Comparison of protein–DNA interactions for specific (*left*) and nonspecific (*right*) binding of *Eco*RV. *Top*: Axial views of the interactions. The proteins and DNA are represented by smooth tubes of the respective backbones with each chain colored differently. The protein residues that contact the DNA (i.e., with nonhydrogen atom centers that are within 3.8 Å of any DNA nonhydrogen atom centers) are shown in licorice, with polar residues in *green*, positively charged (basic) residues in *blue*. *Bottom*: Output of nucplot for the two crystal structures, detailing the protein–DNA interactions [62]

Moreover, while the dimeric protein completely wraps around the DNA in the specific complex, the dimer is much more open in the nonspecific complex. There are more than twice as many water molecules bound to the DNA in the nonspecific complex. More specific/nonspecific pairs of structures of protein–DNA have been determined since then, but the initial summary has remained valid. Nonspecific protein–DNA complexes are looser.

An important starting point for mechanistic knowledge of the sliding state is therefore provided by the very few crystal structures of protein–DNA complexes that appear to capture nonspecific interactions. But how well do crystal structures of "nonspecific" protein–DNA complexes reveal the freely diffusing/sliding state? On one hand, the structures are clearly different from those that form when the specific target is present. On the other hand, there is almost a contradiction between the concept of a crystal structure and that of a nonspecifically bound sliding protein. In fact, various tricks are employed to promote uniformity of complexes, as required for crystallization. For example, the DNA sequence is often very similar to the target sequence, leading to structures that are probably not completely representative of nonspecificity. It is nevertheless hoped that at least some features of the nonspecific state are revealed. In fact, assuming that there are two modes of sliding (i.e., fast and slow search), near-specific structures may turn out to be more representative of the slow mode. Perhaps then, the most relevant structures for understanding of protein sliding are structures of proteins that do not specifically bind anywhere on DNA. A prominent example is the DNA sliding clamps that, not surprisingly, resisted cocrystallization with DNA until very recently. With all the inherent limitations of static structures in the illumination of motion, protein–DNA cocrystal structures may give promising leads in understanding the mechanics of sliding.

Focus on DNA sliding clamps. As mentioned previously, DNA sliding clamps are ring-shaped proteins that encircle DNA helix, topologically trapping DNA inside the ring and providing a mobile platform to which DNA polymerases and other proteins can attach. The fact that sliding clamps are present in all life forms underlies the evolutionary success of this way of keeping other proteins in close association with DNA. The oligomeric state varies among the kingdoms: the bacterial sliding clamp (polIII β subunit or "β clamp") is a dimer, while the eukaryotic and archaeal sliding clamp (PCNA) is trimer. Despite these differences, all sliding clamps have remarkably similar 3D structures featuring pseudo sixfold symmetry. The central channel of sliding clamps is ~35 Å in diameter, significantly larger than the diameter of the DNA double helix in the canonical B-form (~24 Å). Since the first determination of the structure of a DNA sliding clamp, the β clamp, in 1992 [35], these features have raised questions regarding the relative arrangement of the clamp and the DNA. Does the clamp bind directly to DNA or can it "levitate" without making direct contacts with DNA? What is the mechanism of clamp movement/diffusion along the DNA?

Both biochemical and crystallographic studies have shown that the *E. coli* β clamp can bind DNA directly [49]. The crystal structure of the β clamp–DNA complex revealed that the plane of the clamp is not perpendicular to the DNA axis. Instead it is tilted by an angle of about 22° (Fig. 3.5). Since the central channel of the β clamp is wider than the DNA double helix, the tilt allows the DNA to contact both sides of the clamp channel (indeed both protomers of the dimer). The contacts to the DNA are to the phosphate backbone. As we discussed above, the DNA sliding clamp makes an unlikely target for crystallization. Indeed, several aspects of the crystal structure are worth mentioning. First, the DNA is not completely double stranded. It is a "primed site," meaning that it has a four-base single-stranded 5′ overhang, and there are half a dozen contacts of the clamp to the bases (not just the backbone) of the 5′ overhang. Furthermore, due to the crystallographic arrangement, these ssDNA contacts are with neighboring β clamps, not the β clamp penetrated by the dsDNA. Thus, given these "crystallization artifacts," it is fair to ask if β clamp binding to DNA in solution shows similarity to what is observed in the crystal structure.

There are several lines of evidence suggesting that the β clamp-DNA arrangement seen in the crystal does closely reflect the situation in solution. First, a series of molecular dynamics (MD) simulations have been performed in our own lab for the β clamp bound to 18 bp of DNA. This study starts with the DNA centered and aligned to the central axis of the β clamp on DNA. We have found that the clamp quickly adopts a tilted orientation in respect to the DNA, and the tilt angle relative to the central 12 bp is actually slightly larger than that observed in the crystal structure of β–DNA complex. In the simulations, the angle vacillates mostly between 20° and 30°. (A snapshot created at 10 ns is shown in Fig. 3.5). Many of the same residues seen to interact with the DNA in simulations were also identified as contacts by crystallography (Barsky, unpublished results). In addition, our analysis of conserved residues in bacterial *β* subunits (Venclovas, unpublished data) revealed that the majority of the positively charged residues of the β clamp (Arg, Lys, His) interacting with DNA in the simulations appear to be strongly conserved by evolution. Altogether these data are consistent with the β clamp/DNA interaction in the crystal structure.

Having "verified" the plausibility of the crystal structure of the β clamp–DNA complex, we might now ask if there is anything special about how a sliding clamp binds DNA (other than the enclosing topology) that might give a clue to the

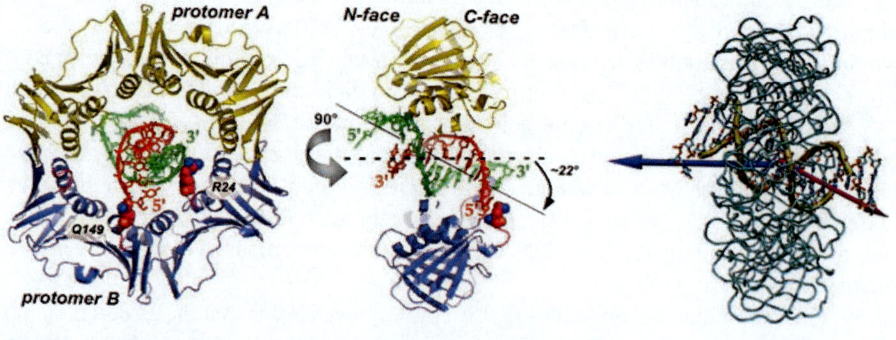

Fig. 3.5 *Front* and *side views* of the β clamp–DNA complex from X-ray crystallography (*left*) [49] and a snapshot at 10 ns from an MD simulation (*right*) showing a 30° angle between the clamp axis and the DNA axis (Barsky, unpublished results)

sliding mechanism. From the point of view of evolutionary conservation, there are several other highly conserved charged residues that either might provide alternative contacts (R176, H175) during β-clamp diffusion on DNA, or (as in the case of K235) might participate in binding in the same mode, but on a longer DNA helix than the one in the crystal structure. The strong evolutionary conservation of positively charged residues inside the inner channel is not surprising considering their anticipated role in binding DNA backbone. However, the surprising finding is that a number of negatively charged residues (Asp, Glu) lining the central channel that are as (or even more) conserved as those with positive charges. Interestingly, these conserved acidic residues appear to be mostly interspersed between positively charged residues that either contact DNA (as seen in the crystal structure) or could potentially contact DNA (Fig. 3.6). Strong conservation suggests a potential role in clamp diffusion, perhaps allowing Arg or Lys side chains to alternate between binding DNA phosphates and forming salt bridges with neighboring Asp and Glu side chains on the clamp. This might significantly lower the activation energy required for breaking charged interactions with DNA phosphates during clamp movement on DNA. Evidence that protein–DNA interactions can be weakened by the "masking" of positively charged residues by forming salt bridges with nearby negatively charged residues has been reported [50].

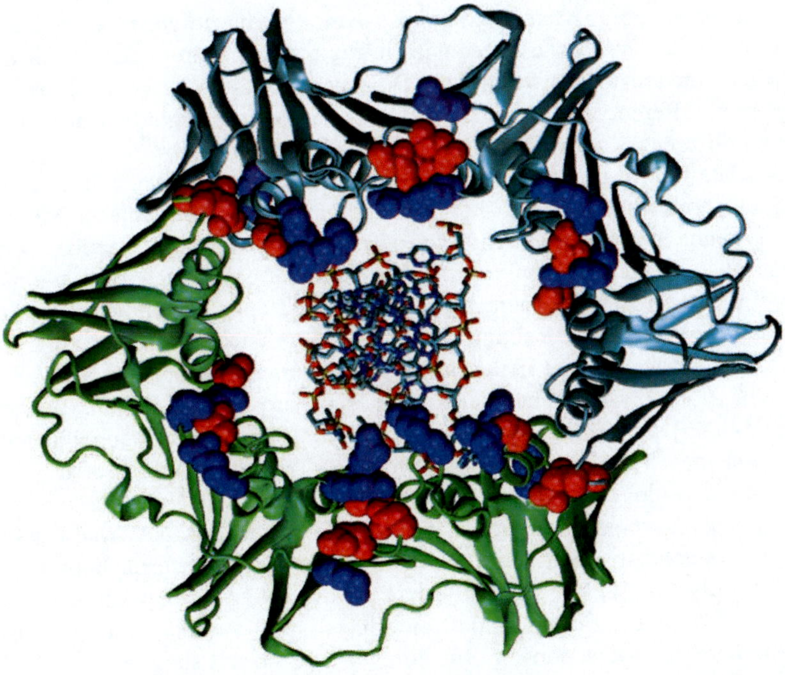

Fig. 3.6 Crystal structure of β clamp/DNA complex (PDB ID: 3bep). The β clamp monomers are shown in differently colored ribbons. The side chains of the most highly conserved positively (Arg, Lys, His) and negatively charged (Asp, Glu) residues lining the inner channel are shown in a space-filled representation in *blue* and *red colors*, respectively

Our MD simulations of the β clamp show that only a small subset of the 24 positively charged residues interact for substantial times with the DNA. For example, in the 18 bp simulation, only nine residues had contact times (ionic bonds or salt bridges of 3 Å or less between the protein and DNA) totaling more than 10% of the simulation, while two pairs of residues (from each side of the dimer) contacted the DNA more than 80% of the time. This supports the idea that key residues dominate the protein–DNA interactions. This also lends some support to the "ice-skating" model of interaction and sliding.

These observations for the β clamp on DNA may be relevant for other DNA sliding clamps. In particular, binding at an angle has been observed for an X-ray crystallographic structure of the yeast PCNA–DNA complex [51] and the structure obtained by electron microscopy for the archaeal DNA ligase–PCNA–DNA complex [52]. Again, while there are reasonable concerns about the relevance of the crystallographic structure, MD simulations have shown that similar to the β clamp, PCNA quickly tilts with respect to the DNA axis [53]. Still, some concerns remain. In the case of PCNA, the angle between clamp and DNA in the crystal structure is substantially larger (40°) than what was reported from the MD simulations (20°). In fact, the DNA is partly disordered and was by necessity modeled as a rigid body, leading the authors to conclude that the orientation of DNA is not definitive [51]. The final structure was termed "an X-ray derived model" (PDB code: 3K4X). This may indicate a highly dynamic complex – exactly what might be expected for a sliding clamp on DNA. The differences in the crystallographic results for PCNA and the β clamp are not easy to explain. Differences might be related to biologically unimportant differences in how the crystals developed, or might be caused by fundamental differences between the two sliding clamps.

From data related to crystal structures, sequence conservation, and modeling, a picture emerges wherein DNA sliding clamps make specific contacts only to the DNA backbone in keeping with the "nonspecific" nature of the interaction. Contacts are transient, and may be supplemented by additional possible contacts and moderated by possible repulsive interactions. Despite the apparent importance of these electrostatic interactions, the sliding of PCNA was only weakly sensitive to the ionic strength – a tenfold increase in the salt concentration led to only a twofold increase in the diffusion constant. The change was in the expected direction. Two recent papers [54, 55] describing crystal structures of PCNA from halophilic archaea (HvPCNA) may be informative in this respect. The inner surface of the archeal PCNA is less positively charged compared to eukaryotic PCNA. The authors note that "strikingly, the positive surface charge considered key to PCNA's role as a sliding clamp is dramatically reduced in the halophilic protein. Instead, bound cations within the solvation shell of HvPCNA may permit sliding along negatively charged DNA by reducing electrostatic repulsion effects" [54]. Perhaps future simulations and experimental observations of diffusion will take up this suggestion.

Finally, we consider the two sliding clamp diffusion modes inferred by Kochaniak et al. who suggested that in the slow mode the proteins follow the DNA helix (undergo rotationally coupled sliding) but not in the fast mode [24]. That the two diffusion modes are due to changes in the protein–DNA interactions is made especially plausible given the stability of the sliding clamp proteins. It is unlikely that sliding clamps undergo the types of conformational changes that are ascribed to other dual-mode sliding proteins. Another idea is that the tilt of the clamp relative to the DNA could affect the diffusion rate. In fact, it may be argued that since the sliding clamp has been observed to slide over 13-nucleotide ssDNA loops and other secondary structures of DNA [56], it need not always remain oriented in a particular way or follow the grooves of the DNA. Nevertheless, whether the change is in helix tracking, in tilt, or both, it is unknown how a particular state would persist while sliding over many base pairs. Perhaps the solution lies in the details of the electrostatics – the particular placement of charges – rearranged so that phosphate backbone would no longer be tracked. Ionic bonds between oppositely charged side chains or long-lived bound ions might accomplish this. Before the structures of DNA bound to the sliding clamps were determined, it had been suggested that the clamps can "ice skate" along the DNA [41]. This still might be true.

3.4.3 Modeling and Simulation

Monte Carlo simulations. Interesting Monte Carlo studies have predicted nonintuitive behavior for proteins sliding on DNA. Dahirel et al. [57] showed by means of Monte Carlo simulations and analytical calculations that there is a counterintuitive repulsion between two oppositely charged macromolecules at a nanometer range. This predicted force is thought to arise from a local increase of the osmotic pressure exerted by ions trapped at the interface. For DNA-binding proteins with concave surfaces, and for realistic protein charge densities, the authors predicted that the DNA–protein interaction free energy has a minimum at a finite surface-to-surface separation. This defines a separation distance at which proteins might easily slide on DNA. When a protein encounters its specific target sequence, the proposed free energy barrier would be overcome by favorable hydrogen bonds, thus enabling sequence recognition. In other Monte Carlo simulations, Barbi et al. show that the consequences of specific site recognition during the process of sliding over nonspecific sites produces "nontrivial" sequence-dependent dynamics that do not necessarily resolve neatly into two modes (or two effective diffusion constants) [40].

Electrostatics calculations and Brownian dynamics. Sun et al. [58] used electrostatic free energy calculations (based on numerical solutions of the Poisson–Boltzmann equation) to take into account both energetic and structural considerations for *Bam*HI restriction endonuclease sliding along DNA. The authors equate nonspecific binding with the sliding state, and they argue from experiments measuring the change in heat capacity and the salt-dependence of binding that the nature of the protein–DNA interactions is dominated by through-solvent electrostatics (no protein–DNA contacts). There are a number of reports that nonspecific binding gives rise to negligible heat capacity change, which Sun et al. and others ascribe mainly to a small change in the hydrophobic contribution to the free energy. This, in turn, implies that there is little change in the hydration of the protein upon binding. Thus, nonspecific binding may imply little contact with DNA, an inference that is at least partly supported by crystal structures. Nonspecific protein binding is known to have a stronger salt-concentration dependence, indicating that sequence nonspecific protein–DNA interactions are dominated by electrostatic forces. Sun et al. apply their modeling to the *Bam*HI–DNA complex. The authors argue that the nonspecific structure of the cocomplex, where just 1 bp was changed from the cognate target sequence, does not represent an electrostatic free energy minimum. To achieve the minimum energy state, the protein would need to be tilted relative to the DNA axis, and moved away from the DNA by ~15 Å. The authors calculated the effect of protein revolution on sliding and found that the protein would need to overcome a barrier of ~2.5 kcal/mol if it did not follow the helical pitch. This barrier would arise from the requirement of crossing from one major groove to the next. The authors also calculated Coulombic interaction energies for charged residues interacting with DNA. Such interactions were predicted to contribute significant favorable free energies (7–11 kcal/mol), with some of the interactions over surprisingly large distances (up to 24 Å). The *Bam*HI diffusion constant has recently been measured (Table 3.1), and it has been inferred that the protein rotates while sliding [31]. It would be interesting to test the prediction of Sun et al. that the specific mutation E51K would affect the *Bam*HI sliding rate [58].

MD simulations. Of all the levels of modeling, by far the most complete and detailed simulations are atomistic molecular dynamics (MD) simulations. We have already discussed some results from our simulations of the β clamp in the section on crystal structures. It is worth considering the feasibility of simulating the actual sliding of proteins on DNA using molecular dynamics (MD) simulations. Currently, state-of-the-art for MD simulations of roughly 100,000 atoms is on the order of 100 ns. Unfortunately, for even the fastest reported diffusion rates ($D=3\times10^{-12}\,m^2/s$), one could expect to observe only about 2 bp of displacement in 100 ns (Table 3.2). Given the flexibility of the DNA and the proteins, 2 bp of overall movement would be difficult to observe unambiguously. Nevertheless, in 500 ns simulations, clear displacements of roughly 5 bp should occur for this diffusion rate. We conclude that the direct observation of sliding on DNA by MD simulation should be feasible in the not-too-distant future.

3.4.4 Proposed Molecular Models of Sliding

Here we briefly speculate on the molecular mechanisms of sliding. In addition to the previously mentioned "ice-skating" model of sliding, we propose two new models: the "inchworm" and "centipede" models. In the inchworm model the protein has two or more binding sites that independently bind the DNA and release it. If the position on the DNA can change between binding events, sliding can occur. In the centipede model the protein has many residues that can form attractive contacts to the DNA. The exchange of these residues can lead to overall movement on the DNA.

Halford postulates that what looks like sliding occurs by the same mechanism as dissociation. "The transfer of the protein from one site to another 1 bp along the DNA is likely to be similar to that for the dissociation of the protein into free solution: both processes involve the same number of bond-breaking events between protein and DNA backbone. For example, a hydrogen bond between a particular functional group on the protein and another on the DNA cannot be maintained while the protein moves 1 bp along the DNA, as this movement, a 3.4 Å translocation and a 34 degree rotation, must extend the length of the hydrogen bond to beyond its breaking point" [59]. By this picture, all true sliding – as opposed to hopping – would be some form of "ice-skating" where no bonds persist, and any exchanging bonds are likely mediated by water molecules. This might be one mechanism of sliding. However, evidence from X-ray crystallography suggests that specific protein–DNA contacts may occur even when the two molecules interact nonspecifically. Thus, there are likely other mechanisms of sliding.

In fact, the picture of sliding as a series of dissociation events inadvertently suggests that true sliding almost certainly requires flexibility, either as global conformational changes (e.g., an "inchworm" model), or as local side chain rearrangements (e.g., a "centipede" model). Even hopping need not happen in a rigid, all-or-none fashion. Protein flexibility could allow the protein to peel off the DNA, much the way dsDNA can fray at the ends. The inchworm and centipede models both involve partial binding and unbinding events in the sliding mechanism.

In the inchworm model in addition to the requirement for separate binding sites, the protein must have internal flexibility to contract and extend so that binding and release events effect translation on the DNA. Note that the requirement for more than one binding sites is similar to what is required for intersegmental transfer (Fig. 3.1). Because intersegmental transfer may involve multimeric protein complexes, each monomer of which can bind DNA, it is not clear if any of the known or suspected intersegmental transfer proteins can slide by the inchworm model. Two known examples are Oct-1 [60] and APOBEC3G, an antiviral restriction factor that acts on foreign single-stranded DNA [61]. Other candidates are the RecA protein and the single-strand binding protein, SSB. It would be interesting to investigate whether any proteins that move by intersegmental transfer can also slide via the inchworm model. Perhaps this can be tested by attaching a pair of FRET labels to the protein and observing it move on stretched DNA where intersegmental transfer is not possible.

In the centipede model the protein has many potential residue contacts with the DNA. As contacts randomly break and reform, the collective effect is a random walk of the protein on the DNA. Many attractive contacts would tend to promote cooperativity and should slow sliding. If, however, many residues are available to bind sites on the DNA (e.g., Lys and Arg residues that can bind phosphates on the DNA backbone), the competition may actually accelerate sliding. Thus, we call this variation of the centipede model a "cooperative/competitive" model of sliding. We propose this model for the DNA sliding clamps in the slow mode. Residues that may inhibit protein–DNA binding, such as the conserved, negatively charged residues of the β clamp, can also accelerate sliding.

3.5 Conclusion

Protein sliding on DNA has been shown to be of intense interest. Experimental methods and theoretical treatments are progressing rapidly toward a mechanistic understanding. The dynamics of protein sliding on DNA are challenging, yet clues are appearing as experimental and simulation methods and resources improve. Given this auspicious situation, it is anticipated that a bumper crop of testable hypotheses will soon emerge regarding all aspects of protein sliding on DNA.

References

1. Berg OG, Winter RB, Vonhippel PH (1981) Diffusion-driven mechanisms of protein translocation on nucleic-acids. 1. Models and theory. Biochemistry 20:6929–6948
2. Adam G, Delbrück M (1968) Reduction of dimensionality in biological diffusion processes. In: Rich A, Davidson N (eds) Structural chemistry and molecular biology. W.H. Freeman and Company, San Francisco, pp 198–215
3. Riggs AD, Bourgeoi S, Cohn M (1970) The lac repressor–operator interaction. 3. Kinetic studies. J Mol Biol 53:401–417
4. Winter RB, Vonhippel PH (1981) Diffusion-driven mechanisms of protein translocation on nucleic-acids. 2. The *Escherichia coli* repressor–operator interaction – equilibrium measurements. Biochemistry 20:6948–6960
5. Winter RB, Berg OG, Vonhippel PH (1981) Diffusion-driven mechanisms of protein translocation on nucleic-acids. 3. The *Escherichia coli* lac repressor–operator interaction – kinetic measurements and conclusions. Biochemistry 20:6961–6977
6. Slutsky M, Mirny LA (2004) Kinetics of protein–DNA interaction: facilitated target location in sequence-dependent potential. Biophys J 87:4021–4035
7. Halford SE, Marko JF (2004) How do site-specific DNA-binding proteins find their targets? Nucleic Acids Res 32:3040–3052
8. Greenleaf WJ, Woodside MT, Block SM (2007) High-resolution, single-molecule measurements of biomolecular motion. Annu Rev Biophys Biomol Struct 36:171–190
9. Herbert KM, Greenleaf WJ, Block SM (2008) Single-molecule studies of RNA polymerase: motoring along. Annu Rev Biochem 77:149–176
10. Wunderlich Z, Mirny LA (2008) Spatial effects on the speed and reliability of protein–DNA search. Nucleic Acids Res 36:3570–3578
11. Laurence TA, Kwon Y, Johnson A, Hollars CW, O'Donnell M, Camarero JA, Barsky D (2008) Motion of a DNA sliding clamp observed by single molecule fluorescence spectroscopy. J Biol Chem 283:22895–22906
12. Maki S, Kornberg A (1988) DNA polymerase III holoenzyme of *Escherichia coli*. 3. Distinctive processive polymerases reconstituted from purified subunits. J Biol Chem 263:6561–6569
13. Yao NY, O'Donnell M (2009) Replisome structure and conformational dynamics underlie fork progression past obstacles. Curr Opin Cell Biol 21:336–343
14. Gowers DM, Halford SE (2003) Protein motion from non-specific to specific DNA by three-dimensional routes aided by supercoiling. EMBO J 22:1410–1418
15. Yao N, Turner J, Kelman Z, Stukenberg PT, Dean F, Shechter D, Pan ZQ, Hurwitz J, Odonnell M (1996) Clamp loading, unloading and intrinsic stability of the PCNA, beta and gp45 sliding clamps of human, *E. coli* and T4 replicases. Genes Cells 1:101–113
16. Redner S (2001) A guide to first-passage processes. Cambridge University Press, Cambridge
17. Austin RH, Karohl J, Jovin TM (1983) Rotational diffusion of *Escherichia coli* RNA polymerase free and bound to deoxyribonucleic acid in nonspecific complexes. Biochemistry 22:3082–3090

18. Gorman J, Chowdhury A, Surtees JA, Shimada J, Reichman DR, Alani E, Greene EC (2007) Dynamic basis for one-dimensional DNA scanning by the mismatch repair complex Msh2–Msh6. Mol Cell 28:359–370
19. Gorman J, Greene EC (2008) Visualizing one-dimensional diffusion of proteins along DNA. Nat Struct Mol Biol 15:768–774
20. Biebricher A, Wende W, Escude C, Pingoud A, Desbiolles P (2009) Tracking of single quantum dot labeled EcoRV sliding along DNA manipulated by double optical tweezers. Biophys J 96:L50–L52
21. Bonnet I, Biebricher A, Porte PL, Loverdo C, Benichou O, Voituriez R, Escude C, Wende W, Pingoud A, Desbiolles P (2008) Sliding and jumping of single EcoRV restriction enzymes on non-cognate DNA. Nucleic Acids Res 36:4118–4127
22. Blainey PC, van Oijent AM, Banerjee A, Verdine GL, Xie XS (2006) A base-excision DNA-repair protein finds intrahelical lesion bases by fast sliding in contact with DNA. Proc Natl Acad Sci USA 103:5752–5757
23. Harada Y, Funatsu T, Murakami K, Nonoyama Y, Ishihama A, Yanagida T (1999) Single-molecule imaging of RNA polymerase–DNA interactions in real time. Biophys J 76:709–715
24. Kochaniak AB, Habuchi S, Loparo JJ, Chang DJ, Cimprich KA, Walter JC, van Oijen AM (2009) Proliferating cell nuclear antigen uses two distinct modes to move along DNA. J Biol Chem 284:17700–17710
25. Lin Y, Zhao T, Jian X, Farooqui Z, Qu X, He C, Dinner AR, Scherer NF (2009) Using the bias from flow to elucidate single DNA repair protein sliding and interactions with DNA. Biophys J 96:1911–1917
26. Tafvizi A, Huang F, Leith JS, Fersht AR, Mirny LA, van Oijen AM (2008) Tumor suppressor p53 slides on DNA with low friction and high stability. Biophys J 95:L1–L3
27. Wang YM, Austin RH, Cox EC (2006) Single molecule measurements of repressor protein 1D diffusion on DNA. Phys Rev Lett 97:4
28. Liu SX, Abbondanzieri EA, Rausch JW, Le Grice SFJ, Zhuang XW (2008) Slide into action: dynamic shuttling of HIV reverse transcriptase on nucleic acid substrates. Science 322:1092–1097
29. Roy R, Kozlov AG, Lohman TM, Ha T (2009) SSB protein diffusion on single-stranded DNA stimulates RecA filament formation. Nature 461:1092–1097
30. Rau DC, Sidorova NY (2010) Diffusion of the restriction nuclease EcoRI along DNA. J Mol Biol 395:408–416
31. Blainey PC, Luo GB, Kou SC, Mangel WF, Verdine GL, Bagchi B, Xie XS (2009) Nonspecifically bound proteins spin while diffusing along DNA. Nat Struct Mol Biol 16:1224–1229
32. Graneli A, Yeykal CC, Robertson RB, Greene EC (2006) Long-distance lateral diffusion of human Rad51 on double-stranded DNA. Proc Natl Acad Sci USA 103:1221–1226
33. Laurence TA, Kwon Y, Yin E, Hollars CW, Camarero JA, Barsky D (2007) Correlation spectroscopy of minor fluorescent species: signal purification and distribution analysis. Biophys J 92:2184–2198
34. Kampmann M (2004) Obstacle bypass in protein motion along DNA by two-dimensional rather than one-dimensional sliding. J Biol Chem 279:38715–38720
35. Kong XP, Onrust R, Odonnell M, Kuriyan J (1992) Three-dimensional structure of the beta-subunit of *Escherichia coli* DNA polymerase III holoenzyme – a sliding DNA clamp. Cell 69:425–437
36. García De La Torre J, Huertas M, Carrasco B (2000) Calculation of hydrodynamic properties of globular proteins from their atomic-level structure. Biophys J 78:719–730
37. Schurr J (1979) The one-dimensional diffusion coefficient of proteins absorbed on DNA hydrodynamic considerations. Biophys Chem 9:413–414
38. Bagchi B, Blainey PC, Xie XS (2008) Diffusion constant of a nonspecifically bound protein undergoing curvilinear motion along DNA. J Phys Chem B 112:6282–6284
39. Zwanzig R (1988) Diffusion in a rough potential. Proc Natl Acad Sci USA 85:2029–2030

40. Barbi M, Place C, Popkov V, Salerno M (2004) A model of sequence-dependent protein diffusion along DNA. J Biol Phys 30:203–226
41. Johnson A, O'Donnell M (2005) Cellular DNA replicases: components and dynamics at the replication fork. Annu Rev Biochem 74:283–315
42. Fay P, Johanson K, McHenry C, Bambara R (1981) Size classes of products synthesized processively by DNA polymerase III and DNA polymerase III holoenzyme of *Escherichia coli*. J Biol Chem 256:976–983
43. Indiani C, McInerney P, Georgescu R, Goodman M, O'Donnell M (2005) A sliding-clamp toolbelt binds high- and low-fidelity DNA polymerases simultaneously. Mol Cell 19:805–815
44. Indiani C, Langston LD, Yurieva O, Goodman MF, O'Donnell M (2009) Translesion DNA polymerases remodel the replisome and alter the speed of the replicative helicase. Proc Natl Acad Sci USA 106:6031–6038
45. Adzuma K (1998) No sliding during homology search by RecA protein. J Biol Chem 273:31565–31573
46. Hu LH, Grosberg AY, Bruinsma R (2008) Are DNA transcription factor proteins Maxwellian Demons? Biophys J 95:1151–1156
47. Jeltsch A, Alves J, Wolfes H, Maass G, Pingoud A (1994) Pausing of the restriction-endonuclease EcoRI during linear diffusion on DNA. Biochemistry 33:10215–10219
48. Jones S, van Heyningen P, Berman HM, Thornton JM (1999) Protein–DNA interactions: a structural analysis. J Mol Biol 287:877–896
49. Georgescu RE, Kim SS, Yurieva O, Kuriyan J, Kong XP, O'Donnell M (2008) Structure of a sliding clamp on DNA. Cell 132:43–54
50. Holbrook JA, Tsodikov OV, Saecker RM, Record MT (2001) Specific and non-specific interactions of integration host factor with DNA: thermodynamic evidence for disruption of multiple IHF surface salt-bridges coupled to DNA binding. J Mol Biol 310:379–401
51. McNally R, Bowman GD, Goedken ER, O'Donnell M, Kuriyan J (2010) Analysis of the role of PCNA-DNA contacts during clamp loading. BMC Struct Biol 10:3
52. Mayanagia K, Kiyonari S, Saito M, Shirai T, Ishino Y, Morikawa K (2009) Mechanism of replication machinery assembly as revealed by the DNA ligase-PCNA–DNA complex architecture. Proc Natl Acad Sci USA 106:4647–4652
53. Ivanov I, Chapados BR, McCammon JA, Tainer JA (2006) Proliferating cell nuclear antigen loaded onto double-stranded DNA: dynamics, minor groove interactions and functional implications. Nucleic Acids Res 34:6023–6033
54. Winter JA, Christofi P, Morroll S, Bunting KA (2009) The crystal structure of *Haloferax volcanii* proliferating cell nuclear antigen reveals unique surface charge characteristics due to halophilic adaptation. BMC Struct Biol 9:55
55. Morgunova E, Gray FC, MacNeill SA, Ladenstein R (2009) Structural insights into the adaptation of proliferating cell nuclear antigen (PCNA) from *Haloferax volcanii* to a high-salt environment. Acta Crystallogr D Biol Crystallogr 65:1081–1088
56. Yao N, Hurwitz J, O'Donnell M (2000) Dynamics of beta and proliferating cell nuclear antigen sliding clamps in traversing DNA secondary structure. J Biol Chem 275:1421–1432
57. Dahirel V, Paillusson F, Jardat M, Barbi M, Victor JM (2009) Nonspecific DNA–protein interaction: why proteins can diffuse along DNA. Phys Rev Lett 102:4
58. Sun J, Viadiu H, Aggarwal AK, Weinstein H (2003) Energetic and structural considerations for the mechanism of protein sliding along DNA in the nonspecific BamHI–DNA complex. Biophys J 84:3317–3325
59. Halford SE (2009) An end to 40 years of mistakes in DNA-protein association kinetics? Biochem Soc Trans 37:343–348
60. Doucleff M, Clore GM (2008) Global jumping and domain-specific intersegment transfer between DNA cognate sites of the multidomain transcription factor Oct-1. Proc Natl Acad Sci USA 105:13871–13876
61. Nowarski R, Britan-Rosich E, Shiloach T, Kotler M (2008) Hypermutation by intersegmental transfer of APOBEC3G cytidine deaminase. Nat Struct Mol Biol 15:1059–1066
62. Luscombe NM, Laskowski RA, Thornton JM (1997) NUCPLOT: a program to generate schematic diagrams of protein–nucleic acid interactions. Nucleic Acids Res 25:4940–4945

Chapter 4
Role of DNA Conformations in Gene Regulation

Ralf Metzler, Bram van den Broek, Gijs J.L. Wuite, and Michael A. Lomholt

The topology of DNA plays an important role in the search of DNA binding proteins for their specific target site on the DNA molecule, this search process being at the heart of gene regulation. Connecting to recent single DNA measurements, we provide evidence for the various mechanisms constituting the facilitated diffusion model, in particular, one-dimensional motion along the DNA chain and intersegmental jumps at points of close contact between chemically remote DNA chain segments, mediated by DNA looping.

4.1 Introduction

Gene regulation in biological cells and viruses comprises the biochemical processes involved in turning the information stored in genes, coding regions on a DNA molecule, into the gene products, mostly proteins [1]. The regulation of a specific gene is often modulated by external or intracellular signals. Commonly these signals influence the binding of certain DNA-binding proteins, so-called transcription factors, to a specific binding, or cognate, site close to the starting sequence of the related gene. The classical example for gene regulation by transcription factors is the regulation by Lac repressor in the lactose metabolism of bacteria: in absence of lactose the Lac repressor is specifically bound at the lacZ gene and prevents the biochemical reading out of the genetic information. When only lactose and no glucose are present the Lac repressor will not bind to the lacZ gene. Instead the catabolite activator protein (CAP) binds to the Lac operon: the gene becomes activated, starting production of lactose-digesting enzymes [2]. The binding of the transcription factor to its cognate site necessitates a successful search of this transcription factor for its cognate site. This search is purely stochastic, that is, based on diffusive processes. Originally this search was assumed to be based on pure three-dimensional diffusion until location of the cognate site.

R. Metzler (✉)
Physics Department T30g, Technical University of Munich, 85747 Garching, Germany
e-mail: metz@ph.tum.de

M.C. Williams and L.J. Maher, III (eds.), *Biophysics of DNA-Protein Interactions*, Biological and Medical Physics, Biomedical Engineering,
DOI 10.1007/978-0-387-92808-1_4, © Springer Science+Business Media, LLC 2011

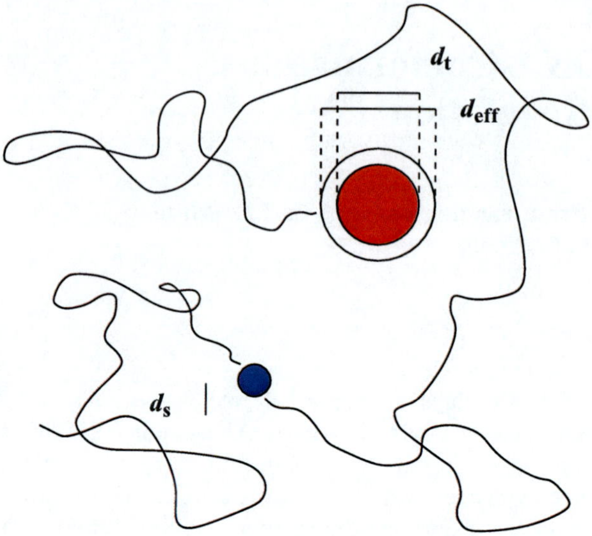

Fig. 4.1 Diffusion-controlled reaction scheme according to the Smoluchowski theory. The searching molecule with diameter d_s diffuses in 3D to the target with diameter d_t. We view the searcher as a point-like particle with diffusivity D_{3d} and the resting target with effective diameter $d_{eff}=d_s+d_t$

The diffusion control of such a search process was originally described by Smoluchowski [3]. Assume that we have a single reaction centre, located at $r=0$ in a three-dimensional liquid environment. This "target" needs to be found by an ensemble of reactants, whose volume density is $n(r,t)$. In Fig. 4.1, we sketch the diffusion path toward the reaction centre of one reactant. If we designate d_t as the diameter of the target and d_s as the diameter of the reactant, we can view the reactant as a point particle and the target as having an effective diameter $d_{eff}=d_s+d_t$. Assuming that the reaction takes place on encounter of the reactant with the target, Smoluchowski's result for the resulting reaction rate can be understood from the solution of the radially symmetric diffusion equation,

$$\frac{\partial n}{\partial t} = D_{3d}\frac{1}{r^2}\frac{\partial}{\partial r}\left(r^2\frac{\partial n}{\partial r}\right) \tag{4.1}$$

in the steady state, $\partial n(r,t)/\partial t=0$. For the reactant density $n(r,t)$, we require that its value is $n(r,t)=n_{bulk}$ far away from the target (i.e., for $r\to\infty$), and that it vanishes on the surface of the target. D_{3d} is the diffusion constant of the reactants. This problem is solved by the steady state density

$$n = n_{bulk}\left(1-\frac{d_{eff}/2}{r}\right). \tag{4.2}$$

The rate constant for this reaction then follows from the stationary flux j_{stat} of reactants onto the effective target surface, $4\pi(d_{eff}/2)^2$, as

$$k_{on}^S = \frac{j_{stat}}{n_{bulk}} = \frac{1}{n_{bulk}}\pi d_{eff}^2 D_{3d}\left.\frac{\partial n}{\partial r}\right|_{r=d_{eff}/2} = 2\pi D_{3d} d_{eff}, \tag{4.3}$$

which is exactly the Smoluchowski result [3]. Note that we normalised the reaction rate such that it becomes a rate constant. This approach is reasonable as typically the steady state is reached before the first particle actually arrives at the target ("rapid equilibrium"). The on-rate k_{on}^{s} has physical dimension $[k_{on}^{s}] = cm^3/s$.

For a typical transcription factor diameter of 5 nm, by Stokes' formula, we obtain $D_{3d} \approx 10^2 \mu m^2/s$. Since the transcription factor can only recognise its cognate site if it is situated right on top of its specific recognition sequence, we assume a typical target site size of one base-pair corresponding to 0.3 nm. Consequently $k_{on}^{s} \approx 10^8 / [(mol/l) \cdot s]$ in biochemical units. Even though this number does not include the time for the binding of the transcription factor itself, it is much slower than some measured on rates: for the Lac repressor, for instance, $k_{on} \approx 10^{10} / [(mol/l) \cdot s]$ was found [4], albeit under unphysiological conditions [5]. The Smoluchowski picture therefore cannot explain the stochastic process underlying the search of some transcription factors for their specific binding site.

Indeed, simple three-dimensional diffusion is not the full answer to the problem: the one-dimensional topology of the DNA chain, embedded in a three-dimensional environment, gives rise to additional stochastic search mechanisms. These are (1) one-dimensional diffusion along the DNA molecule, and (2) so-called intersegmental transfers across contact points of the DNA chain due to DNA looping. During an intersegmental transfer, the transcription factor is transiently bound to both DNA segments. This contrasts (3) intersegmental jumps between nearby DNA segments, which will be treated as a distinct class among the three-dimensional diffusive mechanisms in the following discussion of the role of DNA topology and conformations on gene regulation. We focus on the case of dilute environments typically used in biochemical experiments, such that we can treat the long DNA molecule as a polymer.

We proceed as follows. First, we review evidence for one-dimensional motion along the DNA and discuss the physical mechanism. We then discuss the influence of the full DNA topology and the search mechanisms that DNA coiling effects. We discuss intersegmental transfers, relating them to Lévy flights, and then intersegmental jumps based on recent results from single DNA molecule stretching experiments.

4.2 Nonspecific Binding and 1D Motion Along DNA

Given their specific function, DNA-binding proteins must recognise a specific (cognate) sequence of nucleotides among the megabases contained in the DNA. Without actually opening the double helix, the outside of the DNA can be read by proteins, as the edge of each base pair is exposed at the surface. These patterns are markedly different only in the major groove of the DNA, such that gene regulatory proteins generally bind to the major groove. Apart from single base pair pattern recognition, protein binding is sensitive to special surface features of a DNA region. In bacteria, typical DNA-binding proteins cover around 20 base pairs (Lac repressor: 21, CAP protein: 16; λ repressor CI: 17) [1].

The binding free energy of proteins to DNA involves two contributions [6, 7]: (1) the (average) nonspecific binding free energy due to electrostatic interaction with the

DNA; and (2) additional binding free energy if the sequence of the binding site is sufficiently close to the best (perfectly matching) sequence. The transition from the nonspecific binding is supposed to occur via a conformational change of the regulatory protein from one conformation that permits close contact between the positive charges of the binding protein and the negatively charged DNA backbone to another that allows more sequence-dependent hydrogen bond formation [8]. This is supported by more recent structural studies: While in the nonspecific binding mode, the Lac repressor is bound to DNA in a rather loose and fuzzy way [9], it appears much more ordered in the specific mode. In fact, in the latter case the Lac repressor induces a bend in the DNA [10].[1] In the specific binding case, the additional binding free energy can, to good approximation, be considered independent and additive. Indeed, compared to the specific binding sequence for the protein, each additional base-pair mismatch amounts to the loss of roughly 2 k_BT;[2] the optimal value for the transition between best specific binding to the cognate site and nonspecific binding is approximately 16 k_BT below the specific binding energy [13]. This is quite close to the value of 14 k_BT found for the difference between specific and nonspecific binding for the λ repressor in vivo, see [14] and cited literature therein.

The degree of nonspecific binding can become quite appreciable [15]. For instance, in λ virus-infected *E. coli* cells, the nonspecific binding free energy of the repressor protein CI in vivo was estimated as 7 k_BT, such that in a lysogenic state, approximately 90% of all repressor proteins are indeed nonspecifically bound [14]. A similar value was found for the Lac repressor [16]. Nonspecific binding keeps a binding protein close to the DNA for considerable times. While the exact mechanism is not precisely known, this gives rise to an effective one-dimensional motion along the DNA chain. This 1D diffusion can be quantified by assigning a so-called sliding length: this is the typical length over which the 1D motion persists before dissociation of the protein. The bare sliding length can be relatively short, some 25 bp has been measured [17]. However, once unbound from the DNA, a protein often rebinds immediately and continues the 1D motion along the DNA. The resulting effective sliding length before full detachment from the DNA can be of the order of hundreds of bps [17], and therefore strongly influence the search dynamics.

The existence of sliding motion can be shown by restriction enzymes, that need to keep their orientational direction between two successive binding sites [18]. On the single DNA molecule level, the 1D motion of single-stranded DNA binding proteins was demonstrated in an optical tweezer setup, revealing a one-dimensional diffusion constant $D_{1d} \approx 3.3\times10^{-9}\,\mathrm{cm}^2/\mathrm{s}$ [19, 20]. Using fluorescently labelled binding proteins, the diffusive motion of double-stranded DNA binding proteins along the stretched DNA chain was observed [21], finding values for D_{1d} of the order of $10^{-10}\,\mathrm{cm}^2/\mathrm{s}$. A similar value was obtained in [22], in which 1D motion events could be distinguished from 3D excursions. Remarkably, 1D motion could also be observed in vivo for Lac repressor in an *E. coli* cell, reporting $D_{1d} \approx 4.6\times10^{-10}\,\mathrm{cm}^2/\mathrm{s}$ [23].

---

[1] Similar behaviour is reported for restriction enzymes such as BamHI and EcoRV. See, for example, [11, 12].

[2] Here, $k_B \approx 1.38\times10^{-16}$ erg/K $\approx 3.30\times10^{-24}$ cal/K is the Boltzmann constant, and T denotes absolute temperature.

4.3 The Facilitated Diffusion Model

Nonspecific binding and the effected one-dimensional motion along the DNA chain give rise to a competition between different stochastic mechanisms. This is essential as neither the three-dimensional volume search nor the one-dimensional sliding under the given physiological conditions in the cell are effective search mechanisms: one-dimensional search is very thorough and visits any point on the DNA on its path. However, in one dimension a random walker also frequently returns to previously visited sites. After some time, this one-dimensional search will oversample the DNA, that is, visit sites repeatedly. In fact, the number of sites visited by a one-dimensional random walker grows with the number n of steps like $\mathcal{N}_{\mathrm{dist}}(n) \simeq \sqrt{n}$, corresponding to the expansion of the width of the probability distribution [24]. In contrast, for a random walker progressing in three dimensions it is much more difficult to hit a small target, as discussed for the Smoluchowski model before. This can be anticipated from the fact that the fractal dimension of a Brownian trajectory is 2, and it is therefore unable to fully explore a three-dimensional volume. However, once fully unbound from the DNA, the searching protein can reattach to the DNA at an essentially uncorrelated site on the DNA. It is therefore advantageous that the protein intermittently relocates in three dimensions before resuming its one-dimensional search.

In Fig. 4.2, we sketch different search mechanisms of a protein for its specific target site (labelled O): apart from three-dimensional bulk excursions and one-dimensional sliding, if the protein has two sites for binding DNA it may also transfer directly from one location on the DNA to another, chemically distant segment, when both segments are close in the embedding space, due to looping of DNA. Similarly mediated by DNA looping, a transcription factor with only one DNA binding site may unbind from one DNA segment and immediately rebind to a remote DNA segment, a so-called intersegmental jump (see the discussion below).

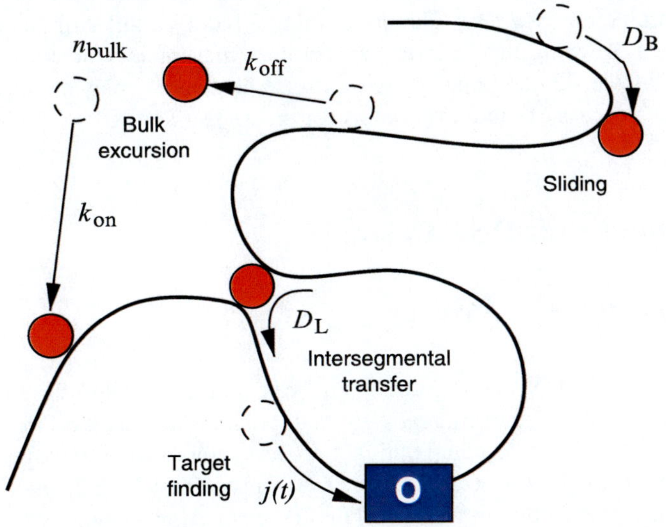

Fig. 4.2 Schematic drawing of the search mechanisms relevant in the effective Lévy flight limit

4.3.1 The Effective Lévy Flight Limit

We may gain some insight into the consequences of the intersegmental transfer mechanism in an approximation to the search for very long chains, neglecting correlations between the unbinding position to the bulk of a transcription factor and its point of rebinding to the DNA. For transcription factors with small nonspecific binding and unbinding rates, its sliding length will be longer than the Kuhn length (twice the persistence length), and we can treat the DNA as an annealed, flexible polymer. From polymer theory, we know that the length ℓ stored in a random loop of such a chain scales like [25]

$$p(\ell) \simeq \ell^{-c}, \tag{4.4}$$

where c is the critical exponent for the contact probability. For a chain in a good, dilute solvent $c \approx 2.2$, while for a Gaussian chain $c = 3/2$. As a transcription factor can use this contact between two DNA segments that are remotely measured along the DNA, but close by in the embedding space, to jump backwards and forwards along the chain, this gives rise to the effective jump length distribution $\lambda(x) \simeq |x|^{-c}$ [26]. Here, x is the one-dimensional coordinate measured along the DNA backbone.

Jump processes governed by a distribution $\lambda(x) \simeq |x|^{-c}$ $(1 < c < 3)$ have diverging variance $\langle x^2 \rangle$ and are called Lévy flights. The number of distinct visited sites for such a Lévy flight grows with the number of jumps n as $\mathcal{N}_{\text{dist}}(n) \simeq n^{1/(c-1)}$ as long as $c > 2$, and as $\simeq n$ when $c < 2$, [24]. Lévy flights with $2 < c < 3$ therefore exhibit less oversampling than a one-dimensional random walker. For $1 < c < 2$, their mixing is as efficient as three-dimensional volume diffusion. When we combine all three mechanisms, one-dimensional search along the DNA, intersegmental transfers across DNA contact points, and three-dimensional diffusion, we see that the Lévy jumps reduce the necessity of decorrelating bulk excursions to arrive at an optimal search efficiency [26]. Having in mind a reduced bulk diffusivity effected by molecular crowding in vivo, we therefore see that intersegmental transfers, or the similar intersegmental jumps to be discussed below, may play an important role in the search process of binding proteins for their specific target.

4.4 Manipulating DNA Coiling

4.4.1 Experimental Strategy

The influence of the DNA configuration on the search rate of DNA binding proteins can indeed be probed experimentally by manipulation of a single DNA molecule [27]. As shown in Fig. 4.3, a dual optical tweezers setup is used to suspend a single DNA chain of length L = 6,538 base pairs (approximately 2.2μm). This DNA contains one specific binding site for EcoRV restriction enzymes, that were in

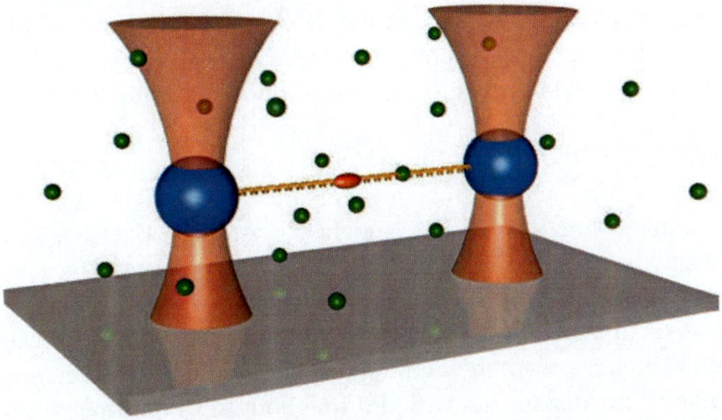

Fig. 4.3 Dual optical tweezers setup. Suspended between two beads held in place by two optical tweezers, the DNA configuration can be deliberately manipulated from an almost completely stretched state to a totally relaxed configuration. At the midpoint of the DNA, a specific binding site for EcoRV enzymes is placed. EcoRV may diffuse in the volume, move along DNA, or perform intersegmental jumps as shown in Fig. 7

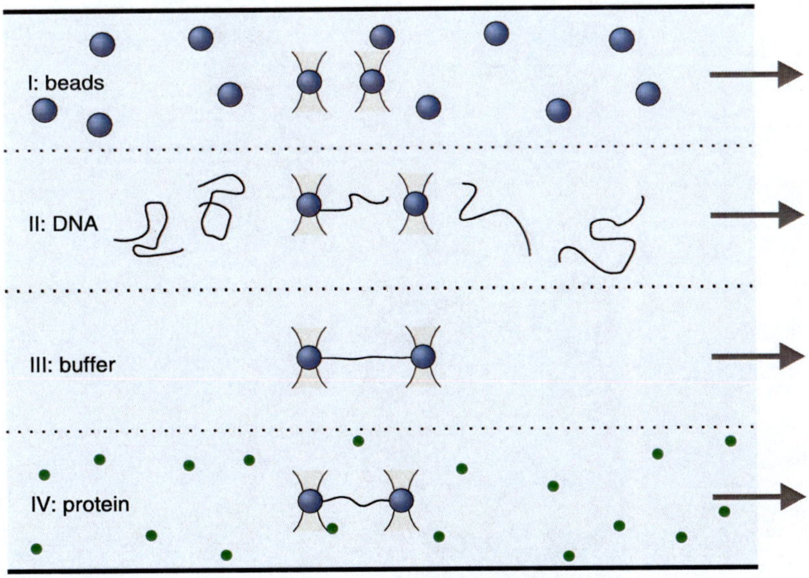

Fig. 4.4 Preparation of the dual tweezers experiment. See text

solution. After locating this site an EcoRV will cut the DNA double strand. Figure 4.4 details how the experiment is prepared. In step I, plastic beads are flushed into a multichannel flow cell, until eventually two beads are trapped in the foci of the two laser beams. These beads are coated with streptavidin. Then, DNA equipped with biotin at the extremities is flushed into the cell, and ultimately one

DNA chain attaches to the two beads in step II. Step III introduces a buffer solution. Finally, in step IV the EcoRV enzymes are introduced, by moving the DNA-tweezers constructs quickly into the enzyme solution, defining the starting point for the actual experiment.

Every second, the DNA is briefly stretched by a force in the range of 5–10 pN, such that the stretching and subsequent relaxation of the DNA chain occurs on a time scale of approximately 25 ms, much shorter than the separation between successive stretching events. As long as the DNA remains intact the stretching leads to a response force by the chain. Once cut this resistance vanishes. Thus, this method allows us to measure the cleavage time of DNA with a 1-s resolution.

After each stretching event the two trapped beads resume a given preset distance from each other. This distance is varied throughout the experiment, allowing us to probe the influence of the degree of DNA coiling on the search efficiency of the restriction enzymes. As shown in Fig. 4.5, we reach three distinct regimes. When the trapped beads are far apart from each other the suspended DNA is almost

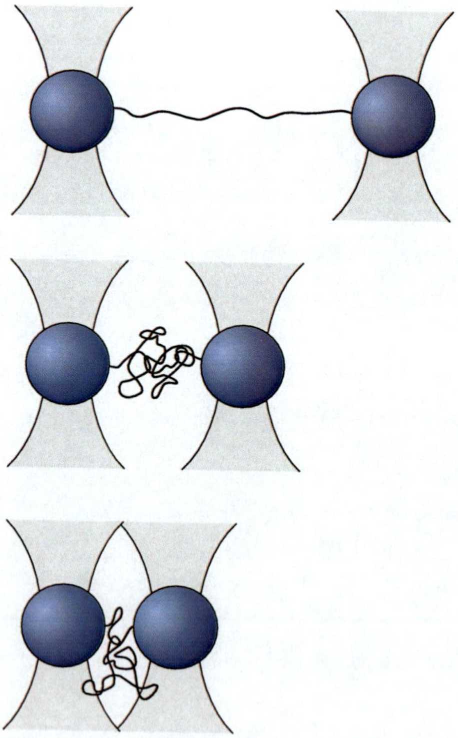

Fig. 4.5 Equilibrium configurations of the DNA chain in the dual tweezers setup. Reducing the distance between the tweezer beads causes the chain to change from an almost stretched configuration to the maximally relaxed coil state, when the distance between the ends of the DNA molecule equals its natural gyration radius. Finally the beads are so close that the suspended DNA is squeezed out of the volume between the beads

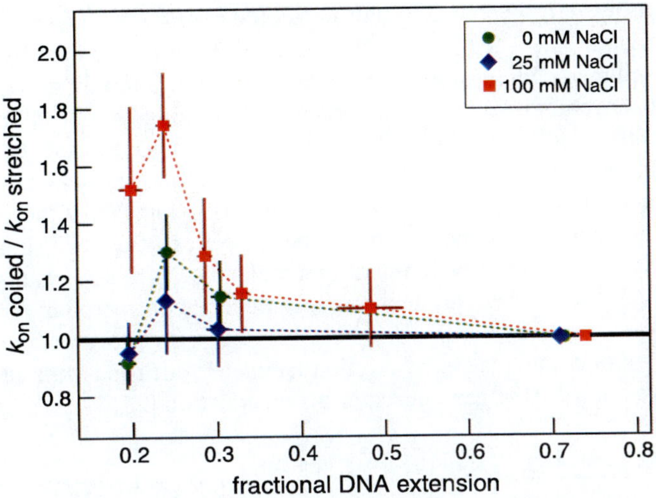

Fig. 4.6 Cleavage rate as function of the fractional extension of the DNA chain, for different salt conditions. Note the distinct maximum at 100 mM NaCl. For each salt concentration the rates are normalised to the respective value in the stretched configuration

stretched. The experiment is repeated for different bead–bead distances, until we reach the maximally relaxed DNA configuration, that is, when the distance becomes comparable to the natural gyration radius of the free chain. The experiment is continued across this point, towards distances at which the DNA is expelled from the volume between the beads.

In Fig. 4.6, we show the measured cleavage rate as function of the fractional DNA extension, that is, the distance between the tweezer beads divided by the DNA length. For physiological salt corresponding to 100 mM NaCl, there occurs a pronounced maximum at a fractional extension of around 0.24. This number corresponds to the maximally random configuration when the DNA size is characterised by the gyration radius of the free DNA chain. The relative increase compared to the almost fully stretched configuration is significant, about 1.7 for the chain length in this experiment. Similar measurements were performed at different NaCl concentrations. Remarkably, the pronounced maximum seen in Fig. 4.6 is decreased significantly at lower NaCl concentration. Also, at 150 mM salt concentration the increase is strongly reduced (data not shown).

4.4.2 *Theoretical Modelling*

EcoRV restriction enzymes only possess one binding pocket for double-stranded DNA. The intersegmental transfer mechanism considered in the original Berg–von Hippel model therefore cannot occur. As we will see, however, a similar mechanism may come into play when the DNA configuration is sufficiently relaxed. This is shown in Fig. 4.7, where we follow an EcoRV restriction enzyme moving along the DNA chain. The effective one-dimensional motion mediated by nonspecific binding combines sliding motion along the DNA and unbinding events followed by immediate rebinding. During its translocation along the DNA backbone, the DNA chain itself will move due to Brownian motion. This may induce looping and form a point of close contact between the location of the restriction enzyme and a remote (foreign) DNA segment. If the enzyme dissociates from the DNA it has a chance to be captured by the foreign segment. We call this mechanism intersegmental jumping. Quite similar to intersegmental transfers this mechanism increases the search efficiency in a way that strongly depends on the configuration of the DNA chain.

To capture in a quantitative description facilitated diffusion with intersegmental jumping, we consider the generalised diffusion equation [28]

$$\frac{\partial n(x,t)}{\partial t} = \left(D_{1d} \frac{\partial^2}{\partial x^2} - k_{off}^{ns} \right) n(x,t) - j(t)\delta(x) + G(x,t) + k_{off}^{ns} \int_{-\infty}^{\infty} dx' \int_0^t dt' \, W_{bulk}(x-x', t-t') n(x', t'). \tag{4.5}$$

for the line density $n(x, t)$ of enzymes on the DNA, measured along the chemical coordinate x along the DNA chain. In (5), the first term on the right hand side represents the one-dimensional motion along the DNA with an effective diffusion constant D_{1d}. The second term is the off rate for proteins unbinding from the DNA into the bulk after nonspecific binding. The third term is the flux of EcoRV restriction enzymes to the specific binding site placed at $x = 0$. These enzymes are being removed by the incorporated δ sink. The associated boundary condition reads $n(x, t) = 0$ at $x = 0$. The last term in (5) is a convolution that relates the unbinding from the DNA at position x' at time t', with the rebinding at x and t. This relocation is mediated by the kernel W_{bulk} that will be considered below. Finally, $G(x, t)$ denotes the virgin flux onto the DNA of enzymes, that have not been bound to DNA previously.

Similar to the derivation of the Smoluchowski result (3), we derive the steady state solution for the binding flux $j_{stat} = k_{1d} n_{eq}^{ns}$ from (5) [28]. Here, we denote by n_{eq}^{ns} the one-dimensional density of proteins on the DNA at nonspecific equilibrium, and k_{1d} dyined through

$$k_{1d}^{-1} = \int_{\infty}^{-\infty} \frac{dq}{2\pi} \frac{1}{D_{1d} q^2 + k_{off}^{ns} (1 - \lambda_{bulk}(q))} \tag{4.6}$$

is a rate constant for the target search on the DNA. We here introduced the Fourier transform $\lambda_{bulk}(q)$ of the distribution $\lambda_{bulk}(x)$ of relocation lengths mediated by bulk excursions. This quantity is defined in terms of the relocation kernel $W_{bulk}(x, t)$ as follows:

$$\lambda_{bulk}(q) = \int_0^{\infty} dt \int_{\infty}^{-\infty} dx e^{iqx} W_{bulk}(x,t). \tag{4.7}$$

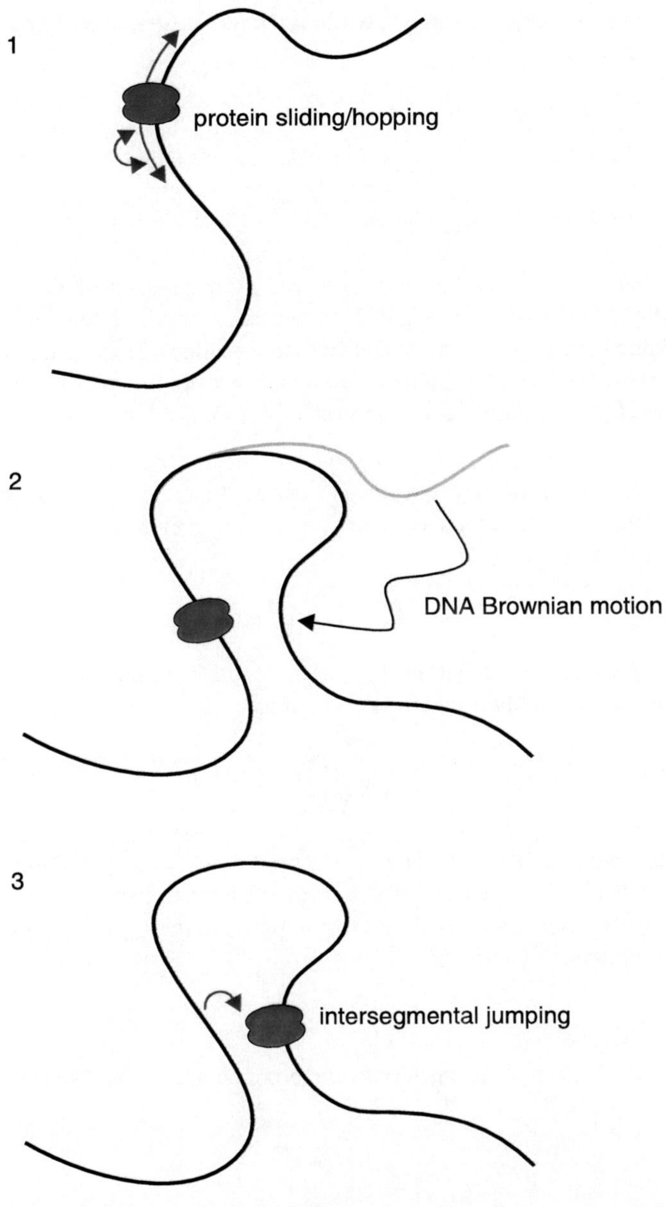

Fig. 4.7 Schematic view of the intersegmental jump mechanism. In *panel 1* an EcoRV enzyme is moving one-dimensionally along the DNA chain, mediated by sliding and hops, short-distance volume excursions with immediate rebinding. While the enzyme is close to the DNA, the chain itself may come in contact to a remote segment due to random looping, as shown in *panel 2*. After dissociation from the DNA, the enzyme has a relatively high chance to be captured by a remote (foreign) segment at the contact point, as sketched in *panel 3*

The desired rate constant with respect to the bulk density of proteins away from the DNA chain is therefore

$$k_{on} = \frac{j_{stat}}{n_{bulk}}. \tag{4.8}$$

4.4.2.1 Straight Rod Configuration

Let us first consider the case of a straight rod configuration of the DNA chain, corresponding to the top panel in Fig. 4.5. In this case, the relocation kernel $W_{bulk}(x, t)$ may be obtained from the cylindrical diffusion equation [28]. We can distinguish two limiting cases based on the nonspecific binding rate of enzymes with the DNA, k_{on}^{ns}, as defined by the equilibrium condition[3] $n_{eq}^{ns}k_{off}^{ns}=n_{bulk}k_{on}^{ns}$:

1. In the high diffusivity case $k_{on}^{ns} \ll D_{3d}$, the rate of nonspecific binding limits the binding of the enzymes: each enzyme after unbinding relocates by a long distance and becomes essentially uncorrelated from the previous binding spot. This effects the search rate

$$k_{on} = 2l_{sl}k_{on}^{ns}. \tag{4.9}$$

Thus, a protein that binds within the sliding length l_{sl} of the target site is able to locate the target. This sliding length is defined as

$$l_{sl} = \sqrt{\frac{D_{1d}}{k_{off}^{ns}}}. \tag{4.10}$$

2. Conversely, when $k_{on}^{ns} \gg D_{3d}$ the enzymes rebind quickly after unbinding from the DNA. This leads to frequent revisits to previously sampled sites (oversampling) and thus to a reduced target search rate compared to the expression given by (9). The asymptotic result becomes

$$k_{on} \sim 4\pi D_{3d} l_{sl}^{eff} / [\ln(l_{sl}^{eff} / r_{int})]^{1/2}, \tag{4.11}$$

where r_{int} is the interaction distance between enzyme and DNA. We can define the effective sliding length

$$l_{sl}^{eff} = \sqrt{\frac{k_{on}^{ns}}{(2\pi D_{3d})}} l_{sl}, \tag{4.12}$$

that can be interpreted as the regular sliding length renormalised by immediate rebinding events [29]. Result (11) is essentially equivalent to the Smoluchowski

---

[3] Note that the dimensions are those of a rate for $[k_{off}^{ns}]=s^{-1}$, a line density $[n_{eq}^{ns}]=cm^{-1}$, and a volume concentration $[n_{bulk}]=cm^{-3}$. Therefore, the nonspecific reaction rate has the dimension of a diffusivity, $[k_{n}^{ns}]=cm^2/s$.

result, the target size being replaced by the effective sliding length, sometimes referred to as the antenna effect. We note that this approach leads to the same result as the original study by Berg and Ehrenberg [29].

4.4.2.2 Coiled DNA Configuration

While the mechanism of intersegmental jumping is quite intuitive, it is not that easy to actually calculate the relocation kernel W_{bulk} in this case. However, we can introduce a number of physically reasonable approximations to interpret the major effects. First, we treat the DNA chain as an array of locally straight segments whose length (Kuhn length) corresponds to two persistence lengths. With the density l_{DNA} of DNA (with dimension of length per volume) around a given segment, we furthermore assume that we are in the dilute limit, and the typical distance $1/\sqrt{l_{DNA}}$ between segments is much larger than the interaction radius r_{int} of DNA and searching enzymes. We obtain the following form of the Fourier transform ($q \leftrightarrow x$) of the relocation kernel,

$$W_{\text{bulk}}(q,t) = W_{\text{bulk}}^{\text{cyl}}(q,t)P_{\text{surv}}^{\text{foreign}}(t) - \frac{dP_{\text{surv}}^{\text{foreign}}(t)}{dt}\delta_{q,0}\int_t^{\infty} W_{\text{bulk}}^{\text{cyl}}(q=0,t')dt'. \tag{4.13}$$

The first term on the right-hand side stands for proteins that return to the original segment without being captured by foreign segments. Those captures occur with probability $1-P_{\text{surv}}^{\text{foreign}}(t)$ and are represented by the second term, in which the factor $\delta_{q,0}$ is a result of the assumption that a capture by foreign DNA segments leads to a long relocation measured along the DNA contour. These captures by foreign segments are exactly the intersegmental jumps.

Extracting values of k_{on} numerically, one finds that the significance of the intersegmental jumps is small when the effective sliding length $l_{\text{sl}}^{\text{eff}}$ is small in comparison to the typical distance between DNA segments: k_{on} will only be slightly larger than the corresponding value for a straight configuration of the DNA. However, the density of DNA segments l_{DNA} is in many situations enhanced compared with the density of a freely fluctuating semiflexible polymer. This enhancement can, for instance, be mediated by supercoiling, attraction between segments induced by multivalent ions, or compaction by various kinds of proteins.

4.4.2.3 Salt Effect

In a quantitative modelling it is not trivial to include the salt dependence, as many model parameters become affected. For low NaCl concentration, the experiment can be described in terms of a local DNA density obtained from the worm-like chain model for DNA with persistence length $\ell_p \approx 50$ nm. At higher

NaCl concentrations and simultaneous presence of divalent Mg^{2+} ions [in our experiments 5 mM Mg^{2+} were used] attractive interactions between DNA segments arise [30, 31]. With reasonable values for this density the observed search rates can be explained, for details, see [27]. Since EcoRV has only one DNA binding site and therefore cannot perform intersegmental transfers, this experiment is a direct proof for the existence of intersegmental jumps. We note that the increase of the search rate effected by intersegmental jumps should be further pronounced when DNA is packaged, for instance, in a cell. A similar increase by a factor of 3–4 was in fact observed in supercoiled versus relaxed plasmids [32].

4.5 Conclusions

Our current understanding of the basic pathways of gene regulation to a large extent reaches back to the work of Lwoff on bacteriophage infection of *E. coli*. In his 1953 review article, Lwoff discusses how, after infection of the host cell, the viral DNA is silently replicated, until eventually it becomes active and leads to production of new viruses [33]. Culminating in the work of Berg and von Hippel, the microscopic stochastic basis of gene regulation, that is, the search of a transcription factor for its specific binding site on the DNA, could be understood in terms of the facilitated diffusion model [6]. With modern single molecule tools, we are now able to look even more closely into the machinery of gene regulation. Thus, we are able to follow single proteins on their pathway, and we can probe the individual mechanisms involved in the regulation process. Here, we focused on the active role played by the DNA molecule itself and showed that its configurational fluctuations assist the search of binding proteins for their specific target site.

In a living cell the situation changes drastically. Even in bacterial cells the DNA is compacted in the centre of the cell, while in eukaryotes the DNA is stored in a hierarchical array in the chromosomes. Moreover the cellular environment is all but dilute. Indeed the cell volume is occupied by larger biomolecules to volume fractions around 40%. These features induce drastic effects to the dynamics of gene regulation. First, the three-dimensional motion of large biopolymers such as mRNAs becomes subdiffusive [34]. For typical binding proteins subdiffusion was also claimed [35, 36]. Even if on relevant time scales there is no subdiffusion, a non-negligible reduction of the bulk diffusivity may arise. Second, packaging and molecular crowding will limit the DNA mobility, and contact points become more permanent, and possibly more specific. Finally, the DNA is decorated with a mixture of binding proteins, both specifically and nonspecifically bound, with interesting effects on the one-dimensional motion along the chain [37]. It is a veritable challenge to come up with a new picture of in vivo gene regulation, and to understand how systems with nanomolar concentrations of a certain transcription factor can be extremely stable.

References

1. Alberts B, Johnson A, Lewis J, Raff M, Roberts K, Walter P (2002) Molecular biology of the cell, 3rd edn. Garland Science, New York
2. Ptashne M (2004) A genetic switch. Cold Spring Harbor Laboratory Press, Cold Spring Harbor, NY
3. Smoluchowski MV (1916) Three presentations on diffusion, molecular movement according to Brown and coagulation of colloid particles. Physikal Zeitschr 17:557–571
4. Riggs AD, Bourgeois S, Cohn M (1970) Lac repressor-operator interaction. 3. Kinetic studies. J Mol Biol 53:401
5. Halford SE (2009) An end to 40 years of mistakes in DNA-protein association kinetics. Biochem Soc Trans 37:344
6. von Hippel PH, Berg OG (1989) Facilitated target location in biological systems. J Biol Chem 264:675
7. Berg OG, Winter RB, von Hippel PH (1981) Diffusion-driven mechanisms of protein translocation on nucleic acids. 1. Models and theory. Biochemistry 20:6929
8. Winter RB, Berg OG, von Hippel PH (1981) Diffusion-driven mechanisms of protein translocation on nucleic acids. 3. The Escherichia coli lac repressor-operator interaction: kinetic measurements and conclusions. Biochemistry 20:6961
9. Kaladimos CG, Biris N, Bonvin AMJJ, Levandovski MM, Guennuegues M, Boelens R, Kaptein R (2004) Structure and flexibility adaptation in nonspecific and specific protein-DNA complexes. Science 305:386
10. Lewis M, Chang G, Horton NC, Kercher MA, Pace HC Schumacher MA, Brennan RG, Lu P (1996) Crystal structure of the lactose operon repressor and its complexes with DNA and inducer. Science 271:1247
11. Viadiu H, Aggarwal AK (2000) Structure of BamHI bound to nonspecific DNA: a model for DNA sliding. Mol Cell 5:889
12. Winkler FK, Banner DW, Oefner C, Tsernoglou D, Brown RS, Heathman SP, Bryan RK, Martin PD, Petratos K, Wilson KS (1993) The crystal-structure of EcoRV endonuclease and of its complexes with cognate and non-cognate DNA fragments. EMBO J 12:1781
13. Gerland U, Moroz JD, Hwa T (2002) Physical constraints and functional characteristics of transcription factor-DNA interaction. Proc Natl Acad Sci U S A 99:12015
14. Bakk A, Metzler R (2004) In vivo non-specific binding of λ CI and Cro repressors is significant. FEBS Lett 563:66
15. von Hippel PH, Revzin A, Gross CA, Wang AC (1974) Non-specific DNA binding of genome regulating proteins as a biological control mechanism: I. The lac operon: equilibrium aspects. Proc Natl Acad Sci U S A 71:4808
16. Kao-Huang Y, Revzin A, Butler AP, O'Conner P, Noble DW, von Hippel PH (1977) Non-specific DNA binding of genome regulating proteins as a biological control mechanism: Measurement of DNA-bound Escherichia coli lac repressor in vivo. Proc Natl Acad Sci U S A 74:4228
17. Stanford NP, Szczelkun MD, Marko JF, Halford SE (2000) One- and three-dimensional pathways for proteins to reach specific DNA sites. EMBO J 19:6546
18. Gowers DM, Wilson GG, Halford SE (2005) Measurement of the contributions of 1D and 3D pathways to the translocation of a protein along DNA. Proc Natl Acad Sci U S A 102:15883
19. Sokolov IM, Metzler R, Pant K, Williams MC (2005) Target search of N sliding proteins on a DNA. Biophys J 89:895
20. Sokolov IM, Metzler R, Pant K, Williams MC (2005) Target search of N sliding proteins on a DNA. Phys Rev E 72:041102
21. Wang YM, Austin RH, Cox EC (2006) Single molecule measurements of repressor protein 1D diffusion on DNA. Phys Rev Lett 97:048302
22. Bonnet I, Biebricher A, Porte P-L, Loverdo C, Benichou O, Voituriez R, Escude C, Wende W, Pingoud A, Desbiolles P (2008) Sliding and jumping of single EcoRV restriction enzymes on non-cognate DNA. Nucleic Acids Res 36:4118

23. Elf J, Li G-W, Xie XS (2007) Probing transcription factor dynamics at the single-molecule level in a living cell. Science 316:1191
24. Hughes BD (1995) Random Walks and Random Environments, Volume 1: Random Walks. Oxford University Press, Oxford, UK
25. Duplantier B (1989) Statistical mechanics of polymer networks of any topology. J Stat Phys 54:581
26. Lomholt MA, Ambjörnsson T, Metzler R (2005) Optimal target search on a fast-folding polymer chain with volume exchange. Phys Rev Lett 95:260603
27. van den Broek B, Lomholt MA, Kalisch S-MJ, Metzler R, Wuite GJL (2008) How DNA coiling enhances the target localization by proteins. Proc Natl Acad Sci U S A 105:15738
28. Lomholt MA, van den Broek B, Kalisch S-MJ, Wuite GJL, Metzler R (2009) Facilitated diffusion with DNA coiling. Proc Natl Acad Sci U S A 106:8204
29. Berg OG, Ehrenberg M (1982) Association kinetics with coupled 3-dimensional and one-dimensional diffusion: chain-length dependence of the association rate to specific DNA sites. Biophys Chem 15:41
30. Lee KC, Borukhov I, Gelbart WM, Liu AJ, Stevens MJ (2004) Effect of mono- and multivalent salts on angle-dependent attractions between charged rods. Phys Rev Lett 93:128101
31. Qiu X et al (2006) Measuring inter-DNA potentials in solution. Phys Rev Lett 96:138101
32. Gowers DM, Halford SE (2003) Protein motion from non-specific to specific DNA by three-dimensional routes aided by supercoiling. EMBO J 22:1410
33. Lwoff A (1953) Lysogeny. Bacteriol Rev 17:269
34. Golding I, Cox EC (2006) Physical nature of bacterial cytoplasm. Phys Rev Lett 96:098102
35. Banks DS, Fradin C (2005) Anomalous diffusion of proteins due to molecular crowding. Biophys J 89:2960
36. Weiss M, Elsner M, Kartberg F, Nilsson T (2004) Anomalous subdiffusion is a measure for cytoplasmic crowding in living cells. Biophys J 87:3518
37. Li GW, Berg OG, Elf J (2009) Effects of macromolecular crowding and DNA looping on gene regulation kinetics. Nat Phys 5:294

Chapter 5
Recognition of Nucleic Acids by Transcription Factor NF-κB

Gourisankar Ghosh, De-Bin Huang, and Tom Huxford

5.1 Introduction

NF-κB is an inducible transcription factor that responds to diverse stress signals by regulating the expression of a large number of genes [29]. At the heart of its orchestrating appropriate cellular responses to potentially toxic stimuli lies the ability of NF-κB to bind specifically and with high affinity to target DNA within the promoters of target genes. In this chapter, we review NF-κB binding to nucleic acids from the perspective of the wealth of structural information that is available on the subject.

5.2 The Rel/NF-κB Family

NF-κB refers to a class of eukaryotic transcription factors formed through the combinatorial assembly of five polypeptide subunits into homo- or heterodimers. The founding member of the NF-κB family subunits is c-Rel, which was discovered in 1984 as a proto-oncogene [79]. Its oncogenic version, v-Rel, is encoded by the reticuloendotheliosis virus strain T (Rev-T) as a recombination product of the avian *c-rel* gene and Rev-A [7, 30, 79]. In v-Rel, 2 N-terminal and 118 C-terminal residues of c-Rel are replaced by viral Envelope-derived residues. Thus, v-Rel is an Env-Rel-Env fusion protein, which is constitutively expressed, resides in the nucleus of infected cells, and confers a distinct gene expression program. Consequently, Rev-T expression of v-Rel causes severe lymphomas and transforms fibroblasts [4, 54]. The discovery and cloning of the *Drosophila* developmental factor Dorsal revealed it to be a protein that shares significant N-terminal sequence homology with c-Rel and v-Rel [73, 74, 79]. Within a few years of these initial discoveries a constitutively active nuclear transcription factor was identified that was found to

G. Ghosh (✉)
Department of Chemistry and Biochemistry, University of California,
9500 Gilman Drive, La Jolla, San Diego, CA 92093-0375, USA
e-mail: gghosh@ucsd.edu

M.C. Williams and L.J. Maher, III (eds.), *Biophysics of DNA-Protein Interactions*, Biological and Medical Physics, Biomedical Engineering,
DOI 10.1007/978-0-387-92808-1_5,

bind with specificity to a 10 bp DNA element within the intronic enhancer of the immunoglobulin *kappa* light chain gene in *B* cells [70, 72]. The factor was consequently named nuclear factor kappa B (NF-κB) and the DNA element that it recognized was dubbed the κB site.

Shortly after its initial discovery it was determined that NF-κB is, in fact, present in all cells. However, and in spite of its original discovery in the nucleus of mature B cells, NF-κB was found to reside predominantly in an inactive state within the cytoplasm of resting cells [69]. This inactive NF-κB was purified and shown to be a trimeric complex consisting of an inhibitor protein, referred to as the inhibitor of *kappaB* binding (IκB), and an NF-κB heterodimer composed of 50 and 65 kDa polypeptides. Treatment of the inactive trimer with the mild detergent deoxycholate was shown to cause dissociation of the inhibitor protein, indicating that it binds NF-κB noncovalently, and convert NF-κB into its active state capable of κB DNA binding [1]. Analysis of the cDNAs encoding the two NF-κB subunits revealed that p50 and p65 (also called RelA) bear homology not only to one another, but also to v-Rel, c-Rel, and Dorsal. The N-terminal portion of approximately 300 amino acid residues in length that is conserved within each of these polypeptides is referred to as the Rel homology region (RHR) [6, 26, 39, 52, 59, 64]. Two additional polypeptides, RelB and p52, were subsequently discovered to contain an N-terminal RHR [5, 51, 65, 67]. Together, RelA, RelB, c-Rel, p50, and p52 constitute the entire mammalian family of NF-κB subunits (Fig. 5.1). During cloning of the p50 and p52

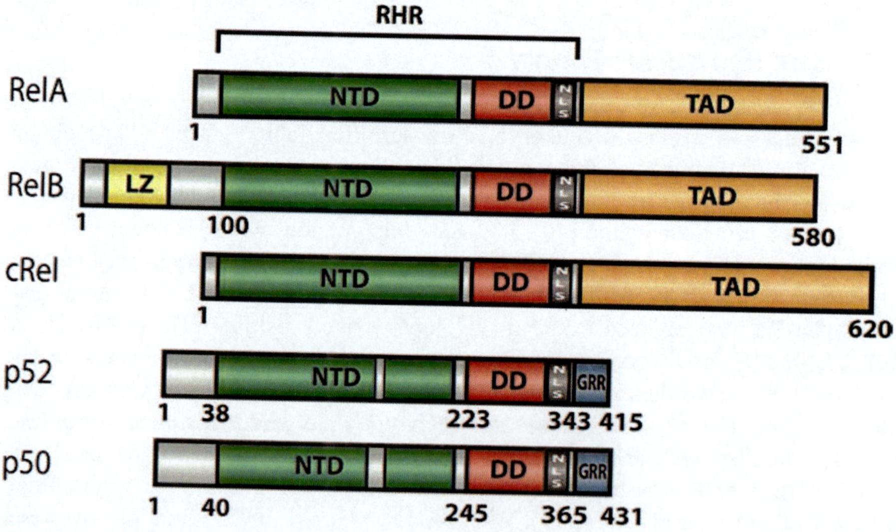

Fig. 5.1 Domain organization schematic of the mammalian NF-κB family subunits. The Rel homology region (RHR) consists of two folded domains referred to as the N-terminal domain (NTD) and dimerization domain (DD), which are linked by a short flexible linker. A nuclear localization signal (NLS) is located at the extreme C-terminus of the RHR. C-terminal to the RHR in RelA, RelB, and c-Rel is a transcription activation domain (TAD). The p50 and p52 subunits lack inherent transcriptional activation potential and contain a glycine-rich region (GRR) C-terminal to the RHR. RelB is unique in that it contains a short leucine zipper motif (LZ) N-terminal to the RHR. Amino acid numbering is taken from the mouse sequences

polypeptides it was discovered that these two subunits represent the mature processed products of the p105 and p100 precursor proteins, respectively. The immature precursors contain p50 or p52 within their N-termini while their C-terminal regions display homology to IκB [2].

The first three-dimensional structures of the RHR were produced by X-ray crystallography of a homodimer of NF-κB p50 subunits bound to two related κB DNA variants. These analyses revealed that the RHR is composed of two immunoglobulin-like folded domains (Fig. 5.2). The N-terminal domain is approximately 200 residues in length and is primarily responsible for sequence specific binding to κB DNA [25, 55]. The C-terminal dimerization domain is around 100 residues in length and is responsible for subunit dimerization. Several basic residues from the dimerization domain also contact κB DNA in a nonspecific manner. A short linker of roughly 10 residues in length connects the two domains.

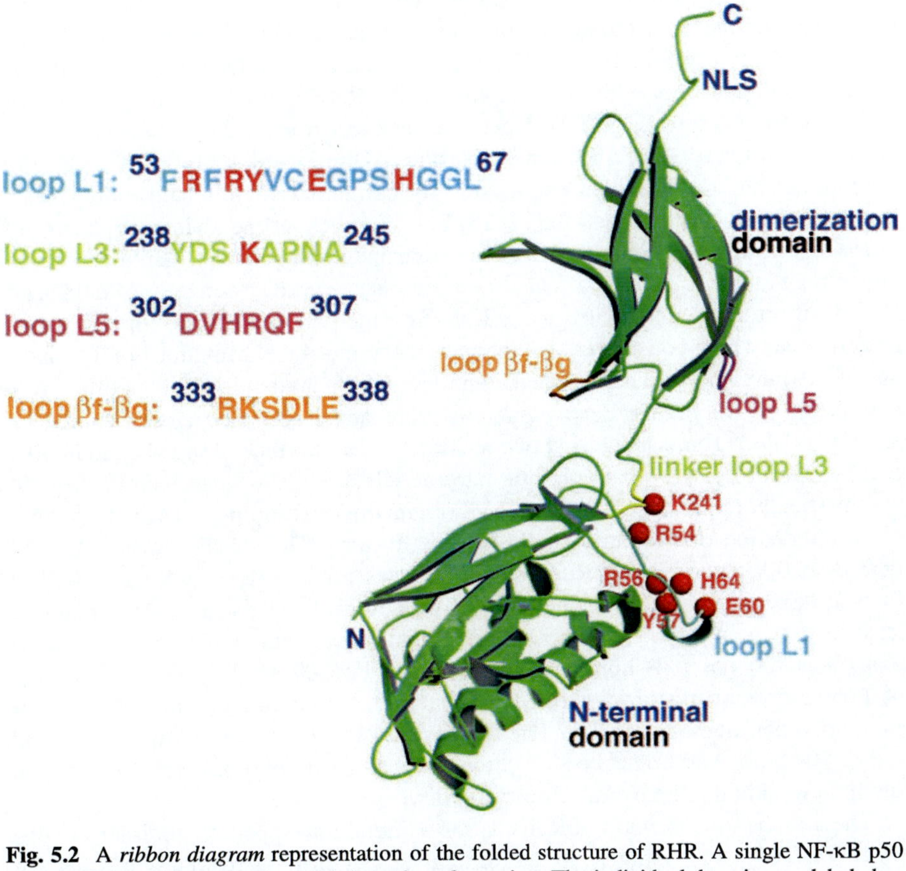

Fig. 5.2 A *ribbon diagram* representation of the folded structure of RHR. A single NF-κB p50 subunit RHR is depicted in its DNA-bound conformation. The individual domains are labeled as well as the linker, NLS, and the N- and C- termini. The amino acid sequences from portions of the molecule discussed in the text are given in single letter code on the *left*. The positions of amino acids that make DNA base-specific contacts with κB DNA are indicated as *red spheres* in the structure. The corresponding amino acids are also highlighted in *red*

The linker contributes one basic residue that makes critical base-specific contacts. It also affords significant motion of the two folded domains relative to one another. Beyond the C-terminal dimerization domain lies 10–20 additional residues that are mostly disordered when the RHR adopts its DNA-bound conformation. Embedded within this flexible polypeptide segment is a type I nuclear localization signal (NLS). Consequently, this region is critical for migration of NF-κB from the cytoplasm to the nucleus. Furthermore, it has been shown that the nuclear localization signal-containing segment is important for NF-κB binding to inhibitor IκB proteins [37, 47].

5.3 NF-κB Dimerization

NF-κB dimers are formed through symmetrical interactions between one beta-sheet from each of the C-terminal immunoglobulin-like dimerization domains of the complex. X-ray crystal structures for the dimerization interface of several NF-κB dimers are known and, with the exception of the RelB homodimer, they all exhibit a similar architecture (Fig. 5.3). Up to 12 residues from each subunit contribute directly to formation of the dimer interface. Dimerization of the NF-κB p50 homodimer has been studied extensively by combined site-directed mutagenesis and biophysical characterization [17, 28, 71]. These experiments indicate that only a few of the amino acids at the interface contribute energetically, while others are either regulatory or do not affect dimer stability. Tyr267, Leu269, Ala308, and Val310 (throughout this review we will use murine p50 numbering) form a hydrophobic core, which constitutes the hot spot of the interface. Mutations of Tyr, Leu, and Val to alanine significantly weaken p50 homodimer stability. The methyl group side chain of Ala307 fits neatly into this core and larger side chains cannot be accommodated. There are at least two residues at the interface, Asp254 and Phe307, that function to discriminate binding partner selection. The side chain of Asp254 symmetrically opposes another Asp254 from its dimer-forming p50 subunit partner in an interaction that is unfavorable (Fig. 5.3b, top). The substitution of Asp254 with Asn200 in the NF-κB RelA subunit allows for a perfect hydrogen bond to form at the dimer interface of the p50:RelA heterodimer. This serves to explain, at least in part, why the NF-κB p50:RelA heterodimer forms with greater stability than does the p50:p50 homodimer. Similarly, Phe307 slightly destabilizes the p50:p50 homodimer as evidenced by the fact that its mutation to Ala yields a more stable p50:p50 homodimer. This Phe307 position is substituted to Val in RelA and c-Rel. The bulky and branched side chains maintain a suboptimal distance between subunits weakening the overall dimer stability.

The dimerization domain of RelB forms a domain-swapped homodimer in vitro (Fig. 5.3) [35]. Surprisingly, all the critical dimer-forming residues in RelB are either identical or homologous to other family members. Therefore, it is difficult to rationalize by simple observation alone just why the interfacial residues in the RelB homodimer do not adopt the same side-by-side dimerization architecture exhibited

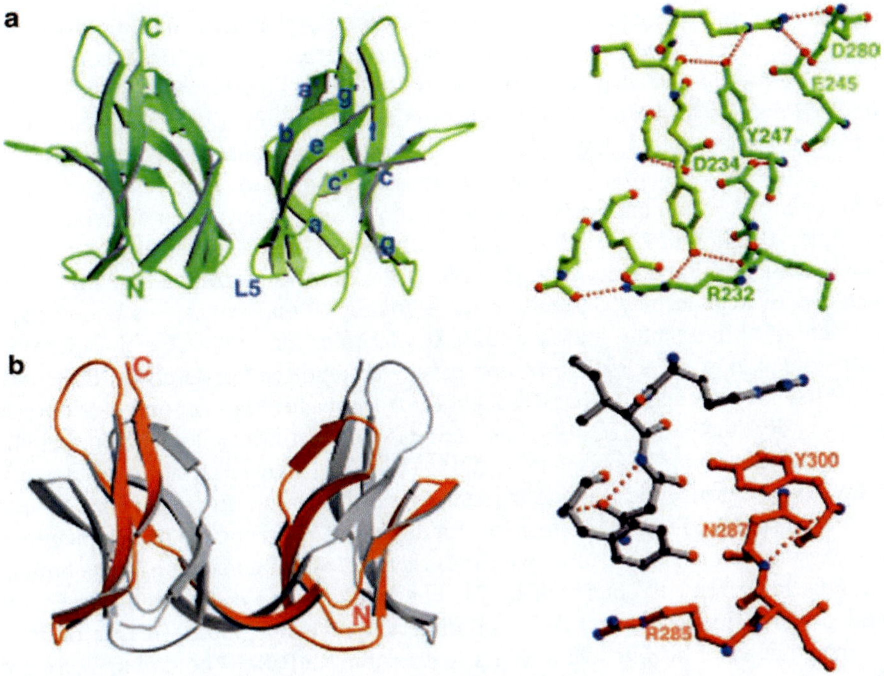

Fig. 5.3 The NF-κB dimer interface. (**a**) On the *left* is depicted an overall view of the NF-κB p50:p50 homodimer DD as a *ribbon diagram*. The individual beta-strands are labeled (a–g') and the amino- and carboxy-termini are labeled (N and C, respectively). On the *right* in *ball-and-stick representation* are shown the amino acid residues that participate in inter-subunit hydrogen bond contacts at the dimer interface. *Dotted lines* denote individual hydrogen bonds. (**b**) The *left panel* contains a *ribbon diagram* representation of the RelB homodimer DD. The two identical intertwined RelB subunits are represented in *orange* and *grey*. On the *right* is depicted the same segment of the open interface of the intertwined RelB dimer is shown as in p50. Note the absence of hydrogen bonds at the RelB:RelB dimer interface

by the other NF-κB dimers. One possibility is that residues outside of the dimer interface indirectly play a role in guiding the proper formation of NF-κB dimers. Indeed, we have observed that the folding stability of the RelB dimerization domain as well as the N-terminal immunoglobulin-like domain is low, leading to rapid degradation of the entire protein in cells. RelB preferentially forms heterodimers with p50 and p52 in vivo. Both RelB:p50 and RelB:p52 exist as regular side-by-side dimers and their interactions at the dimer interface are conserved. These observations suggest that energy gained from proper interactions at the interface also helps stabilize folding of the RelB dimerization domain [22, 53]. Consistent with this domain stability model of side-by-side dimerization, specific mutations at the p50:p50 dimer interface aimed at disrupting dimerization resulted in a domain-swapped dimer [15].

Taken together these structural observations suggest that domain folding stability plays an important role in NF-κB dimerization. Interestingly, the RelB homodimer is not observed in vivo as it degrades rapidly [48]. This further supports the idea that the RelB domain fold is unstable and that in the absence of p50 or p52 binding partners it undergoes rapid degradation by cellular proteases. Consistent with this notion, investigators have identified several mutations in other NF-κB subunits that result in their failure to participate in direct interactions across the dimer interface. For example, mutation of Cys216 to Ala destabilizes RelA homodimer formation [23]. The cysteine residue projects into the core of the RelA dimerization domain and its mutation to alanine may destabilize RelA folding altering dynamics and, hence, interactions at the subunit interface [32]. Similar roles for noninterfacial residues in the modulation of NF-κB dimerization have been reported for v-Rel and RelB [46, 57, 66]. One report showed that RelB and RelA form a heterodimer only when RelA is phosphorylated at Ser276 [49]. This serine is part of a surface loop βf-βg that lies outside of the dimer interface (Fig. 5.3). Therefore, phosphorylation must work indirectly to enhance the RelB:RelA heterodimer formation. In conclusion, NF-κB subunits display differential preferences for dimerization among the family members and the selection of dimerization partner is dictated by amino acid positions both at the dimer interface and outside of it. The physiological consequences of dimer formation selectivity are that some NF-κB dimers, such as the p50:RelA heterodimer, are more abundant in cells while others, such as the RelB:RelB homodimer, are not observed at all or form only under specialized conditions. As we continue to discuss below, the various NF-κB homo- and heterodimers also select for slightly different κB DNA target sequences.

5.4 κB DNA Response Elements

NF-κB dimers are DNA-binding transcription factors that recognize DNA with sequence specificity [69, 72]. One of the surprises at the time of its discovery was that the NF-κB DNA recognition element was short – only 10 bp long. Subsequent to its identification, the NF-κB binding site was found to be included in genomes of several animal viruses, such as human immunodeficiency virus (HIV-1), SV40, and cytomegalovirus (CMV), as well as located within the promoters of multiple human and mouse genes, such as the MHC class I and interleukin-6 [36, 42–44]. Comparison of these sequences led to the identification of a consensus element for NF-κB binding. The critical feature of this consensus is the presence of a 5′-GG at both ends while the central part of the sequence displays greater variation. Now, 20 years since the first report of the consensus NF-κB binding site and after the discovery of hundreds of its variants, the consensus sequence for NF-κB binding remains: 5′-GGGRNWYYCC-3′, where R=A or G; N=A, C, G, or T; W=A or T; and Y=C or T (Table 5.1). All NF-κB binding sequences that in principle follow this consensus are known as κB sites or κB DNA.

The expression of several hundred genes is now known to be regulated by NF-κB and the ability of the κB DNA sites within these genes to drive transcription

Table 5.1 A short list of κB DNA sites recognized by NF-κB dimers

a

Site	Sequence	Positions
	5'GGGRNWYYCC	-5 -4 -3 -2 -1 0 +1 +2 +3 +4
HIV	5'GGGACTTTCC	-5 -4 -3 -2 -1 0 +1 +2 +3 +4
iFNβ	5'GGGAAATTCC	-5 -4 -3 -2 -1 0 +1 +2 +3 +4
NFKB2	5'GGGGCTTTCC	-5 -4 -3 -2 -1 0 +1 +2 +3 +4
IL1β	5'GGGAAAATCC	-5 -4 -3 -2 -1 0 +1 +2 +3 +4
CCL11	5'GGGAGATTCC	-5 -4 -3 -2 -1 0 +1 +2 +3 +4
kB-55	5'GGGAAATTCC	-5 -4 -3 -2 -1 0 +1 +2 +3 +4

b

Site	Sequence	Positions
MHC	5'GGGGAATCCCC	-5 -4 -3 -2 -1 0 +1 +2 +3 +4 +5
NF-kB1	5'GGGAAGCCCCC	-5 -4 -3 -2 -1 0 +1 +2 +3 +4 +5
MMP9	5'GGGGTTGCCCC	-5 -4 -3 -2 -1 0 +1 +2 +3 +4 +5

c

Site	Sequence	Positions
IL-8	5'GGAAATTCC	-4 -3 -2 -1 0 +1 +2 +3 +4
IL-2	5'GGAATTTCT	-4 -3 -2 -1 0 +1 +2 +3 +4
IRF-7	5'GGAAACTCC	-4 -3 -2 -1 0 +1 +2 +3 +4
TNC	5'GGGAATTCC	-4 -3 -2 -1 0 +1 +2 +3 +4

d

Site	Sequence	Positions
kB-33	5'GGAAATTTCC	-4 -3 -2 -1 0 +1 +2 +3 +4 +5
CCR7	5'GGGGCTTTTT	-5 -4 -3 -2 -1 0 +1 +2 +3 +4
CCND1	5'GGGGAGTTTT	-5 -4 -3 -2 -1 0 +1 +2 +3 +4

(a) κB sites preferably bound by the NF-κB heterodimers with p50 or p52 as one of the subunits. Consensus sequence is at the top
(b) κB sites preferably bound by p50 and p52 homxodimers
(c) κB sites bound by RelA and c-Rel homodimers
(d) κB sites showing only half site specificity. The ideal half sites are marked by highlighting the signature G nucleotides in color. The nonideal half sites are marked by highlighting disallowed bp at position +3 in color

has been confirmed experimentally. All NF-κB target genes contain at least one κB DNA site within their promoters. Most often these sequences are found within 500 bp upstream of the transcription start site. However, in a few cases the κB sites are located several kbp up- or downstream. Identification of the κB DNA sites that drive transcription of some chemokine genes such as ICAM, IL-8, and GM-CSF clearly revealed a shortened κB site. In these cases, one 5′-G is absent leading to identification of a new κB DNA subclass with the 9 bp consensus: 5′-HGGRNWYYCC-3′ (H=A, C, or T) (Table 5.1c). Finally, a few of the functional κB sites contain an extra C (underlined) at the 3′ end resulting in an 11 bp consensus: 5′-GGGRNWYYCC<u>C</u>-3′. The significance of an absent 5′-G or additional 3′-C was not originally obvious. With very few exceptions, κB DNA sites are nonsymmetric. Interestingly, many of these asymmetric κB DNA sites are known targets of NF-κB homodimers rather than heterodimers in vivo.

5.5 NF-κB-DNA Recognition

A detailed molecular understanding of the NF-κB:DNA recognition mechanism began with the elucidation of first X-ray structures of NF-κB p50:p50 RHR homodimer complexes to two different κB sites [25, 55]. Subsequently, structures of six other NF-κB dimers (p52:p52, p50:RelA, p50:RelB, p52:RelB, RelA:RelA, and cRel:cRel) bound to nine different κB sites have been reported [3, 11–14, 16, 20, 22, 31, 53]. Together these structures have provided a wealth of information on DNA recognition strategies employed by various NF-κB dimers (Fig. 5.4). In general, each κB DNA is recognized as two half sites and each monomer within a functional NF-κB dimer confines itself to only one of the two DNA half sites in terms of sequence-specific contacts with DNA bases. However, each monomer also reaches across to the other half site and participates in nonspecific contacts with the DNA backbone. In addition, nonspecific contacts extend beyond the length of the target site by at least three more bp on either side. Thus, despite its 10 bp consensus binding sequence, the NF-κB p50:RelA heterodimer contacts a 16 bp long DNA sequence of which the central 10 bp it binds with sequence specificity. The central bp within any κB DNA site, which does not participate in any specific contact with the protein, coincides with the pseudo-dyad axis of the protein and separates the two half sites resulting in pseudo-symmetric DNA targets. For the originally described 10 bp κB sites this central bp asymmetrically divides 5- and 4-bp half sites.

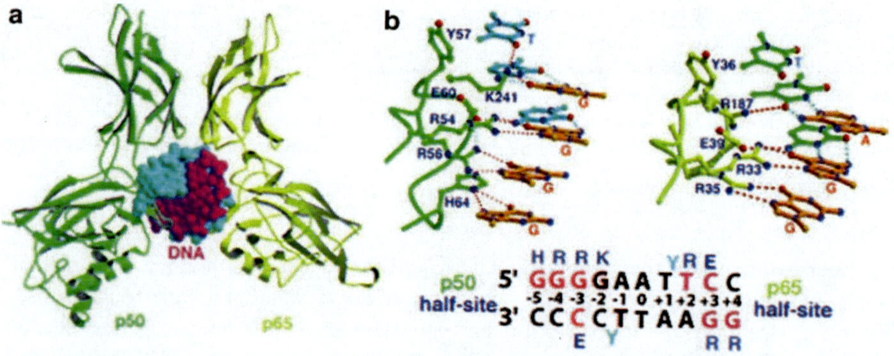

Fig. 5.4 The X-ray structure of NF-κB p50:RelA heterodimer bound to a κB DNA. (**a**) Overall structure of the heterodimer bound to DNA. The structure is oriented with a pseudo-twofold symmetry axis running vertically in the plane of the page. This positions the DNA such that it is viewed down its long axis. The DNA double helix is shown as a space-filling model with its two strands in *cyan* and *magenta* while the p50:RelA heterodimer is depicted as a *ribbon diagram* with the subunits colored *green* and *yellow*, respectively. (**b**) On the *top* is shown a close-up view of the detailed hydrogen bonding contacts between DNA and p50 subunit (*left*) and RelA subunit (*right*). On the *bottom* is a *schematic representation* of the original κB DNA site and the base contacts made by p50 residues at the 5′-half site and RelA residues at the 3′-half site. The bases are numbered and the central bp is assigned the number 0. Amino acid residues that make direct hydrogen bonds with DNA are colored *dark blue*

The NF-κB p50 and p52 subunits in most heterodimers such as p50:RelA or p52:RelA occupy the 5-bp half sites while the remaining subunit (RelA, RelB, or c-Rel) of the dimer binds the 4-bp half site. On the other hand, NF-κB homodimers of either p50 and p52 subunits ideally recognize a 11-bp κB site wherein a central bp separates two 5 bp half sites (Table 5.1b). It follows then that NF-κB homodimers of RelA, c-Rel, and RelB recognize 9 bp κB sites composed of two 4 bp half sites divided by the central bp (Table 5.1c). Although 10 bp κB sites predominate in cells, these 11 and 9 bp κB sites also exist and function as the preferred sites for NF-κB homodimer binding as predicted by the structural analyses.

5.5.1 Recognition at the 5′-GG Elements

A constellation of conserved amino acid residues from within structured loops that connect elements of secondary structure directly contact bases within κB DNA. In p50 (murine) these residues are Arg54, Arg56, Tyr57, Glu60, His64, and Lys241 (Fig. 5.4b). The first five residues derive from loop L1 of the N-terminal domain, whereas Lys241 lies within the interdomain linker (loop L3) (Fig. 5.1). The two Arg, the Tyr, and the Glu are invariant in all NF-κB subunits. His64 directly contacts the G:C bp at position ±5. The substitution of Ala in c-Rel, RelA, and RelB at the position homologous to His64 in p50 and p52 gives rise to the differences in half site length observed by these two classes of NF-κB subunits as the Ala cannot compensate for the loss of this base-specific contact. The G:C at position ±4 is contacted by one of the two invariant arginines (Arg56 in murine p50) from loop L1, whereas the second arginine contacts the G:C bp at position ±3. The invariant glutamate contacts the paired C at this position. This G:C bp contact by the Arg–Glu pair is a conserved feature in all NF-κB:DNA complexes. Recognition of both nucleotides at this position suggests a more important role of the G:C bp at position ±3 than the G:C bp at the ±5 and ±4 positions. Mutation of this G:C bp to any other bp (C:G, A:T, and T:A) dramatically affects DNA binding affinity.

Structural comparison of loop L1 of p50 in three distinct states – κB DNA bound, RNA aptamer bound (discussed later), and unbound – reveals its important characteristics. The C-terminal portion of the loop is flexible and can contact the DNA backbone of nucleotides flanking the κB sequence. The N-terminal portion of this loop contributes all of the DNA base-contacting residues mentioned and above appears to be rigidly structured. The stability of this segment of the loop is illustrated by its nearly identical conformations in the three ligand-bound states of p50. Moreover, the same region of loop L1 exhibits similar conformations in each of the other NF-κB subunits that have been studied crystallographically. Conserved hydrophobic amino acids within this loop are positioned toward the core of the N-terminal domain. This stabilizes the loop in a rigid conformation that projects DNA-contacting residues into an arrangement that is ideal for hydrogen bonding with nucleic acid bases. Furthermore, some of these DNA-contacting residues contact one another further stabilizing the loop L1 conformation and allowing them all to act as a unit. In p50, Glu60 bridges Arg54 and Arg56 as together they contact DNA as a unit. In RelA and c-Rel, a similar Glu brings together one of the two Arg residues form the loop L1 and the Arg from the interdomain linker. These cooperative interactions between the amino acids side chains not only maintain a properly oriented conformation of the functional groups primed to contact DNA but also contribute to differential base pair selectivity.

Another important difference between p50 and p52 and the remaining NF-κB subunits within loop L1 is a tripeptide segment bracketed by two glycines (Fig. 5.2). In p50 and p52 the sequence of the tripeptide is Pro–Ser–His. The last residue is His64 that was previously described to directly contact the G nucleotide base at the ±5 position. The analogous tripeptide sequence in RelA, RelB, and c-Rel is Arg–Ser–Ala. The presence of an Arg within this short element gives these subunits the potential to make additional direct base-specific contacts with DNA. Indeed, in the structure of the NF-κB p52:RelB heterodimer in complex with DNA the RelB subunit employs this Arg to contact a G:C bp at the +5 position. Intriguingly, in other structures the Ser of this tripeptide has been found to make contact with the G at the –5 position. Therefore, RelA, c-Rel, and RelB are, in fact, capable of directly contacting a 5-bp half site. However, the contacts mediated by Arg–Ser–Ala are not so stable as those mediated by the Pro–Ser–His of p50 or p52. This suggests that the His64-mediated contact by p50 and p52 is preferred to contacts by Arg or Ser for the same base. It is likely that conformational restriction imposed by Pro62 in p50 primes His64 for optimal binding of +5 G bases.

5.5.2 *Recognition at the Inner Variable Positions*

Base pairs at positions ±2 and ±1 in κB DNA exhibit more variability than do the more peripheral bases. A semiconserved Arg residue from loop L3 (the interdomain linker loop), present only in RelA and c-Rel, contacts the T of the opposite stand of an A:T bp at position +2. A Lys residue (Lys241) present in p50 and p52 can interact with both T of the A:T bp or a G of the G:C bp at the same position. This is due to greater flexibility and charge density of the ε-amino group of the Lys side chain as compared to the guanidino group of the Arg. Base pairs at position ±1 in κB DNA do not participate in any hydrogen bonding contacts in the cases of RelA and c-Rel. In the cases of p50 and p52 the linker Lys can mediate contacts at this position depending on the DNA sequence. The residue corresponding to this L3 Arg in RelA and c-Rel is Lys274 in RelB. Interestingly, this Lys does not contact DNA but projects inward to make an ion pair with Asp272 (Fig. 5.5). It is possible that this explains why RelB might tolerate greater κB DNA sequence diversity at the inner positions.

The ±1 bp does appear to play a role in stacking interactions with the invariant Tyr (Tyr57 in murine p50 and Tyr36 in murine RelA) of loop L1 (Fig. 5.6a). Indeed, the phenolic ring of the tyrosine residue makes stacking interactions with bases at

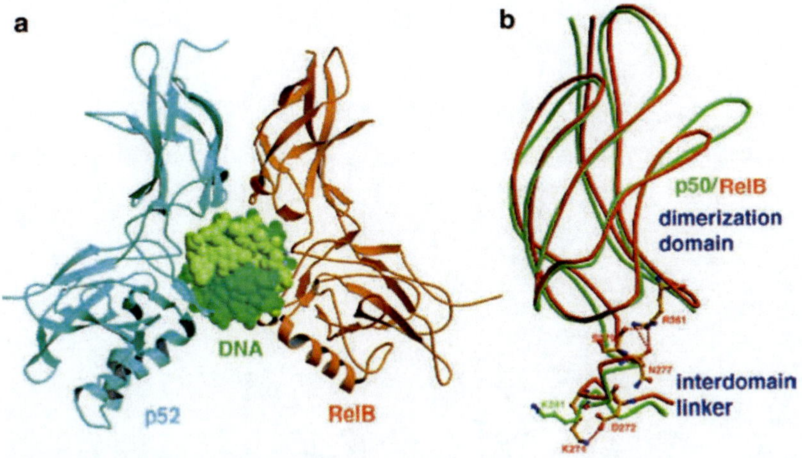

Fig. 5.5 κB DNA recognition by the NF-κB p52:RelB heterodimer. (**a**) A *ribbon diagram* representation of the X-ray structure of the p52:RelB heterodimer bound to a κB DNA (5′-CGGGAATTCCC-3′). The DNA is represented by *space-filling model* with strands colored *green* and *yellow*. (**b**) Superposition of the dimerization domains of p50 (*green*) and RelB (*red*) illustrates the conformational differences of the linker regions (Loop 3) between the two subunits. Lys274 (K274) of RelB is oriented differently than the corresponding Lys residue (K241) of p50. Consequently, Lys274 of RelB is not involved in DNA contact

both ±1 and ±2 of the same strand. This stacking is favored by the presence of two successive thymines (TT) where their exocyclic five-methyl groups favor the interaction. Although a Phe at the same position could substitute for Tyr and maintain stacking interactions, Tyr also participates in hydrogen bonds through its hydroxyl group making it an absolutely required residue for DNA recognition. Either two C bases or any combination of T and C can also be accommodated at these positions and an A or G at either position is unfavorable. The critical role played by this invariant Tyr that stacks against the inner core residues is illustrated by the overrepresentation of the sequence AAATT or AATTT at the central five positions of the 9-bp κB sequences recognized by RelA and c-Rel homodimers. Narrowing of the minor groove within the center of the κB DNA is observed in all NF-κB:DNA complexes and appears to be a result of the squeeze experienced by both strands of DNA due to Tyr stacking on both sides at this region. Furthermore, van der Waals contacts at the center of the DNA may influence the conformation of the flanking bp. The side chain of the conserved Tyr57 in murine p50 is held tightly through van der Waals stacking interactions with a Phe, which in turn is held in a stable conformation through a "ring stack" that runs through the core of the NTD (Fig. 5.6b). In all NF-κB subunits a similar stack of amino acid side chains positions the homologous Tyr such that is poised to add one ring to the stack by properly positioning itself upon DNA bases at ±1 and ±2 positions. Finally, in addition to favorable contacts between the Tyr and the TT dinucleotide, an AT bp at the central position (position 0) may also favor protein–DNA recognition.

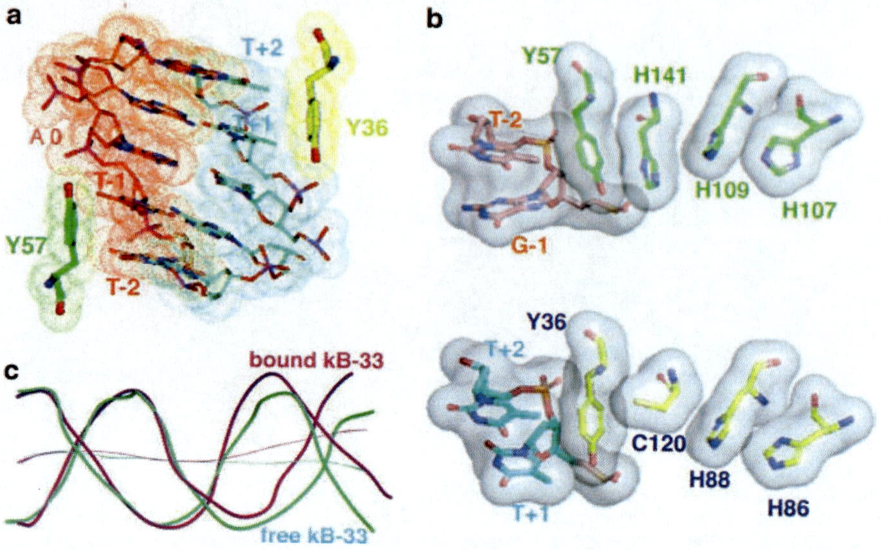

Fig. 5.6 Protein–DNA recognition at the central region of κB DNA. (**a**) The stacking of Y57 (p50) to T−1 and T−2, and Y36 (RelA) to the T+1 and T+2 are shown. DNA and protein side chains are shown as *ball-and-stick models* within van der Waals spheres. The two DNA strands are colored *orange* and *cyan*. Coordinates from the NF-κB p50–RelA heterodimer complexed to an IFN-κB site were used to make this figure. (**b**) Conformational stability of Y57 and Y36 is maintained through series of amino acid side chain stacking interactions through the NTD. (**c**) Structural differences of the DNA backbone in its NF-κB bound (*magenta*) and unbound (*green*) states are depicted. The curvature of the B-DNA helical axis is also shown as a *thin line* down the center of the DNA double helix. The sequence of κB-33 DNA is 5′-GGAAATTTCC-3′

5.5.3 The Role of Protein–Protein Interactions in Stabilizing NF-κB:DNA Complexes

Intra- and intermolecular protein–protein interactions distal from the protein: DNA interface can drastically affect binding affinity of the complex. Both protein:DNA interfacial and protein–protein interactions are interrelated, and could be affected by the DNA conformation. One illustration of this point is the βf-βg loop of the NF-κB dimerization domain, which projects toward κB DNA but does not directly contact it. Two conserved acidic residues (Asp267 and Glu269 in chicken c-Rel) are located within this loop and reside near the DNA in the c-Rel:IL2-CDRE complex. These residues would be expected to repel DNA and weaken binding [31]. A conserved Arg of loop L1 that precedes the two DNA-contacting Arg residues is observed to directly contact these two acidic residues in the X-ray crystal structure. These intrasubunit protein–protein contacts augment DNA binding in two ways: through neutralization of negative change, and by holding the DNA-contacting residues in place. This βf-βg loop is a particularly intriguing example of a distal modifier of DNA binding because it also contains two basic residues (Fig. 5.2). Mutation of either of these dramatically reduces DNA-binding affinity [18, 66]. More importantly, the βf-βg loop contains a critical serine, which in RelA has been shown to be extremely important for gene regulation. Cells expressing RelA bearing the Ser276Ala mutant show dramatically reduced transcriptional activity. This serine has been shown to undergo phosphorylation by the cAMP-dependent protein kinase (PKA) and its posttranslational modification is essential for RelA transcriptional activity [18, 62, 78, 82]. Although the RelA Ser276Ala appears to bind DNA, defects in DNA affinity cannot be completely ruled out in light of the importance of other residues in the same loop in the protein:DNA complex formation.

5.6 Structural Features of Unbound κB DNA

One of the most fascinating regulatory aspects of protein–DNA recognition involves "invisible" differences in stability of specific DNA base pairs. Local DNA sequence differences can affect the orientation of amino acid residues at the protein–DNA interface resulting in differential binding affinity and/or overall conformation of the complex. For instance, dynamics of the TA base step is different than the AA base step. The effect of these subtle DNA dynamic and conformational differences is often difficult to measure accurately. However, it can be inferred by measuring differences in affinity and/or kinetics of binding in solution, and possibly through the elucidation of structures of complexes with mutant DNA.

Although the dynamics of DNA due to a TA base step might be essential for specific recognition in some protein:DNA complexes, the TA base step is rarely observed in κB sequences. Although both GGAATTTCC and GGAAATTCC sequences are optimum targets for the RelA or c-Rel homodimers from the standpoint of hydrogen bonding contacts, the GGAATATCC sequence is not an ideal κB sequence, though it potentially could accommodate all of the same favorable hydrogen bonding interactions. This is likely for two reasons: first, the stable stacking interaction between tyrosine and thymine described above is not possible and second, the TA step is expected to induce dynamics of the neighboring bp resulting in reduced binding affinity. The presence of a G:C or C:G bp at the central position is fairly uncommon among κB sites although this position does not directly contact protein. A G:C or C:G bp could also potentially affect the conformation/dynamics of the neighboring sites. For instance, the narrow width of the minor groove at the center, an universal feature of A:T/T:A centered κB sites observed to date might be difficult to achieve with a G:C/C:G bp at the center.

The X-ray structure of a free κB DNA allowed for comparison of the structural transitions of the same DNA upon binding to a RelA:RelA homodimer [33] (Fig. 5.6c). The sequence of the central part of the DNA was identical to another DNA sequence whose X-ray structure is available [19]. The features of the central AAATTT sequence in both DNA are similar with a small roll angle and consequently these DNA are closer to ideal B form. RelA induces a smooth bending around the central AT-rich sequence [12, 13]. Moreover, the free κB DNA exhibits a wider minor grove and deeper major groove. NF-κB binding results in narrowing of the minor groove and widening of the major groove where direct protein contacts are observed.

NMR analysis of both wild-type and mutant HIV κB DNA sequences showed some key differences in their conformations. Although the base changes were introduced at the 5′-junction in the mutant κB, differences in conformation within the κB core were observed. Whereas, the phosphates at the boundary between the 5′-end of the wild-type HIV κB site and flanking sequence existed in a BI–BII equilibrium, the same phosphates in the mutant κB site exclusively exhibited the BI conformation [76]. Interestingly, the same phosphates in the X-ray structure of free κB DNA showed BII conformation. These mutant HIV κB DNA used in the NMR studies showed reduced NF-κB binding suggesting that the flanking sequence negatively affects DNA binding by changing the preferred phosphate conformation to BII whereas NF-κB preferentially binds to DNA with phosphates in the BI conformation [75, 76].

Molecular dynamics simulations of the IL-2 κB (AGAAATTCC) site embedded within a 22-mer duplex DNA over a 1-μs time scale has revealed striking structural transitions. [56]. At around 0.7 μs of simulation the central A:T bp becomes completely sheared and the bases undergo cross-stand stacking. This phenomenon leads to the flipping of neighboring thymine at position –1, leaving its paired adenine free. These structural changes might be linked to sequence. For example, interruption of an otherwise AT-rich central sequence by a G:C bp may not result in similar structural changes. In all, these studies on free κB DNA strongly suggest that sequence-dependent DNA conformations play a vital role in NF-κB recognition. Moreover, sequence variations in the flanking regions can also affect the conformation of the κB DNA in such a way that NF-κB binding affinity is altered. Consistent with this, we have observed that the overall conformations of all NF-κB:DNA complexes studied thus far exhibit unexpected differences.

5.7 Biochemical Studies of NF-κB:DNA Complexes

Many κB DNA sites have been tested for their binding by two of the early identified dimers, p50:p50 and p50:RelA. Most of these early studies used electrophoretic mobility shift assay (EMSA). Recent studies have relied upon surface plasmon biosensor-based assay (SPR) and fluorescence-based solution methods. In general, these studies agree well in terms of relative binding affinities of p50:p50 and p50:RelA dimers for different κB sites although the absolute affinity values differ considerably. EMSA analyses revealed extremely high affinities between HIV, MHC, and IFN κB DNA sites and NF-κB p50:RelA heterodimers with K_D (dissociation equilibrium constant) values reported to be in the pM range [21, 81]. Interestingly, SPR, which is capable of monitoring complex assembly and disassembly on a surface in real time, revealed an unexpected binding mechanism. Affinity measured from rates of association and dissociation ($K_D = k_{off}/k_{on}$) indicated pM binding affinity while SPR experiments measured at equilibrium detected a

K_D value in the nanomolar range [27]. One possible explanation for this discrepancy could be that a relatively high nonspecific binding affinity of NF-κB for DNA flanking the κB site influences overall binding rates. Despite these nonspecific interactions, NF-κB ultimately occupies the specific κB DNA sites, which is what is detected under equilibrium conditions. In support of this model, affinity measurements performed at equilibrium by solution-based fluorescence polarization assays also yield nM range binding affinities [61]. In all cases, however, NF-κB binding to DNA was determined to be highly salt and pH sensitive. These experiments are conducted with *Escherichia coli* expressed proteins truncated to encode only the RHR. It is, therefore, possible that the presence of the transcriptional activation domain and/or modification counteracts this salt sensitivity. It is also possible that inside cells and within the presence of other proteins and counterions, NF-κB is not as salt sensitive as observed in vitro. Further experiments are required to resolve the thermodynamic and kinetic mechanisms of NF-κB:DNA recognition using pure native proteins.

Studies by Kunsch et al. have enhanced our understanding on κB site selectivity for p50, RelA, and c-Rel from pools of random DNA sequences using in vitro selection methods [40]. Using this approach they identified different κB DNA site consensus sequences for NF-κB p50:p50, RelA:RelA, and c-Rel:c-Rel homodimers. The consensus sequences derived previously from structural analyses of p50 and RelA monomers on κB DNA half sites fit well with this selection analysis [40]. However, the consensus sequence for c-Rel deviates significantly from the consensus derived from physiological κB sites and also from known physiological c-Rel binding sites. More recently, Udalova and coworkers determined comparative affinity of a large number of κB DNA for p50:p50, p50:RelA, and p52:p52 dimers [58, 77]. A total of 256 different DNA site variants were prepared by changing the internal six positions (GGRRNNYYCC). One of the key observations made in these studies was that at least one of the two half sites must contain G:C at the ±3 positions (5′-GGGXX-3′ or 5′-XCCC-3′) to have any appreciable binding for the p50 homodimer. This clearly establishes the critical importance of the G:C bp at ±3 positions. The significance of this position is due to the recognition of both bases by protein side chains as described previously. This study also confirms the importance of TT over CC at +1 and +2 positions. As expected, a G or C is not favored at these positions.

5.8 NF-κB-RNA Aptamer Recognition

Using the SELEX approach, RNA aptamers have been identified that selectively bind NF-κB p50:p50 and RelA:RelA homodimers [9, 10, 41, 80]. These RNA aptamers are highly specific and do not interact with the non-cognate NF-κB dimers with any appreciable affinity. Furthermore, they do not bear any sequence similarity with the consensus κB sites. Finally, the RNA molecules are asymmetric. It was, therefore, not immediately clear how these nucleic acids might be recognized by symmetrical NF-κB homodimers.

The X-ray structure of the NF-κB p50:p50 homodimer bound to a 29-mer RNA aptamer served to resolve this puzzle. The structure revealed that each p50 monomer interacts with one RNA molecule [34]. Each RNA molecule binds the NTD and the linker region of one p50 monomer for recognition. The relative flexibility of the NTD with respect to the dimerization domain allows each NTD to recognize an RNA molecule without any hindrance from the other subunit of the dimer (Fig. 5.7a). Overall, the p50:RNA complex appears more open where the NTD:RNA subcomplexes are splayed away from each other. The two p50 subunits remain held together through their dimerization domains only. In the DNA-bound complexes, by comparison, the N-terminal domains are forced to approach one another as they each interact with one half site of the same κB DNA molecule. The proximity of all four domains (two NTDs and two DDs) in the DNA complex endows additional protein–protein contacts between the NTD and DD of each subunit.

In spite of the sequence differences between the RNA aptamer and DNA, and differences in the overall binding conformation of the NF-κB p50:p50 homodimer, close inspection of the p50:RNA complex revealed remarkable similarities in the mechanism of nucleic acid recognition (Fig. 5.7b, c). The secondary structure of the 29-mer RNA aptamer is an internal loop flanked by a stem on one side and stem-loop on the other. In its p50-bound state the RNA adopts an irregularly bent, long helical structure with bending occurring within the internal loop region. The tetraloop of the hairpin is highly structured with only one nucleotide free of any interactions. Sequence-specific recognition primarily involves bases of the internal loop, which is presented to NF-κB as a duplex. The G18G19U20U21G22G23 sequence of one strand interacts with His64, Agr56, Arg54, Lys241, and Tyr57 of loops L1 and L3 in a manner reminiscent of p50:DNA complexes. Of these nucleotides, G22 and G23 are from the internal loop. Tertiary interactions such as cross-strand stacking hold these two guanines in a conformation amenable to protein binding. Cys59 and Glu60 contact RNA bases A10, A11, and C12 that occupy positions in the reverse strand of this double helical region. Interestingly, the conformation of loop L1 is almost identical in both DNA and RNA complexes suggesting that any structural change must occur at the level of RNA to accommodate the protein side chains in proper order [24, 34]. Although the total number of contacts in the p50:RNA complex is greater than those observed in p50:DNA complexes, the core of the interface exhibits remarkably similar chemistry in both types of complexes.

The three-dimensional structure of the free RNA aptamer has been done using NMR which further provides insights into how p50 recognizes the RNA [63]. The RNA aptamer sequence has evolved to display both preorganized features and protein-induced structural changes essential for high affinity RNA-p50 recognition. The features of the major groove of the internal loop distorted helix are remarkably similar to DNA duplexes and are different from typical A-form RNA duplexes. The cross-strand stacking of guanines in the internal loop appear to be the major determinant for these distinct major groove features. This preformed RNA major groove in the internal loop facilitates protein binding, which upon binding to protein undergoes further structural changes such as a kink in the internal loop and base unstacking in the tetra loop [63].

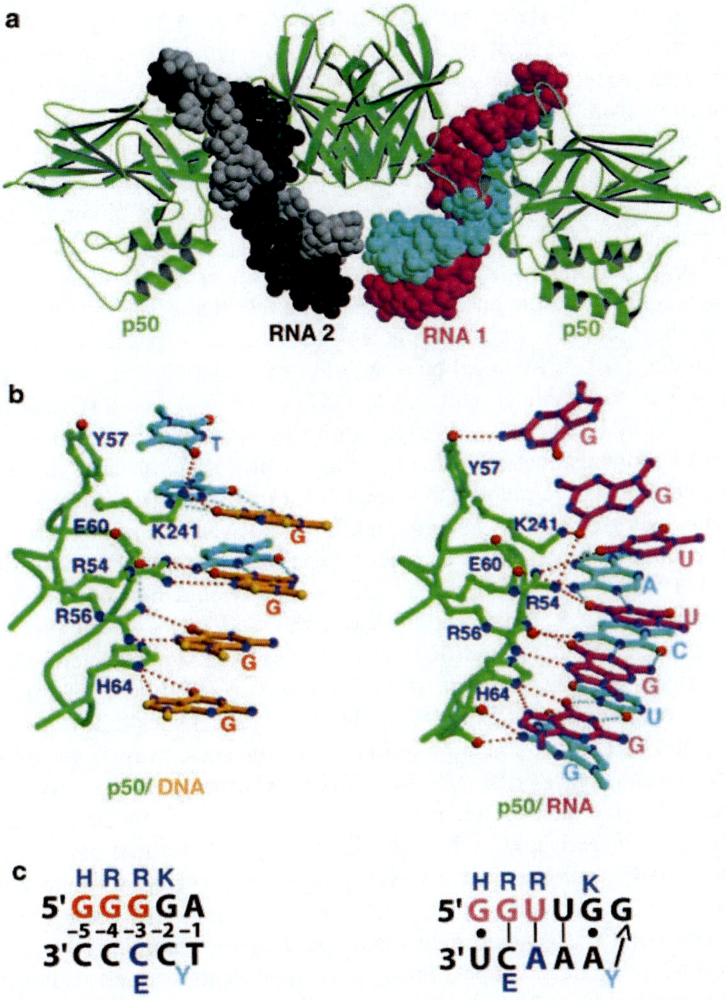

Fig. 5.7 RNA aptamer binding by NF-κB p50:p50 homodimer. (**a**) The overall structure of the NF-κB p50:p50 homodimer bound to two molecules of a specific RNA aptamer. As previously, NF-κB is represented as a *ribbon diagram* (*green*) while the two single-stranded RNA molecules are depicted as *space filling models* (*black/grey and magenta/cyan*). (**b**) Comparison of DNA (*left*) and RNA (*right*) binding by the residues of loops L1 of p50 (*green ball-and-stick*). Hydrogen bonds are depicted as *dashed lines*. (**c**) κB DNA and RNA recognition sequences in the two complexes

5.9 DNA Binding by NF-κB in the Cell

A string of recent reports describes the binding of NF-κB dimers to regulatory DNA sequences on the whole genome scale [8, 45, 50, 68]. Although these reports contradict one another with regards to the precise number of NF-κB binding sequences and their locations within the genome, the studies reveal some important observations.

For instance, they all show that NF-κB dimers bind a large number of DNA sequences and that most of these sequences are present within the promoters (±500 bp with respect to transcription start sites). Intriguingly, several DNA fragments isolated from chromatin immunoprecipitation (ChIP) experiments do not contain a stringently defined NF-κB consensus sequence. When these newly identified NF-κB binding sequences are analyzed, many are found to exhibit half site consensus. This suggests that NF-κB dimers are capable of binding, albeit with lower affinity, to DNA with one ideal half site even if the other half site is not (Table 5.1d). Such binding events might happen under the conditions where the NF-κB concentration in the nucleus is high. The fact that at any given time a large fraction of the genome is bound to nucleosomes suggests that accessible genomic DNA is limited [38]. If the local concentration of active NF-κB is relatively high, however, it is a reasonable assumption that NF-κB dimers would be able to bind a wide spectrum of DNA sites that show high to low sequence similarity with the consensus. A rough estimation is that the number of RelA molecules in the nucleus of stimulated cells is around 100,000 (50% of a total of 200,000 RelA molecules enter the nucleus after any given stimulus). If all other NF-κB monomers were also present in similar numbers and were present in the same cell at the same time then total number of NF-κB monomers in the nucleus would be 500,000. Thus, the concentration of NF-κB dimers could be as high as ~0.5 μM. At such a concentration most of the high- and low-affinity available κB sites are expected to be recognized by NF-κB. An interesting question that arises from NF-κB:DNA recognition studies in vivo is that if most experimentally identified κB sites follow the κB DNA consensus and, therefore, are high-affinity binding sites, why have low-affinity binding sites never been previously identified experimentally? An answer to this puzzle might be that all experimentally identified κB sites are located within the promoters of genes activated by NF-κB dimers. Therefore, these newly identified, low-affinity, half-site consensus κB sites might be involved in gene regulation through some other mechanism such gene repression and/or epigenetics.

There are many unanswered questions regarding the status of NF-κB in vivo. For example, how do different posttranslational modifications such as phosphorylation or acetylation alter DNA recognition by NF-κB [60]. One particular example, phosphorylation of Ser276 of RelA and homologous residues in other NF-κB subunits, has been introduced. But it has been shown that phosphorylation is essential for DNA binding by RelA to some, but not all, κB DNA sites. Several sites in RelA also undergo acetylation and the impact of these modifications in DNA binding has not been thoroughly established. Second, identification of the low-affinity sites and their physiological role in NF-κB regulation remains to be clarified. Finally, one cannot rule out the possibility that NF-κB might be involved in RNA binding in vivo. This possibility is even more feasible when one considers that the amount of NF-κB is high in the nucleus, that most of the genome is masked by histones in the form of nucleosomes, and that diverse RNA sequences can target specific NF-κB subunits. Although it is currently unknown whether NF-κB:RNA binding has any consequence on physiological regulation, the ability of NF-κB to be targeted as a consequence of its robust nucleic acid recognition potential is unquestionable.

Acknowledgments The authors thank Vivien Wang for reading the manuscript and for aid in the preparation of figures. Research in the laboratory of GG is supported by the NIH (grants GM085490 and AI064326). TH is supported by American Cancer Society grant RSG-08-287-01-GMC.

References

1. Baeuerle PA, Baltimore D (1988) Activation of DNA-binding activity in an apparently cytoplasmic precursor of the NF-kappa B transcription factor. Cell 53:211–217
2. Baldwin AS Jr (1996) The NF-kappa B and I kappa B proteins: new discoveries and insights. Annu Rev Immunol 14:649–683
3. Berkowitz B, Huang DB, Chen-Park FE, Sigler PB, Ghosh G (2002) The x-ray crystal structure of the NF-kappa B p50.p65 heterodimer bound to the interferon beta -kappa B site. J Biol Chem 277:24694–24700
4. Bose HR Jr (1992) The Rel family: models for transcriptional regulation and oncogenic transformation. Biochim Biophys Acta 1114:1–17
5. Bours V, Burd PR, Brown K, Villalobos J, Park S, Ryseck RP, Bravo R, Kelly K, Siebenlist U (1992) A novel mitogen-inducible gene product related to p50/p105-NF-kappa B participates in transactivation through a kappa B site. Mol Cell Biol 12:685–695
6. Bours V, Villalobos J, Burd PR, Kelly K, Siebenlist U (1990) Cloning of a mitogen-inducible gene encoding a kappa B DNA-binding protein with homology to the rel oncogene and to cell-cycle motifs. Nature 348:76–80
7. Breitman ML, Lai MM, Vogt PK (1980) The genomic RNA of avian reticuloendotheliosis virus REV. Virology 100:450–461
8. Bunting K, Rao S, Hardy K, Woltring D, Denyer GS, Wang J, Gerondakis S, Shannon MF (2007) Genome-wide analysis of gene expression in T cells to identify targets of the NF-kappa B transcription factor c-Rel. J Immunol 178:7097–7109
9. Cassiday LA, Maher LJ 3rd (2001) In vivo recognition of an RNA aptamer by its transcription factor target. Biochemistry 40:2433–2438
10. Cassiday LA, Maher LJ 3rd (2003) Yeast genetic selections to optimize RNA decoys for transcription factor NF-kappa B. Proc Natl Acad Sci U S A 100:3930–3935
11. Chen FE, Huang DB, Chen YQ, Ghosh G (1998) Crystal structure of p50/p65 heterodimer of transcription factor NF-kappaB bound to DNA. Nature 391:410–413
12. Chen YQ, Ghosh S, Ghosh G (1998) A novel DNA recognition mode by the NF-kappa B p65 homodimer. Nat Struct Biol 5:67–73
13. Chen YQ, Sengchanthalangsy LL, Hackett A, Ghosh G (2000) NF-kappaB p65 (RelA) homodimer uses distinct mechanisms to recognize DNA targets. Structure 8:419–428
14. Chen-Park FE, Huang DB, Noro B, Thanos D, Ghosh G (2002) The kappa B DNA sequence from the HIV long terminal repeat functions as an allosteric regulator of HIV transcription. J Biol Chem 277:24701–24708
15. Chirgadze DY, Demydchuk M, Becker M, Moran S, Paoli M (2004) Snapshot of protein structure evolution reveals conservation of functional dimerization through intertwined folding. Structure 12:1489–1494
16. Cramer P, Larson CJ, Verdine GL, Müller CW (1997) Structure of the human NF-kappaB p52 homodimer-DNA complex at 2.1 A resolution. EMBO J 16:7078–7090
17. Day YS, Bacon SL, Hughes-Thomas Z, Blackburn JM, Sutherland JD (2002) Mutational analysis of the kinetics and thermodynamics of transcription factor NF-kappaB homodimerisation. Chembiochem 3:1192–1199
18. Dong J, Jimi E, Zhong H, Hayden MS, Ghosh S (2008) Repression of gene expression by unphosphorylated NF-kappaB p65 through epigenetic mechanisms. Genes Dev 22:1159–1173
19. Edwards KJ, Brown DG, Spink N, Skelly JV, Neidle S (1992) Molecular structure of the B-DNA dodecamer d(CGCAAATTTGCG)2. An examination of propeller twist and minor-groove water structure at 2.2 A resolution. J Mol Biol 226:1161–1173

20. Escalante CR, Shen L, Thanos D, Aggarwal AK (2002) Structure of NF-kappaB p50/p65 heterodimer bound to the PRDII DNA element from the interferon-beta promoter. Structure 10:383–391
21. Fujita T, Nolan GP, Ghosh S, Baltimore D (1992) Independent modes of transcriptional activation by the p50 and p65 subunits of NF-kappa B. Genes Dev 6:775–787
22. Fusco AJ, Huang DB, Miller D, Wang VY, Vu D, Ghosh G (2009) NF-kappaB p52:RelB heterodimer recognizes two classes of kappaB sites with two distinct modes. EMBO Rep 10:152–159
23. Ganchi PA, Sun SC, Greene WC, Ballard DW (1993) A novel NF-kappa B complex containing p65 homodimers: implications for transcriptional control at the level of subunit dimerization. Mol Cell Biol 13:7826–7835
24. Ghosh G, Huang DB, Huxford T (2004) Molecular mimicry of the NF-kappaB DNA target site by a selected RNA aptamer. Curr Opin Struct Biol 14:21–27
25. Ghosh G, van Duyne G, Ghosh S, Sigler PB (1995) Structure of NF-kappa B p50 homodimer bound to a kappa B site. Nature 373:303–310
26. Ghosh S, Gifford AM, Riviere LR, Tempst P, Nolan GP, Baltimore D (1990) Cloning of the p50 DNA binding subunit of NF-kappa B: homology to rel and dorsal. Cell 62:1019–1029
27. Hart DJ, Speight RE, Cooper MA, Sutherland JD, Blackburn JM (1999) The salt dependence of DNA recognition by NF-kappaB p50: a detailed kinetic analysis of the effects on affinity and specificity. Nucleic Acids Res 27:1063–1069
28. Hart DJ, Speight RE, Sutherland JD, Blackburn JM (2001) Analysis of the NF-kappaB p50 dimer interface by diversity screening. J Mol Biol 310:563–575
29. Hayden MS, Ghosh S (2008) Shared principles in NF-kappaB signaling. Cell 132:344–362
30. Hoelzer JD, Franklin RB, Bose HR Jr (1979) Transformation by reticuloendotheliosis virus: development of a focus assay and isolation of a nontransforming virus. Virology 93:20–30
31. Huang DB, Chen YQ, Ruetsche M, Phelps CB, Ghosh G (2001) X-ray crystal structure of proto-oncogene product c-Rel bound to the CD28 response element of IL-2. Structure 9:669–678
32. Huang DB, Huxford T, Chen YQ, Ghosh G (1997) The role of DNA in the mechanism of NFkappaB dimer formation: crystal structures of the dimerization domains of the p50 and p65 subunits. Structure 5:1427–1436
33. Huang DB, Phelps CB, Fusco AJ, Ghosh G (2005) Crystal structure of a free kappaB DNA: insights into DNA recognition by transcription factor NF-kappaB. J Mol Biol 346:147–160
34. Huang DB, Vu D, Cassiday LA, Zimmerman JM, Maher LJ 3rd, Ghosh G (2003) Crystal structure of NF-kappaB (p50)2 complexed to a high-affinity RNA aptamer. Proc Natl Acad Sci U S A 100:9268–9273
35. Huang DB, Vu D, Ghosh G (2005) NF-kappaB RelB forms an intertwined homodimer. Structure (Camb) 13:1365–1373
36. Isräel A, Le Bail O, Hatat D, Piette J, Kieran M, Logeat F, Wallach D, Fellous M, Kourilsky P (1989) TNF stimulates expression of mouse MHC class I genes by inducing an NF kappa B-like enhancer binding activity which displaces constitutive factors. EMBO J 8:3793–3800
37. Jacobs MD, Harrison SC (1998) Structure of an IkappaBalpha/NF-kappaB complex. Cell 95:749–758
38. Jiang C, Pugh BF (2009) Nucleosome positioning and gene regulation: advances through genomics. Nat Rev Genet 10:161–172
39. Kieran M, Blank V, Logeat F, Vandekerckhove J, Lottspeich F, Le Bail O, Urban MB, Kourilsky P, Baeuerle PA, Israel A (1990) The DNA binding subunit of NF-kappa B is identical to factor KBF1 and homologous to the rel oncogene product. Cell 62:1007–1018
40. Kunsch C, Ruben SM, Rosen CA (1992) Selection of optimal kappa B/Rel DNA-binding motifs: interaction of both subunits of NF-kappa B with DNA is required for transcriptional activation. Mol Cell Biol 12:4412–4421
41. Lebruska LL, Maher LJ 3rd (1999) Selection and characterization of an RNA decoy for transcription factor NF-kappa B. Biochemistry 38:3168–3174

42. Lenardo MJ, Baltimore D (1989) NF-kappa B: a pleiotropic mediator of inducible and tissue-specific gene control. Cell 58:227–229
43. Lenardo MJ, Kuang A, Gifford A, Baltimore D (1989) Purified bovine NF-kappa B recognizes regulatory sequences in multiple genes expressed during activation of T- and B-lymphocytes. Haematol Blood Transfus 32:411–415
44. Libermann TA, Baltimore D (1990) Activation of interleukin-6 gene expression through the NF-kappa B transcription factor. Mol Cell Biol 10:2327–2334
45. Lim CA, Yao F, Wong JJ, George J, Xu H, Chiu KP, Sung WK, Lipovich L, Vega VB, Chen J, Shahab A, Zhao XD, Hibberd M, Wei CL, Lim B, Ng HH, Ruan Y, Chin KC (2007) Genome-wide mapping of RELA(p65) binding identifies E2F1 as a transcriptional activator recruited by NF-kappaB upon TLR4 activation. Mol Cell 27:622–635
46. Liss AS, Bose HR Jr (2002) Mutational analysis of the v-Rel dimerization interface reveals a critical role for v-Rel homodimers in transformation. J Virol 76:4928–4939
47. Malek S, Huang DB, Huxford T, Ghosh S, Ghosh G (2003) X-ray crystal structure of an IkappaBbeta x NF-kappaB p65 homodimer complex. J Biol Chem 278:23094–23100
48. Marienfeld R, Berberich-Siebelt F, Berberich I, Denk A, Serfling E, Neumann M (2001) Signal-specific and phosphorylation-dependent RelB degradation: a potential mechanism of NF-kappaB control. Oncogene 20:8142–8147
49. Marienfeld R, May MJ, Berberich I, Serfling E, Ghosh S, Neumann M (2003) RelB forms transcriptionally inactive complexes with RelA/p65. J Biol Chem 278:19852–19860
50. Martone R, Euskirchen G, Bertone P, Hartman S, Royce TE, Luscombe NM, Rinn JL, Nelson FK, Miller P, Gerstein M, Weissman S, Snyder M (2003) Distribution of NF-kappaB-binding sites across human chromosome 22. Proc Natl Acad Sci U S A 100:12247–12252
51. Mercurio F, Didonato J, Rosette C, Karin M (1992) Molecular cloning and characterization of a novel Rel/NF-kappa B family member displaying structural and functional homology to NF-kappa B p50/p105. DNA Cell Biol 11:523–537
52. Meyer R, Hatada EN, Hohmann HP, Haiker M, Bartsch C, Rothlisberger U, Lahm HW, Schlaeger EJ, van Loon AP, Scheidereit C (1991) Cloning of the DNA-binding subunit of human nuclear factor kappa B: the level of its mRNA is strongly regulated by phorbol ester or tumor necrosis factor alpha. Proc Natl Acad Sci U S A 88:966–970
53. Moorthy AK, Huang DB, Wang VY, Vu D, Ghosh G (2007) X-ray structure of a NF-kappaB p50/RelB/DNA complex reveals assembly of multiple dimers on tandem kappaB sites. J Mol Biol 373:723–734
54. Mosialos G, Gilmore TD (1993) v-Rel and c-Rel are differentially affected by mutations at a consensus protein kinase recognition sequence. Oncogene 8:721–730
55. Müller CW, Rey FA, Sodeoka M, Verdine GL, Harrison SC (1995) Structure of the NF-kappa B p50 homodimer bound to DNA. Nature 373:311–317
56. Mura C, McCammon JA (2008) Molecular dynamics of a kappaB DNA element: base flipping via cross-strand intercalative stacking in a microsecond-scale simulation. Nucleic Acids Res 36:4941–4955
57. Nehyba J, Hrdlickova R, Bose HR Jr (1997) Differences in kappaB DNA-binding properties of v-Rel and c-Rel are the result of oncogenic mutations in three distinct functional regions of the Rel protein. Oncogene 14:2881–2897
58. Nijnik A, Mott R, Kwiatkowski DP, Udalova IA (2003) Comparing the fine specificity of DNA binding by NF-kappaB p50 and p52 using principal coordinates analysis. Nucleic Acids Res 31:1497–1501
59. Nolan GP, Ghosh S, Liou HC, Tempst P, Baltimore D (1991) DNA binding and I kappa B inhibition of the cloned p65 subunit of NF-kappa B, a rel-related polypeptide. Cell 64: 961–969
60. Perkins ND (2006) Post-translational modifications regulating the activity and function of the nuclear factor kappa B pathway. Oncogene 25:6717–6730
61. Phelps CB, Sengchanthalangsy LL, Malek S, Ghosh G (2000) Mechanism of kappa B DNA binding by Rel/NF-kappa B dimers. J Biol Chem 275:24392–24399

62. Reber L, Vermeulen L, Haegeman G, Frossard N (2009) Ser276 phosphorylation of NF-kB p65 by MSK1 controls SCF expression in inflammation. PLoS ONE 4:e4393
63. Reiter NJ, Maher LJ 3rd, Butcher SE (2008) DNA mimicry by a high-affinity anti-NF-kappaB RNA aptamer. Nucleic Acids Res 36:1227–1236
64. Ruben SM, Dillon PJ, Schreck R, Henkel T, Chen CH, Maher M, Baeuerle PA, Rosen CA (1991) Isolation of a rel-related human cDNA that potentially encodes the 65-kD subunit of NF-kappa B. Science 254:11
65. Ryseck RP, Bull P, Takamiya M, Bours V, Siebenlist U, Dobrzanski P, Bravo R (1992) RelB, a new Rel family transcription activator that can interact with p50-NF-kappa B. Mol Cell Biol 12:674–684
66. Ryseck RP, Novotny J, Bravo R (1995) Characterization of elements determining the dimerization properties of RelB and p50. Mol Cell Biol 15:3100–3109
67. Schmid RM, Perkins ND, Duckett CS, Andrews PC, Nabel GJ (1991) Cloning of an NF-kappa B subunit which stimulates HIV transcription in synergy with p65. Nature 352:733–736
68. Schreiber J, Jenner RG, Murray HL, Gerber GK, Gifford DK, Young RA (2006) Coordinated binding of NF-kappaB family members in the response of human cells to lipopolysaccharide. Proc Natl Acad Sci U S A 103:5899–5904
69. Sen R, Baltimore D (1986) Inducibility of kappa immunoglobulin enhancer-binding protein Nf-kappa B by a posttranslational mechanism. Cell 47:921–928
70. Sen R, Baltimore D (1986) Multiple nuclear factors interact with the immunoglobulin enhancer sequences. Cell 46:705–716
71. Sengchanthalangsy LL, Datta S, Huang DB, Anderson E, Braswell EH, Ghosh G (1999) Characterization of the dimer interface of transcription factor NFkappaB p50 homodimer. J Mol Biol 289:1029–1040
72. Singh H, Sen R, Baltimore D, Sharp PA (1986) A nuclear factor that binds to a conserved sequence motif in transcriptional control elements of immunoglobulin genes. Nature 319:154–158
73. Stephens RM, Rice NR, Hiebsch RR, Bose HR Jr, Gilden RV (1983) Nucleotide sequence of v-rel: the oncogene of reticuloendotheliosis virus. Proc Natl Acad Sci U S A 80:6229–6233
74. Steward R (1987) Dorsal, an embryonic polarity gene in Drosophila, is homologous to the vertebrate proto-oncogene, c-rel. Science 238:692–694
75. Tisne C, Delepierre M, Hartmann B (1999) How NF-kappaB can be attracted by its cognate DNA. J Mol Biol 293:139–150
76. Tisne C, Hartmann B, Delepierre M (1999) NF-kappa B binding mechanism: a nuclear magnetic resonance and modeling study of a GGG CTC mutation. Biochemistry 38:3883–3894
77. Udalova IA, Mott R, Field D, Kwiatkowski D (2002) Quantitative prediction of NF-kappa B DNA-protein interactions. Proc Natl Acad Sci U S A 99:8167–8172
78. Vermeulen L, De Wilde G, Van Damme P, Vanden Berghe W, Haegeman G (2003) Transcriptional activation of the NF-kappaB p65 subunit by mitogen- and stress-activated protein kinase-1 (MSK1). EMBO J 22:1313–1324
79. Wilhelmsen KC, Eggleton K, Temin HM (1984) Nucleic acid sequences of the oncogene v-rel in reticuloendotheliosis virus strain T and its cellular homolog, the proto-oncogene c-rel. J Virol 52:172–182
80. Wurster SE, Maher LJ 3rd (2008) Selection and characterization of anti-NF-kappaB p65 RNA aptamers. RNA 14:1037–1047
81. Zabel U, Schreck R, Baeuerle PA (1991) DNA binding of purified transcription factor NF-kappa B. Affinity, specificity, Zn2+ dependence, and differential half-site recognition. J Biol Chem 266:252–260
82. Zhong H, Voll RE, Ghosh S (1998) Phosphorylation of NF-kappa B p65 by PKA stimulates transcriptional activity by promoting a novel bivalent interaction with the coactivator CBP/p300. Mol Cell 1:661–671

Chapter 6
Dynamics and Mechanism of DNA-Bending Proteins in Binding Site Recognition

Anjum Ansari and Serguei V. Kuznetsov

6.1 Introduction: Protein–DNA Interactions

The three-dimensional shape of biological macromolecules (proteins, DNA, and RNA), is determined by a myriad of "weak" noncovalent interactions (ionic, hydrophobic, van der Waals, and hydrogen bonds), each of which can be disrupted by thermal fluctuations, leading to constantly changing conformations accessible to the macromolecule [1]. These conformational fluctuations are essential to biology and are central to molecular recognition, in which two or more interacting macromolecules rely on complementary shapes and charge distributions to form a multitude of weak intermolecular bonds that lead to higher-order complexes. An overarching goal in molecular biophysics is to elucidate the underlying energetics of these interactions, by measuring the dynamics of conformational fluctuations in the macromolecular complexes.

Protein–DNA interactions regulate many important biological functions such as DNA replication, transcription, recombination, gene regulation, and repair. Proteins that bind to DNA in a sequence-nonspecific manner play a role in chromosome packaging and in stabilizing single-stranded intermediates. Alternatively, many cellular functions are initiated by the binding of proteins that recognize specific DNA sequences. These site-specific proteins typically bind their target sites with thousand- or even million-fold higher affinities relative to random DNA sequences.

Two puzzles have particularly intrigued researchers in this field. The first seeks to understand how site-specific DNA-binding proteins efficiently search for their binding sites in DNA, among a sea of nonspecific sites. This question came to the forefront with the discovery by Riggs et al. [2] that *lac* repressor protein could locate its target site on long DNA 100–1,000-fold faster than expected from three-dimensional diffusion and random collisions between protein and DNA. These and other measurements suggested "facilitated" diffusion, together with hopping between sites, as mechanisms for accelerated target location. Thus, a protein is

A. Ansari (✉)
Physics Department, M/C 273, University of Illinois at Chicago,
845 W. Taylor St., Chicago, IL, 60607 USA
e-mail: ansari@uic.edu

M.C. Williams and L.J. Maher, III (eds.), *Biophysics of DNA-Protein Interactions*, Biological and Medical Physics, Biomedical Engineering,
DOI 10.1007/978-0-387-92808-1_6,

imagined to bind nonspecifically anywhere along the DNA, slide along the DNA in a one-dimensional search for its binding site, then dissociate and randomly collide with another segment of the DNA, to continue its search [3–7].

The second related question asks how a sequence-specific DNA-binding protein actually determines that it has reached its target site on genomic DNA. Structural and thermodynamic studies of a wide range of protein–DNA complexes have shown that the DNA structure is often deformed in the complex, and the protein also adjusts its conformation, to facilitate favorable interactions [8, 9]. These concerted rearrangements in protein and DNA conformation are believed to be key to the underlying induced-fit mechanism proposed for the recognition of DNA binding sites. Details of the recognition mechanism remain elusive. In particular, very little is known about the dynamics of the conformational rearrangements that lead to precise recognition.

Kalodimos et al. [10] provided the first tantalizing glimpse of *lac* repressor, a sequence-specific DNA-binding (and bending) protein, bound to a nonspecific site where the DNA was observed to be unbent. Their NMR study indicated large-scale conformational fluctuations at the protein–DNA interface in the nonspecific complex, occurring on micro to millisecond time scales. This study suggested that the nonspecific complex is highly dynamical in which the protein and DNA sample a range of conformations prior to "falling into" a free energy minimum characterizing the specific complex. Smaller fluctuations were observed in the specific complex, indicating a tighter fit. This enhanced flexibility in the nonspecific complex is believed to be important for efficient search of the specific binding site on genomic DNA [10–12].

6.1.1 Binding Site Recognition: Direct Versus Indirect Readout

Structural studies of protein–DNA complexes have implicated two underlying mechanisms by which proteins discriminate between potential target sites. One is "direct readout," in which specific amino-acid residues of the protein make contacts (e.g., hydrogen bonds) with specific bases (A, T, G, or C). Proteins that use direct readout typically bind to the major groove of DNA where the pattern of hydrogen-bond donors and acceptors is distinct for each Watson–Crick base pair (bp) [13]. Thus, members of this class of proteins recognize their binding sites as a result of direct examination of the hydrogen-bonding pattern exposed in the major groove, and which is unique for each kind of base pair. Examples of protein motifs with DNA-binding domains (DBDs) that primarily employ direct readout are the helix-turn-helix motif in, for example, bacteriophage λ-repressor [14], 434 repressor [15], and bacterial Trp repressor [16], the basic leucine zipper (bZIP) motif in yeast GCN4 protein [17], and the zinc finger motifs in TFIIIA, a transcriptional regulator from *Xenopus* [18].

Alternatively, a significant fraction of site-specific DNA-binding proteins recognizes the minor groove of DNA where the pattern of hydrogen bond donors and acceptors is similar for all Watson–Crick base pairs, and symmetric, making it difficult for the protein to distinguish A from T, and G from C [19]. Furthermore, the G:C pair is distinguished from the A:T pair only by one additional centrally located donor group. Thus, for the minor groove-binding proteins, there is a lack of direct sequence information. Such proteins recognize their target sites with high fidelity via "indirect readout." In this mechanism, sequence-dependent variations in the mechanical properties of DNA, such as local DNA flexibility/"bendability" and changes in major or minor groove width, as well as differences in the stacking or twisting parameters, play important roles in the recognition process. Thus, minor groove-binding proteins are believed to "read" DNA sequence by the ease with which they can bend, twist, or deform the potential binding sites to match their binding interface. Quite often, proteins use a combination of both direct and indirect readout to recognize their binding sites.

In protein–DNA complexes causing DNA bending, the DNA deformation may be spread over the length of the DNA, as when DNA is wrapped around histone octamers to form nucleosomes for chromosome packaging [20, 21]. Alternatively, the DNA may be severely kinked at one or two sites, with the intervening segments left relatively undistorted. These sharp kinks are often stabilized by the intercalation of protein side chains between adjacent Watson–Crick pairs whose stacking interactions are disrupted [22]. Examples of specific binding proteins that introduce sharp, localized distortions in DNA are bacterial nucleoid-associated architectural proteins from the integration host factor (IHF)/HU family that kink DNA at two sites to form a "U-turn" [23, 24]; *lac* repressor, a tetrameric bacterial gene regulatory protein that simultaneously binds DNA at two sites and loops the intervening DNA [25]; TATA-binding protein (TBP), a eukaryotic basal transcription factor that recognizes a TATA element in the promoter sequence of DNA to initiate transcription, with severe DNA distortion at two sites to make a ~108° bend [26]; restriction enzymes that bind to specific sites with DNA kinking by ~50° prior to cleavage [27]; and DNA repair proteins that recognize and bind to lesions in DNA, stabilizing a ~60° kink at the lesion site [28, 29].

6.1.2 Sequence-Dependent DNA Flexibility and Bendability

The energetic cost of deforming DNA is often estimated using a continuum worm-like chain description for duplex DNA, in which the intrinsic rigidity is described in terms of a characteristic persistence length. For length scales much less than the persistence length, the polymer is assumed to behave like an elastic rod with harmonic bending energy potential, while for length scales much longer than the persistence length, it adopts configurations of a random coil [30]. Persistence length measurements of DNA have been made with a wide range of techniques such as rotational diffusion measurements on DNA oligomers that probe the shortening of the end-to-end distance in the elastic rod regime [31–33]; ligase-catalyzed cyclization of DNA molecules with cohesive ends that probe ring-closure probabilities [34–36] and, more recently, micromanipulation experiments on long single molecules of DNA that probe the entropic spring constant of polymeric DNA under applied force [37–39]. These measurements have yielded persistence length values of 50–60 nm (~150–180 bp), with essentially no ionic-strength dependence observed in the range of 10 mM–1 M Na^+ ions [40].

On length scales comparable to the target sites of sequence-specific DNA-binding proteins, the deformability of DNA depends on the local sequence, and this variation from the average is a key feature that is recognized by DNA-bending proteins. Hogan et al. [33] demonstrated that both the bending stiffness and the torsional constant of DNA fragments are strongly dependent on base composition, with persistence length values ranging from 125 bp for poly(dA).poly(dT) to 500 bp for poly(dG).poly(dC), and the torsional constant varying 40-fold in the range of sequences they studied. DNA cyclization studies showed that a DNA sequence with high histone octamer affinity exhibits twofold smaller bending constant and ~35% smaller torsional constant [41], and that alternating dA–dT sequences are ~30% more flexible in comparison with random sequences [42].

It is now well established that the deformability of DNA varies enormously along the length of genomic DNA, reflecting variations in the DNA sequence [43, 44]. This is consistent with the observation that nucleosome positioning sequences are not randomly distributed along the length of DNA [43, 45]. Statistical analysis of natural and synthetic nucleosome positioning sequences has revealed a pattern of high occurrence of AA/TT/TA dinucleotide steps phased 10 bp apart, alternating with GC dinucleotide steps shifted 5 bp away, indicating that certain sequence arrangements are more easily deformed into tight circles than others [46–48].

Two examples illustrate very nicely the correlation between sequence-dependent variations in the bendability of DNA, and the binding affinity of site-specific proteins. The first is an early study on phage 434 repressor protein, in which the high resolution structure of the protein bound to its cognate sequence showed that the binding site was bent and overtwisted, with the central 4 bp of the cognate site making no contact with the protein [49]. Nevertheless, the binding affinity could be reduced 50-fold by sequence variations in this segment [50]. Hogan and Austin [51] found a correlation between theoretical estimates of bending and twisting constants for each of the sequences, and the measured binding affinities for the repressor protein. A similar correlation was observed between the binding affinity of the E2 regulatory protein from the human papillomavirus (HPV) genome for DNA substrates containing varying E2 binding sites, and their intrinsic curvature and flexibility parameters [52, 53]. Cocrystal structures of the DBD of E2 show that the DNA substrates are bent by 45°, and, as in the case of the 434 repressor protein, there are no direct contacts between the E2 DBD and the central 4-bp spacer sequence within the binding site [54]. Maher and coworkers used electrophoretic mobility measurements on DNA molecules containing three distinct E2-binding sites appropriately phased relative to A-tracts to determine the intrinsic curvature at the site of the E2 spacer sequence [52, 55]. These authors found high intrinsic curvature (~18°) for a spacer sequence with ~30-fold higher affinity for E2 DBD in comparison with a low affinity spacer sequence, which exhibited essentially no intrinsic curvature, while an intermediate affinity sequence showed ~11° curvature [52]. Crothers and coworkers carried out high-throughput cyclization based on a Forster resonance energy transfer (FRET) assay, combined with a statistical mechanical theory of cyclization probabilities to extract the bending and torsional constants, from which they calculated the probability that each of the DNA substrate would adopt the bent conformation, even in the absence of the E2 protein [42, 53]. These results demonstrated remarkable agreement between the calculated probabilities and the more than three orders of magnitude variation in the binding affinities for 15 of the 16 different spacer sequences studied [53].

6.1.3 *Beyond the Continuum Worm-like Chain Description*

DNA segments comparable in length to the persistence length of DNA can wrap nearly twice around histone octamers in the nucleosome. This observation led Klug and Crick [56] to propose, more than 30 years ago, that the energetic cost of forming small circles could be reduced if DNAs were to kink at regular intervals, rather than be uniformly distorted throughout. Evidence of kinked DNA in nucleosomes was reported by Austin and coworkers [57]. High-resolution nucleosome crystal structures show smooth bending of DNA segments that are bent into the major groove, while DNA segments that are bent into the minor groove are either kinked, at six 5′-CA-3′ steps out of a total of ten such steps, or smoothly bent but with the bases alternately shifted to accommodate the tight bend [21].

Recent interest in developing theoretical descriptions of the polymer properties of DNA beyond the semiflexible, worm-like chain model came from cyclization studies of Widom and coworkers [58, 59], in which they reported that ~100-bp -long DNA fragments can form circles with 3–5 orders of magnitude higher probabilities than expected for a worm-like chain, especially DNA sequences that bind tightly to nucleosomes. These studies led to theoretical extensions of the continuum, elastic worm-like chain model that included sharp bending or kinking of DNA, either from local melting to produce highly flexible single-stranded regions [60], or from sharp kinks at sites of low stacking energy [61, 62].

Vologodskii and coworkers revisited cyclization efficiency measurements of DNA fragments 105–130 bp in length and demonstrated that the worm-like chain model remains adequate for predicting cyclization probabilities of these fragments [62]. Their experimental study provided a technical explanation for the apparently unusually high cyclization probabilities observed by Widom and coworkers, although the strong sequence dependence observed in the original study remains enigmatic. Another recent study, carried out on 200-bp-long fragments, showed nearly fivefold variation in cyclization probability as a result of eight nucleotide changes designed to locally melt more easily, highlighting the ability of certain sequences to form circles more easily than others [63]. Biochemical assays to probe localized disruptions of the regular DNA double-helical structure upon forming minicircles [64, 65] as well as cryoelectron microscopy images of such minicircles [65] indicated no direct evidence for sharp kinks in 85–86-bp circles

[64] or 94-bp circles [65], although evidence for kinked DNA was observed for much smaller circles of 64–65 bp [64]. Deviations from the simple worm-like chain description have also been observed in recent experiments that reported enhanced flexibility of DNA on short length scales, down to ~15 bp [66, 67], indicating a "softening" of the harmonic bending energy potential implicit in the worm-like chain description [66].

6.1.4 Free Energy Cost of Kinking DNA

Obtaining accurate estimates of the free energy cost of kinking DNA has been problematic because of the difficulty in separating the stacking free energy from the base-pairing energies that contribute to the stability of duplex DNA. Based on thermal melting of long DNA duplexes, the free energy cost associated with disruption of a single Watson–Crick base pair is estimated to be about ~1 kcal/mol for an AT pair, and ~2 kcal/mol for a CG pair, in the presence of 100 mM monovalent cations [68–70]. Another estimate of base-pair disruption, ~6–9 kcal/mol, comes from NMR measurements of imino proton exchange that probe the time scales for opening and closing of a single base pair inside long duplex DNA [71]. The wide range of base pair opening times observed in the NMR measurements, especially for A–T pairs, reflects the sequence-context dependent cost of breaking stacking interactions that accompany base pair opening, and that may be the more dominant contribution to the energetic cost of base-pair disruption. Base stacking has been shown to be more important than Watson–Crick pairing for the rigidity of duplex DNA [72, 73].

Estimates of stacking free energies for the ten distinct dinucleotide steps have been obtained primarily from analysis of thermal melting of DNA oligomers of varying sequences and lengths, and unified by SantaLucia [74]. More recently, Frank-Kamenetskii and coworkers have applied a novel approach to obtain the stacking free energies, from measurements of the electrophoretic mobility of a series of 300 bp-long DNA molecules with a nick in the sugar–phosphate backbone on one of the strands, sandwiched between all combinations of dinucleotide steps [70]. The deviations of the observed mobility for each of the nicked duplexes in comparison with intact DNA were modeled as arising from equilibria between the stacked (straight) and unstacked (bent) conformations at the nicked position, to obtain directly the ten independent stacking free energies. Olson and coworkers took advantage of the available database of known protein–DNA complex structures to develop sequence-dependent empirical energy functions that describe DNA deformability at each of the dinucleotide steps [75, 76]. They obtained average values, and deviations from the average, of six step parameters that describe the local stacking geometry between neighboring base pairs. A global parameter from this analysis that can be readily compared with stacking free energies is the volume of conformational space that is thermally accessible at each step (Fig. 6.1). Despite

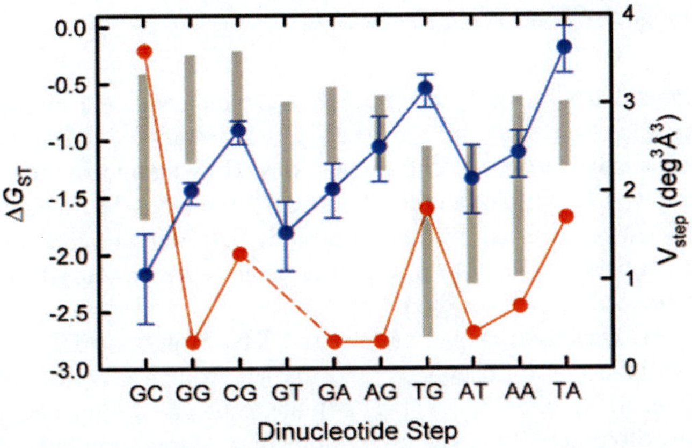

Fig. 6.1 Comparison of nearest-neighbor (nn) stacking free energies (ΔG_{ST}) and volumes of conformational space (V_{step}) for the ten different dinucleotide steps of duplex DNA. The nn stacking free-energy parameters, extracted from thermal denaturation experiments on oligonucleotide duplexes, are shown as *vertical gray bars*. The length of each *vertical bar* indicates the range of the stacking parameters from seven independent research groups, obtained under different salt conditions and for varying lengths of duplex DNA, and unified by SantaLucia [74]. Stacking free energies from electrophoretic mobility measurements (*filled blue circle*) on DNA fragments containing a nick in the sugar–phosphate backbone, between all possible combinations of dinucleotide steps are from [70]. The volume of thermally accessible conformational space for each dinucleotide step (*filled red circle*), obtained from fluctuations and correlations of base step parameters in DNA–protein crystal complexes, is from [75]. The volume for the GT step was not reported in [75], as indicated by the *dashed red line*. The estimates of stacking free energies from the thermal melting and the gel mobility assays, in general, correlate well with the volume of accessible conformational space at each dinucleotide step, with perhaps the exception of the GC step

some variations in the estimates of sequence-dependent deformability obtained from the different approaches, some trends emerge, with the pyrimidine–purine (Y–R) steps being the most easily deformable.

6.2 Dynamics of Protein Binding and DNA Bending

Characterization of the rates at which proteins bind to, and bend, DNA provides insights into reaction intermediates, and is a natural starting point to probe the molecular mechanism of binding site recognition. These studies have begun to yield answers to the following questions: What is the sequence of molecular rearrangements in the protein–DNA complex and where are the bottlenecks in the transition from nonspecific to specific complex? What role does sequence-dependent DNA flexibility play in the mechanism by which minor groove DNA-bending proteins recognize their target sites? Does the protein first bind to straight DNA and then induce DNA distortion as a second step, or does the protein capture thermally accessible predistorted DNA conformations? Does DNA bending rate play a role in the recognition mechanism?

6.2.1 Stopped-Flow Measurements

Most experimental data on the dynamics of DNA bending and induced conformational changes in proteins have come from stopped-flow measurements. These studies have yielded partial answers to three central questions: How distinct are the bimolecular association rates from diffusion-controlled rates? Do DNA-bending proteins bind and bend DNA in a concerted manner, or do they first bind and then bend DNA, in a sequential manner? How do DNA sequence changes that modify DNA flexibility affect observed binding and bending rates?

In stopped-flow measurements, protein and DNA solutions are mixed together on time scales of a few milliseconds, and the subsequent kinetics are monitored with one of several probes [77–81]: (1) fluorescent anisotropy decay of single-labeled DNA oligomers to measure bimolecular association, (2) DNase I or hydroxyl-radical "footprinting" to measure protection of DNA by bound protein, and (3) FRET between donor and acceptor fluorophores attached at opposite ends of a short DNA oligomer to measure DNA bending.

Early stopped-flow studies on the binding of the TBP to promoter sites revealed two salient results. First, the binding of promoter substrates to TBP followed second-order behavior, with bimolecular association rates in the range $(0.5–2)\times10^6\ M^{-1}s^{-1}$ [77, 82, 83], significantly smaller than the $\sim10^8–10^9\ M^{-1}s^{-1}$ rate expected for diffusion-controlled bimolecular association between protein and a short DNA oligomer [3, 84]. Second, FRET measurements to probe the bending kinetics of labeled DNA substrates also followed second-order behavior with similar association rates, thus ruling out a reaction mechanism of formation of a stable "encounter complex" followed by slow DNA bending [78]. To account for the slower than diffusion limited association rate, the authors of this study suggested a mechanism in which the protein and DNA are in rapid equilibrium with a transient encounter complex (with negligible steady-state population), which is separated from the final, bent complex by a large activation barrier [78]. Subsequent kinetic studies by Brenowitz and Parkhurst and coworkers, carried out over a range of temperatures, showed significant deviations from single-exponential decay under pseudo-first order conditions [78, 85, 86]. These studies provided evidence for a complex reaction pathway between the transient encounter complex and the final bent complex, with multiple steps consisting of partially or fully bent intermediate states. Some of these intermediate states have been directly observed in a single-molecule study described below [87].

Bimolecular association rates that are significantly smaller than diffusion-controlled rates, $\sim3\times10^6\ M^{-1}s^{-1}$, have also been reported for the binding of a DNA repair protein, *Taq* MutS, to DNA substrates with single-T insertions [88]. Plausible mechanisms that can account for the relatively slow bimolecular association rates include the possibility that the unbound DNA samples a distribution of DNA conformations, and the protein binds to the fraction that is partially prebent [78], or alternatively, that the protein fluctuates between distinct conformations, with only a subset that is able to bind DNA [89].

Perona and coworkers monitored binding of the restriction enzyme *Eco*RV to its cognate DNA site, with stopped-flow anisotropy measurements, and the bending of the DNA in complex, with FRET [80]. In the presence of divalent ions, these authors observed identical second-order rates with both sets of probes, indicating simultaneous DNA binding and bending events [80]. The bimolecular association rate for the *Eco*RV–DNA complex was ~10^8 $M^{-1}s^{-1}$ [80, 89]. The binding and bending of a cognate DNA site by bacterial DNA methyltransferase *M.Eco*RI also showed similar rates for the bimolecular association step (~3×10^7 $M^{-1}s^{-1}$) when monitored with fluorescence anisotropy and FRET [90].

Brenowitz and Crothers and coworkers used time-resolved hydroxyl radical footprinting to obtain the time scales on which different regions of a cognate DNA substrate are protected from cleavage when bound to IHF [79]. As discussed below, the IHF–DNA interface extends over ~35 bp, with the DNA bent at two sites and wrapped around the protein in three roughly equal segments. Simultaneous protection of the three segments of DNA was observed, again indicating concerted binding and bending for the IHF–DNA complex, at least at the ~30-ms time resolution of these measurements [79]. The pseudo-first order association kinetics as well as the dissociation kinetics showed deviations from single-exponential decay suggesting a multistep mechanism with evidence for rapid, facilitated diffusion along the DNA to locate the target site, followed by concerted bending. The bimolecular association rate for this complex was found to be ~5×10^8 $M^{-1}s^{-1}$ [81].

An underlying conclusion from the initial stopped-flow studies was that protein binding and bending appeared to occur simultaneously [77–80, 90]. Evidence for separate binding and bending events that occur sequentially, with a stable intermediate state with protein bound to straight DNA, first came from single-molecule micromanipulation studies of DNA cleavage by *Eco*RV [91]. More recently, Sugimura and Crothers [81] demonstrated sequential binding then bending of a cognate site by IHF in a stopped-flow study that also yielded, for the first time, the rates of DNA bending. These studies, described below, have opened up the possibility of such a sequential mechanism for other DNA-bending proteins as well, with the bending step possibly occurring faster than the time resolution of stopped-flow techniques.

6.2.2 *Single-Molecule Measurements*

Micromanipulation experiments that probe the mechanical response of single molecules of λ DNA to applied force have emerged as a powerful tool to probe protein–DNA interactions, and have begun to unravel the role that nucleoid-associated nonspecific DNA-bending proteins play in the compaction of DNA [92–96]. Single-molecule studies have also begun to yield dynamical information on DNA bending by site-specific proteins [87, 91, 97, 98].

Wuite and coworkers [91] probed the mechanism of sequence recognition by *Eco*RV by monitoring the rate at which *Eco*RV cleaved a single molecule of DNA containing the specific site, stretched between two optical traps (Fig. 6.2). At high values of the applied tension, the rate of DNA bending appeared to be the rate-limiting step in the cleavage reaction. Extrapolation to zero force suggested DNA-bending rates between 35 and 1,000 s^{-1}. A particularly interesting observation in this study was the bimolecular association rate of ~4×10^6 $M^{-1}s^{-1}$ for finding the target site

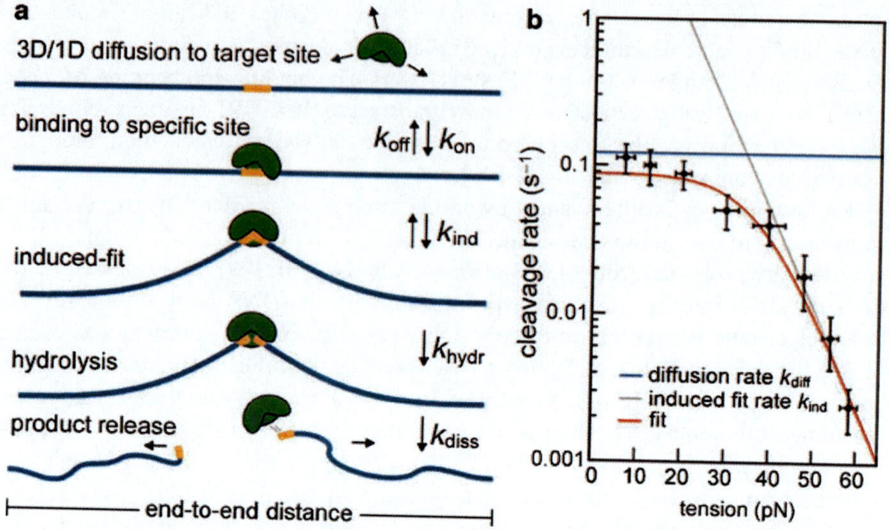

Fig. 6.2 Micromanipulation measurements of the force dependence of the cleavage rate of DNA by a type II restriction endonuclease *Eco*RV. (**a**) A schematic of the reaction pathway for the restriction enzyme that includes binding to the target site, DNA bending, the hydrolysis step that cleaves the DNA, and product release. A long DNA molecule (*blue*) contains the cognate site for the restriction enzyme (*orange*) and is shown to be mechanically stretched. k_{on} and k_{off} are the association and dissociation rates, respectively, for binding to the target site, k_{ind} is the rate of DNA bending and induced conformational changes in the protein, k_{hydr} is the hydrolysis rate, and k_{diss} is the rate at which the product is released. (**b**) The measured cleavage rate for *Eco*RV, obtained by Wuite and coworkers [91], is plotted versus DNA tension. The force dependence of the cleavage reaction was determined as follows: a DNA molecule containing the *Eco*RV restriction site was captured between two beads held by two optical traps; one trap was moved relative to the other, to apply tension on the DNA, and the force was measured by the displacement of the other bead. A series of measurements of the time delay between when the restriction enzymes were flowed in and when the DNA was cut yielded the cleavage rate as a function of the applied tension. The rate at which DNA bends in the complex was assumed to be the rate-limiting step at high values of the applied tension, shown as the *gray line* (**b**). Extrapolation of $k_{ind}(F)$ to zero force yielded bending rates from 35 to 1,000 s^{-1}. Figure reprinted from [91], with permission from Oxford University Press

on stretched DNA, smaller than the ~10^8 $M^{-1}s^{-1}$ obtained from stopped-flow measurements with short DNA oligomers [80, 89], and significantly smaller than the ~10^{10} $M^{-1}s^{-1}$ observed for target location by *lac* repressor on long, randomly coiled DNA [99]. This study illustrated that stretched DNA may inhibit bimolecular rate enhancement by protein hopping from one nonspecific site to another, a mechanism that, together with one-dimensional sliding, allows rapid location of target sites on randomly coiled DNA [7].

Dixit et al. [97] detected local bending and shortening of a 76-bp DNA oligomer as a result of IHF binding to a specific site, by optically monitoring the displacement of micron-size bead tethered to a surface by DNA. Reinhard et al. [98] applied a novel technique in which a pair of gold nanoparticles, attached at either ends of a DNA segment containing the restriction site for *Eco*RV, was used as "plasmon rulers" to monitor both DNA cleavage by *Eco*RV (detected by a sharp change in the plasmon coupling signal), as well as subtle changes in DNA conformation just prior to the cleavage. Tolic-Norrelykke et al. [87] detected TBP binding and dissociation events at the single-molecule level by video microscopy images of the Brownian motion of beads tethered by DNA oligomers containing a TATA-box binding site. The range of dynamics of the bead varied between straight DNA and DNA bent by TBP, with evidence for partially bent intermediate states that ties in with the multistep kinetic scheme proposed from bulk studies of Parkhurst and coworkers [78, 85].

6.2.3 Laser Temperature-Jump to Probe Protein–DNA Interactions

The stopped-flow and single-molecule studies discussed above suggest DNA-bending rates, in the presence of bound protein, in excess of ~100 s^{-1}, thus requiring the development of fast experimental techniques. In the last decade, laser temperature-jump (T-jump) has emerged as a powerful complement to stopped-flow and single-molecule studies, whereby the conformational dynamics of biomolecules can be monitored on time scales of a few nanoseconds to several milliseconds. Typically, ~10 ns laser pulses in the near infrared (~1.54–1.9 μm), at a wavelength where water absorbs, produce a rapid ~10°C increase in the temperature of the sample, within the pulse width of the laser [100–102]. The T-jump perturbs the equilibrium distribution of biomolecular conformations, and the system responds by adjusting the distribution of conformations to the new, higher temperature. The rate at which the system relaxes to its new equilibrium is monitored with spectroscopic probes sensitive to the average biomolecular conformation. The T-jump eventually decays with a half-life of ~100–300 ms. Laser T-jump has been widely used by several groups to investigate fast dynamics in the folding of proteins [100, 103–107] and nucleic acids [108–111].

Complementary techniques, such as fluorescence correlation spectroscopy (FCS) on labeled samples, allow dynamics in proteins and nucleic acids to be investigated with microsecond time resolution [112–115]. However, a distinct advantage of the T-jump technique is that intrinsic fluorescence of proteins can be monitored without interference from extrinsic labels [105, 106, 116–118]. Furthermore, the development of a variety of fluorescent analogs of DNA bases allows the measurement of local changes in the twist or flexibility of DNA upon protein binding [88, 119–121]. Thus, the T-jump technique combines superior time resolution with exquisite sensitivity to monitor subtle changes in the conformation of biomolecules.

Kuznetsov et al. [118, 122] extended the application of laser T-jump to measuring the dynamics of protein–DNA interactions. The first such study probed the previously unresolved wrapping/unwrapping dynamics of single-stranded DNA (ssDNA) around a single-strand binding (SSB) protein, an essential step in all DNA metabolic processes. These experiments showed that this step occurs on a time scale of tens of microseconds [118]. This study set an upper limit on how fast SSB can translocate along DNA during replication, recombination, and repair, and held out promise for similar studies on other protein–DNA systems with unresolved dynamics. The second study, on the bending/unbending dynamics of DNA in complex with IHF [122, 123], is discussed below.

6.3 Integration Host Factor: A Case Study

The IHF/HU family consists of closely related prokaryotic DNA-binding proteins that induce large bends in DNA as a result of localized distortions, making them a particularly interesting system to probe the role of DNA bendability in the recognition mechanism. These proteins are functionally analogous to both the histones and the HMG box proteins of eukaryotes, and aid in chromosome compaction as well as play specific architectural roles. All prokaryotes appear to have at least one IHF/HU homolog. Cocrystal structures of three proteins from this family, the IHF from *Escherichia coli* [23], the histone-like protein from *Anabaena* (AHU) [124], and Hbb from the Lyme disease-causing spirochete *Borrelia burgdorferi*, were determined by Rice and coworkers [24]. All three proteins bind to the minor groove and, when sequence specificity is involved, recognize their binding sites primarily by the indirect readout mechanism [24, 125–129].

6.3.1 Specific and Nonspecific Binding Modes of the IHF/HU Family of Proteins

IHF is an essential protein in several cellular processes including site-specific recombination, transcription, DNA replication, and λ phage packaging [130–133]. It binds with low or sub nanomolar affinities to several specific binding sites on λ phage DNA. These binding sites share a consensus sequence WATCARNNNNTTR (W refers to A or T, N is any base, and R refers to a purine, A or G) located on one half of the ~35 bp DNA segment that wraps around the protein [134]. Some IHF sites also contain an A-tract consisting of 4–6 adenines located about 8–9 bp upstream of the consensus region [135, 136].

HU functions within the cell by facilitating the formation of multiprotein complexes mediated by DNA looping or bending [137–140], and is also implicated in DNA repair [141, 142]. It has no sequence specificity and binds with moderate affinity (~100 nM) to random DNA sequences. However, it shows strong preference for DNA segments containing structural distortions such as nicks, gaps, bulges, or branches [128, 143, 144]. Hbb is a recently characterized protein that, like IHF, plays a role in DNA compaction as well as assembly of higher-order nucleoprotein complexes. Hbb exhibits strong preferences for particular DNA sequences [24, 145].

In their nonspecific binding modes, these proteins play a role in chromosome compaction [146]. The ability of IHF and HU to compact DNA has been investigated using micromanipulation techniques as well as with atomic force microscopy [92, 93, 147, 148]. Force-extension measurements on λ DNA bound to nonspecific DNA-bending proteins from the IHF/HU family, as well as the HMGB family, revealed the formation of very tight complexes at protein concentrations above a certain threshold. Under these conditions, no spontaneous dissociation of the proteins was observed [93]. These results are in contrast to what is expected from dissociation constant measurements on proteins bound to short DNA oligomers and suggest some kind of cooperativity between bound proteins, perhaps mediated by underlying distortions of the DNA structure that stabilize a filament [93]. Above a threshold concentration, HU also demonstrates the unique ability to stiffen DNA [93, 147].

6.3.2 In the Specific Binding Mode, IHF Bends DNA in a U-Turn

Proteins in the IHF/HU family have the same overall fold, with a largely α-helical body capped by β-sheets that extend into two β-ribbon "arms" that are disordered in the absence of DNA [23, 127, 149–152] (Fig. 6.3). In the complex, the β-arms wrap around the DNA and lie in the minor groove, and the DNA is severely kinked at two sites separated by 9 bp, which results in a large "U-turn" bend in the DNA. A highly conserved proline residue at the tip of each arm induces or stabilizes DNA bending by intercalating between base pairs at the site of the kinks. The DNA segments on either side of the kink closely "hug" the protein, making several contacts with positively charged residues on the sides of the protein. The cocrystal structure of IHF, obtained with a 35-bp DNA duplex containing the H′ binding site from phage λ, showed that ~26 positively charged residues on the surface of the protein make ionic interactions with DNA in the wrapped complex [23, 124]; in the Hbb complex with 35-bp duplex DNA, the number of ionic contacts is ~32 [24]. The cocrystal structure of AHU was obtained with a 19-bp duplex [127], which is not long enough to show the interactions that might be operative in stabilizing longer sequences. However, interplay between intrinsic DNA flexibility and the ability to make more extended contacts with the positively charged surface of the protein is believed to be an important factor that contributes to the dramatically increased affinity of all three proteins for sequences with inserted distortions [124]. This may be relevant for the role of HU in recognizing and wrapping DNA repair intermediates [153–157].

6.3.3 Ion Release in the IHF–H′ Complex Is Coupled to Protein Conformational Changes

The polyelectrolyte nature of DNA results in the condensation of high concentration of cations in the vicinity of the double helix [158–160]. Release of these counterions when proteins bind to DNA is responsible for a favorable entropic contribution to the

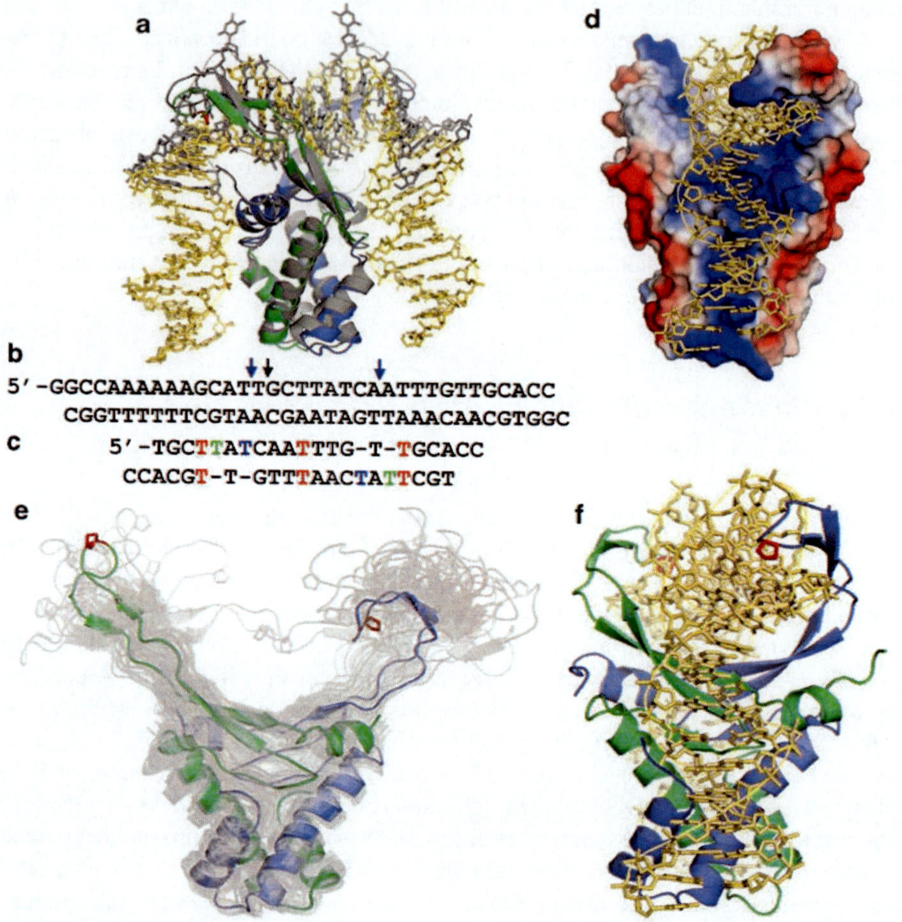

Fig. 6.3 Cocrystal structures of the IHF/HU family of DNA-bending proteins. (**a**) Superposition of the IHF–DNA and AHU–DNA cocrystal structures. The α- and β-chains of the IHF protein are shown in *blue* and *green*, respectively, and the IHF DNA in *gold* (PDB code: 1IHF). Conserved proline residues are shown in *red*. AHU and the bound DNA are shown in *gray* (PDB code: 1P71). (**b**) Sequence of the DNA substrate in the cocrystal structure with IHF, which contains the H′ binding site from bacteriophage λ. The H′ sequence was nicked at the position indicated by the *black arrow*, to facilitate crystal packing. The DNA is observed to be sharply kinked at two sites in complex with IHF, at two locations indicated by the *blue arrows*. (**c**) The sequence of the DNA substrate in the cocrystal structure with AHU containing three T•T mismatches (shown in *red*) and four unpaired T bases (*green bases* are stacked in the crystal structure, while the *blue bases* are flipped out). (**d**) Side view of electrostatic surface representation of IHF. The localization of positive and negative charges is shown in *blue* and *red*, respectively. (**e**) A superposition of all 25 NMR structures of the HU protein from *Bacillus stearothermophilus* (PDB code: 1HUE) is shown in *gray*, with one of these structures highlighted, with the α- and β-chains shown in *blue* and *green*, respectively, and conserved proline residues shown in *red*. (**f**) The structure of IHF–H′ complex (1IHF) rotated by 90° relative to the view (**a**). The β-ribbon arms of the proteins of this family are more flexible in the absence of a bound DNA. In the presence of a DNA substrate, they wrap around the DNA and lie in the minor groove, and the prolines intercalate between the kinks in the DNA

stability of protein–DNA complexes, resulting in a linear dependence of the logarithm of the binding affinity versus log[salt]. The slope of this linear dependence can yield an estimate of the net number of ions released when the complex is formed [161]. For the IHF–H′ complex, this slope is ~8–9 [79, 123, 162], which is significantly smaller than the number of closely associated counterions expected to be released from DNA, a number closer to ~20 [162, 163]. Moreover, this slope depends on the type and concentration of anions [164], as was previously observed in the case of ssDNA binding to SSB [160]. These studies indicate that the uptake and release of anions by the protein also can contribute to the observed salt dependence of protein–nucleic acid interactions [165].

Two additional observations made with isothermal titration calorimetry (ITC) measurements of Saecker and Record and coworkers indicate unusual thermodynamics of the IHF–H′ complex. First, H′ binding to IHF is accompanied by unusually large negative enthalpy, entropy, and heat capacity changes, with $\Delta H^0_{sp} \approx -19$ kcal/mol and $T\Delta S^0_{sp} \approx -9$ kcal/mol at ~20°C, and $\Delta C_{sp} \approx -2$ kcal/K, in 100 mM KCl [162]. Second, the magnitudes of ΔH^0_{sp}, ΔS^0_{sp}, and ΔC_{sp} becoming less negative with increasing [KCl] [162]. Negative enthalpy, entropy, and heat capacity changes that depend on the salt concentration have also been observed in the nonspecific binding of SSB to poly(dT) [165], and the nonspecific binding of mammalian high mobility group protein A2 (HMGA2) to poly(dA–dT)$_2$ [166], in contrast to what is expected if the observed salt-dependence is solely from the entropic release of cations associated with the free DNA.

Negative enthalpy and entropy changes for the IHF–H′ complex are unexpected. In a comparison of the enthalpy and entropy changes for a wide range of DNA-binding proteins, Jen-Jacobson et al. [167] pointed out a correlation between the enthalpy changes and the extent of DNA distortions in the complex: complexes with minimal distortion of the DNA typically exhibited favorable enthalpy change compensated by an unfavorable entropy change, while complexes in which the DNA was bent or severely distorted exhibited an unfavorable enthalpy change, compensated by a favorable entropy change. In another study, Privalov et al. [168] showed that minor groove-binding proteins exhibit positive enthalpy changes, while major groove-binding proteins exhibit negative enthalpy changes. They attributed the unfavorable enthalpy changes for the minor groove-binding proteins to the displacement of highly ordered water from the A/T rich minor groove. IHF stands out as a notable exception in that, despite binding to minor groove regions that are generally A/T rich and severely distorting DNA, it exhibits large negative enthalpy and entropy changes upon H′ binding.

To explain the unusual thermodynamics of the IHF–H′ complex, Saecker and Record and coworkers proposed that the surface of IHF may be characterized by a network of intraprotein salt bridges near the DNA wrapping surface, which become protein–DNA salt bridges as a result of conformational changes in the protein upon DNA binding. The large negative enthalpy and entropy changes were attributed to the hydration of the negatively charged protein residues that are exposed when the salt bridges are exchanged [162, 163]. The dependence of the binding enthalpy on salt could be attributed to the effect of salt on the stability of the protein [162, 169], and to contributions from differential (salt dependent) ion–protein binding [165].

6.4 Time-Resolved FRET Measurements on IHF–H′

Binding of IHF to its cognate site is accompanied by DNA bending, wrapping of the β-ribbon arms around the DNA, intercalation of prolines to stabilize the kinks, formation of electrostatic contacts between the DNA sugar–phosphate backbone and the positively charged wrapping surface of the protein, release of ions, conformational rearrangements in the protein, and formation of specific contacts between the protein and the bent DNA that depend on the sequence of DNA in the consensus regions. What are the time- scales for each of these steps in the recognition mechanism? The kinetics measurements described below provide a glimpse into the sequence and time scales of molecular rearrangements along the transition pathway.

6.4.1 Sequential Binding then Bending Observed for the IHF–H′ Complex

Do proteins bind and bend DNA in a concerted manner, or is the binding and bending sequential, with an intermediate state in which the DNA is essentially straight or only partially bent? Evidence for sequential binding and bending first came from the study by Wuite and coworkers [91], on the rate of DNA cleavage by *Eco*RV. More recently, stopped-flow measurements of Sugmura and Crothers [81], together with laser T-jump studies of Kuznetsov et al. [122], provided the first direct observation of the DNA-bending rates in the IHF–H′ complex [170].

The large bend in the H′ substrate when bound to IHF makes it very amenable to biophysical studies of the dynamics in the complex. The two ends of a 35-bp long DNA segment containing the H′ binding site are ~100 Å apart in free DNA and ~50 Å apart in the IHF–H′ complex [23]. Thus, time-resolved FRET measurements on labeled DNA oligomers provide a very sensitive measure of changes in the end-to-end distance and hence the bending/unbending dynamics [81, 171, 172]. Sugimura and Crothers [81] carried out stopped-flow FRET measurements on the IHF–H’ complex and demonstrated that, under pseudo-first order conditions with IHF concentration above the dissociation constant of the complex and in excess over the DNA oligomer concentration, the observed relaxation rates at low IHF concentrations scaled linearly with the concentration, as expected for a bimolecular process, but saturated at high IHF concentrations (Fig. 6.5c). They interpreted the limiting value of the observed relaxation rates as arising from a concentration-independent, unimolecular process and identified it as the DNA bending step, which occurred on time scales less than ~10 ms at room temperature. Their measurements left open the possibility that the stopped-flow technique might be observing only

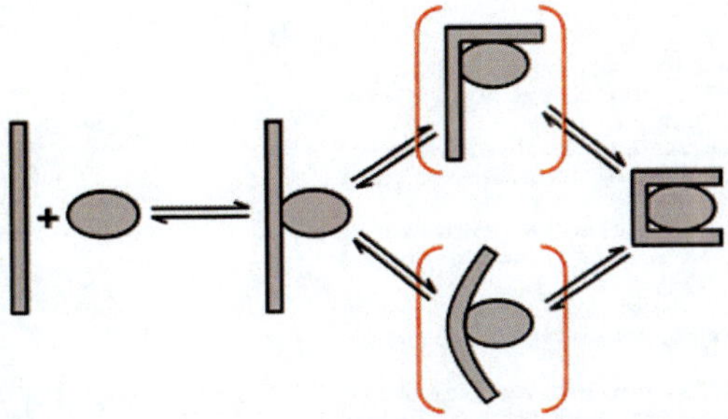

Fig. 6.4 Sequential binding then bending mechanism proposed for the IHF–DNA complex formation. IHF first binds nonspecifically to straight DNA, and then bends DNA to form the specific complex. Two plausible pathways exist for the bending step. Each includes an intermediate partially bent state (shown inside *red brackets*): (1) bending of first one arm of the DNA and then the other (*top*) or (2) a partial bending of DNA, perhaps from the wrapping of the β-arms of IHF around DNA

the end of a multistep DNA-bending pathway following the initial binding event, with perhaps sequential bending of first one flanking arm and then the other, or a weakly bent intermediate state (Fig. 6.4).

Kuznetsov et al. [122] used laser T-jump methods to probe the DNA bending/unbending kinetics in the IHF–H′ complex with a time resolution of a few microseconds. This work extended the relaxation kinetics measurements to shorter time scales and a wider temperature range (Fig. 6.5d). These authors obtained relaxation rates for the unimolecular bending/unbending process that were in excellent agreement with the bending rates obtained from stopped-flow studies, providing direct support for the sequential binding and bending model.

6.4.2 *DNA Bending Occurs on Time Scales Similar to Single Base Pair Disruption*

Stopped-flow and T-jump studies of DNA bending rates during formation of the IHF–H′ complex suggested a ~13 kcal/mol activation energy for the DNA bending step (Fig. 6.5d), consistent with previous estimates of the enthalpic cost for disrupting base-stacking interactions in DNA [173, 174]. An intriguing revelation from these measurements was that the time scales and activation energy for DNA bending were within the range of the time scales and activation energies for the opening of a single, internal A:T base pair in B-DNA, obtained from NMR measurements of imino proton exchange in nucleic acids [71, 175, 176]. These results suggested the possibility that transient thermal disruption of base pairing and/or stacking, at weak points along the DNA, may lead to spontaneous bending/kinking of DNA. If the resulting kink occurs in the presence of bound protein and at a site appropriate for specific binding, then this thermal fluctuation would be sufficient to overcome the free energy barrier. Favorable interactions between the bent DNA and the protein, facilitated by conformational

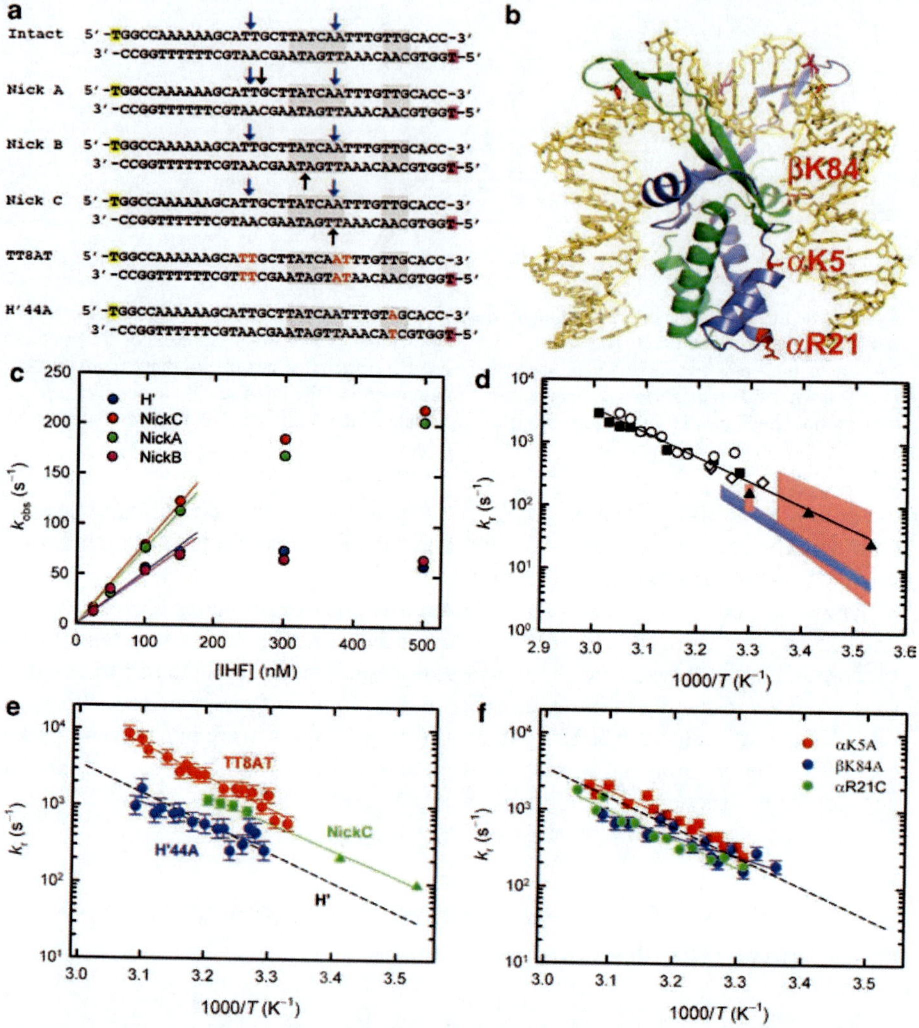

Fig. 6.5 Kinetics measurements on the IHF–H′ complex using stopped flow and laser T-jump. (**a**) Several variants are shown for the 35-bp long DNA oligomer containing the H′ binding site of IHF, labeled with fluorescein and TAMRA on the 5′ ends of the DNA strands, with the consensus regions indicated in *gray*. *Blue arrows* denote locations of the two kinks in the IHF–H′ cocrystal structure, and *black arrows* denote positions of nicks in the sugar–phosphate backbone, in Nick A, Nick B, and Nick C sequences. Nicked substrates were made without the dangling phosphoryl group at the 3′-end. The TT8AT sequence contains two tandem mismatches indicated in *red*, TT•TT and AT•TA, separated by 8 bp. The H′44A sequence has a T-to-A substitution, shown in *red*, in the TTR consensus region of H′. (**b**) The IHF–H′ cocrystal structure (PDB code: 1IHF) is shown with location of three positively charged residues, shown in *red*, that were substituted for a neutral residue: (1) αK5A, with Lys→Ala substitution at position 5 of the α-chain, (2) βK84A, with Lys→Ala substitution at position 84 of the β-chain, and (3) αR21C, with Arg→Cys substitution at position 21 of the α-chain. The crystal structure was obtained with Nick A. The location of the nicks in the Nick B and Nick C sequences are indicated in *pink*, at the *center and right-hand side* of the H′ structure, respectively. (**c**) The observed relaxation rate k_{obs} versus IHF concentration,

rearrangements in the complex, would then lead to a stable complex. The substantial positive activation energy observed for the bending step, despite the large negative enthalpy change obtained in the thermodynamics of forming the complex, supports the picture that protein conformational rearrangements that contribute to the negative enthalpy change occur after the rate-limiting step that characterizes the bottleneck in the transition.

The picture that emerges is the following: the protein slides along the DNA in search of its target site, probing the sequence-dependent thermal fluctuations of the underlying DNA. When the protein reaches its target site, these random conformations adopted by the DNA match the binding interface of the protein. The bound protein recognizes the appropriate transiently bent or kinked DNA conformation, and captures this structure to form the tightly bound complex. Spontaneous bending/kinking of DNA in the absence of a bound protein has been implicated in recent studies [59–61, 67, 177–179], raising the question as to whether the DNA is capable of undergoing thermal fluctuations in which distorted conformations are energetically accessible [66, 122, 180–183]. The importance of thermal disruption of base pairs has been implicated in the recognition of promoter sequence by RNA polymerase [182, 183]. Similarly, thermal stretching fluctuations may be playing a role in the binding of RecA to DNA, with RecA binding preferentially to stretched DNA [180].

In the case of IHF, it can be debated whether disruption of a base pair is necessary for kinking DNA or whether it is sufficient to disrupt stacking interactions [184]. The cocrystal structures of IHF–H′ and HU–DNA show severe distortion of the DNA helix parameters at the site of the kinks, but no evidence of disruption of the Watson–Crick base pairs. Furthermore, the NMR time scales for the opening and closing of base pairs depend strongly on the sequence context (Fig. 6.5d) and may represent disruption of the stacking between bases, especially for an A:T base pair, as suggested by the very wide range of opening times reported for opening of an A:T pair in comparison with a G:C pair (Fig. 6.5d) [71].

Fig. 6.5 (continued) obtained from stopped-flow measurements, is plotted for intact H′ (*filled blue circle*), Nick A (*filled green circle*), Nick B (*filled pink circle*), and Nick C (*filled red circle*) substrates at 20°C. (Data from [81, 190]). (**d**) The relaxation rates, k_r, are plotted versus inverse temperature as obtained from three sets of T-jump measurements (*filled circle*, *open diamond*, and *open circle*) under conditions where the T-jump perturbation does not dissociate the IHF–H′ complex (data from [122, 185]). The plateau values of k_{obs} at high [IHF], obtained from stopped-flow measurements, are also shown (*filled triangle*; data from [81]). The *shaded areas* represent the range of base pair opening rates from imino proton exchange measurements (*pink*: A:T opening rates; *blue*: C:G opening rates; data from [71]). The *pink vertical bar* represents opening rates measured for an A:T pair in the H′ sequence, at the site of the kink (data from [176]). (**e**) The relaxation rates k_r, obtained from T-jump measurements on IHF–NickC (*filled green circle*), IHF–TT8AT (*filled red circle*), and IHF–H′44A (*filled blue circle*), are plotted versus inverse temperature. Plateau values of k_{obs} at high [IHF], from stopped-flow measurements on IHF–NickC, are also shown (*filled green triangle*). (**f**) The relaxation rates k_r, obtained from T-jump measurements on αK5A-H′ (*filled red circle*), βK84A-H′ (*filled blue circle*), and αR21C-H′ (*filled green circle*), are plotted versus inverse temperature. The continuous lines in (**d**)–(**f**) are Arrhenius fits to the relaxation rates. The *dashed black lines* in (**e**) and (**f**) are the Arrhenius fit for the IHF–H′ relaxation rates, reproduced from (**d**)

6.4.3 *Insight into the Transition State Ensemble Probed with Ionic Strength Dependence on the DNA-Bending Rate*

The thermodynamic complexity revealed in the ITC measurements of the IHF–H′ complex [162] was further probed by Vivas et al. [123], in an attempt to parse the observed ionic strength dependence on the binding affinity into the individual microscopic steps of bimolecular association/dissociation and unimolecular bending/unbending. T-jump measurements in the range of 100–500 mM KCl showed a nonlinear dependence of the observed relaxation rates on increasing [salt]. Below ~250 mM KCl, the observed relaxation was unimolecular, with the relaxation rate nearly independent of [KCl]. Above ~300 mM KCl, dissociation of the IHF–DNA complex became significant, and the observed relaxation process included contributions from the bimolecular association/dissociation step, with the relaxation rate decreasing with increasing [KCl], with a slope of ~−2.3 in a log(rate) versus log([KCl]) plot [123]. Based on stopped-flow measurements, Dhavan et al. [79] reported a weak salt dependence for the on-rates, with a slope of ~−2 and a stronger salt dependence for the off-rates, with a slope of ~4.

A unified description of the salt dependence of the equilibrium binding affinities and relaxation rates for the IHF–H′ complex was obtained in terms of the salt dependence of the microscopic rates corresponding to the bimolecular association/dissociation and unimolecular bending/unbending [185]. This analysis indicated that approximately five of the total 8–9 ions are released in the bimolecular step, with the bulk of the ionic effects appearing in the microscopic step corresponding to the disruption of the nonspecific complex [185]. This result is unexpected, since the extent to which the H′ substrate makes contact with the protein in the fully wrapped specific complex is significantly higher than in the nonspecific complex. A plausible explanation for the net release of more ions in the bimolecular step than during the unimolecular bending of DNA and wrapping around the protein is that net counterion release during the unimolecular step is reduced as a result of uptake and release of ions by the protein during conformational rearrangements and salt bridge exchange in the protein–DNA complex.

6.5 Role of DNA Bendability in the Recognition Mechanism

A thermodynamic linkage between increased probability for cyclization and increased binding affinity by *E. coli* protein CAP [186, 187] and HPV E2 protein [53] have demonstrated that if protein binding results in DNA bending, then the protein will display a higher affinity for easily bent or "prebent" DNA. Another example of enhanced affinity for prebent DNA comes from binding studies of promoter sequences by TBP [188, 189]. The question persists: is this enhanced affinity reflected in an increase in the rate of DNA bending, or a decrease in the rate of DNA unbending? Is the DNA already bent in the transition state ensemble?

The role of sequence-dependent DNA flexibility in the ability of IHF to recognize its cognate site and the nature of the transition state ensemble were further examined by introducing (1) nicks in the sugar–phosphate backbone and (2) mismatches to create internal loops, at or near the site of the kinks in the IHF–H′ complex (Fig. 6.5a).

6.5.1 Nicks in DNA at or near the Site of the Kinks Increase the Bending Rate

Sugimura and Crothers introduced nicks in the sugar–phosphate backbone (Fig. 6.5a) at (1) position A (located 1 bp removed from the site of the kink near the A-tract in the H′ sequence), (2) position B (located in between the two kink sites), and (3) position C (located at the site of the other kink). Nick A substrate bound with nearly the same affinity for IHF as the intact H′, Nick B bound with approximately twofold higher affinity [81], and Nick C with approximately fourfold higher affinity [128, 190]. The stopped-flow measurements showed nearly fourfold increase in the bending rates for both Nick A and Nick C, while the bending rates for Nick B were unchanged in comparison with the intact H′ substrate (Fig. 6.5c). Thus, making the DNA more flexible at the site of either of the kinks accelerates DNA bending by reducing the energetic cost for kinking DNA, while a nick where the DNA is not bent in complex leaves the bending rate unchanged. Although both Nick A and Nick C enhanced the bending rates, only Nick C showed an enhanced binding affinity, indicating that the increased flexibility of the Nick A substrate is fortuitously compensated elsewhere in the bent complex in such a way as to accelerate the unbending/dissociation rate by roughly the same factor.

The effect of nicks on the binding affinity and the bending rates were modest, and the activation enthalpy for the bending of the Nick C substrate was very similar to that for the intact H′ DNA (Fig. 6.5e). These results are consistent with previous studies on the structure of nicked DNA that indicate essentially straight DNA in solution, highlighting the importance of stacking in maintaining the helical conformation [70, 191–193].

The cocrystal structure of the IHF–H′ complex was obtained with the Nick A substrate (Fig. 6.3b), to facilitate crystal packing [23]. A nick at this site was shown not to affect significantly the binding affinity [81, 171], or the bent conformation in complex with IHF [171]. Rice and coworkers obtained the cocrystal structure of IHF bound to a doubly nicked H′ substrate, with nicks at positions A and C [128], as well as the cocrystal structure of AHU bound to a tight-binding oligomer with phased single-T insertions, spaced approximately the same distance apart as the kinks in the IHF–H′ structure [124] (Fig. 6.3c). An overlay of these structures (Fig. 6.6) provided a picture of how the nick at position C in H′ relieves the strain in the complex in a manner very similar to DNA substrates with single-T insertions

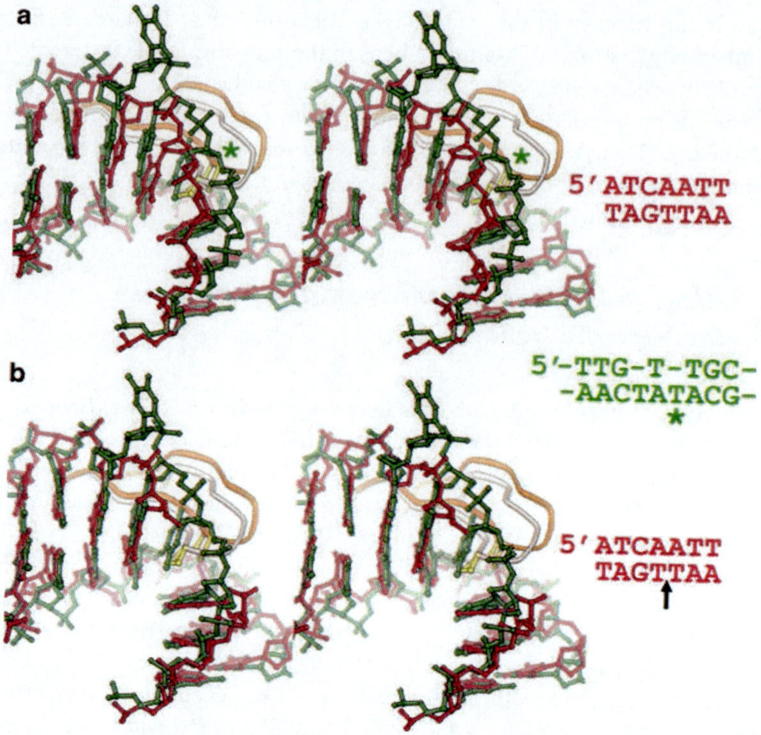

Fig. 6.6 Comparison of DNA structures in complex with IHF and AHU. (**a**) A stereoview of a superposition of kinked DNA from the IHF–DNA (*pink*; PDB code: 1IHF) and the AHU–DNA (*green*; PDB code: 1P78) structures shown in Fig. 6.3a, zoomed in on the site of the kink located away from the nick in the DNA substrate for IHF. The tips of the β-ribbon arms of IHF and AHU are shown in *white* and *gold*, respectively, with the intercalating proline residues in *yellow*. The DNA substrate for the AHU complex (Fig. 6.3c) has two single-T insertions near the site of each kink, one of which is shown flipped out in the crystal structure, and the other, marked by an *asterisk*, is intercalated. (**b**) A stereoview of a superposition of kinked DNA from another IHF–DNA structure (*pink*; PDB code: 2HT0), with doubly nicked DNA at positions indicated by Nick A and Nick C in Fig. 6.5a, and the AHU–DNA structure (*green*; PDB code: 1P78). Nick C is directly at the site of the kink. This picture illustrates that the strain in the kinked DNA backbone is relieved in a similar manner when there is a nick or an extra T at the site of the kink. Figure reprinted from [128], with permission from Elsevier

in complex with AHU [128]. Based on the structural information and binding affinity measurements, Rice and coworkers suggested that the higher binding affinity of IHF for the Nick C substrate is the result of a release of ~0.8 kcal/mol of backbone strain in the bent complex, owing to a nick directly at the site of the kink at position C [128]. The observation that the bending rate for Nick C increases by nearly the same factor as the binding affinity, in comparison with the intact H′ substrate, suggests that the bending strain in the DNA substrate is already relieved in the transition state, thus also lowering the free energy barrier for DNA bending by the same amount.

6.5.2 How Flexible/Bendable Are Inserted Mismatches at the Site of the Kinks?

Grove and coworkers [125, 126] carried out a detailed study of the effect of sequence-dependent variations in the flexibility of DNA on binding affinity for DNA-bending proteins. The binding of IHF, HU, and the eukaryotic HMGB1 was measured for DNA substrates whose flexibility was enhanced by mismatches that were inserted at various sites along the DNA. When compared to perfect DNA duplexes, all three proteins showed higher affinity for DNA duplexes containing tandem mismatches at two separate sites, creating four nucleotide (4-nt) loops. Optimal increase in affinity was observed when the 4-nt loops were separated by 8–9 bp [126]. As an example, a tenfold increase in binding affinity was obtained for the TT8AT variant of the H′ sequence, with two mismatches separated by 8 bp; one mismatch at the TT site coincident with the kink on the A-tract side of the H′ sequence, and another mismatch at the AT site located at the edge of the WATCAR consensus sequence, and shifted 1 bp in the 3′ direction from the site of the other kink, so as not to disturb the consensus region (Fig. 6.5a).

T-jump measurements on the IHF–TT8AT complex yielded an approximately sevenfold increase in the DNA-bending rates (Fig. 6.5e). Therefore, as in the case of the Nick C substrate, nearly all the increase in the stability of the IHF–TT8AT complex (~1.4 kcal/mol) is reflected in a corresponding smaller free energy barrier for bending/kinking DNA. These results demonstrate that any weakening of the base-pairing/stacking interactions at or near the site of the kinks enhances the binding affinities primarily because of the ease with which the DNA can be deformed, and which appears in accelerated bending rates. An interesting experiment to further probe the contribution of enhanced DNA flexibility to the bending step would involve design of H′ substrates with single-T insertions near the kink sites. The ~1,000-fold increase in binding affinity of AHU for DNA substrates with single-T insertions, reported by Swinger and Rice [128], is already an indication that weakening the stacking interactions at the site of the kinks may significantly accelerate the bending rates.

6.5.3 IHF–H′ Interactions that Stabilize the Bent Complex are Made After the DNA Bending Step

What role does IHF play in the DNA bending step? What interactions between IHF and H′, that stabilize the bent complex, are already present in the transition state ensemble separating the nonspecific from the specific complex? One strategy to probe the nature of the transition state ensemble is to perturb the complex by modifying the DNA or protein and observing the effects on transition rates. Any perturbation of stabilizing interactions that are formed after the system passes through the rate-limiting transition state should have no effect on the observed DNA bending rate (as long as the flexibility of DNA at the site of the kinks remains unmodified). On the other hand, any perturbation of interactions that are already formed in the transition state should affect the bending rate by the same factor as the binding affinity. Such a strategy would be analogous to the ϕ-value analysis pioneered by Fersht and coworkers to probe the transition state ensemble for protein folding [194], and which has also been applied recently to examine the nature of the encounter complex in the binding of the protein HPV16 E2C to DNA [195].

For this purpose, a particularly interesting modification would be the H′44A sequence in which a single mutation of A for T at position 44 of the TTR consensus region of H′ in phage λDNA (Fig. 6.5a), results in a ~100-fold decrease in the binding affinity [196]. Crystallographic studies from the Rice group indicate that the TTG trinucleotide in the original H′ consensus sequence results in an over-twist of the DNA helix at the TG step that allows a chain of salt bridges (ionic interactions) to be made with three amino acid residues of IHF [196]. In the TAG sequence, the twist is spread out more evenly, thus disrupting the specific ionic interactions. This result illustrates how sequence-dependent variation in the "twistability" of DNA is exploited by IHF to recognize a part of its binding site. Kinetics measurements on the IHF–H′44A complex showed that, despite the significant decrease in the binding affinity of the H′44A substrate relative to H′, the bending rates remained unchanged (Fig. 6.5e), indicating that the specific interactions that IHF makes with the TTR consensus site of H′ are not yet made in the transition state ensemble [185].

Similar conclusions were drawn from measurements of H′ bending kinetics in complex with IHF mutants that perturb electrostatic interactions between the positively charged wrapping surface of IHF and the bent DNA, by substitution of positively charged residues for neutral residues at locations where IHF makes close contact with the DNA backbone. Kinetics measurements on three such mutants (Fig. 6.5b), αK5A (Lys →Ala at position 5 of the α-chain), βK84A (Lys →Ala at position 84 of the β-chain), and αR21C (Arg →Cys at position 21 of the α-chain), yielded less than twofold change in the DNA-bending rates (Fig. 6.5f), despite nearly 60-fold decrease in binding affinity of αK5A and βK84A for H′, again indicating that each of these contacts are made after the transition state ensemble [185].

6.6 Recognition of Cognate Site by IHF Occurs in Two Steps

The microsecond sensitivity of the T-jump technique has begun to reveal two distinct phases in the kinetics traces for the IHF–H′ system (Fig. 6.7a), with a previously undetected fast phase appearing at ~100 μs, and a slower phase between ~1 and 10 ms (Fig. 6.7d) [185]. Thus, the mechanism by which IHF recognizes and binds tightly to its cognate site appears to involve at least two steps (Fig. 6.7c). The origin of the fast phase remains to be fully investigated. Kinetics measurements on the TT8AT substrate (Fig. 6.7e), together with two additional substrates, the TT-loop (with the 4-nt loop located on only one kink site), and the AT-loop (with the 4-nt

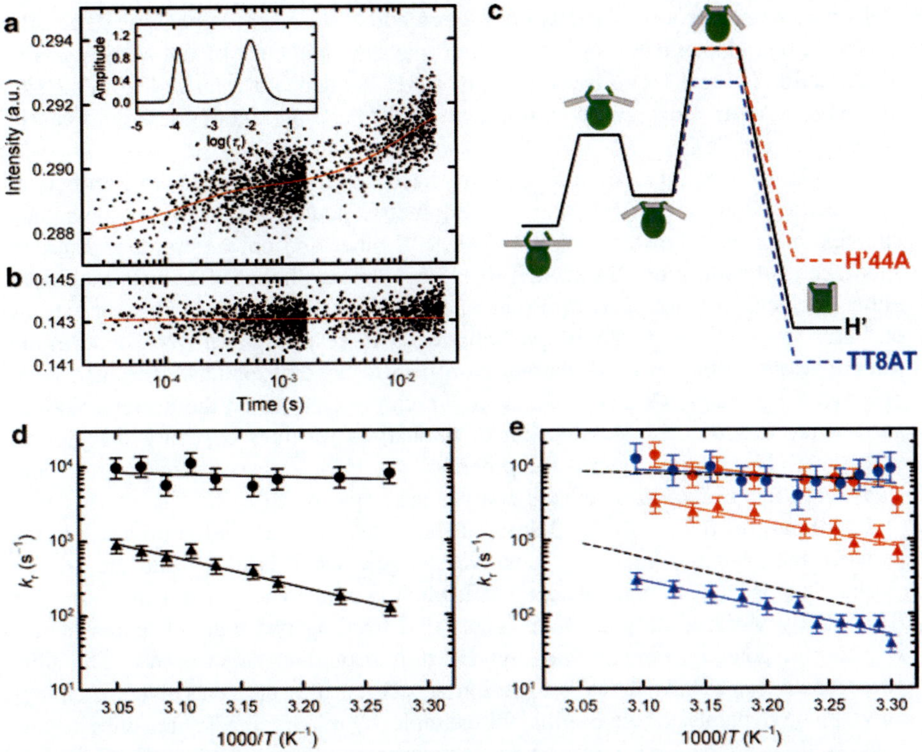

Fig. 6.7 Biphasic kinetics observed for the unimolecular bending step in the IHF–H′ complex. (**a**) Relaxation kinetics trace in response to a ~7°C T-jump (from 25°C to 32°C) for the IHF–H′ complex, probed by fluorescence changes of the donor (fluorescein) at 520 nm, is plotted versus time on a logarithmic scale. The *red line* represents a fit to the relaxation kinetics in terms of a distribution of relaxation times (data from [185]). (*Inset*) The distribution of relaxation times ($\tau_r = k_r^{-1}$) that best describes the relaxation kinetics shows two distinct peaks at ~100 μs and ~1 ms. (**b**) Measurements on a control sample, with donor-only labeled strand of DNA, in the absence of IHF, show no kinetics in the time range of 30 μs–10 ms. The *red line* represents a horizontal fit to the data points. (**c**) A free energy schematic illustrating two steps in the bending kinetics. The first, rapid step, appears to be independent of variations in the H′ sequence. The second, slow step, is unaffected by the H′44A mutation but gets faster for the TT8AT substrate, by roughly the same factor as the increase in binding affinity. (**d, e**) Relaxation rates for the two components (*circles*: fast phase; *triangles*: slow phase) are plotted as a function of inverse temperature for IHF–H′ (*filled black circle, filled black triangle*), IHF–TT8AT (*filled red circle, filled red triangle*), and IHF–H′44A (*filled blue circle, filled blue triangle*). The continuous lines represent Arrhenius fits to the two separate components for each sample. The *dashed black lines* in (**e**) represent the Arrhenius fits to the two components of the IHF–H′ complex, reproduced from (**d**)

loop located on the other kink site), provide a clue. All three variants of the H′ substrate showed two phases [185]. However, in each case, only the slower phase was accelerated by the insertion of the 4-nt loop, whether on one kink site or the other or both, while the fast phase remained unchanged [185]. These results suggest that the two steps do not reflect sequential bending of the two flanking arms of

DNA but, instead, from a partially bent intermediate state, whose bending dynamics appear to be independent of DNA flexibility at the kink sites. In H′44A, where the modification in the TTR consensus site of the H′ sequence is removed from the site of the kinks, neither of the two components in the bending kinetics are significantly affected (Fig. 6.7e).

One plausible explanation for the rapid phase is that it corresponds to a sequence-nonspecific bending of DNA, due to asymmetric neutralization of the phosphate charges on one face of the DNA [197–199]. Maher and coworkers demonstrated spontaneous bending of DNA when the charge on one helical face was modified, either by incorporation of neutral phosphate analogs [200] or by tethering cations on bases to form ions pairs with phosphates [201], resulting in a DNA bend in the direction toward the reduced charge. Nonspecific bending of DNA has also been observed in protein–DNA complexes, in which the charges on the protein binding surface are modified by substitution of uncharged residues by charged residues, with the DNA bending toward the protein if the excess charges are positive, and away from the protein if the excess charges are negative [202–205].

In the case of IHF, both the β-arms of the protein as well as the protein surface between the arms are lined with positive charges, and it is likely that the protein partially bends DNA even in its nonspecific binding mode, with perhaps the β-arms of IHF weakly wrapped around DNA. This rapid bending phase could represent fast scanning of potential binding sites by IHF during random target search. The slow phase would then be the ultimate recognition step. Direct measurements of the wrapping/unwrapping dynamics of the β-arms, for example, by introducing Trp residues at suitable positions on the arms, could be a key experiment to resolve the origin of the fast phase, and to add insight into DNA binding site recognition and bending by IHF.

6.7 Concluding Remarks

This chapter has discussed recent progress in monitoring the DNA binding and bending dynamics in site-specific recognition by IHF, an architectural protein from the eubacterial family of DNA-bending proteins ubiquitous in prokaryotes. This family of proteins provides a paradigm for investigating indirect readout of DNA sites and the mechanics of DNA bending. IHF recognizes several sites on phage λ DNA, and sharply bends the DNA at its cognate site by nearly 180° over about 35 bp, creating two kinks in the bent DNA stabilized by intercalation of highly conserved proline residues, located on two β-ribbon arms that wrap around the DNA in the complex.

Stopped-flow measurements on the IHF–H′ complex, together with laser T-jump studies, provided the first direct observation of the unimolecular bending kinetics of DNA. These measurements resolved a long-standing question as to whether binding and bending of DNA by DNA-bending proteins must occur in a concerted manner, or can be sequential, and demonstrated distinct binding and bending steps. DNA bending in the IHF–H′ complex occurs on the same time scale as thermal disruption of a single base pair in B-DNA, indicating that spontaneous bending/kinking of DNA by thermal disruption of base-pairing/stacking interactions may be the rate-limiting step in the DNA conformational change responsible for high-affinity recognition of the specific binding site. Modifications in the H′ sequence that increase its flexibility at the site of the kinks accelerate the bending rates, as expected because of the reduced energetic cost of bending DNA. On the other hand, modification in a consensus region well removed from the site of the kink, designed to perturb specific protein–DNA contacts, leaves the bending rates unchanged despite a 100-fold decrease in the binding affinity. Similarly, IHF mutants have been prepared with neutral residues replacing positively charged residues at sites where the flanking DNA arms on either side of the kinks make close contact with the protein. These mutants do not affect DNA bending rates, although the binding affinity decreases ~60-fold.

T-jump measurements also revealed a previously undetected rapid phase in the bending kinetics, occurring on time scales of ~100 ms. In contrast to the relaxation rates for the slow phase, which are affected by modifications in the DNA that make it more flexible at the site of the kinks, the relaxation rates for the fast phase appear to be unaffected. An attractive possibility, that remains to be explored, is that this rapid phase corresponds to the wrapping and unwrapping of the β-arms of the protein in a nonspecific binding mode, as IHF scans potential binding sites.

IHF–H′ is the only protein–DNA complex for which the rates of the unimolecular bending step have been reported within the time resolution of stopped -flow. It is possible that the rather severe bend induced in DNA by IHF, together with the fact that the DNA must kink at two sites, makes for slow (millisecond) recognition of its binding site. It is unknown how quickly DNA bends during formation of complexes with other DNA-bending proteins. An outstanding question remains: how fast is the recognition step in comparison with the time required for protein scanning in the vicinity of a potential binding site. Direct visualization of protein sliding on DNA has yielded one-dimensional diffusion constants in the range of $\sim 2\times10^3$ to 3×10^6 bp^2/s [206–210], which correspond to stepping times per bp of ~200 ns to ~500 μs. Sequence-specific DNA-binding proteins must recognize their binding sites faster than they diffuse away. The experimental tools at hand now present an exciting opportunity to gain insight into the time scales of binding site recognition.

Acknowledgements We thank the present and former members of the Ansari laboratory: Paula Vivas, Yogambigai Velmurugu and Ranjani Narayanan, for their help and for many discussions. We are grateful to James Maher and Phoebe Rice for critical reading of the manuscript and for their incisive comments and suggestions. We are especially indebted to Phoebe Rice for her offer and help in making all illustrations of protein structures shown in this chapter. A.A acknowledges support from the National Science Foundation (MCB-0721937). S.V.K. acknowledges support from the American Heart Association (AHA 0730254N).

References

1. Alberts B, Bray D, Lewis J, Raff M, Roberts K, Watson JD (1989) Molecular biology of the cell. Garland, New York
2. Riggs AD, Bourgeois S, Cohn M (1970) The lac repressor-operator interaction. 3. Kinetic studies. J Mol Biol 53:401–417

3. Berg OG, Winter RB, von Hippel PH (1981) Diffusion-driven mechanisms of protein translocation on nucleic acids. 1. Models and theory. Biochemistry 20:6929–6948
4. Winter RB, Berg OG, von Hippel PH (1981) Diffusion-driven mechanisms of protein translocation on nucleic acids. 3. The *Escherichia coli* lac repressor–operator interaction: kinetic measurements and conclusions. Biochemistry 20:6961–6977
5. von Hippel PH, Berg OG (1989) Facilitated target location in biological systems. J Biol Chem 264:675–678
6. Stanford NP, Szczelkun MD, Marko JF, Halford SE (2000) One- and three-dimensional pathways for proteins to reach specific DNA sites. EMBO J 19:6546–6557
7. Halford SE, Marko JF (2004) How do site-specific DNA-binding proteins find their targets? Nucleic Acids Res 32:3040–3052
8. Spolar RS, Record MT Jr (1994) Coupling of local folding to site-specific binding of proteins to DNA. Science 263:777–784
9. Garvie CW, Wolberger C (2001) Recognition of specific DNA sequences. Mol Cell 8:937–946
10. Kalodimos CG, Biris N, Bonvin AM, Levandoski MM, Guennuegues M, Boelens R, Kaptein R (2004) Structure and flexibility adaptation in nonspecific and specific protein-DNA complexes. Science 305:386–389
11. von Hippel PH (2004) Biochemistry. Completing the view of transcriptional regulation. Science 305:350–352
12. Slutsky M, Mirny LA (2004) Kinetics of protein-DNA interaction: facilitated target location in sequence-dependent potential. Biophys J 87:4021–4035
13. Rice PA (2008) Introduction. In: Correll CC, Rice PA, Correll CC (eds) Protein-nucleic acid interactions. Royal Society of Chemistry, Cambridge
14. Beamer LJ, Pabo CO (1992) Refined 1.8 A crystal structure of the lambda repressor-operator complex. J Mol Biol 227:177–196
15. Aggarwal AK, Rodgers DW, Drottar M, Ptashne M, Harrison SC (1988) Recognition of a DNA operator by the repressor of phage 434: a view at high resolution. Science 242:899–907
16. Otwinowski Z, Schevitz RW, Zhang RG, Lawson CL, Joachimiak A, Marmorstein RQ, Luisi BF, Sigler PB (1988) Crystal structure of trp repressor/operator complex at atomic resolution. Nature 335:321–329
17. Ellenberger TE, Brandl CJ, Struhl K, Harrison SC (1992) The GCN4 basic region leucine zipper binds DNA as a dimer of uninterrupted alpha helices: crystal structure of the protein–DNA complex. Cell 71:1223–1237
18. Foster MP, Wuttke DS, Radhakrishnan I, Case DA, Gottesfeld JM, Wright PE (1997) Domain packing and dynamics in the DNA complex of the N-terminal zinc fingers of TFIIIA. Nat Struct Biol 4:605–608
19. Lawson CL, Berman HM (2008) Indirect readout of DNA sequence by proteins. In: Rice PA, Correll CC (eds) Protein-nucleic acid interactions. Royal Society of Chemistry, Cambridge
20. Luger K, Mader AW, Richmond RK, Sargent DF, Richmond TJ (1997) Crystal structure of the nucleosome core particle at 2.8 A resolution. Nature 389:251–260
21. Richmond TJ, Davey CA (2003) The structure of DNA in the nucleosome core. Nature 423:145–150
22. Werner MH, Gronenborn AM, Clore GM (1996) Intercalation, DNA kinking, and the control of transcription. Science 271:778–784
23. Rice PA, Yang S, Mizuuchi K, Nash HA (1996) Crystal structure of an IHF-DNA complex: a protein-induced DNA U-turn. Cell 87:1295–1306
24. Mouw KM, Rice PA (2007) Shaping the *Borrelia burgdorferi* genome: crystal structure and binding properties of the DNA-bending Hbb. Mol Microbiol 63:1319–1330
25. Kalodimos CG, Bonvin AM, Salinas RK, Wechselberger R, Boelens R, Kaptein R (2002) Plasticity in protein-DNA recognition: lac repressor interacts with its natural operator 01 through alternative conformations of its DNA-binding domain. EMBO J 21:2866–2876
26. Kim Y, Geiger JH, Hahn S, Sigler PB (1993) Crystal structure of a yeast TBP/TATA-box complex. Nature 365:512–520

27. Winkler FK, Banner DW, Oefner C, Tsernoglou D, Brown RS, Heathman SP, Bryan RK, Martin PD, Petratos K, Wilson KS (1993) The crystal structure of EcoRV endonuclease and of its complexes with cognate and non-cognate DNA fragments. EMBO J 12:1781–1795
28. Obmolova G, Ban C, Hsieh P, Yang W (2000) Crystal structures of mismatch repair protein MutS and its complex with a substrate DNA. Nature 407:703–710
29. Lamers MH, Perrakis A, Enzlin JH, Winterwerp HH, de Wind N, Sixma TK (2000) The crystal structure of DNA mismatch repair protein MutS binding to a G x T mismatch. Nature 407:711–717
30. Howard J (2001) Mechanics of motor proteins and the cytoskeleton. Sinauer Associates, Sunderland
31. Hogan M, Dattagupta N, Crothers DM (1978) Transient electric dichroism of rod-like DNA molecules. Proc Natl Acad Sci USA 75:195–199
32. Hagerman PJ (1981) Investigation of the flexibility of DNA using transient electric birefringence. Biopolymers 20:1503–1535
33. Hogan M, LeGrange J, Austin B (1983) Dependence of DNA helix flexibility on base composition. Nature 304:752–754
34. Shore D, Langowski J, Baldwin RL (1981) DNA flexibility studied by covalent closure of short fragments into circles. Proc Natl Acad Sci USA 78:4833–4837
35. Levene SD, Crothers DM (1986) Ring closure probabilities for DNA fragments by Monte Carlo simulation. J Mol Biol 189:61–72
36. Crothers DM, Drak J, Kahn JD, Levene SD (1992) DNA bending, flexibility, and helical repeat by cyclization kinetics. Methods Enzymol 212:3–29
37. Bustamante C, Marko JF, Siggia ED, Smith S (1994) Entropic elasticity of lambda-phage DNA. Science 265:1599–1600
38. Bustamante C, Smith SB, Liphardt J, Smith D (2000) Single-molecule studies of DNA mechanics. Curr Opin Struct Biol 10:279–285
39. Marko JF, Siggia ED (1995) Stretching DNA. Macromolecules 28:8759–8770
40. Baumann CG, Smith SB, Bloomfield VA, Bustamante C (1997) Ionic effects on the elasticity of single DNA molecules. Proc Natl Acad Sci USA 94:6185–6190
41. Roychoudhury M, Sitlani A, Lapham J, Crothers DM (2000) Global structure and mechanical properties of a 10-bp nucleosome positioning motif. Proc Natl Acad Sci USA 97:13608–13613
42. Zhang Y, Crothers DM (2003) High-throughput approach for detection of DNA bending and flexibility based on cyclization. Proc Natl Acad Sci USA 100:3161–3166
43. Satchwell SC, Drew HR, Travers AA (1986) Sequence periodicities in chicken nucleosome core DNA. J Mol Biol 191:659–675
44. Widom J (2001) Role of DNA sequence in nucleosome stability and dynamics. Q Rev Biophys 34:269–324
45. Drew HR, Travers AA (1985) DNA bending and its relation to nucleosome positioning. J Mol Biol 186:773–790
46. Lowary PT, Widom J (1997) Nucleosome packaging and nucleosome positioning of genomic DNA. Proc Natl Acad Sci USA 94:1183–1188
47. Lowary PT, Widom J (1998) New DNA sequence rules for high affinity binding to histone octamer and sequence-directed nucleosome positioning. J Mol Biol 276:19–42
48. Segal E, Fondufe-Mittendorf Y, Chen L, Thastrom A, Field Y, Moore IK, Wang JP, Widom J (2006) A genomic code for nucleosome positioning. Nature 442:772–778
49. Anderson JE, Ptashne M, Harrison SC (1987) Structure of the repressor-operator complex of bacteriophage 434. Nature 326:846–852
50. Koudelka GB, Harrison SC, Ptashne M (1987) Effect of non-contacted bases on the affinity of 434 operator for 434 repressor and Cro. Nature 326:886–888
51. Hogan ME, Austin RH (1987) Importance of DNA stiffness in protein-DNA binding specificity. Nature 329:263–266
52. Zimmerman JM, Maher LJ 3rd (2003) Solution measurement of DNA curvature in papillomavirus E2 binding sites. Nucleic Acids Res 31:5134–5139

53. Zhang Y, Xi Z, Hegde RS, Shakked Z, Crothers DM (2004) Predicting indirect readout effects in protein-DNA interactions. Proc Natl Acad Sci USA 101:8337–8341
54. Kim SS, Tam JK, Wang AF, Hegde RS (2000) The structural basis of DNA target discrimination by papillomavirus E2 proteins. J Biol Chem 275:31245–31254
55. Hardwidge PR, Zimmerman JM, Maher LJ 3rd (2000) Design and calibration of a semi-synthetic DNA phasing assay. Nucleic Acids Res 28:E102
56. Crick FH, Klug A (1975) Kinky helix. Nature 255:530–533
57. Hogan ME, Rooney TF, Austin RH (1987) Evidence for kinks in DNA folding in the nucleosome. Nature 328:554–557
58. Cloutier TE, Widom J (2004) Spontaneous sharp bending of double-stranded DNA. Mol Cell 14:355–362
59. Cloutier TE, Widom J (2005) DNA twisting flexibility and the formation of sharply looped protein-DNA complexes. Proc Natl Acad Sci USA 102:3645–3650
60. Yan J, Marko JF (2004) Localized single-stranded bubble mechanism for cyclization of short double helix DNA. Phys Rev Lett 93:108108
61. Wiggins PA, Phillips R, Nelson PC (2005) Exact theory of kinkable elastic polymers. Phys Rev E 71:021909
62. Du Q, Smith C, Shiffeldrim N, Vologodskaia M, Vologodskii A (2005) Cyclization of short DNA fragments and bending fluctuations of the double helix. Proc Natl Acad Sci USA 102:5397–5402
63. Forties RA, Bundschuh R, Poirier MG (2009) The flexibility of locally melted DNA. Nucleic Acids Res 37:4580–4586
64. Du Q, Kotlyar A, Vologodskii A (2008) Kinking the double helix by bending deformation. Nucleic Acids Res 36:1120–1128
65. Demurtas D, Amzallag A, Rawdon EJ, Maddocks JH, Dubochet J, Stasiak A (2009) Bending modes of DNA directly addressed by cryo-electron microscopy of DNA minicircles. Nucleic Acids Res 37:2882–2893
66. Wiggins PA, van der Heijden T, Moreno-Herrero F, Spakowitz A, Phillips R, Widom J, Dekker C, Nelson PC (2006) High flexibility of DNA on short length scales probed by atomic force microscopy. Nat Nanotechnol 1:137–141
67. Yuan C, Chen H, Lou XW, Archer LA (2008) DNA bending stiffness on small length scales. Phys Rev Lett 100:018102
68. Frank-Kamenetskii F (1971) Simplification of the empirical relationship between melting temperature of DNA, its GC content and concentration of sodium ions in solution. Biopolymers 10:2623–2624
69. Wartell RM, Benight AS (1982) Fluctuational base-pair opening in DNA at temperatures below the helix-coil transition region. Biopolymers 21:2069–2081
70. Protozanova E, Yakovchuk P, Frank-Kamenetskii MD (2004) Stacked–unstacked equilibrium at the nick site of DNA. J Mol Biol 342:775–785
71. Coman D, Russu IM (2005) A nuclear magnetic resonance investigation of the energetics of basepair opening pathways in DNA. Biophys J 89:3285–3292
72. Mills JB, Hagerman PJ (2004) Origin of the intrinsic rigidity of DNA. Nucleic Acids Res 32:4055–4059
73. Yakovchuk P, Protozanova E, Frank-Kamenetskii MD (2006) Base-stacking and base-pairing contributions into thermal stability of the DNA double helix. Nucleic Acids Res 34:564–574
74. SantaLucia J Jr (1998) A unified view of polymer, dumbbell, and oligonucleotide DNA nearest-neighbor thermodynamics. Proc Natl Acad Sci USA 95:1460–1465
75. Olson WK, Gorin AA, Lu XJ, Hock LM, Zhurkin VB (1998) DNA sequence-dependent deformability deduced from protein-DNA crystal complexes. Proc Natl Acad Sci USA 95:11163–11168
76. Olson WK, Zhurkin VB (2000) Modeling DNA deformations. Curr Opin Struct Biol 10:286–297
77. Perez-Howard GM, Weil PA, Beechem JM (1995) Yeast TATA binding protein interaction with DNA: fluorescence determination of oligomeric state, equilibrium binding, on-rate, and dissociation kinetics. Biochemistry 34:8005–8017

78. Parkhurst KM, Brenowitz M, Parkhurst LJ (1996) Simultaneous binding and bending of promoter DNA by the TATA binding protein: real time kinetic measurements. Biochemistry 35:7459–7465
79. Dhavan GM, Crothers DM, Chance MR, Brenowitz M (2002) Concerted binding and bending of DNA by *Escherichia coli* integration host factor. J Mol Biol 315:1027–1037
80. Hiller DA, Fogg JM, Martin AM, Beechem JM, Reich NO, Perona JJ (2003) Simultaneous DNA binding and bending by EcoRV endonuclease observed by real-time fluorescence. Biochemistry 42:14375–14385
81. Sugimura S, Crothers DM (2006) Stepwise binding and bending of DNA by *Escherichia coli* integration host factor. Proc Natl Acad Sci USA 103:18510–18514
82. Hoopes BC, LeBlanc JF, Hawley DK (1992) Kinetic analysis of yeast TFIID-TATA box complex formation suggests a multi-step pathway. J Biol Chem 267:11539–11547
83. Petri V, Hsieh M, Brenowitz M (1995) Thermodynamic and kinetic characterization of the binding of the TATA binding protein to the adenovirus E4 promoter. Biochemistry 34:9977–9984
84. Berg OG, Ehrenberg M (1982) Association kinetics with coupled three- and one-dimensional diffusion. Chain-length dependence of the association rate of specific DNA sites. Biophys Chem 15:41–51
85. Parkhurst KM, Richards RM, Brenowitz M, Parkhurst LJ (1999) Intermediate species possessing bent DNA are present along the pathway to formation of a final TBP-TATA complex. J Mol Biol 289:1327–1341
86. Delgadillo RF, Whittington JE, Parkhurst LK, Parkhurst LJ (2009) The TATA-binding protein core domain in solution variably bends TATA sequences via a three-step binding mechanism (dagger). Biochemistry 48:1801–1809
87. Tolic-Norrelykke SF, Rasmussen MB, Pavone FS, Berg-Sorensen K, Oddershede LB (2006) Stepwise bending of DNA by a single TATA-box binding protein. Biophys J 90:3694–3703
88. Jacobs-Palmer E, Hingorani MM (2007) The effects of nucleotides on MutS-DNA binding kinetics clarify the role of MutS ATPase activity in mismatch repair. J Mol Biol 366:1087–1098
89. Erskine SG, Baldwin GS, Halford SE (1997) Rapid-reaction analysis of plasmid DNA cleavage by the EcoRV restriction endonuclease. Biochemistry 36:7567–7576
90. Hopkins BB, Reich NO (2004) Simultaneous DNA binding, bending, and base flipping: evidence for a novel M.EcoRI methyltransferase-DNA complex. J Biol Chem 279:37049–37060
91. van den Broek B, Noom MC, Wuite GJ (2005) DNA-tension dependence of restriction enzyme activity reveals mechanochemical properties of the reaction pathway. Nucleic Acids Res 33:2676–2684
92. Ali BM, Amit R, Braslavsky I, Oppenheim AB, Gileadi O, Stavans J (2001) Compaction of single DNA molecules induced by binding of integration host factor (IHF). Proc Natl Acad Sci USA 98:10658–10663
93. Skoko D, Wong B, Johnson RC, Marko JF (2004) Micromechanical analysis of the binding of DNA-bending proteins HMGB1, NHP6A, and HU reveals their ability to form highly stable DNA-protein complexes. Biochemistry 43:13867–13874
94. McCauley M, Hardwidge PR, Maher LJ 3rd, Williams MC (2005) Dual binding modes for an HMG domain from human HMGB2 on DNA. Biophys J 89:353–364
95. Skoko D, Yoo D, Bai H, Schnurr B, Yan J, McLeod SM, Marko JF, Johnson RC (2006) Mechanism of chromosome compaction and looping by the *Escherichia coli* nucleoid protein Fis. J Mol Biol 364:777–798
96. Zhang J, McCauley MJ, Maher LJ 3rd, Williams MC, Israeloff NE (2009) Mechanism of DNA flexibility enhancement by HMGB proteins. Nucleic Acids Res 37:1107–1114
97. Dixit S, Singh-Zocchi M, Hanne J, Zocchi G (2005) Mechanics of binding of a single integration-host-factor protein to DNA. Phys Rev Lett 94:118101
98. Reinhard BM, Sheikholeslami S, Mastroianni A, Alivisatos AP, Liphardt J (2007) Use of plasmon coupling to reveal the dynamics of DNA bending and cleavage by single EcoRV restriction enzymes. Proc Natl Acad Sci USA 104:2667–2672

99. Riggs AD, Bourgeois S, Newby RF, Cohn M (1968) DNA binding of the lac repressor. J Mol Biol 34:365–368
100. Dyer RB, Gai F, Woodruff WH, Gilmanshin R, Callender RH (1998) Infrared studies of fast events in protein folding. Acc Chem Res 31:709–716
101. Hofrichter J (2001) Laser temperature-jump methods for studying folding dynamics. In: Murphy KP (ed) Methods in molecular biology. Humana Press, Totowa, NJ
102. Kubelka J (2009) Time-resolved methods in biophysics. 9. Laser temperature-jump methods for investigating biomolecular dynamics. Photochem Photobiol Sci 8:499–512
103. Munoz V, Thompson PA, Hofrichter J, Eaton WA (1997) Folding dynamics and mechanism of beta-hairpin formation. Nature 390:196–199
104. Gruebele M, Sabelko J, Ballew R, Ervin J (1998) Laser temperature jump induced protein refolding. Acc Chem Res 31:699–707
105. Thompson PA, Munoz V, Jas GS, Henry ER, Eaton WA, Hofrichter J (2000) The helix-coil kinetics of a heteropeptide. J Phys Chem B 104:378–389
106. Qiu L, Pabit SA, Roitberg AE, Hagen SJ (2002) Smaller and faster: the 20-residue Trp-cage protein folds in 4 micros. J Am Chem Soc 124:12952–12953
107. Hauser K, Krejtschi C, Huang R, Wu L, Keiderling TA (2008) Site-specific relaxation kinetics of a tryptophan zipper hairpin peptide using temperature-jump IR spectroscopy and isotopic labeling. J Am Chem Soc 130:2984–2992
108. Ansari A, Kuznetsov SV, Shen Y (2001) Configurational diffusion down a folding funnel describes the dynamics of DNA hairpins. Proc Natl Acad Sci USA 98:7771–7776
109. Proctor DJ, Ma H, Kierzek E, Kierzek R, Gruebele M, Bevilacqua PC (2004) Thermodynamics and kinetics of YNMG RNA hairpins: incorporation of 8-bromoguanosine leads to stabilization by enhancement of the folding rate. J Am Chem Soc 43:14004–14014
110. Brauns EB, Dyer RB (2005) Time-resolved infrared spectroscopy of RNA folding. Biophys J 89:3523–3530
111. Kuznetsov SV, Ren C, Woodson SA, Ansari A (2008) Loop dependence of the stability and dynamics of nucleic acid hairpins. Nucleic Acids Res 36:1098–1112
112. Bonnet G, Tyagi S, Libchaber A, Kramer FR (1999) Thermodynamic basis of the enhanced specificity of structured DNA probes. Proc Natl Acad Sci USA 96:6171–6176
113. Kim HD, Nienhaus GU, Ha T, Orr JW, Williamson JR, Chu S (2002) Mg2+-dependent conformational change of RNA studied by fluorescence correlation and FRET on immobilized single molecules. Proc Natl Acad Sci USA 99:4284–4289
114. Altan-Bonnet G, Libchaber A, Krichevsky O (2003) Bubble dynamics in double-stranded DNA. Phys Rev Lett 90:138101
115. Li G, Levitus M, Bustamante C, Widom J (2005) Rapid spontaneous accessibility of nucleosomal DNA. Nat Struct Mol Biol 12:46–53
116. Kubelka J, Eaton WA, Hofrichter J (2003) Experimental tests of villin subdomain folding simulations. J Mol Biol 329:625–630
117. Qiu L, Hagen SJ (2004) A limiting speed for protein folding at low solvent viscosity. J Am Chem Soc 126:3398–3399
118. Kuznetsov SV, Kozlov AG, Lohman TM, Ansari A (2006) Microsecond dynamics of protein-DNA interactions: direct observation of the wrapping/unwrapping kinetics of single-stranded DNA around the *E. coli* SSB tetramer. J Mol Biol 359:55–65
119. Hawkins ME, Pfleiderer W, Balis FM, Porter D, Knutson JR (1997) Fluorescence properties of pteridine nucleoside analogs as monomers and incorporated into oligonucleotides. Anal Biochem 244:86–95
120. Martin GT, Ujvari A, Liu C (2003) Evaluation of fluorescence spectroscopy methods for mapping melted regions of DNA along the transcription pathway. Methods Enzymol 371:13–33
121. Wojtuszewski K, Hawkins ME, Cole JL, Mukerji I (2001) HU binding to DNA: evidence for multiple complex formation and DNA bending. Biochemistry 40:2588–2598
122. Kuznetsov SV, Sugimura S, Vivas P, Crothers DM, Ansari A (2006) Direct observation of DNA bending/unbending kinetics in complex with DNA-bending protein IHF. Proc Natl Acad Sci USA 103:18515–18520

123. Vivas P, Kuznetsov SV, Ansari A (2008) New insights into the transition pathway from nonspecific to specific complex of DNA with *Escherichia coli* integration host factor. J Phys Chem B 112:5997–6007
124. Swinger KK, Lemberg KM, Zhang Y, Rice PA (2003) Flexible DNA bending in HU-DNA cocrystal structures. EMBO J 22:3749–3760
125. Grove A, Galeone A, Mayol L, Geiduschek EP (1996) On the connection between inherent DNA flexure and preferred binding of hydroxymethyluracil-containing DNA by the type II DNA-binding protein TF1. J Mol Biol 260:196–206
126. Grove A, Galeone A, Mayol L, Geiduschek EP (1996) Localized DNA flexibility contributes to target site selection by DNA-bending proteins. J Mol Biol 260:120–125
127. Swinger KK, Rice PA (2004) IHF and HU: flexible architects of bent DNA. Curr Opin Struct Biol 14:28–35
128. Swinger KK, Rice PA (2007) Structure-based analysis of HU-DNA binding. J Mol Biol 365:1005–1016
129. Aeling KA, Opel ML, Steffen NR, Tretyachenko-Ladokhina V, Hatfield GW, Lathrop RH, Senear DF (2006) Indirect recognition in sequence-specific DNA binding by *Escherichia coli* integration host factor: the role of DNA deformation energy. J Biol Chem 281:39236–39248
130. Nash HA, Robertson CA (1981) Purification and properties of the *Escherichia coli* protein factor required for lambda integrative recombination. J Biol Chem 256:9246–9253
131. Winkelman JW, Hatfield GW (1990) Characterization of the integration host factor binding site in the ilvPG1 promoter region of the ilvGMEDA operon of *Escherichia coli*. J Biol Chem 265:10055–10060
132. Polaczek P, Kwan K, Liberies DA, Campbell JL (1997) Role of architectural elements in combinatorial regulation of initiation of DNA replication in *Escherichia coli*. Mol Microbiol 26:261–275
133. Xin W, Feiss M (1993) Function of IHF in lambda DNA packaging. I. Identification of the strong binding site for integration host factor and the locus for intrinsic bending in cosB. J Mol Biol 230:492–504
134. Goodrich JA, Schwartz ML, McClure WR (1990) Searching for and predicting the activity of sites for DNA binding proteins: compilation and analysis of the binding sites for *Escherichia coli* integration host factor (IHF). Nucleic Acids Res 18:4993–5000
135. Hales LM, Gumport RI, Gardner JF (1994) Determining the DNA sequence elements required for binding integration host factor to two different target sites. J Bacteriol 176:2999–3006
136. Hales LM, Gumport RI, Gardner JF (1996) Examining the contribution of a dA+dT element to the conformation of *Escherichia coli* integration host factor-DNA complexes. Nucleic Acids Res 24:1780–1786
137. Lavoie BD, Chaconas G (1993) Site-specific HU binding in the Mu transpososome: conversion of a sequence-independent DNA-binding protein into a chemical nuclease. Genes Dev 7:2510–2519
138. Lavoie BD, Shaw GS, Millner A, Chaconas G (1996) Anatomy of a flexer-DNA complex inside a higher-order transposition intermediate. Cell 85:761–771
139. Aki T, Adhya S (1997) Repressor induced site-specific binding of HU for transcriptional regulation. EMBO J 16:3666–3674
140. Lyubchenko YL, Shlyakhtenko LS, Aki T, Adhya S (1997) Atomic force microscopic demonstration of DNA looping by GalR and HU. Nucleic Acids Res 25:873–876
141. Boubrik F, Rouviere-Yaniv J (1995) Increased sensitivity to gamma irradiation in bacteria lacking protein HU. Proc Natl Acad Sci USA 92:3958–3962
142. Li S, Waters R (1998) *Escherichia coli* strains lacking protein HU are UV sensitive due to a role for HU in homologous recombination. J Bacteriol 180:3750–3756
143. Kamashev D, Balandina A, Rouviere-Yaniv J (1999) The binding motif recognized by HU on both nicked and cruciform DNA. EMBO J 18:5434–5444
144. Castaing B, Zelwer C, Laval J, Boiteux S (1995) HU protein of *Escherichia coli* binds specifically to DNA that contains single-strand breaks or gaps. J Biol Chem 270:10291–10296

145. Kobryn K, Naigamwalla DZ, Chaconas G (2000) Site-specific DNA binding and bending by the *Borrelia burgdorferi* Hbb protein. Mol Microbiol 37:145–155
146. Johnson JB, Stella S, Heiss JK (2008) Bending and compaction of DNA by proteins. In: Rice PA, Correll CC (eds) Protein-nucleic acid interactions. Royal Society of Chemistry, Cambridge
147. van Noort J, Verbrugge S, Goosen N, Dekker C, Dame RT (2004) Dual architectural roles of HU: formation of flexible hinges and rigid filaments. Proc Natl Acad Sci USA 101:6969–6974
148. Sagi D, Friedman N, Vorgias C, Oppenheim AB, Stavans J (2004) Modulation of DNA conformations through the formation of alternative high-order HU-DNA complexes. J Mol Biol 341:419–428
149. Tanaka I, Appelt K, Dijk J, White SW, Wilson KS (1984) 3-A resolution structure of a protein with histone-like properties in prokaryotes. Nature 310:376–381
150. Vis H, Mariani M, Vorgias CE, Wilson KS, Kaptein R, Boelens R (1995) Solution structure of the HU protein from *Bacillus stearothermophilus*. J Mol Biol 254:692–703
151. Boelens R, Vis H, Vorgias CE, Wilson KS, Kaptein R (1996) Structure and dynamics of the DNA binding protein HU from *Bacillus stearothermophilus* by NMR spectroscopy. Biopolymers 40:553–559
152. White SW, Wilson KS, Appelt K, Tanaka I (1999) The high-resolution structure of DNA-binding protein HU from *Bacillus stearothermophilus*. Acta Crystallogr D Biol Crystallogr 55:801–809
153. Fernandez S, Rojo F, Alonso JC (1997) The *Bacillus subtilis* chromatin-associated protein Hbsu is involved in DNA repair and recombination. Mol Microbiol 23:1169–1179
154. Kamashev D, Rouviere-Yaniv J (2000) The histone-like protein HU binds specifically to DNA recombination and repair intermediates. EMBO J 19:6527–6535
155. Grove A, Saavedra TC (2002) The role of surface-exposed lysines in wrapping DNA about the bacterial histone-like protein HU. Biochemistry 41:7597–7603
156. Grove A (2003) Surface salt bridges modulate DNA wrapping by the type II DNA-binding protein TF1. Biochemistry 42:8739–8747
157. Kamau E, Tsihlis ND, Simmons LA, Grove A (2005) Surface salt bridges modulate the DNA site size of bacterial histone-like HU proteins. Biochem J 390:49–55
158. Record MT Jr, Anderson CF, Lohman TM (1978) Thermodynamic analysis of ion effects on the binding and conformational equilibria of proteins and nucleic acids: the roles of ion association or release, screening, and ion effects on water activity. Q Rev Biophys 11:103–178
159. Manning GS (1978) The molecular theory of polyelectrolyte solutions with applications to the electrostatic properties of polynucleotides. Q Rev Biophys 11:179–246
160. Lohman TM, Mascotti DP (1992) Thermodynamics of ligand-nucleic acid interactions. Methods Enzymol 212:400–424
161. Bloomfield VA, Crothers DM, Tinoco IJ (2000) Nucleic acids: structures, properties, and functions. University Science Books, Sausalito, CA
162. Holbrook JA, Tsodikov OV, Saecker RM, Record MT Jr (2001) Specific and non-specific interactions of integration host factor with DNA: thermodynamic evidence for disruption of multiple IHF surface salt-bridges coupled to DNA binding. J Mol Biol 310:379–401
163. Saecker RM, Record MT Jr (2002) Protein surface salt bridges and paths for DNA wrapping. Curr Opin Struct Biol 12:311–319
164. Vander Meulen KA, Saecker RM, Record MT Jr (2008) Formation of a wrapped DNA-protein interface: experimental characterization and analysis of the large contributions of ions and water to the thermodynamics of binding IHF to H′ DNA. J Mol Biol 377:9–27
165. Lohman TM, Overman LB, Ferrari ME, Kozlov AG (1996) A highly salt-dependent enthalpy change for *Escherichia coli* SSB protein-nucleic acid binding due to ion-protein interactions. Biochemistry 35:5272–5279
166. Cui T, Wei S, Brew K, Leng F (2005) Energetics of binding the mammalian high mobility group protein HMGA2 to poly(dA-dT)2 and poly(dA)-poly(dT). J Mol Biol 352:629–645

167. Jen-Jacobson L, Engler LE, Jacobson LA (2000) Structural and thermodynamic strategies for site-specific DNA binding proteins. Structure 8:1015–1023
168. Privalov PL, Dragan AI, Crane-Robinson C, Breslauer KJ, Remeta DP, Minetti CA (2007) What drives proteins into the major or minor grooves of DNA? J Mol Biol 365:1–9
169. Dragan AI, Read CM, Makeyeva EN, Milgotina EI, Churchill ME, Crane-Robinson C, Privalov PL (2004) DNA binding and bending by HMG boxes: energetic determinants of specificity. J Mol Biol 343:371–393
170. Khrapunov S, Brenowitz M, Rice PA, Catalano CE (2006) Binding then bending: a mechanism for wrapping DNA. Proc Natl Acad Sci USA 103:19217–19218
171. Lorenz M, Hillisch A, Goodman SD, Diekmann S (1999) Global structure similarities of intact and nicked DNA complexed with IHF measured in solution by fluorescence resonance energy transfer. Nucleic Acids Res 27:4619–4625
172. Hillisch A, Lorenz M, Diekmann S (2001) Recent advances in FRET: distance determination in protein-DNA complexes. Curr Opin Struct Biol 11:201–207
173. Holbrook JA, Capp MW, Saecker RM, Record MT Jr (1999) Enthalpy and heat capacity changes for formation of an oligomeric DNA duplex: interpretation in terms of coupled processes of formation and association of single-stranded helices. Biochemistry 38:8409–8422
174. Bosch D, Campillo M, Pardo L (2003) Binding of proteins to the minor groove of DNA: what are the structural and energetic determinants for kinking a basepair step? J Comput Chem 24:682–691
175. Gueron M, Leroy JL (1995) Studies of base pair kinetics by NMR measurement of proton exchange. Methods Enzymol 261:383–413
176. Dhavan GM, Lapham J, Yang S, Crothers DM (1999) Decreased imino proton exchange and base-pair opening in the IHF-DNA complex measured by NMR. J Mol Biol 288:659–671
177. Yan J, Kawamura R, Marko JF (2005) Statistics of loop formation along double helix DNAs. Phys Rev E Stat Nonlin Soft Matter Phys 71:061905
178. Lankas F, Lavery R, Maddocks JH (2006) Kinking occurs during molecular dynamics simulations of small DNA minicircles. Structure 14:1527–1534
179. Maher LJ 3rd (2006) DNA kinks available...if needed. Structure 14:1479–1480
180. Leger JF, Robert J, Bourdieu L, Chatenay D, Marko JF (1998) RecA binding to a single double-stranded DNA molecule: a possible role of DNA conformational fluctuations. Proc Natl Acad Sci USA 95:12295–12299
181. deHaseth PL, Zupancic ML, Record MT Jr (1998) RNA polymerase-promoter interactions: the comings and goings of RNA polymerase. J Bacteriol 180:3019–3025
182. deHaseth PL, Nilsen TW (2004) Molecular biology. When a part is as good as the whole. Science 303:1307–1308
183. Young BA, Gruber TM, Gross CA (2004) Minimal machinery of RNA polymerase holoenzyme sufficient for promoter melting. Science 303:1382–1384
184. Travers A (2005) DNA dynamics: bubble 'n' flip for DNA cyclisation? Curr Biol 15:R377–379
185. Vivas P (2009) Mechanism of integration host factor, a DNA-bending protein, probed with laser temperature-jump, in Physics. University of Illinois at Chicago, Chicago
186. Kahn JD, Crothers DM (1992) Protein-induced bending and DNA cyclization. Proc Natl Acad Sci USA 89:6343–6347
187. Kahn JD, Yun E, Crothers DM (1994) Detection of localized DNA flexibility. Nature 368:163–166
188. Parvin JD, McCormick RJ, Sharp PA, Fisher DE (1995) Pre-bending of a promoter sequence enhances affinity for the TATA-binding factor. Nature 373:724–727
189. Starr DB, Hoopes BC, Hawley DK (1995) DNA bending is an important component of site-specific recognition by the TATA binding protein. J Mol Biol 250:434–446
190. Sugimura S (2005) Kinetic and steady-state studies of binding and bending of lambda phage DNA by integration host factor. Ph.D. thesis, Chemistry Department, Yale University, New Haven, CT

191. Aymami J, Coll M, van der Marel GA, van Boom JH, Wang AH, Rich A (1990) Molecular structure of nicked DNA: a substrate for DNA repair enzymes. Proc Natl Acad Sci USA 87:2526–2530
192. Mills JB, Cooper JP, Hagerman PJ (1994) Electrophoretic evidence that single-stranded regions of one or more nucleotides dramatically increase the flexibility of DNA. Biochemistry 33:1797–1803
193. Kozerski L, Mazurek AP, Kawecki R, Bocian W, Krajewski P, Bednarek E, Sitkowski J, Williamson MP, Moir AJ, Hansen PE (2001) A nicked duplex decamer DNA with a PEG(6) tether. Nucleic Acids Res 29:1132–1143
194. Fersht A (1999) Structure and mechanism in protein science. W.H. Freeman & Co, New York
195. Ferreiro DU, Sanchez IE, de Prat Gay G (2008) Transition state for protein-DNA recognition. Proc Natl Acad Sci USA 105:10797–10802
196. Lynch TW, Read EK, Mattis AN, Gardner JF, Rice PA (2003) Integration host factor: putting a twist on protein-DNA recognition. J Mol Biol 330:493–502
197. Mirzabekov AD, Rich A (1979) Asymmetric lateral distribution of unshielded phosphate groups in nucleosomal DNA and its role in DNA bending. Proc Natl Acad Sci USA 76:1118–1121
198. Manning GS, Ebralidse KK, Mirzabekov AD, Rich A (1989) An estimate of the extent of folding of nucleosomal DNA by laterally asymmetric neutralization of phosphate groups. J Biomol Struct Dyn 6:877–889
199. Williams LD, Maher LJ 3rd (2000) Electrostatic mechanisms of DNA deformation. Annu Rev Biophys Biomol Struct 29:497–521
200. Strauss JK, Maher LJ 3rd (1994) DNA bending by asymmetric phosphate neutralization. Science 266:1829–1834
201. Strauss JK, Roberts C, Nelson MG, Switzer C, Maher LJ 3rd (1996) DNA bending by hexamethylene-tethered ammonium ions. Proc Natl Acad Sci USA 93:9515–9520
202. Strauss-Soukup JK, Maher LJ 3rd (1998) Electrostatic effects in DNA bending by GCN4 mutants. Biochemistry 37:1060–1066
203. McDonald RJ, Kahn JD, Maher LJ 3rd (2006) DNA bending by bHLH charge variants. Nucleic Acids Res 34:4846–4856
204. Leonard DA, Rajaram N, Kerppola TK (1997) Structural basis of DNA bending and oriented heterodimer binding by the basic leucine zipper domains of Fos and Jun. Proc Natl Acad Sci USA 94:4913–4918
205. Ramirez-Carrozzi VR, Kerppola TK (2001) Long-range electrostatic interactions influence the orientation of Fos-Jun binding at AP-1 sites. J Mol Biol 305:411–427
206. Blainey PC, van Oijen AM, Banerjee A, Verdine GL, Xie XS (2006) A base-excision DNA-repair protein finds intrahelical lesion bases by fast sliding in contact with DNA. Proc Natl Acad Sci USA 103:5752–5757
207. Wang YM, Austin RH, Cox EC (2006) Single molecule measurements of repressor protein 1D diffusion on DNA. Phys Rev Lett 97:048302
208. Tafvizi A, Huang F, Leith JS, Fersht AR, Mirny LA, van Oijen AM (2008) Tumor suppressor p53 slides on DNA with low friction and high stability. Biophys J 95:L01–03
209. Bonnet I, Biebricher A, Porte PL, Loverdo C, Benichou O, Voituriez R, Escude C, Wende W, Pingoud A, Desbiolles P (2008) Sliding and jumping of single EcoRV restriction enzymes on non-cognate DNA. Nucleic Acids Res 36:4118–4127
210. Gorman J, Chowdhury A, Surtees JA, Shimada J, Reichman DR, Alani E, Greene EC (2007) Dynamic basis for one-dimensional DNA scanning by the mismatch repair complex Msh2-Msh6. Mol Cell 28:359–370

Chapter 7
Studies of Sequence-Nonspecific HMGB DNA-Binding Proteins

L. James Maher, III

7.1 The DNA Persistence Length Problem in Brief

DNA is the genetic material in most organisms. Extremely long DNA molecules (1 m per haploid genome in the case of humans) must be packaged, replicated, repaired, and transcribed within living cells [1]. Remarkably, DNA is among the stiffest of all natural polymers, both in terms of resistance to bending and to twisting. What is the origin of this DNA stiffness and what mechanisms do cells use to manage it? These are the questions that motivate our work in this area.

The stiffness of purified DNA is perhaps most directly measured by ring closure experiments catalyzed by DNA ligase [2–5]. The cyclization probability reflects the relative concentration of one DNA terminus in the vicinity of the other and is quantified as the *J* factor (units of molarity). The length dependence of the *J* factor is predicted well by application of the worm-like chain model (7.1) over a wide range of DNA lengths [6–12]:

$$\langle R^2 \rangle = 2PL\left[1-\frac{P}{L}\left(1-e^{-\frac{L}{P}}\right)\right] \tag{7.1}$$

where L is the contour length, $<R^2>$ is the mean squared end-to-end distance, and P is the polymer persistence length (~140 bp or 47.6 nm for DNA under physiological conditions).

Plots of the theoretical dependence of the *J* factor on DNA chain length are informative. An example is shown in Fig. 7.1a. Three important conclusions are evident from this plot. First, the maximum likelihood of DNA cyclization occurs at ~500 bp, and the probability of cyclization falls dramatically for shorter sequences. Second, the length dependence of cyclization rapidly diminishes for chains longer than 500 bp. Third, the twist inflexibility causes a striking sinusoidal oscillation in *J* factor when DNA end alignment is considered. This twist inflexibility rapidly

L.J. Maher, III (✉)
Department of Biochemistry and Molecular Biology, Mayo Clinic College of Medicine, Rochester, MN, USA
e-mail: maher@mayo.edu

M.C. Williams and L.J. Maher, III (eds.), *Biophysics of DNA-Protein Interactions*, Biological and Medical Physics, Biomedical Engineering,
DOI 10.1007/978-0-387-92808-1_7,

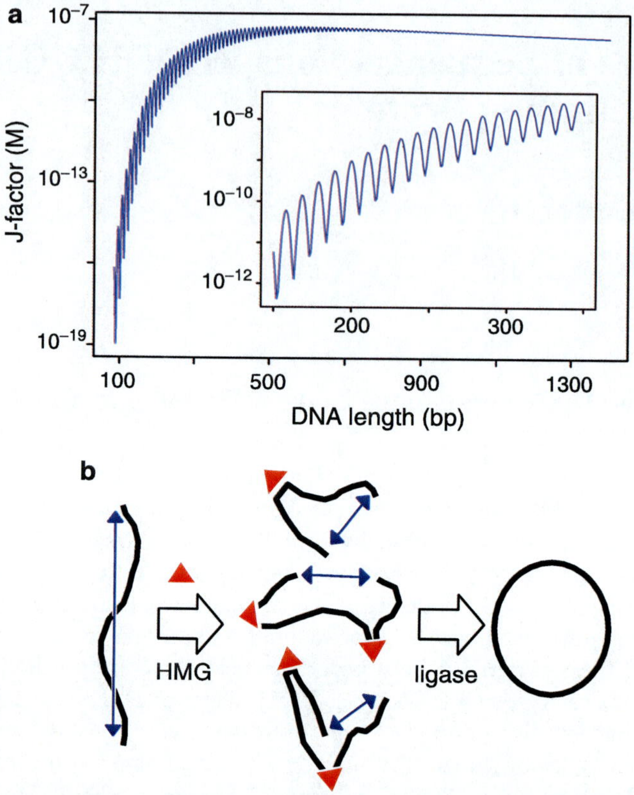

Fig. 7.1 Conceptual role of HMGB proteins in modulating apparent DNA stiffness. (**a**) Length-dependent DNA flexibility predicted by the worm-like chain (WLC) model in DNA cyclization experiments. The *J* factor measures end-to-end concentration sensitive to helical alignment. The inset expands a portion of the length range. (**b**) Schematic illustration of the potential of HMGB proteins (*triangles*) to induce a random ensemble of collapsed DNA structures with enhanced *J* factors. Members of this ensemble cyclize more rapidly than naked DNA. Adapted from Zimmerman and Maher [62]

decays for lengths above 500 bp. These results express the fact that DNA molecules of length P have an average R/L ratio of $(2/e)^{0.5}$ (i.e., ~0.9) and shorter molecules are increasingly rod like. In contrast, long DNA molecules behave in a more flexible manner, with end-to-end distance scaling as $(2PL)^{0.5}$. Thus, short DNAs are rod like (rigid) and long DNAs are threadlike (flexible). This result has crucial implications for DNA function. Active compaction of even long thread-like DNA is required in viruses and cells. Although the 1-m human genome is predicted to spontaneously collapse into a coil with end-to-end distance of ~400 μm, the diameter of the human cell nucleus is ~7 μm, requiring an additional ~60-fold compaction. More importantly, gene regulation and recombination often requires interaction between proteins bound to DNA sites separated by a few hundred base pairs or less.

Over these distances, the rod-like character and twist inflexibility of DNA are dramatic and should to create large energy barriers to protein–protein contacts.

A compelling puzzle has emerged when the in vitro and in vivo physical properties of DNA are compared. DNA appears to be more flexible in vivo than in vitro. Classic experiments have shown DNA looping to be the mechanism of repression in the *Escherichia coli lac* operon. Detailed analysis led Record et al. to estimate DNA bend and twist flexibilities in vivo that are sevenfold and twofold higher, respectively, than in vitro [13–15]. A recent updated analysis by Zhang et al. places the discrepancies at 1.6-fold and threefold, respectively [16]. What is the origin of this apparent DNA "softening" in vivo? It seems unlikely that DNA has intrinsically different physical properties in bacteria. Rather, it is hypothesized that the apparent difference is due to factors such as negative supercoiling and DNA-binding proteins present in vivo. Travers [17], Crothers [18], and Johnson [19–21] originally proposed that sequence-nonspecific architectural proteins are capable of transient binding and kinking of DNA and contribute to reduced apparent DNA persistence length (Fig. 7.1b). Such factors could facilitate the tight bending of DNA required for looping in recombination and gene control. Johnson et al., as well as others [22–24], identified specific architectural proteins capable of these effects, including the heat unstable (HU) protein in bacteria and high-mobility group B (HMGB) proteins in yeast, insects, and vertebrates [25].

7.2 HMGB Background

7.2.1 HMGB In Vitro Characteristics and Properties

High-mobility group (HMG) proteins are abundant eukaryotic chromatin-associated proteins named for their small size and high mobility on polyacrylamide gels. The HMG proteins encompass three broad families that are unrelated in sequence and structure [26–29]. First, HMGA family members contain AT-hook motifs. Second, HMGN proteins contain a nucleosome-binding domain. Finally, HMGB family members (the focus of this chapter) can be sequence-specific or nonspecific and contain highly conserved HMG "box" motifs. Sequences and structures of representative HMGB proteins are illustrated in Fig. 7.2 and compared with the unrelated *E. coli* HU protein [30, 31]. The structure of a designed two-box HMGB protein complex with DNA has also recently been reported [32]. Interestingly, DNA bending by two-box HMG proteins can be less than by single-box proteins because the bends induced by the two-box protein may be out of phase. While different in structure, HU and HMGB proteins both cause strong bending of DNA with little or no sequence specificity. Both proteins combine the action of one or more amino acid side chains as intercalative "wedges" in the minor groove with additional favorable electrostatic interactions and asymmetric charge neutralization (Fig. 7.2b, see Nhp6A). The HU protein is abundant with ~25,000 heterodimers per cell during exponential growth. HU appears to bind and kink DNA in the *E. coli GAL* repression loop in vivo and

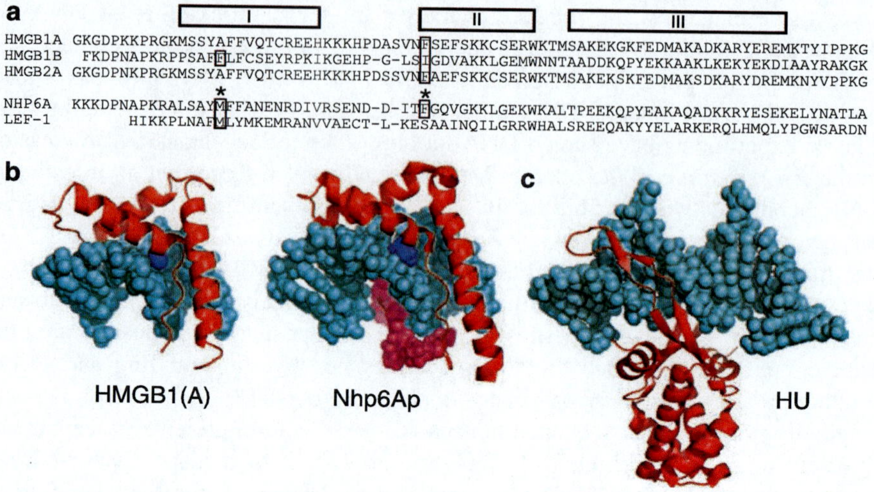

Fig. 7.2 HMGB protein sequence and structure. (**a**) Conserved HMG box sequences from mammalian double-box proteins HMGB1 and HMGB2, yeast single-box protein Nhp6Ap, and the mammalian sequence-specific HMGB single-box protein LEF-1. Conserved alpha-helical segments are indicated by numbered *rectangles* (*above*). Primary intercalating residues that interact with the DNA minor groove are boxed and highlighted by *asterisks*. (**b**) Structures of HMGB1 box A interacting with platinum-cross-linked DNA (*left*, pdb code 1ckt [34]), and yeast Nhp6A (*center*, pdb code 1j5n [30]). HMGB intercalating residues are shown in *dark blue*, and the cationic leader of Nhp6Ap is shown in *magenta*. (**c**) The bacterial architectural protein HU (pdb code 1p78 [31, 83]). Adapted from Zimmerman and Maher [62]

in vitro [22–24]. HMGB proteins are also abundant with ten different HMG box proteins in yeast and the similar HMGB1 and HMGB2 proteins of mammals present at one HMGB protein per ~15 nucleosomes. Sequence-nonspecific HMGB proteins contain one or two HMG boxes. Each box encodes amino acids that form three alpha helices to engage DNA as an "L"-shaped structure (Fig. 7.2b). This module widens the DNA minor groove through partial intercalation of one or more amino acid side chains between base pair stacks (Fig. 7.2a, boxes) [25, 33]. These perturbations cause strong DNA bending away from the engaged DNA surface. The HMG box A of mammalian HMGB1 (Fig. 7.2b, left) is shown bound to a DNA segment distorted by platinum cross-linking [34]. Figure 7.2b (right) depicts the structure of the yeast single-box protein Nhp6Ap complexed with unmodified DNA [35]. Putative intercalating residues are shown in dark blue. Nhp6Ap also contains a strongly cationic leader sequence (Fig. 7.2b, right, magenta) positioned in the major groove to asymmetrically neutralize the compressed sugar–phosphate backbone. This interaction apparently contributes significantly to DNA bending [36]. Thus, DNA bending by HMGB proteins involves a combination of minor groove widening, amino acid wedging, and asymmetric charge neutralization. HMGB proteins appear to be partially unstructured when not bound to DNA and electrostatic forces have been shown to dominate van der Waals and hydrogen bonding interactions in both DNA binding and bending by sequence-nonspecific HMGB proteins [36]. While strong sequence-nonspecific bending by the prokaryotic nucleoid protein HU involves minor groove interactions, a completely different scaffold is involved (Fig. 7.2b, left).

Though similar, subtle features of "A box" and "B box" domains of two-box mammalian HMGB proteins can be distinguished [25]. The A box domain differs in the shape and orientation of helix I and the identity of potential intercalating residues. Isolated A box domains are less effective at DNA bending [20]. The sequence of human HMGB2(box A) [hereafter, HMG box domains A and/or B are indicated in parentheses, e.g., HMGB2(A)] is illustrated in Fig. 7.2a. Whereas the intercalating phenylalanine in helix II is present in both box A and box B domains of two-box HMGB proteins, the alanine in helix I does not intercalate strongly [25]. Boxes A and B are also distinguished by the lengths and geometries of helices I and II, with helix I being short and straight in box A, but bent in box B [25]. The single-box HMGB protein Nhp6Ap conforms to the box B family (Fig. 7.2). Nhp6Ap is distinguished from HMGB2(A) by the identity of the intercalating residue in helix I (methionine in Nhp6Ap), a longer helix I, and a highly cationic leader sequence. Characteristics of HMGB proteins have been reviewed [17, 18, 28, 29, 37–40]. Early results characterized HMGB proteins as specific for binding distorted DNA, as in cruciform structures and chemical cross-links [34, 41, 42].

Despite the in vitro evidence that HMGB proteins can enhance apparent DNA flexibility by sequence-nonspecific DNA kinking, the actual in vivo functions of HMGB proteins remain unknown. Just as deletion of genes encoding the *E. coli* HU protein causes a nucleoid structure defect and other problems [21, 43, 44], deletion of mammalian HMGB1 causes neonatal death in mice due to hypoglycemia of unknown origin [45]. The *Saccharomyces cerevisiae* single-HMG-box proteins Nhp6Ap and Nhp6Bp result from apparent gene duplication [46, 47]. Deletion of these genes causes a temperature-sensitive growth phenotype in yeast that has been shown to result from the role of Nhp6 proteins in transcription of the U6 small nuclear RNA by RNA polymerase III [48, 49].

Besides their role as architectural factors, HMGB proteins have been proposed as catalytic facilitators in transcription factor binding [19, 50–52], components of the FACT complex that facilitates transcription through chromatin [53], cofactors in DNA repair [54], and modifiers of chromatin structure [40]. Remarkably, mammalian HMGB proteins are also recognized as extracellular hormones in cardiac pathology [55]. Thus, much remains to be learned of these proteins. This work summarized here has provided some new insights.

7.2.2 *The Mystery of HMGB Function In Vivo*

The actual roles for HMGB proteins in living cells remain uncertain and controversial. Some possibilities are illustrated in Fig. 7.3. The original studies of Johnson et al. [19, 20] emphasize the ability of HMGB proteins to replace HU in supporting tightly bend DNA structures required for recombination (Fig. 7.3a). The fact that HMGB proteins could improve the growth and morphology of HU mutant *E. coli* also suggested a potential role for HMGB proteins in general DNA compaction [21]. Interestingly (perhaps surprisingly), these authors showed that the yeast HMGB proteins Nhp6A/B also enhanced Gal4-dependent gene transcription by a mechanism that seemed to involve functions of the *GAL1* core promoter rather than facilitation of DNA looping between the core promoter and Gal4p-binding site [19]. On the other hand, a fascinating recent study of factors that restrict gene activation to short distances in yeast revealed that mutation of the HMGB gene *SPT2* caused gene activation to occur over longer distances, consistent with a role of Spt2p in reducing average loop distances by enhancing DNA flexibility [56]. Spt2p has also been implicated as a chromatin component that participates in proper recruitment of the machinery responsible for mRNA cleavage and polyadenylation [57].

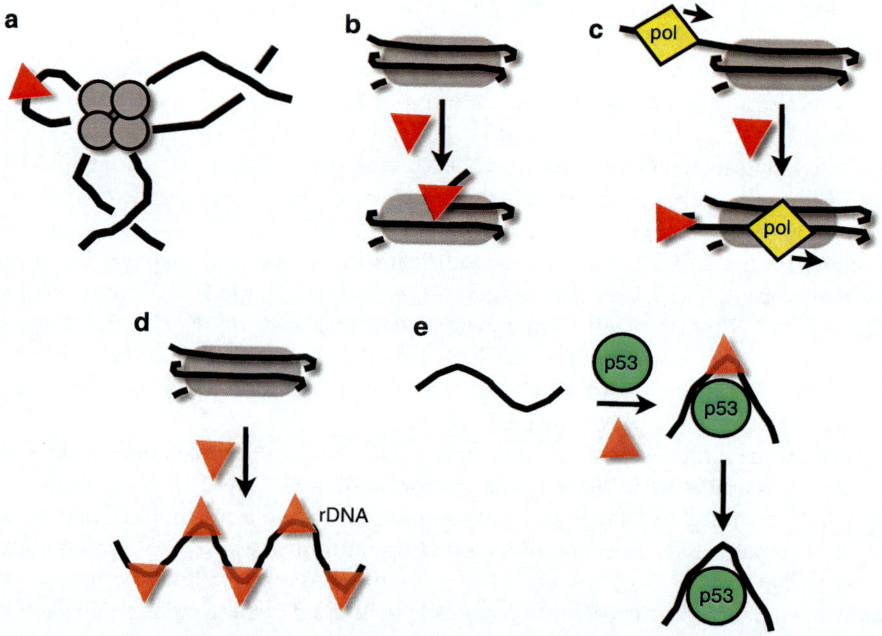

Fig. 7.3 Proposed roles for HMGB proteins. **(a)** As originally discovered by Paull et al. [20], mammalian HMGB1 and HMGB2 (*red triangle*) can substitute for bacterial HU during *in vitro* DNA inversion recombination by Hin recombinase (*gray circles*). **(b)** As proposed by Travers [40], HMGB proteins could deform DNA near the nucleosome entry point. The result could be loosened chromatin structure and facilitation of nucleosome migration and transcription factor binding. **(c)** Possible role of yeast Nhp6A in altering core DNA/histone interactions to facilitate transcription through chromatin [58]. **(d)** Possible reorganization of normal euchromatin into an alternate chromatin structure containing the yeast HMGB protein Hmo1p at highly transcribed ribosomal gene clusters (rDNA) [61]. **(e)** "Catalytic" role for HMGB proteins in facilitating transcription factor binding, noting the example of p53 [51]

The preference of HMGB proteins for distorted DNA suggests the possibility that nucleosome binding is involved in the dominant HMGB function. Travers has presented a thoughtful proposal [40] for HMGB involvement in "priming" of nucleosomes for displacement by sliding (Fig. 7.3b). Yeast Nhp6 proteins are also known to be components of the FACT complex that enhances RNA polymerase transcription through chromatin. This observation and subsequent experiments have suggested a rather different role (Fig. 7.3c) for HMGB protein binding to core (rather than entry point) nucleosomal DNA during modification of chromatin structure [58].

Recent studies of the yeast single-HMG-box Hmo1 protein [59] reveal an entirely different possibility for this HMGB protein. Approximately 10% of the yeast genome and 60% of all yeast RNA transcripts encode ribosomal RNAs. The peculiar chromatin structure at active ribosomal genes appears to be devoid of histones, but is enriched in Hmo1p [60, 61]. This intriguing result suggests that HMGB proteins could replace histone octamers in the formation of an alternative "chromatin" of unknown structure (Fig. 7.3d). Finally, there is evidence that HMGB proteins can facilitate transcription factor or repair protein binding to DNA [19, 50–52, 54]. In at least one curious case involving p53, the role appears to be catalytic (Fig. 7.3e).

7.3 Overview of Recent Contributions

7.3.1 HMGB Mechanism from the Single-Molecule Perspective

Some of the *in vitro* properties of purified HMGB proteins were illustrated in our recent experiments to test the hypothesis that HMGB interactions with DNA are transient. The study used conventional ensemble experiments [62]. We tested our hypothesis by adding HMGB proteins to create distorted DNAs with rapidly interconverting conformations. We showed that optimal HMGB concentrations indeed enhance ligase-mediated cyclization (Fig. 7.4a). Using timed competition with unlabeled competitor DNA, electrophoretic gel mobility shift experiments showed that HMGB proteins are rapidly exchanging on and off the DNA such that all the proteins had exchanged onto unlabeled competitor within 60 s (Fig. 7.4b). Thus, HMGB binding was found to be both weak and transient under conditions where DNA cyclization is strongly enhanced. During the course of this work, we also detected novel complexes in which HMGB proteins simultaneously bind more than one DNA duplex when elevated concentrations of DNA are present (Fig. 7.4c). This interesting result is observed for both single-box and double-box proteins, suggesting that HMGB proteins can bridge DNA segments that cross forming nodes [62].

We recently applied ensemble experiments to understand the importance of amino acid residues that distinguish box A-like and box B-like HMGB proteins [63]. As described above, box A and box B homology domains have subtle sequence differences such that box B domains bend DNA strongly, while DNA

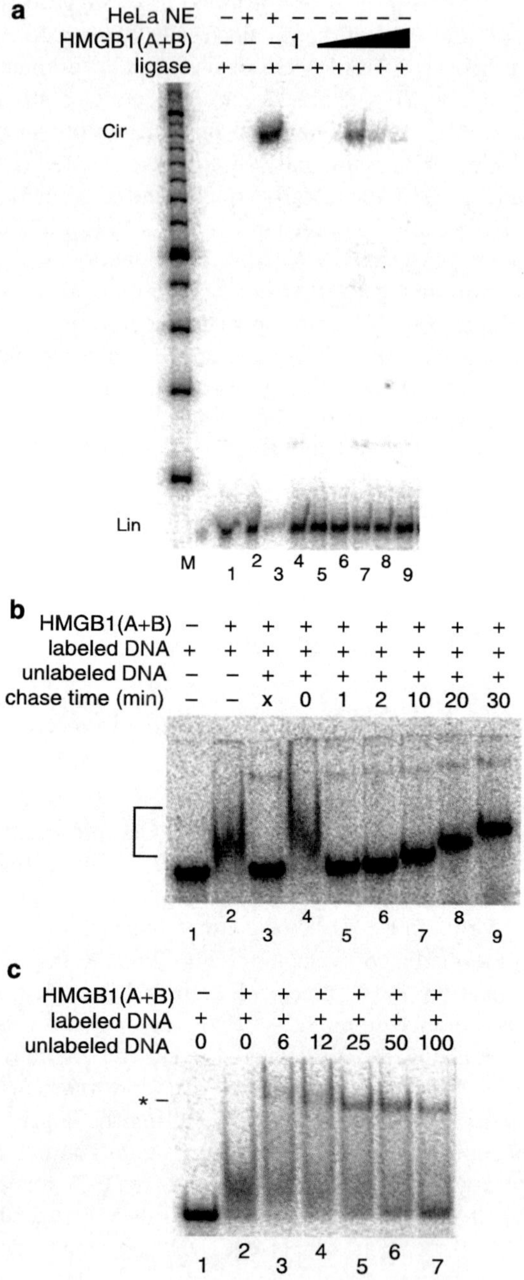

Fig. 7.4 In vitro activities of rat HMGB1. (**a**) Enhancement of ligase-mediated cyclization of 195-bp radiolabeled DNA ("Lin") to form *circles* ("Cir"). Heat-treated HeLa cell nuclear extract ("HeLa NE") and purified recombinant rat HMGB1(A + B) enhance cyclization in a ligase and protein concentration-dependent manner. Lanes 5–9 correspond to HMGB1(A + B) concentrations of 1.6, 8, 40, 200, and 1,000 nM, respectively. (**b**) Transient nature of HMGB–DNA complexes

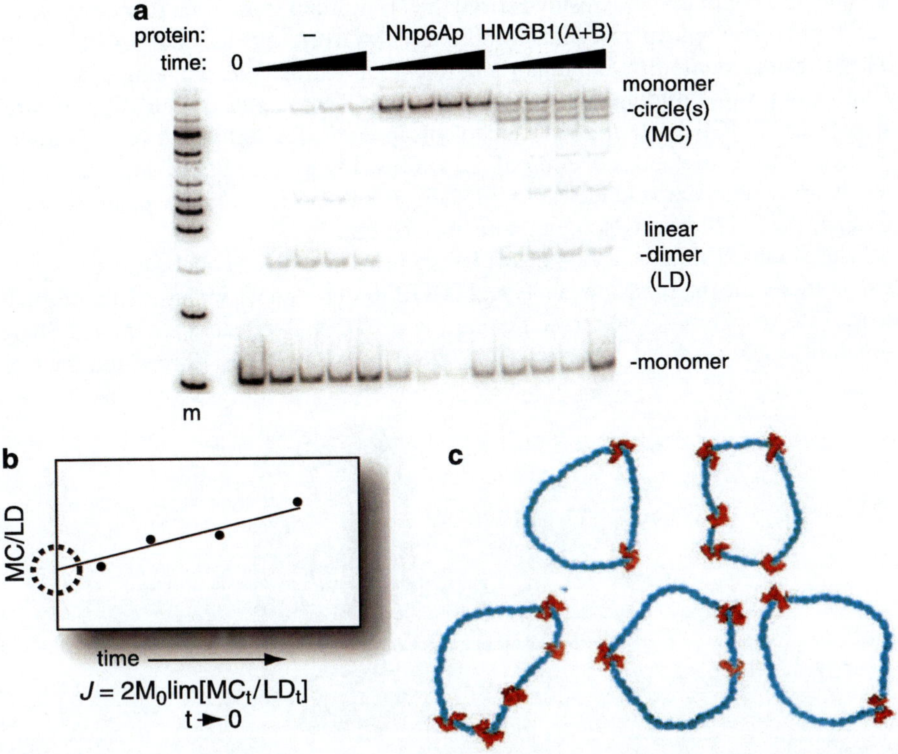

Fig. 7.5 Cyclization kinetics and modeling of DNA flexibility enhancement by HMGB proteins. (**a**) Effects of Nhp6Ap and HMGB1(A+B) on DNA ligase-mediated cyclization kinetics of 200-bp DNA. (**b**) Approach to calculate experimental J factors from cyclization data. (**c**) Modeling by Monte Carlo simulation to explore cyclic DNA chain conformations (cyan) whose probabilities are enhanced by HMGB proteins (Nhp6A proteins in this example; *red*)

bending by isolated box A domains is weaker. We applied DNA T4 ligase-mediated cyclization kinetics assays in vitro (Fig. 7.5a) to show that an isolated HMG box A domain derived from human HMGB2 folded poorly and did not enhance apparent DNA flexibility in this in vitro assay. The application of formal cyclization kinetics analysis (Fig. 7.5b) is important to distinguish enhancement of apparent DNA flexibility from simple enhancement of end joining. We found that substitution of

Fig. 7.4 (continued) revealed by binding to 0.5 nM labeled DNA followed by addition of 1 μM unlabeled competitor and electrophoresis after the indicated competition times. HMGB1(A+B)-DNA complexes (lanes 2 and 4) completely exchange with unlabeled DNA within 1 min of competition (lane 5). (**c**) Interaction of 0.5 nM labeled duplex DNA (lane 1) with 400 nM HMGB1(A+B) alone (lane 2) or with addition of the indicated nanomolar concentrations of unlabeled duplex DNA (lanes 3–7). The novel complex involving HMGB1(A+B) protein bridging labeled and unlabeled DNA duplexes is indicated (*asterisks*). Adapted from Zimmerman and Maher [62] a small number of cationic residues from the N-terminal leader of a functional yeast box B protein, Nhp6Ap, rescued the ability of HMGB2A to enhance DNA flexibility. These results demonstrated important roles for cationic leader amino acids of Nhp6A in protein folding, DNA interaction, and DNA bending. Our experimental studies of HMGB protein enhancement of apparent flexibility are being supplemented with theoretical modeling of the results. Figure 7.5c shows an example of results of Monte Carlo modeling of HMGB effects on DNA ring closure (Rueter, Czapla, Peters, Maher, Olson, unpublished results).

The Maher, Williams, and Israeloff laboratories have been collaborating to apply single-molecule methods to analyze HMGB protein mechanisms. Our original work [64] involved single-molecule analysis of HMGB effects on apparent DNA flexibility using optical tweezers (Fig. 7.6a) to measure the forces required to

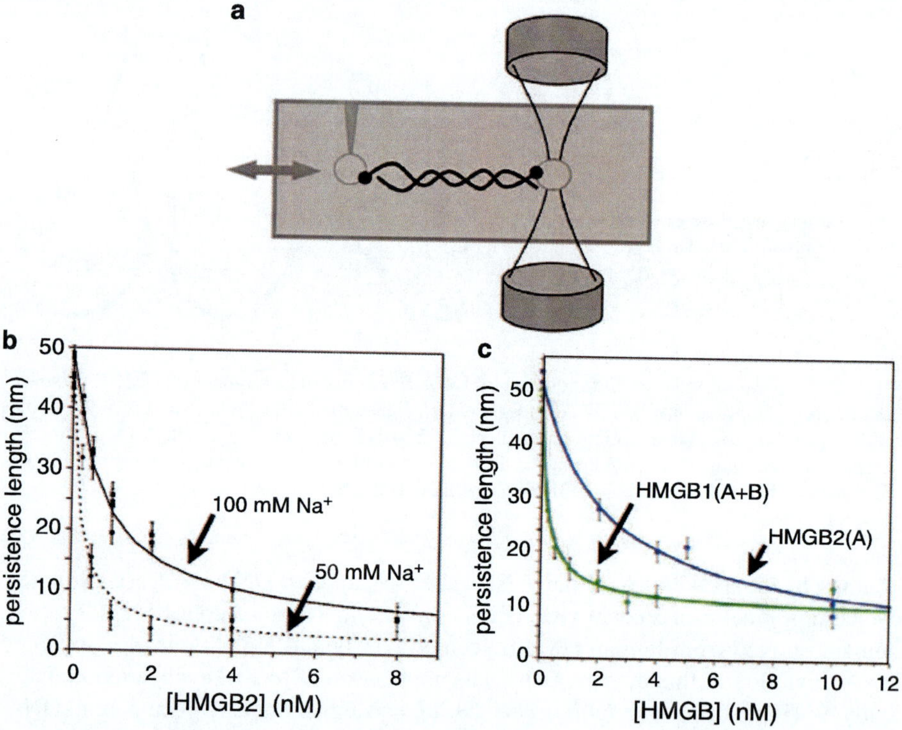

Fig. 7.6 Single-molecule force spectroscopy studies of HMGB effects on apparent DNA persistence length. (**a**) Diagram of optical tweezers instrumentation with laser trap focused on the bead at *right*, with tension exerted by movement of the bead at *left*. Phage λ DNA is captured by its termini. Force-extension data are collected and fit to the WLC model to extract the apparent DNA persistence length in the absence or presence of DNA-binding proteins. (**b**) Decrease in apparent DNA persistence length induced by human HMGB2(A) protein. The naked DNA persistence length was ~45 nm (132 bp), vs. ~3 nm (9 bp) for 8 nM HMGB1 in 50 mM Na^+. (**c**) Decrease in apparent DNA persistence length in the presence of human HMGB2(A) compared to rat HMGB1(A+B). Adapted from McCauley et al. [64, 66]

stretch single-DNA molecules. Parameters describing DNA flexibility, including contour length and persistence length, were revealed. In the presence of nanomolar concentrations of isolated HMG box A from HMGB2, DNA showed a decrease in its persistence length, where the protein induced an average DNA bend angle of $114 \pm 21°$ for 50 mM Na^+. The DNA contour length increased from 0.341 ± 0.003 to 0.397 ± 0.012 nm/bp, independent of salt concentration. In 50 mM Na^+, the protein did not unbind even at high DNA extension, while in 100 mM Na^+, the protein unbound only below concentrations of 2 nM. These observations supported a flexible hinge model for non-cooperative HMG binding at low protein concentrations. These studies are consistent with the results of Marko et al. [65]. However, at higher protein concentrations, a cooperative filament mode was observed instead of hinge binding. This mode was uniquely characterized by high-force optical tweezers experiments [64]. A key result from this work was the effect of the HMG domain on DNA persistence length (Fig. 7.6b).

We continued this work by studying the effects of single- and double-HMGB box proteins on single-phage lambda DNA molecules [66]. A dual beam optical tweezers was used to extend double-stranded DNA in the absence and presence of a single-box derivative of human HMGB2 [HMGB2(A)] and a double-box derivative of rat HMGB1 [HMGB1(A+B)]. The single-box domain was observed to reduce the persistence length of the double helix, generating sharp DNA bends with an average bend angle of $99 \pm 9°$, and at very high concentrations also stabilized double-stranded DNA. The double-box motif contains two consecutive HMG box domains joined by a flexible tether. This protein also reduced the DNA persistence length, inducing an average bending angle of $77 \pm 7°$ and stabilizing dsDNA at much lower concentrations. These results suggested that the single- and double-box proteins both increase DNA flexibility and stability, but the effects are achieved at lower concentrations for the double-box protein at constant monovalent salt concentration. A key result is shown in Fig. 7.6c.

We recently published atomic force microscopy (AFM) data [67] relevant to the mechanism of HMGB enhancement of apparent flexibility. Details of induced DNA bending were studied by examining complexes of double-stranded DNA with HMGB2(A) and HMGB1(A+B) using AFM in tapping mode in air (Fig. 7.7a, b). DNA end-to-end distances and local DNA bend angle distributions (Fig. 7.7c) were analyzed for protein complexes deposited on a mica surface. For HMGB2(A) binding, we found a mean induced DNA bend angle of 78° with a standard deviation of 23°, while HMGB1(A+B) binding gave a mean of 67° and a standard deviation of 21°. These results were consistent with analysis of the observed global persistence length changes derived from end-to-end distance measurements, and with results of DNA stretching experiments. The distributions of bend angles induced by both proteins were moderately broad. This result is inconsistent with either a static kink model or a freely flexible hinge model. Thus, the HMGB mechanism for enhancement of apparent DNA flexibility must be intermediate, differing from that of the *E. coli* HU protein, which in previous studies showed a very broad angle distribution consistent with a flexible hinge model.

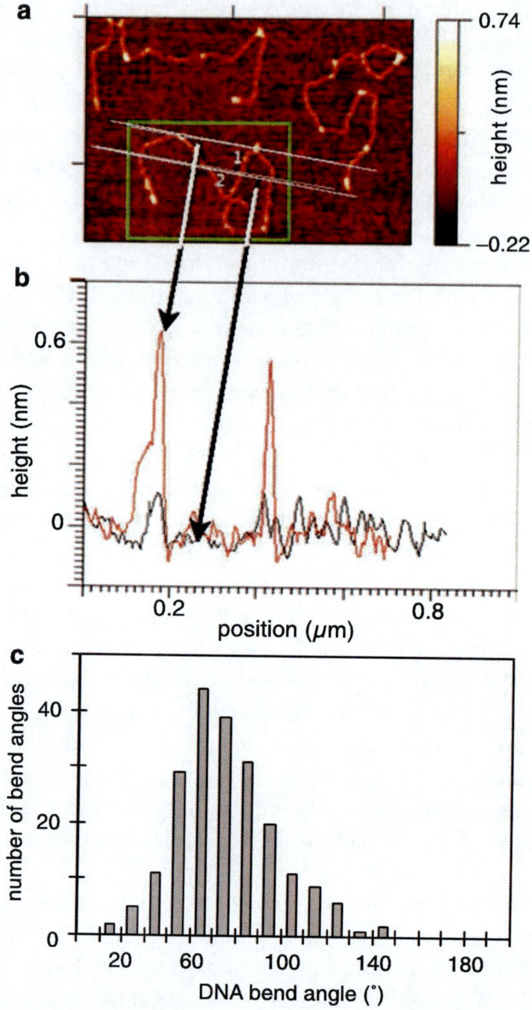

Fig. 7.7 AFM analysis of DNA bend angles induced by HMGB2(A). Linear plasmid DNA is deposited from dilute Mg^{+2} solution onto mica and dried. Imaging was by tapping. (**a**) Image of 0.09 nM DNA in the presence of 3.1 nM HMGB2(A) colored to show height information. (**b**) Calculated height information based on (**a**), distinguishing bound HMGB proteins (*upper line* that transects protein/DNA complexes) from DNA alone (*lower line*). (**c**) Distribution of HMGB protein-induced DNA bend angles showing a moderately broad angle distribution inconsistent with either a flexible hinge or a static kink model of HMGB mechanism. Adapted from Zhang et al. [67]

7.3.2 HMGB Mechanism in an *E. coli* DNA Looping Model

Can the effects of HMGB proteins on apparent DNA stiffness be studied quantitatively within living cells? An in vivo *lac* repression looping assay [13–15, 43, 68–71] was originally conceived (independently) by Record et al. and Müller-Hill et al.

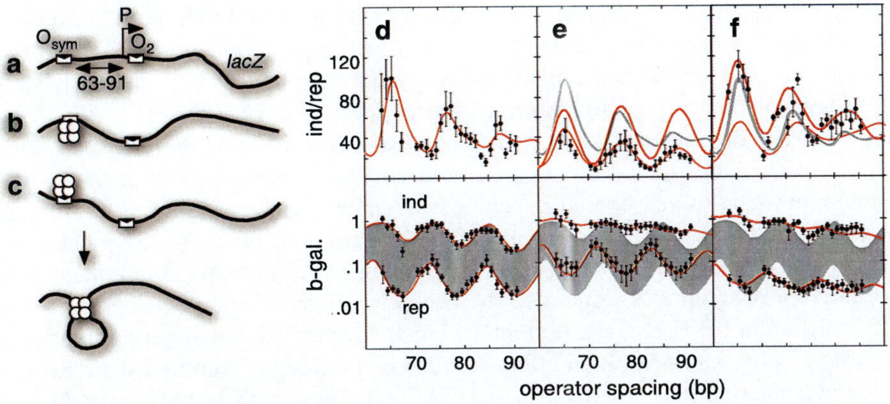

Fig. 7.8 Lac repressor looping assays of DNA flexibility in living *E. coli* cells. (**a**) Simplified *lac* promoter (P)-reporter (*lacZ*) constructs placed on F′ episome. Strong (O_{sym}) and weak (O_2) operators are spaced between 63 and 91 bp. (**b**) Out-of-phase spacings loop poorly due to DNA twist inflexibility. (**c**) In-phase spacings allow looping and gene repression. (**d**) Reporter gene activity under inducing (ind) or repressing (rep) conditions shown as repression ratio (*above*) or normalized reporter gene activity (*below*). The irregular pattern in *upper panel* is due to dephasing of sinusoidal patterns in *lower panel*. (**e**) Repression looping is inhibited in the absence of the architectural protein HU (data points and *red lines*). (**f**) Rescue of DNA looping by a variant of the yeast Nhp6A protein (data points and *red lines*) in the absence of the architectural protein HU. Data from panel D are shown in *gray* for comparison. Adapted from Becker et al. [73]

in their studies of constraints on DNA repression loops in living bacteria due to inflexibility of the repressor protein and looped DNA. As shown in Fig. 7.8a, the *lacZ* reporter gene is placed downstream from a simple promoter that overlaps a very weak lac repressor-binding site (O_2 operator). This gene is poorly repressed, but repression can be dramatically enhanced by placement of a strong operator (O_{sym}) upstream of the promoter. It was found that analysis of *lacZ* repression as a function of operator spacing provided a very sensitive assay of the bending and twisting flexibility of DNA in vivo [1, 13, 16, 43, 68, 72]. When the operators are out of phase (Fig. 7.8b), looping is energetically unfavorable due to the need for DNA twisting. When in-phase (Fig. 7.8c), looping is facilitated because the local concentration of repressor at O_2 by virtue of proximity to O_{sym} is actually higher than the concentration of free repressor.

The *E. coli lac* looping system is amenable to quantitative analysis (e.g., Fig. 7.8d, e). Here, the lower panels show the β-galactosidase (product of the *lacZ* reporter gene) activity extracted from living cells as a function of operator spacing for constructs inserted into the large F′ episome of *E. coli*. Cells are analyzed for β-galactosidase activity in the absence (uninduced) or presence (induced) of IPTG, a lactose analog that reduces affinity for lac repressor for DNA. The ratio of the induced:repressed activities (the so-called repression ratio) is shown in the upper panels. Thermodynamic modeling is then possible by several approaches [13, 14, 16, 43, 68, 72, 73]. Fitting the oscillating plots of the repression dependence on operator spacing [13, 43, 68] suggests that the in vivo apparent DNA torsional modulus is ~fourfold lower than the accepted range of approximately $2\text{–}4 \times 10^{-19}$ erg cm for naked DNA in dilute solution [74]. DNA-bending stiffness did not appear to affect looping for loop sizes of 60–90 bp [43, 68]. Such short loops are highly unfavorable in naked DNA [75], though exactly how unfavorable has been the subject of recent debate [12, 76, 77].

We originally recreated an in vivo *E. coli lac*-operon-based system for measuring apparent DNA bend and twist flexibility in vivo [43]. The dependence of reporter gene repression on *lac* operator spacing reflects the in vivo energetics of forming the small DNA repression loop involving bending and twisting strain. The key results of our original work were that (1) binding of inducer alters both the affinity of lac repressor for DNA and also the geometry of residual repression loops and (2) deletion of the *E. coli* HU protein drastically destabilizes small repression loops (compare Fig. 7.8d and e), an effect that can be partially overcome by the expression of a heterologous HMGB protein (Fig. 7.8f). These results showed for the first time that the inherent inflexibility of DNA restrained looping and modulated by HU in vivo. Strikingly, analysis of these data has suggested that the in vivo physical properties of the repressor loop DNA in the absence of HU more closely match the behavior of naked DNA in solution [16].

The fact that eukaryotic HMGB proteins can complement bacterial HU deficit [21, 43, 63, 73] is actually quite remarkable. HMGB and HU proteins share no sequence homology and interact with the DNA minor groove using very different strategies (Fig. 7.2b). Bacterial chromatin lacks nucleosomes, and it seems unlikely that bacterial proteins can guide HMGB proteins to DNA-binding sites. In spite of this seemingly inhospitable environment, HMGB proteins and variants are capable of addressing fundamental and universal challenges of DNA compaction as they rescue DNA looping in cells lacking HU.

We took advantage of this in vivo system to explore effects of other architectural proteins [78] on apparent DNA flexibility in living *E. coli* cells [68]. After building appropriate genetic deletion strains, we tested DNA looping in bacteria lacking the nucleoid proteins HU, IHF, or H-NS. We confirmed that deletion of HU inhibits looping and that quantitative modeling suggests residual looping even in the induced operon. Deletion of IHF had little effect. Interestingly, DNA looping was strongly *enhanced* in the absence of H-NS, and we proposed an explanatory model involving reduced interhelix bridging in the absence of H-NS. Chloroquine titration, psoralen cross-linking, and supercoiling-sensitive reporter assays showed that the effects of nucleoid proteins on looping were not correlated with their effects on either total or unrestrained supercoiling. These results suggested that host nucleoid proteins can directly facilitate or inhibit DNA looping in bacteria.

Our initial experiments had suggested that some eukaryotic HMGB proteins can functionally substitute for HU in facilitating DNA looping in our *lac* assay strains [43]. We then further explored the extent to which HMGB proteins and derivatives could complement the DNA looping defect in *E. coli*-lacking HU protein. We found that derivatives of yeast HMGB protein Nhp6Ap could rescue DNA looping in *E. coli*-lacking HU, in some cases facilitating looping to a greater extent than is observed in *E. coli* expressing normal levels of HU protein. Nhp6Ap-induced changes in the DNA length dependence of repression efficiency suggest that Nhp6Ap altered DNA twist in vivo. In contrast, human HMGB2-box A derivatives did not rescue looping [68].

More recently, we applied the *E. coli* in vivo DNA looping assay in combination with in vitro biochemical experiments to address a question related to HMGB structure and function. HMGB proteins are composed of one or two conserved HMG box domains, each forming three alpha helices that fold into a sequence-nonspecific DNA-binding module recognizing the DNA minor groove. As described above, box A and box B homology domains have subtle sequence differences such that box B domains bend DNA strongly, while DNA bending by isolated box A domains is weaker. Both box A and box B domains preferentially bind to distorted DNA structures. Single-box-HMGB proteins resemble box B. We showed using DNA cyclization kinetics assays (in vitro) and *E. coli* DNA looping assays (in vivo) that an isolated HMG box A domain derived from human HMGB2 folds poorly and does not enhance apparent DNA flexibility in these assays [63]. Surprisingly, substitution of a small number of cationic residues from the N-terminal leader of functional yeast box B protein Nhp6Ap conferred the ability to enhance DNA flexibility. These results demonstrated important roles for cationic leader amino acids in HMGB folding, DNA interaction, and DNA bending [63]. These data reinforce the concept that asymmetric charge neutralization plays a role in the mechanism of Nhp6Ap.

7.4 Future Directions in the Study of HMGB Proteins

Despite years of study (and multiple excellent review articles), we remain ignorant of both the fundamental DNA-bending mechanism of HMGB proteins and their true functions in the complex biology of the nucleus. Considering their small size, simple structure, and remarkable abilities to alter apparent DNA physical properties, the HMGB proteins remain attractive targets for biophysics and molecular biology. A variety of informative experiments can be imagined for the future. Some of these approaches are ongoing through collaborations in which we are fortunate to be involved. Examples are discussed below.

With respect to the fundamental mechanisms of DNA bending by HMGB proteins, time-resolved ensemble and single-molecule fluorescence resonance energy transfer (FRET) experiments have the potential to reveal details of protein-binding and DNA-bending kinetics. These data will allow insight into whether there are kinetic intermediates in the kinking of DNA by HMGB proteins. Does the HMGB protein transiently occupy linear DNA and then engage in bending only after the DNA stochastically samples a kinked geometry through thermal energy or are DNA binding and kinking concerted and inseparable? FRET experiments could also supplement AFM [67] to reveal the extent of DNA flexibility at the site of protein binding. Single-molecule experiments (optical tweezers, AFM, and other methods) also have the unique potential to reveal how HMGB proteins and mutants change both the structure and physical properties of naked DNA and reconstituted chromatin in vitro.

Because the biological roles of HMGB proteins remain unclear, there are tremendous opportunities for molecular biology to shed new light. Techniques such as chromatin immunoprecipitation and chromatin endogenous cleavage [79] offer the potential to map various HMGB proteins on chromatin in situ within living cells. Not only can these methods illuminate the various roles of HMGB proteins in eukaryotes but can also help us to understand whether the ability of HMGB proteins to rescue DNA looping in HU-deficient bacteria is due to direct protein binding to the repression loop. Other techniques, such as chromosome conformation capture [80, 81], provide powerful tools to understand how DNA flexibility and looping are influenced by perturbation of architectural proteins (including endogenous or exogenous HMGB proteins) in mammalian cells, yeast, and bacteria. From an engineering perspective, artificial targeting of HMGB proteins to new chromosomal sites offers intriguing prospects for the artificial control of gene expression. Such targeting could involve direct fusion between a sequence-nonspecific HMGB protein and a sequence-specific targeting factor [82] or development of a noncovalent HMGB recruitment domain to attract the architectural factor to the engineered protein. DNA flexibility and loop geometry are likely to be important in control of gene expression in both prokaryotes and eukaryotes. Building tools to artificially and gene specifically influence these parameters is an exciting challenge.

7.5 Summary

Discovered as small, chromatin-associated eukaryotic proteins, the sequence-nonspecific HMGB proteins are intriguing in their ability to alter the apparent physical properties of DNA in vitro and in vivo. Both ensemble biochemical analyses and single-molecule experiments are revealing details of the DNA-binding and DNA-bending mechanisms of these small proteins. A host of biological functions have been suggested for the HMGB proteins, necessitating application of new assays within living cells. Even after many years of study, much remains to be learned about these simple, powerful, and enigmatic proteins.

Acknowledgments The author is delighted to acknowledge the seminal contributions of present and past students (Julie Soukup, Phil Hardwidge, Andy Rodrigues, Eric Ross, Laura Cassiday, Anne Keating, Robert Dean, Joe Azok, Tessa Davis, Bob McDonald, and Justin Peters), staff scientists (Claudia McDonald, Matt Ferber, Jeff Zimmerman, and Emily Rueter) and collaborators (Jason Kahn, Michael Fried, Stephen Levine, Gerald Manning, Wilma Olson, Luke Czapla, Luis Marky, Anjum Ansari, Ivan Rasnik, Alex Vologodskii, Darrin York, Barry Gold, Rob Phillips, Yitzhak Tor, Yuan Ping Pang, Peter Privalov, Larry Parkhurst, Bill Kirk, Mark Williams, Loren Williams, Nathan Israeloff, Chris Switzer, and Udayan Mohanty). The exceptional skills and intellectual contributions of Nicole Becker are deeply appreciated. It was the teaching of Tom Record that first inspired the author's interest in this field. Work on HMGB proteins in the author's lab has been funded by the Mayo Foundation for Medical Education and Research and by NIH grant GM75965.

References

1. Garcia HG, Grayson P, Han L, Inamdar M, Kondev J, Nelson PC, Phillips R, Widom J, Wiggins PA (2007) Biological consequences of tightly bent DNA: the other life of a macromolecular celebrity. Biopolymers 85:115–130
2. Shimada J, Yamakawa H (1984) Ring-closure probabilities for twisted wormlike chains. Application to DNA. Macromolecules 17:689–698
3. Shore D, Baldwin RL (1983) Energetics of DNA twisting. I. Relation between twist and cyclization probability. J Mol Biol 170:957–981
4. Shore D, Langowski J, Baldwin RL (1981) DNA flexibility studied by covalent closure of short fragments into circles. Proc Natl Acad Sci U S A 78:4833–4837
5. Vologodskaia M, Vologodskii A (2002) Contribution of the intrinsic curvature to measured DNA persistence length. J Mol Biol 317:205–213
6. Ariel G, Andelman D (2003) Persistence length of a strongly charged rodlike polyelectrolyte in the presence of salt. Phys Rev E 67:11805–11814
7. Barrat JL, Joanny JF (1993) Persistence length of polyelectrolyte chains. Europhys Lett 24:333–338
8. Hagerman PJ (1992) Straightening out the bends in curved DNA. Biochim Biophys Acta 1131:125–132
9. Maher LJ III (1998) Mechanisms of DNA bending. Curr Opin Chem Biol 2:688–694
10. Schellman JA, Harvey SC (1995) Static contributions to the persistence length of DNA and dynamic contributions to DNA curvature. Biophys Chem 55:95–114
11. Williams LD, Maher LJ III (2000) Electrostatic mechanisms of DNA deformation. Annu Rev Biophys Biomol Struct 29:497–521
12. Du Q, Smith C, Shiffeldrim N, Vologodskaia M, Vologodskii A (2005) Cyclization of short DNA fragments and bending fluctuations of the double helix. Proc Natl Acad Sci U S A 102:5397–5402
13. Bellomy G, Mossing M, Record M (1988) Physical properties of DNA *in vivo* as probed by the length dependence of the *lac* operator looping process. Biochemistry 27:3900–3906
14. Law SM, Bellomy GR, Schlax PJ, Record MT Jr (1993) In vivo thermodynamic analysis of repression with and without looping in lac constructs. Estimates of free and local lac repressor concentrations and of physical properties of a region of supercoiled plasmid DNA in vivo. J Mol Biol 230:161–173
15. Mossing MC, Record MT Jr (1986) Upstream operators enhance repression of the *lac* promoter. Science 233:889–892
16. Zhang Y, McEwen AE, Crothers DM, Levene SD (2006) Analysis of in-vivo LacR-mediated gene repression based on the mechanics of DNA looping. PLoS ONE 1:e136
17. Travers AA, Ner SS, Churchill MEA (1994) DNA chaperones: a solution to a persistence problem. Cell 77:167–169
18. Crothers DM (1993) Architectural elements in nucleoprotein complexes. Curr Biol 3:675–676
19. Paull TT, Carey M, Johnson RC (1996) Yeast HMG proteins NHP6A/B potentiate promoter-specific transcriptional activation *in vivo* and assembly of preinitiation complexes *in vitro*. Genes Dev 10:2769–2781
20. Paull TT, Haykinson MJ, Johnson RC (1993) The nonspecific DNA-binding and -bending proteins HMG1 and HMG2 promote the assembly of complex nucleoprotein structures. Genes Dev 7:1521–1534
21. Paull TT, Johnson RC (1995) DNA looping by *Saccharomyces cerevisiae* high mobility group proteins NHP6A/B. J Biol Chem 270:8744–8754
22. Aki T, Adhya S (1997) Repressor induced site-specific binding of HU for transcriptional regulation. EMBO J 16:3666–3674
23. Lewis DE, Adhya S (2002) In vitro repression of the gal promoters by GalR and HU depends on the proper helical phasing of the two operators. J Biol Chem 277:2498–2504

24. Lia G, Bensimon D, Croquette V, Allemand JF, Dunlap D, Lewis DE, Adhya S, Finzi L (2003) Supercoiling and denaturation in Gal repressor/heat unstable nucleoid protein (HU)-mediated DNA looping. Proc Natl Acad Sci U S A 100:11373–11377
25. Thomas JO, Travers AA (2001) HMG1 and 2, and related 'architectural' DNA-binding proteins. Trends Biochem Sci 26:167–174
26. Bianchi ME, Beltrame M (2000) Upwardly mobile proteins. Workshop: the role of HMG proteins in chromatin structure, gene expression and neoplasia. EMBO Rep 1:109–114
27. Bustin M (2001) Revised nomenclature for high mobility group (HMG) chromosomal proteins. Trends Biochem Sci 26:152–153
28. Grosschedl R, Giese K, Pagel J (1994) HMG domain proteins: architectural elements in the assembly of nucleoprotein structures. Trends Genet 10:94–100
29. Hock R, Furusawa T, Ueda T, Bustin M (2007) HMG chromosomal proteins in development and disease. Trends Cell Biol 17:72–79
30. Masse JE, Wong B, Yen YM, Allain FH, Johnson RC, Feigon J (2002) The *S. cerevisiae* architectural HMGB protein NHP6A complexed with DNA: DNA and protein conformational changes upon binding. J Mol Biol 323:263–284
31. Swinger KK, Lemberg KM, Zhang Y, Rice PA (2003) Flexible DNA bending in HU-DNA cocrystal structures. EMBO J 22:3749–3760
32. Stott K, Tang GS, Lee KB, Thomas JO (2006) Structure of a complex of tandem HMG boxes and DNA. J Mol Biol 360:90–104
33. Klass J, Murphy F IV, Fouts S, Serenil M, Changela A, Siple J, Churchill ME (2003) The role of intercalating residues in chromosomal high-mobility-group protein DNA binding, bending and specificity. Nucleic Acids Res 31:2852–2864
34. Ohndorf UM, Rould MA, He Q, Pabo CO, Lippard SJ (1999) Basis for recognition of cisplatin-modified DNA by high-mobility-group proteins. Nature 399:708–712
35. Allain FH, Yen YM, Masse JE, Schultze P, Dieckmann T, Johnson RC, Feigon J (1999) Solution structure of the HMG protein NHP6A and its interaction with DNA reveals the structural determinants for non-sequence-specific binding. EMBO J 18:2563–2579
36. Dragan AI, Read CM, Makeyeva EN, Milgotina EI, Churchill ME, Crane-Robinson C, Privalov PL (2004) DNA binding and bending by HMG boxes: energetic determinants of specificity. J Mol Biol 343:371–393
37. Baxevanis AD, Landsman D (1995) The HMG-1 box protein family: classification and functional relationships. Nucleic Acids Res 23:1604–1613
38. Bianchi ME, Beltrame M (1998) Flexing DNA: HMG-box proteins and their partners. Am J Hum Genet 63:1573–1577
39. Wolffe AP (1999) Architectural regulations and Hmg1. Nat Genet 22:215–217
40. Travers AA (2003) Priming the nucleosome: a role for HMGB proteins? EMBO Rep 4:131–136
41. Churchill ME, Changela A, Dow LK, Krieg AJ (1999) Interactions of high mobility group box proteins with DNA and chromatin. Methods Enzymol 304:99–133
42. Jung Y, Lippard SJ (2003) Nature of full-length HMGB1 binding to cisplatin-modified DNA. Biochemistry 42:2664–2671
43. Becker NA, Kahn JD, Maher LJ III (2005) Bacterial repression loops require enhanced DNA flexibility. J Mol Biol 349:716–730
44. Megraw TL, Chae CB (1993) Functional complementarity between the HMG1-like yeast mitochondrial histone HM and the bacterial histone-like protein HU. J Biol Chem 268:12758–12763
45. Calogero S, Grassi F, Aguzzi A, Voigtlander T, Ferrier P, Ferrari S, Bianchi ME (1999) The lack of chromosomal protein HMG1 does not disrupt cell growth but causes lethal hypoglycaemia in newborn mice. Nature Genet 22:276–280
46. Kolodrubetz D, Burgum A (1990) Duplicated *NHP6* genes of *Saccharomyces cerevisiae* encode proteins homologous to bovine high mobility group protein 1. J Biol Chem 265:3234–3239
47. Kolodrubetz D, Kruppa M, Burgum A (2001) Gene dosage affects the expression of the duplicated *NHP6* genes of *Saccharomyces cerevisiae*. Gene 272:93–101

48. Kruppa M, Kolodrubetz D (2001) Mutations in the yeast Nhp6 protein can differentially affect its in vivo functions. Biochem Biophys Res Commun 280:1292–1299
49. Kruppa M, Moir RD, Kolodrubetz D, Willis IM (2001) Nhp6, an HMG1 protein, functions in *SNR6* transcription by RNA polymerase III in *S. cerevisiae*. Mol Cell 7:309–318
50. Aidinis V, Bonaldi T, Beltrame M, Santagata S, Bianchi ME, Spanopoulou E (1999) The RAG1 homeodomain recruits HMG1 and HMG2 to facilitate recombination signal sequence binding and to enhance the intrinsic DNA-bending activity of RAG1-RAG2. Mol Cell Biol 19:6532–6542
51. Jayaraman L, Moorthy NC, Murthy KG, Manley JL, Bustin M, Prives C (1998) High mobility group protein-1 (HMG-1) is a unique activator of p53. Genes Dev 12:462–472
52. Laser H, Bongards C, Schüller J, Heck S, Johnsson N, Lehming N (2000) A new screen for protein interactions reveals that the *Saccharomyces cerevisiae* high mobility group proteins Nhp6A/B are involved in the regulation of the *GAL1* promoter. Proc Natl Acad Sci U S A 97:13732–13737
53. LeRoy G, Orphanides G, Lane WS, Reinberg D (1998) Requirement of RSF and FACT for transcription of chromatin templates in vitro. Science 282:1900–1904
54. Prasad R, Liu Y, Deterding LJ, Poltoratsky VP, Kedar PS, Horton JK, Kanno S, Asagoshi K, Hou EW, Khodyreva SN, Lavrik OI, Tomer KB, Yasui A, Wilson SH (2007) HMGB1 is a cofactor in mammalian base excision repair. Mol Cell 27:829–841
55. Lotze MT, Tracey KJ (2005) High-mobility group box 1 protein (HMGB1): nuclear weapon in the immune arsenal. Nat Rev Immunol 5:331–342
56. Dobi KC, Winston F (2007) Analysis of transcriptional activation at a distance in *Saccharomyces cerevisiae*. Mol Cell Biol 27:5575–5586
57. Hershkovits G, Bangio H, Cohen R, Katcoff DJ (2006) Recruitment of mRNA cleavage/polyadenylation machinery by the yeast chromatin protein Sin1p/Spt2p. Proc Natl Acad Sci U S A 103:9808–9813
58. Rhoades AR, Ruone S, Formosa T (2004) Structural features of nucleosomes reorganized by yeast FACT and its HMG box component, Nhp6. Mol Cell Biol 24:3907–3917
59. Kamau E, Bauerle KT, Grove A (2004) The *Saccharomyces cerevisiae* high mobility group box protein HMO1 contains two functional DNA binding domains. J Biol Chem 279:55234–55240
60. Hall DB, Wade JT, Struhl K (2006) An HMG protein, Hmo1, associates with promoters of many ribosomal protein genes and throughout the rRNA gene locus in *Saccharomyces cerevisiae*. Mol Cell Biol 26:3672–3679
61. Merz K, Hondele M, Goetze H, Gmelch K, Stoeckl U, Griesenbeck J (2008) Actively transcribed rRNA genes in *S. cerevisiae* are organized in a specialized chromatin associated with the high-mobility group protein Hmo1 and are largely devoid of histone molecules. Genes Dev 22:1190–1204
62. Zimmerman J, Maher LJ III (2008) Transient HMGB protein interactions with B-DNA duplexes and complexes. Biochem Biophys Res Commun 371:79–84
63. Sebastian NT, Bystry EM, Becker NA, Maher LJ III (2009) Enhancement of DNA flexibility in vitro and in vivo by HMGB box A proteins carrying box B residues. Biochemistry 48:2125–2134
64. McCauley M, Hardwidge PR, Maher LJ III, Williams MC (2005) Dual binding modes for an HMG domain from human HMGB2 on DNA. Biophys J 89:353–364
65. Skoko D, Wong B, Johnson RC, Marko JF (2004) Micromechanical analysis of the binding of DNA-bending proteins HMGB1, NHP6A, and HU reveals their ability to form highly stable DNA-protein complexes. Biochemistry 43:13867–13874
66. McCauley MJ, Zimmerman J, Maher LJ III, Williams MC (2007) HMGB binding to DNA: single and double box motifs. J Mol Biol 374:993–1004
67. Zhang J, McCauley MJ, Maher LJ III, Williams MC, Israeloff NE (2009) Mechanism of DNA flexibility enhancement by HMGB proteins. Nucleic Acids Res 37:1107–1114
68. Becker NA, Kahn JD, Maher LJ III (2007) Effects of nucleoid proteins on DNA repression loop formation in *Escherichia coli*. Nucleic Acids Res 35:3988–4000

69. Kramer H, Niemoller M, Amouyal M, Revet B, von Wilcken-Bergmann B, Müller-Hill B (1987) Lac repressor forms loops with linear DNA carrying two suitably spaced lac operators. EMBO J 6:1481–1491
70. Muller J, Oehler S, Müller-Hill B (1996) Repression of lac promoter as a function of distance, phase and quality of an auxiliary lac operator. J Mol Biol 257:21–29
71. Oehler S, Eismann ER, Kramer H, Müller-Hill B (1990) The three operators of the lac operon cooperate in repression. EMBO J 9:973–979
72. Zhang Y, McEwen AE, Crothers DM, Levene SD (2006) Statistical-mechanical theory of DNA looping. Biophys J 90:1903–1912
73. Becker NA, Kahn JD, Maher LJ III (2008) Eukaryotic HMGB proteins as replacements for HU in *E. coli* repression loop formation. Nucleic Acids Res 36:4009–4021
74. Bryant Z, Stone MD, Gore J, Smith SB, Cozzarelli NR, Bustamante C (2003) Structural transitions and elasticity from torque measurements on DNA. Nature 424:338–341
75. Ross ED, Hardwidge PR, Maher LJ III (2001) HMG proteins and DNA flexibility in transcription activation. Mol Cell Biol 21:6598–6605
76. Cloutier TE, Widom J (2004) Spontaneous sharp bending of double-stranded DNA. Mol Cell 14:355–362
77. Maher LJ III (2006) DNA kinks available...if needed. Structure 14:1479–1480
78. Ali Azam T, Iwata A, Nishimura A, Ueda S, Ishihama A (1999) Growth phase-dependent variation in protein composition of the *Escherichia coli* nucleoid. J Bacteriol 181:6361–6370
79. Schmid M, Durussel T, Laemmli UK (2004) ChIC and ChEC: genomic mapping of chromatin proteins. Mol Cell 16:147–157
80. Ansari A, Hampsey M (2005) A role for the CPF 3′-end processing machinery in RNAP II-dependent gene looping. Genes Dev 19:2969–2978
81. Dekker J, Rippe K, Dekker M, Kleckner N (2002) Capturing chromosome conformation. Science 295:1306–1311
82. Tang LJ, Li J, Katz DS, Feng JA (2000) Determining the DNA bending angle induced by non-specific high mobility group-1 (HMG-1) proteins: a novel method. Biochemistry 39:3052–3060
83. Swinger KK, Rice PA (2004) IHF and HU: flexible architects of bent DNA. Curr Opin Struct Biol 14:28–35

Chapter 8
DNA Interactions with Single-Stranded DNA Binding Proteins and Retroviral Nucleic Acid Chaperones by Force Spectroscopy

Mark C. Williams and Ioulia Rouzina

8.1 Introduction

In this chapter, we compare and contrast the biophysical properties of two important classes of nonsequence-specific nucleic acid binding proteins: single-stranded DNA binding proteins (SSBs) from bacteriophages and nucleocapsid proteins (NCs) from retroviruses. The SSBs comprise the primary noncatalytic component of the DNA replication machinery, functioning to bind and protect all available single-stranded DNA (ssDNA) at the DNA replication fork. The NC proteins are necessary components of the RNA reverse transcription complex, playing roles analogous to those of SSBs in DNA replication. The primary function of NC is to facilitate RNA and DNA refolding into low-energy conformations, a property referred to as nucleic acid chaperone activity. This function is necessary for almost every step of retroviral reverse transcription. Application of single molecule DNA force spectroscopy methods has advanced the characterization of bacteriophage SSB and retroviral NC interactions with single-stranded (ss) and double-stranded (ds) DNA molecules beyond what was known from conventional solution studies. By applying single molecule techniques to wild type and mutant NC and SSB proteins from several biological systems, we have been able to illustrate a continuous spectrum of properties that distinguish proteins as effective nucleic acid chaperones or ssDNA stabilizers.

8.2 Single-Stranded DNA Binding Proteins

Single-stranded DNA binding (SSB) proteins are essential DNA replication proteins that participate in DNA replication in all three domains of life [1]. The most basic function of SSBs is to bind single-stranded DNA produced transiently during DNA

M.C. Williams (✉)
Department of Physics, Northeastern University, Boston, MA, USA
e-mail: mark@neu.edu

M.C. Williams and L.J. Maher, III (eds.), *Biophysics of DNA-Protein Interactions*, Biological and Medical Physics, Biomedical Engineering,
DOI 10.1007/978-0-387-92808-1_8,

replication or repair processes. Binding protects ssDNA from nucleases and inhibits the formation of secondary structure within the ssDNA. However, SSBs are also known to interact with other replication, recombination, and repair proteins, and may directly facilitate these processes, perhaps even recruiting additional proteins to ssDNA [2]. SSBs have the common feature that they bind ssDNA through an oligonucleotide/oligosaccharide (OB) fold, but SSBs from different replication systems vary significantly in the number of OB folds, the mechanism of interaction with other proteins, the multiplicity of subunits in the binding unit, and in cooperativity. Here, we discuss the simplest SSB proteins, derived from bacteriophage replication systems. Despite the simple replication system of the bacteriophages and the presence of only one OB fold in each monomeric binding unit, the DNA binding mechanisms of bacteriophage SSBs are only beginning to be understood.

The most well-studied SSB is the bacteriophage T4 gene 32 protein (gp32), which was the first SSB to be isolated [3]. gp32 plays an essential role in DNA replication, recombination, and repair [4]. gp32 also binds preferentially to ssDNA relative to double-stranded DNA (dsDNA), and displays high cooperativity only when binding to ssDNA. This cooperative binding allows gp32 to form clusters along ssDNA, enhancing ssDNA protection from nucleases and preventing secondary structure formation. gp32 has three major structural domains, as determined by proteolytic cleavage [5]. The central, core domain (residues 22–253) contains the DNA binding site, while the N-terminal domain (residues 1–21) is involved in homotypic protein–protein interactions that lead to highly cooperative ssDNA binding. The C-terminal domain (CTD) (residues 254–301) is involved in heterotypic protein–protein interactions, which may help to regulate its DNA affinity or to recruit other proteins to ssDNA bound by gp32. The acidic CTD also diminishes the ability of the protein to bind nucleic acids, a property that will be described in detail later [6–9].

Like T4, bacteriophage T7 serves as a model replication system that can be used to understand how the complex DNA replication process is coordinated with relatively few proteins. In fact, T7 only requires four proteins to form the T7 replisome: DNA polymerase with its processivity factor thioredoxin, the helicase/primase, and the SSB protein gp2.5 [10]. Like T4 gp32, T7 gp2.5 also binds to ssDNA with a single OB fold, and it binds preferentially to ssDNA. However, T7 gp2.5 differs in that it binds ssDNA with little or no cooperativity and forms a dimer in solution [11], whereas gp32 is monomeric. Dimerization involves strong electrostatic interactions between the acidic C-termini of monomers and the DNA binding site of its dimer partner. Thus, in order for gp2.5 to bind DNA, this dimerization interaction must be broken [12, 13]. gp2.5 interacts with other components of the T7 replisome and is also involved in T7 DNA replication and recombination [14–21].

The SSB proteins from bacteriophage T4 and T7 locate ssDNA binding sites through weak dsDNA binding followed by rapid diffusion along dsDNA in one dimension. Both proteins dissociate from ssDNA slowly, on the time scales of tens of seconds to several minutes. Such slow dissociation is likely necessary for the proteins to perform their primary sequestering functions. On the other hand, such slow binding kinetics may also inhibit recombination events, in which the protein must be actively removed to promote strand annealing or other nucleic acid rearrangements. In T4, slow binding kinetics is overcome by recombination mediator proteins such as UvsY, which facilitate the removal of T4 gp32 from ssDNA during recombination [22].

8.3 Nucleic Acid Chaperone Proteins from Retroviruses

In contrast to the relatively complex bacteriophage recombination system, nucleic acid recombination and rearrangement in retroviruses is facilitated primarily by the nucleocapsid (NC) protein. NC proteins are generally small, cationic proteins of 50–100 amino acids, less than one-third the size of the bacteriophage SSBs. All orthoretroviral NC proteins contain one or two zinc fingers as a primary structural feature, in which the zinc is coordinated by a CCHC motif [23, 24]. NC coats nucleic acids in a manner reminiscent of SSBs, but is more readily displaced from the nucleic acid by other proteins. We do not fully understand the mechanism by which NC proteins bind with high affinity to ssDNA yet facilitate nucleic acid rearrangement. During the initial retroviral assembly process, the NC protein is a small component of a polyprotein called Gag. The latter contains three major proteins: NC, capsid, and matrix. When the virus is assembled, NC binds and packages two identical copies of genomic RNA [25–27]. The matrix protein on the opposite end binds to the cell membrane, creating a spherical structure with NC and RNA on the inside and the membrane on the outside. Later, during retroviral maturation required for infectivity, Gag is cleaved by the retroviral protease and the overall structure of the virus is rearranged. NC, capsid, and matrix are released, and the structure of the NC-bound RNA also changes in a process called dimerization [28–30]. NC is released from Gag as NCp15, but is then further cleaved to NCp9 and then to NCp7, mature NC. These precursor forms of NC have different DNA binding characteristics [31, 32], and there is some evidence that they may play specific roles during HIV-1 replication [33–35]. Because dimerization requires rearrangement of RNA secondary structure, this process likely requires the nucleic acid chaperone activity of NC. It is not yet clear whether Gag itself can facilitate significant secondary structure rearrangements, or if rearrangements such as dimerization require NC to be first cleaved from Gag.

After the retrovirus infects the host cell, its genomic RNA must be copied into dsDNA so that the viral genomic DNA can be inserted into the host cell genome. The process of copying the ssRNA into proviral dsDNA is called reverse transcription, and is performed by the enzyme reverse transcriptase (RT) with the participation of NC, as will be discussed in more detail below. The process of inserting viral dsDNA into the host genome is referred to as integration, performed by the enzyme integrase with the participation of NC [33, 36]. After integration, viral RNA is transcribed and packaged or translated into proteins and more viruses are assembled to continue the viral life cycle [37].

NC is required for multiple steps in reverse transcription, including tRNA primer annealing and strand transfer events. In these processes, NC serves primarily as a nucleic acid chaperone. This function is best understood through a specific example. Reverse transcription begins with a single molecule of single-stranded RNA. However, this lowest energy form of this RNA involves intramolecular base pairs, creating a complex structure. Initiation of reverse transcription requires an annealed primer with a free 3′ terminus to act as nucleophile. In HIV-1, the primer is provided by tRNA^{Lys,3}, which is annealed to a complementary primer binding site on the HIV-1 genomic RNA with the assistance of NC [38–40]. Reverse transcription then proceeds from this point until the end of the template is reached. As this occurs, the RNaseH domain of RT degrades the RNA that has been copied. In order for reverse transcription to continue, the new ssDNA strand must be transferred to the other end of the genome (presumably brought into proximity by folding in three dimensions) and annealed to a complementary RNA sequence. This seemingly unlikely process is termed minus-strand transfer (Fig. 8.1). Remarkably, stable secondary structures (over 20 bp each) are formed by both the DNA that has been polymerized by RT and the complementary RNA. Without NC, these two structures cannot anneal to form the DNA–RNA hybrid duplex required for completion of reverse transcription. NC is required for obligatory and random strand transfer events observed during reverse transcription [41–45] and other nucleic acid refolding events [46].

Because NC is an essential component of every stage of the HIV-1 life cycle, it is an intriguing anti-HIV drug target [47–49]. The most obvious candidates for

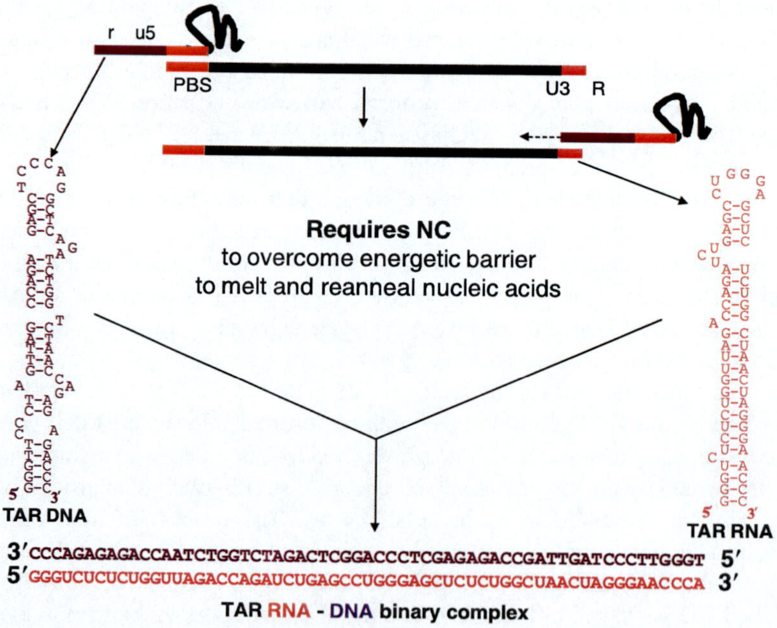

Fig. 8.1 Schematic diagram of minus-strand transfer in HIV-1. In this process, a DNA hairpin containing 21 bp is annealed to the TAR RNA structure, which contains 24 bp, to form a 98-bp DNA–RNA hybrid structure

HIV-1 antagonists are compounds that target the zinc finger domains of NC. Such agents initially showed great promise, but were not pursued due to toxicity [50]. More sophisticated methods are needed to target NC and Gag. This will require a detailed mechanistic understanding of NC function in retroviral replication. Here, we describe a detailed biophysical mechanism for the nucleic acid chaperone function of NC. We identify specific regions of the protein responsible for this activity, and we compare the activity of HIV-1 NC with that of NC proteins from other retroviruses. Finally, we compare the DNA interactions of retroviral NC proteins with those of bacteriophage SSBs, which exhibit many similar characteristics, despite their significantly different functions during replication.

8.4 Single Molecule Force Spectroscopy Studies of Bacteriophage T4 and T7 SSB Proteins

Figure 8.2a shows a schematic diagram of an optical tweezers instrument used for stretching single DNA molecules. In these experiments, a single bacteriophage λ genomic DNA (48,500 bp) is labeled on each end with several biotin molecules. Only one strand is labeled at each DNA terminus so the DNA is free to rotate as it is stretched. Similar experiments can be done with magnetic tweezers and atomic force microscopes. All three instruments allow one to stretch single DNA molecules and measure force–extension curves with nanometer or better position resolution and piconewton (pN) or better force resolution [51].

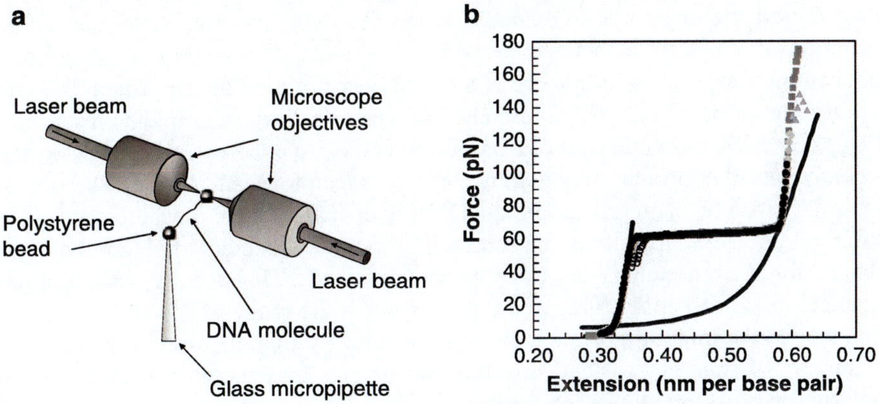

Fig. 8.2 (**a**) Schematic diagram of dual beam optical tweezers instrument for stretching single DNA molecules. (**b**) Typical DNA stretching (*various solid symbols*) and relaxation (*open symbols*) data from single λ DNA molecules in 100 mM Na^+ solution (10 mM HEPES, pH 7.5). The *left solid line* is a theoretical curve representing the wormlike chain model for a stiff polymer corresponding to dsDNA, while the *right solid line* is a theoretical curve representing the freely jointed chain model for a more flexible polymer, corresponding to ssDNA. When the approximately constant force transition exhibited by the data at ~65 pN is reversible, the area between the stretching data and the ssDNA stretching curve represents the DNA melting free energy

As DNA is stretched to its B-form contour length, the required force begins to increase rapidly as the randomly coiled DNA is straightened and the helical form itself begins to stretch [52, 53]. At forces near 60 pN, an abrupt phase transition occurs. In this regime, DNA can be stretched to 1.7 times its B-form contour length with very little increase in force (Fig. 8.2b) [54–56]. It has been demonstrated that this transition, referred to as the DNA overstretching transition, is the force analog of thermal melting. The DNA is converted from dsDNA to ssDNA, and the fraction of melted DNA increases as the length of the construct increases through the transition [57, 58]. For example, solution conditions such as high and low pH, high temperature, and low salt, which lower the DNA thermal melting temperature, also correspondingly lower the overstretching force [59–61]. In addition, it has been shown that the fraction of DNA base pairs exposed to solution corresponds to the position in the transition, proving that the DNA is melted during the process [62]. Finally, small molecules and proteins that destabilize [63] and stabilize [56, 64] DNA in thermal melting correspondingly decrease or increase the required overstretching force [65, 66]. Therefore, the DNA overstretching transition should properly be considered as a DNA force-induced melting transition.

8.5 Single Molecule Measurements of Equilibrium ssDNA Binding by SSB Proteins

When DNA is stretched in the presence of single-stranded DNA binding proteins, the required melting force is reduced. When the experiment is performed under conditions that protein association and dissociation occurs more rapidly than force-induced melting, the melting force is a direct measurement of the protein binding affinity. However, when DNA is stretched at typical pulling rates in the presence of T7 gp2.5 or T4 gp32, this process is not reversible, as shown in Fig. 8.3. In the presence of either protein, the melting force is indeed lower than in the absence of protein. However, upon relaxation, the DNA does not reanneal, as indicated by the different relaxation and stretching forces. Theoretically, one could obtain the equilibrium force by stretching the DNA extremely slowly. However, it is much more practical to instead apply force at a typical rate (in the range of 25–100 nm/s) and then stop at the midpoint of the melting transition. By then measuring the force as a function of time at constant position, one obtains an exponential decay of the melting force. At long times (on the order of 30 min), the force eventually becomes essentially constant, revealing the equilibrium DNA melting force. The equilibrium protein–ssDNA binding affinity, K_{ss} ,can be obtained by measuring the equilibrium melting force as a function of concentration and fitting that dependence to a simple thermodynamic model [7]:

$$F_m = F_m^0 - \frac{2k_B T}{n_{ss}\,\Delta x}\ln(1+K_{ss}c) \tag{8.1}$$

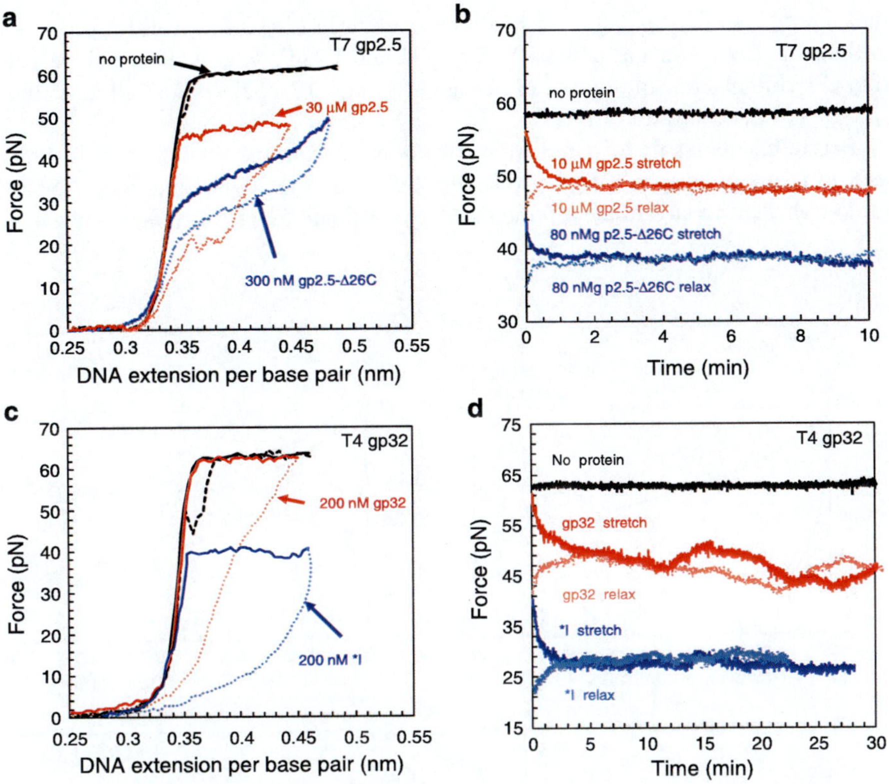

Fig. 8.3 (**a**) Stretching (*solid line*) and relaxation (*dotted line*) curves for λ-DNA in the absence of protein (*black*) and in the presence of 30 μM T7 gp2.5 (stretch – *red*, relax – *pink*) or in the presence of 300 nM T7 gp2.5-Δ26C (stretch – *blue*, relax – *light blue*). Data are taken in 10 mM HEPES pH 7.5, 50 mM [Na^+] and with a pulling rate of 25 nm/s. (**b**) DNA stretching force at constant position as a function of time in the absence of protein (*black*) and in the presence of 10 μM T7 gp2.5 (*red*) or 80 nM gp2.5-Δ26C (*blue*). The *light colored curves* show the time dependence of DNA reannealing, in which the DNA molecule has been melted by force and relaxed back to the same position in the presence of 10 μM gp2.5 (*pink*) and 80 nM gp2.5-Δ26C (*light blue*). Data are taken in 10 mM HEPES pH 7.5, 25 mM [Na^+]. (**c**) Stretching (*solid line*) and relaxation (*dotted line*) curves for λ-DNA in the absence of protein (*black*) and in the presence of 200 nM gp32 (*red*) or 200 nM *I (*blue*). Data are taken in 10 mM HEPES pH 7.5, 100 mM [Na^+] with a pulling rate of 100 nm/s. (**d**) DNA stretching force as a function of time at constant position in the absence of protein (*black*) and in the presence of 200 nM gp32 (*red*) or 200 nM *I (*blue*) as well as the time dependence of DNA reannealing in which the DNA molecule has been melted by force and relaxed back to the same position in the presence of 200 nM gp32 (*pink*) or 200 nM *I (*light blue*). Figure is adapted from [67] and is used with permission from the publisher (Institute of Physics)

Here F_m is the measured equilibrium melting force at a given protein concentration, F_m^0, is the DNA melting force in the absence of protein, n_{ss}, is the ssDNA binding site size in nucleotides, Δx is the change in length of DNA upon melting at a given force, and k_B is Boltzmann's constant. Thus, by fitting F_m vs. C, one obtains the measurements of K_{ss} and n_{ss}. Such fits are shown in Fig. 8.4 as solid lines, along with the measured data as points, for T4 gp32 and T7 gp2.5, as well as their respective C-terminal truncation mutants, T4 gp32 *I and T7 gp2.5-Δ26C, all as a function of salt concentration [67].

Figure 8.5 shows the quantitative results derived from the fits of Fig. 8.4, shown as K_{ss} for the wild-type (WT) gp32 and gp2.5 in Fig. 8.4a, and as K_{ss} for the CTD-truncated version of these proteins, T4 gp32 *I and T7 gp2.5-Δ26C, in Fig. 8.4b.

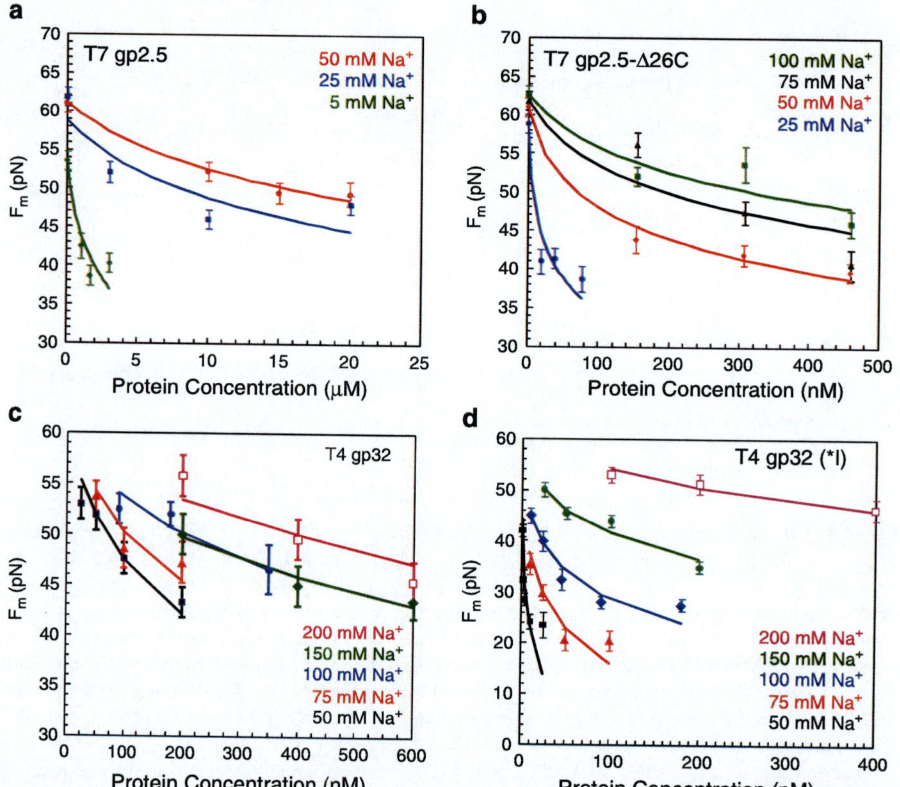

Fig. 8.4 (**a**) Measured DNA equilibrium melting force as a function of protein concentration for T7 gp2.5 for solutions of 5 mM Na^+ (*diamond*), 25 mM Na^+ (*square*), and 50 mM Na^+ (*circle*). (**b**) Measured DNA equilibrium melting force as a function of protein concentration for T7 gp2.5-Δ26C for solutions of 25 mM Na^+ (*circle*), 50 mM Na^+ (*diamond*), 75 mM Na^+ (*triangle*), and 100 mM Na^+ (*square*). (**c**) Measured DNA equilibrium melting force as a function of protein concentration for T4 gp32 for solutions of 50 mM Na^+ (*square*), 75 mM Na^+ (*triangle*), 100 mM Na^+ (*circle*), 150 mM Na^+ (*diamond*), and 200 mM Na^+ (*open square*). (**d**) Measured DNA equilibrium melting force as a function of protein concentration for the T4 gp32 C-terminal truncate *I for solutions of 50 mM Na^+ (*square*), 75 mM Na^+ (*triangle*), 100 mM Na^+ (*diamond*), 150 mM Na^+ (*circle*), and 200 mM Na^+ (*open square*). For all parts, lines were fitted to data using (8.1) and a χ^2 analysis. Figure is taken from [67] and is used with permission from the publisher (Institute of Physics)

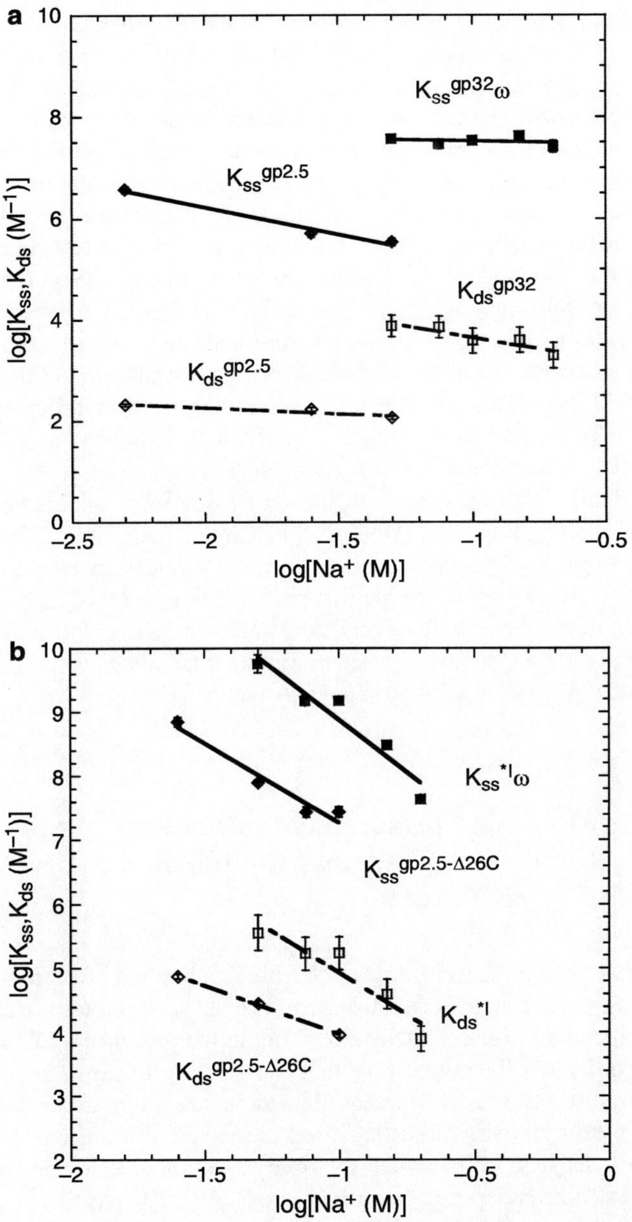

Fig. 8.5 (**a**) Measured equilibrium association constants for T4 gp32 (*square*) and T7 gp2.5 (*diamond*) to ssDNA (*filled symbol*) or dsDNA (*open symbols*). (**b**) Measured (*symbols*) equilibrium association constants for T4 gp32 C-terminal truncate *I (*square*) or T7 gp2.5-Δ26C (*diamond*) to ssDNA (*filled symbol*) or dsDNA (*open symbol*). Lines are fits to the data. Figure is modified from [67] and is used with permission from the publisher (Institute of Physics)

The K_{ds} results are discussed below. For both T4 and T7 proteins, the wild-type SSB shows very weak salt dependence, while the C-terminal truncation shows a much stronger salt dependence. This change may reflect the high negative charge of the C terminus in both cases. The C-terminal tail interacts electrostatically with the cationic DNA binding site in the core of the protein. In the case of T4 gp32, the C-terminal tail binds directly to the DNA binding site of the same molecule, inhibiting ssDNA binding. In the case of T7 gp2.5, the C-terminal tail binds to the ssDNA binding site of its dimer partner, facilitating dimerization and inhibiting ssDNA binding. In both cases, removal of the C terminus exposes the DNA binding site on the protein core and facilitates ssDNA binding by the protein. This interpretation, which explains the difference between the wild type and C-terminal truncates of both proteins, is supported by a wealth of bulk biochemical data on T4 gp32 [68–72] and T7 gp2.5 [14, 15, 73–77]. However, a complete understanding of the salt-dependent ssDNA binding and helix-destabilization properties of these well-studied model SSB proteins was not available until these force spectroscopy experiments [6–9, 13, 78]. Two unique aspects of single molecule force spectroscopy were important in this case. First, thermal melting studies of DNA binding proteins are limited by the thermal stability of the protein. Workers must use short oligonucleotides or use conditions of very low salt to melt DNA without melting protein. Force-induced melting avoids this issue. Second, force spectroscopy experiments allow direct control of the double- or single-stranded state of the DNA molecule in real time, allowing control of the rate of protein reaction with the ssDNA or dsDNA lattice.

8.6 Single Molecule Measurement of the Rate of SSB Association with ssDNA and Equilibrium dsDNA Binding by SSB Proteins

In addition to allowing measurement of equilibrium binding of proteins to ssDNA, single molecule force spectroscopy also permits analysis of protein–ssDNA binding kinetics. As shown in Fig. 8.3, DNA stretching in the presence of T4 gp32 and T7 gp2.5 displays hysteresis, suggesting that the binding reaction is not reversible under the conditions tested. The nonequilibrium nature of the experiments suggests that protein binding is slow and that the kinetics of binding can be measured. To pursue this analysis, we exploited the observation that the force at which force-induced melting occurs depends on pulling rate [6, 78]. The physics behind this rate dependence can be described by a simple model, in which the transition occurs when the rate of pulling (nm/s) is equal to the rate at which proteins can bind to ssDNA exposed on the end of the DNA molecule due to thermal fluctuations. The rate of fluctuational opening of ssDNA at the ends of the molecule decreases exponentially with the free energy of base pair opening, which is of the order of $2k_BT$ under the solution conditions used here. In order for a protein to bind, a number of bases equal to the binding site size of the protein must be exposed. In the case of both T7 gp2.5 and T4 gp32, the site size is $n_{ss} \approx 7$, so the rate of protein binding to ssDNA in the absence of force is negligible. However, in the presence of force, the free energy of melting decreases significantly such that fluctuational opening of the base pairs at the end of the molecule becomes significant. The condition for the pulling rate v to equal the fluctuational opening rate is given by:

$$F_k(v) = \frac{k_B T}{n_{ss}\,\Delta x} \ln\left(\frac{v}{2n_{ss}\,\Delta x\, k_a}\right) + F_m^0. \tag{8.2}$$

Here, $F_k(v)$ is the pulling rate-dependent observed melting force in the presence of protein, Δx is the change in length of DNA upon conversion from dsDNA to protein-bound ssDNA, k_a is the rate at which proteins encounter a new ssDNA site, and the factor of 2 reflects the possibility of ssDNA site opening at either end of the molecule. The other variables have the same meanings as in (8.1).

Measurements of $F_k(v)$ vs. in (v) for different concentrations of T7 gp2.5-Δ26C and T4 gp32 *I are shown in Fig. 8.6a, c, respectively. For both proteins, the slope of the plot is constant and independent of protein concentration, and the same holds for the wild-type proteins (data not shown). Based on (8.2), the slope can be used to directly determine n_{ss} because the parameters preceding the natural logarithm are known. We obtain $n_{ss} \approx 7$, in agreement with bulk measurements of the binding site size for these proteins. In addition to allowing the estimation of binding site size, the condition $F_k(v) = F_m^0$ gives the rate at which proteins encounter binding sites at the ends of the molecule. These rates can be calculated theoretically from the "first passage time" for the protein [79, 80]. The results, shown in Fig. 8.4b, d, were initially surprising because they do not depend linearly on protein concentration and they exceed the theoretical 3D diffusion limit for colliding with the two available binding sites on the molecule.

The surprisingly rapid kinetics with which SSB proteins find stretch-induced ssDNA binding sites can be explained by recognizing that these proteins need not diffuse from bulk solution, but can be captured from local dsDNA to which they are prebound. This hypothesis explains the fact that the apparent bimolecular rate of SSB-ssDNA association exceeds its 3D diffusion limit because this rate is in fact not bimolecular, but rather depends on the square of the protein concentration in solution (Fig. 8.6b, d [81]). Indeed, if sufficient protein is prebound to dsDNA, then the time required to encounter a new ssDNA site is determined only by 1D sliding of the protein on dsDNA, a function of the square of bulk protein concentration. It has been shown that the rate under these conditions can be expressed as [6]:

$$k_a = (2\Theta / n_{ds})^2\, k_s \tag{8.3}$$

Here, k_a is the observed rate of finding the binding site from (8.2), k_s is the sliding rate along the DNA molecule, which is a property of the protein related to its 1D diffusion constant, and n_{ds} and Θ are the binding site size and fractional saturation of the protein on the dsDNA lattice. The fractional saturation of the DNA lattice is determined by the binding affinity of the protein for dsDNA, which in turn is estimated by the McGhee–von Hippel binding isotherm:

$$\Theta = K_{ds}\, n_{ds}\, C \frac{(1-\Theta)^{n_{ds}}}{(1-\Theta+\Theta/n_{ds})^{n_{ds}-1}} \tag{8.4}$$

By combining (8.3) and (8.4), we then fit the data in Fig. 8.6b, d with three fitting parameters K_{ds}, n_{ds}, and k_s. We obtain typical values for the latter quantities and

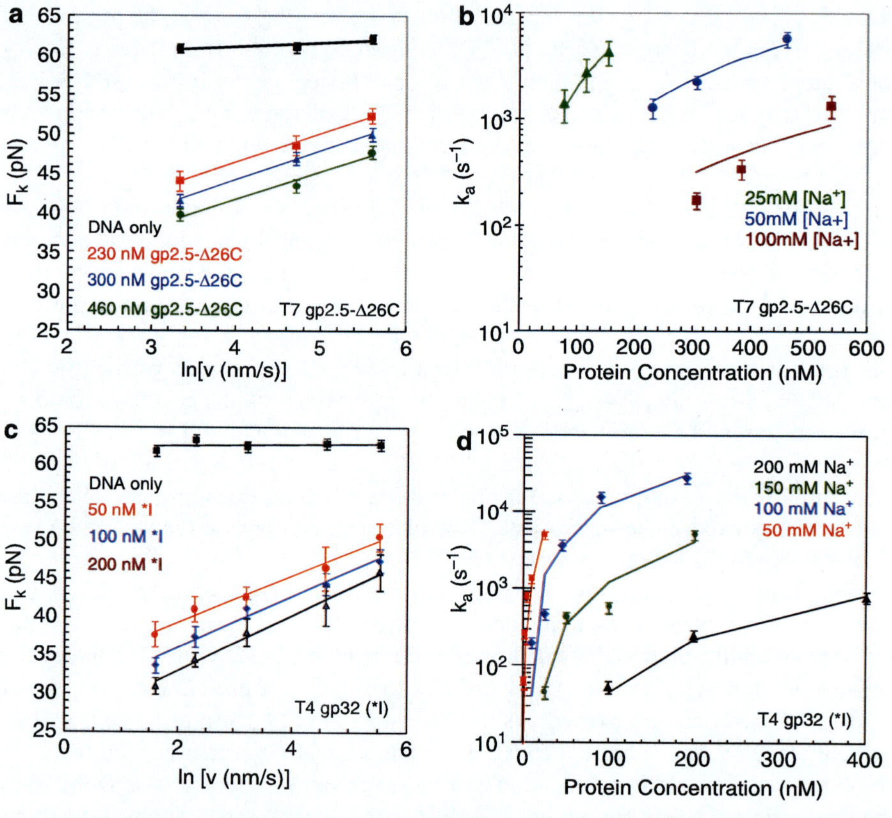

Fig. 8.6 (**a**) Measured nonequilibrium DNA melting force, F_k, as a function of the rate of pulling, v, in 50 mM Na^+ in the absence of protein (*black square*), in the presence of 230 nM gp2.5-Δ26C (*red square*), 300 nM gp2.5-Δ26C (*blue triangle*), and 460 nM gp2.5-Δ26C (*green circle*). (**b**) Protein binding rate (k_a) as a function of protein concentration for gp2.5-Δ26C in 25 mM Na^+ (*green triangle*), 50 mM Na^+ (*blue circle*), and 100 mM Na^+ (*brown square*). (**c**) Measured nonequilibrium DNA melting force, F_k, as a function of the rate of pulling, v. Data are shown in 100 mM Na^+ and in the absence of protein (*black square*), in the presence of 50 nM *I (*red circle*), 100 nM *I (*blue diamond*), and 200 nM *I (*black triangle*). (**d**) Protein binding rate (k_a) as a function of protein concentration for *I in 50 mM Na^+ (*red circle*), 100 mM Na^+ (*blue diamond*), 150 mM Na^+ (*green circle*), and 200 mM Na^+ (*black triangle*). All data were taken in 10 mM HEPES pH 7.5 and fits to the data are shown as *solid lines*. Figure is taken from [67] and is used with permission from the publisher (Institute of Physics)

are most interested in the values for K_{ds}, the equilibrium association constant for protein binding to dsDNA. These values are plotted in Fig. 8.5, along with the corresponding values for K_{ss}, the equilibrium constant for binding to ssDNA.

We have outlined single molecule force spectroscopy methods for quantifying equilibrium binding of ssDNA binding proteins to both ssDNA and dsDNA. The results, summarized in Fig. 8.5, reveal the mechanism by which these proteins interact with both forms of DNA. All the SSB proteins studied, T7 gp2.5, T4 gp32, and their respective C-terminal truncates, bind to dsDNA and diffuse in one dimension along the DNA molecule until they encounter available ssDNA binding sites. This relatively weak sequence nonspecific and primarily electrostatic prebinding of SSBs to dsDNA is likely a very general mechanism facilitating SSB association with ssDNA under in vivo conditions, when ssDNA exists only transiently at boundaries between dsDNA and ssDNA. Interestingly, a similar mechanism of 1D diffusion on dsDNA is also common for facilitating the search by specific dsDNA binding proteins for their unique dsDNA binding sites [82–87]. All four proteins discussed here (gp32 and gp2.5 as well as their CTD truncates presented in Fig. 8.5) have ssDNA binding affinities that exceed dsDNA binding affinity by about four orders of magnitude. All therefore share a similar ability to destabilize dsDNA in equilibrium. Interestingly, while gp32 achieves this binding preference for ssDNA through strong cooperative binding, T7 gp2.5 does not exhibit significant cooperative binding [11].

The ssDNA and dsDNA binding by T7 gp2.5 and T4 gp32 wild-type proteins is only weakly salt dependent, while DNA binding by their C-terminal truncates is strongly salt dependent. This is explained in both cases by a model in which the CTD modulates DNA binding by interacting with its own DNA binding site (or that of its dimer partner in the case of T7 gp2.5), thus inhibiting DNA interactions. A schematic diagram of this model for these CTD interactions is shown in Fig. 8.7. This model explains the weak salt dependence of DNA binding by the wild-type SSBs because lower salt causes an increase in affinity of the C terminus for the protein, thereby offsetting any increase in DNA binding affinity for the protein's DNA binding site in low salt. Alternatively, this process can be viewed in the framework of counterion condensation, in which a specific number of counterions must be released for charged molecules to bind to DNA. In the case of T4 gp32, approximately three counterions are released when its C-terminal truncate binds DNA in low salt, as determined from the slope of the log(K) vs. log[Na^+] graph shown in Fig. 8.5 for *I binding to both ssDNA and dsDNA. In contrast, for wild-type T4 gp32, the slope is very small for both ssDNA and dsDNA. Therefore, the net number of counterions released is approximately zero in this reaction. However, according to the model, the DNA binding site is the same in both cases. In order for the net counterions released to be zero, the CTD must bind three counterions when it is exposed to solution as the wild-type protein binds DNA in low salt. Such compensatory ion uptake was previously observed in other protein–DNA interactions [88]. Therefore, this model provides a compelling explanation for the unusual behavior of the wild-type protein at low salt. In high salt, the slope of the binding curve in Fig. 8.5 for both wild-type T4 gp32 and *I is about 7. This suggests that seven counterions are released upon binding, which can be explained by recognizing that

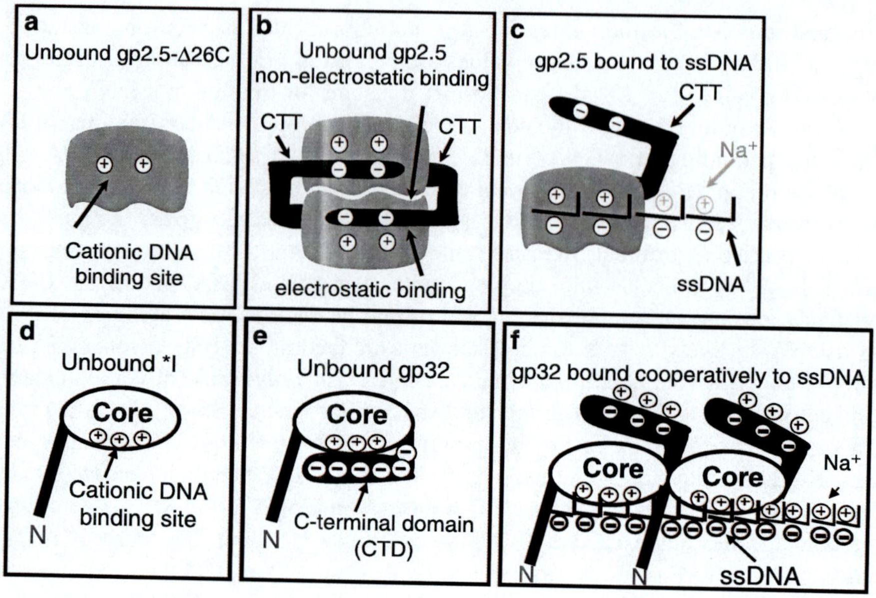

Fig. 8.7 Models for regulation of DNA binding by the C-terminal regions of T4 and T7 single-stranded DNA binding proteins. (**a**) Model for T7 gp2.5-Δ26C. This protein has a dimer interface and cationic DNA binding site. (**b**) Model for T7 gp2.5 dimerization. In this model, the C terminus of each dimer partner binds to the DNA binding site of the other monomer, thus occluding DNA binding. (**c**) Model for DNA binding by T7 gp2.5. In this model, the dimer interaction must be broken so that the C terminus is exposed to solution, allowing DNA binding. (**d**) Model for the T4 gp32 C-terminal truncate *I. This protein contains an available cationic DNA binding site. (**e**) Model for T4 gp32 in low salt. Under these solution conditions, the C terminus binds to the DNA binding site of the monomer, which occludes DNA binding by the protein. (**f**) Model for DNA-bound T4 gp32. In this model, the C-terminal domain must be unbound from the core domain and exposed to solution in order for the protein to bind DNA. Figure is taken from [67] and is used with permission from the publisher (Institute of Physics)

the acidic C terminus of T4 gp32 will already be exposed to solution in high salt and will already have counterions bound to its C terminus. Therefore, in high salt, the CTD does not participate in the reaction and does not affect the salt dependence of protein–DNA binding. This explains why T4 gp32 and *I have similar salt dependences, but it does not explain the fact that the slope of the graph increases to 7 in high salt. The latter might be explained by postulating that anions (in this case, Cl^- ions) can bind to the positively charged DNA binding site of the protein at high anion concentration. This model is summarized in Fig. 8.7. In reference [9], all the available single molecule and bulk DNA binding data for T4 gp32 and *I in different salt concentrations with different anions and DNA substrates are fit to this model with a total of four fitting parameters, fully quantifying and validating the schematic model presented in Fig. 8.7a–c. A similar schematic model describes the salt-dependent binding of T7 gp2.5 to ssDNA and dsDNA. Here, there is an added complication that the CTD is involved in a dimerization interaction, which must be broken before DNA binding can occur. Thus, as the salt concentration is lowered, the dimerization interaction becomes stronger. In the absence of the C terminus and any strong dimerization interaction, lower salt increases the DNA binding affinity, as was the case for T4 gp32 *I. This additional interaction can also be described in terms of counterion condensation, in which two counterions are released upon T7 gp2.5-Δ26C binding to DNA. The model in Fig. 8.7d, e is quantified in [13] and compared with available bulk data. T7 gp2.5 is not as well studied as T4 gp32, so less bulk data are available for comparison with single molecule results. Nonetheless, the available data for T7 gp2.5 and T7 gp2.5-Δ26C are quantitatively consistent with the model in Fig. 8.7d, e.

It is important to note that for both bacteriophage T4 and T7, the CTD is known to interact with other replication proteins. It is therefore possible that DNA binding can be regulated by these interactions. Alternatively, the C terminus may serve to recruit other replication proteins to the DNA replication fork, as the C terminus becomes available after DNA binding by SSB protein. CTD regulation of DNA binding can be controlled by solution conditions or by the presence of CTD binding proteins. This mechanism has the potential to strongly alter the nature of SSB interactions with both ssDNA and dsDNA.

8.7 Single Molecule Force Spectroscopy Studies of Retroviral Nucleic Acid Chaperone Proteins

Because nucleic acid chaperone proteins facilitate the rearrangement of nucleic acid secondary structure [89], they are expected to bind strongly to both dsDNA and ssDNA (or RNA). On the other hand, these proteins must destabilize nucleic acid base pairing to some extent in order to facilitate rearrangements. Initial single molecule experiments on HIV-1 NC confirmed that this protein destabilizes DNA secondary structure, lowering the overall free energy of DNA melting [90, 91]. The effect of HIV-1 NC on DNA force-induced melting is illustrated in Fig. 8.8. Figure 8.8a shows ssDNA and dsDNA stretching curves. By calculating the area between the two curves, we find that the DNA melting free energy is about $2k_BT$(1 k_BT is about 0.6 kcal/mol). In contrast, in the presence of 10 nM HIV-1 NC, which is near saturated binding, the overall DNA melting free energy (averaged over the entire sequence of this DNA molecule) is reduced to about 1 k_BT. Thus, HIV-1 NC destabilizes the DNA helix without melting the DNA.

Figure 8.8c can be contrasted with Fig. 8.3a, c. The latter show the significant hysteresis observed in DNA force–extension experiments with T7 gp2.5 and T4 gp32 protein. An important difference is observed between HIV-1 NC and SSB proteins. When the DNA is relaxed after stretching in the presence of HIV-1 NC, the force–extension curve almost exactly follows the original profile. This shows that NC rapidly dissociates from ssDNA, allowing the two strands to reanneal and form dsDNA. In contrast, Fig. 8.3a, c imply that SSB proteins prevent DNA reannealing until the extension is relaxed well below the DNA contour length.

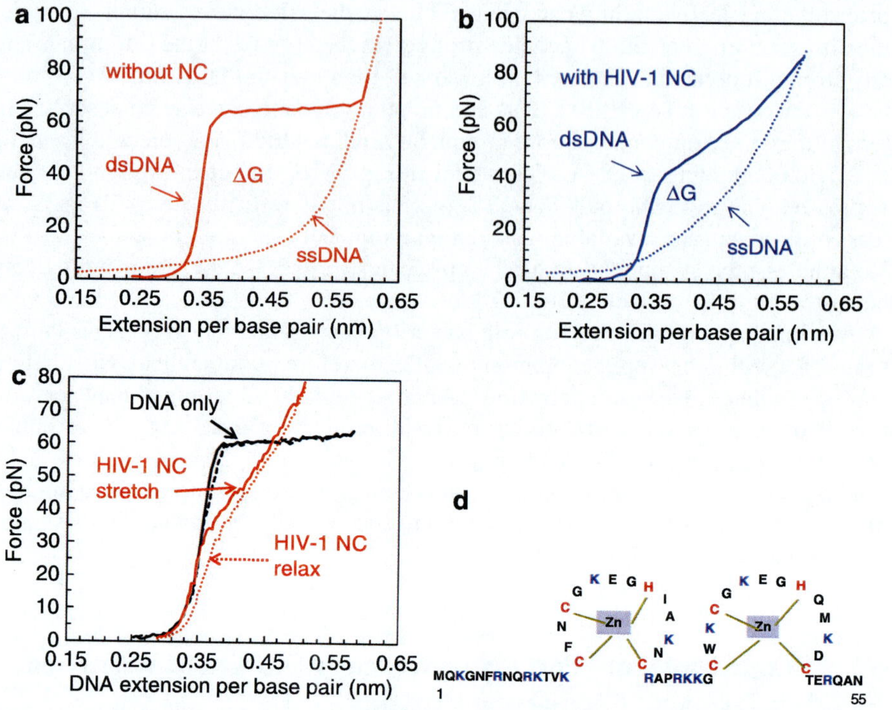

Fig. 8.8 (**a**) dsDNA stretching (*solid line*) and ssDNA stretching (*dashed line*, data from [55], modeled using the freely jointed chain model) illustrating the free energy of DNA melting in the absence of protein (50 mM Na^+, 10 mM HEPES, pH 7.5). (**b**) dsDNA stretching (*solid line*) and ssDNA stretching (*dashed line*, data from [91]) illustrating the free energy of DNA melting in the presence of 7 nM HIV-1 NC (50 mM Na^+, 10 mM HEPES, pH 7.5). (**c**) dsDNA stretching (*solid lines*) and relaxation (*dashed lines*) in the absence of protein (*black*) and in the presence of 10 nM HIV-1 NC (*red*) [98]. Data were taken in 50 mM Na^+, 10 mM HEPES, pH 7.5. (**d**) Primary structure of HIV-1 NC showing the 55 amino acids that make up the protein as well as a schematic shape of the two zinc fingers coordinated by the standard CCHC motif. Positively charged residues are shown in *blue*

Thus, SSB proteins exhibit slow ssDNA dissociation kinetics, while HIV-1 NC exhibits rapid ssDNA dissociation kinetics [92].

Which interactions of HIV-1 NC (Fig. 8.8d) are important for rapid binding kinetics during nucleic acid rearrangements? To answer this question, we examined several HIV-1 NC mutants. Examples are shown in Fig. 8.9. The first mutant (2–1 mutant, Fig. 8.9a) has the order of the zinc fingers switched. This mutation has two significant effects (Fig. 8.9c). First, the mutation weakens the overall affinity of the protein for DNA, such that higher protein concentrations are required to observe significant effects on DNA stretching. Second, and most important, this simple mutation destroys the rapid DNA interaction kinetics observed for wild-type NC. Simply duplicating the first zinc finger to form the 1–1 mutant results in a protein with both slower interaction kinetics and weaker binding affinity. However, this

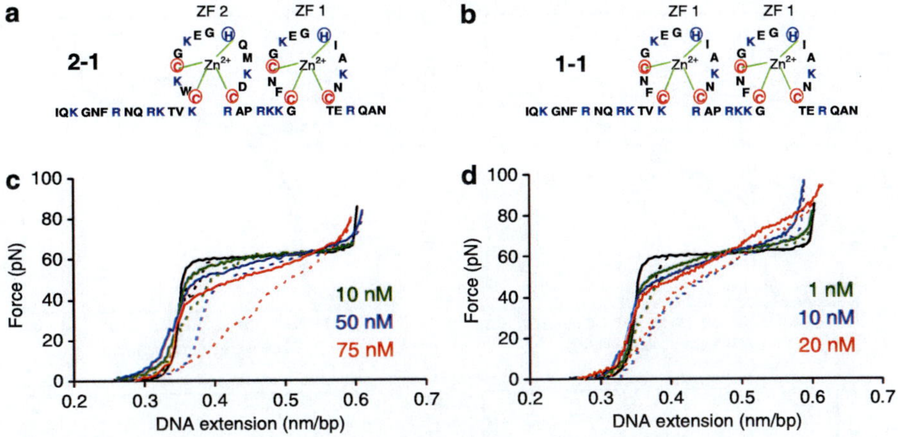

Fig. 8.9 (**a**) Primary structure of HIV-1 NC mutant 2–1, in which the order of the zinc fingers is reversed. (**b**) Primary structure of HIV-1 NC mutant 1–1, in which the first zinc finger is duplicated. (**c**) DNA stretching (*solid lines*) and relaxation (*dashed lines*) in the presence of the 2–1 mutant, which shows significant hysteresis not seen with the wild-type protein. (**d**) DNA stretching (*solid lines*) and relaxation (*dashed lines*) in the presence of the 1–1 mutant, which shows more hysteresis than the wild-type protein, but less than the 2–1 mutant

mutant was unique among those tested in its relatively rapid interaction kinetics and its efficient minus-strand transfer activity in in vitro experiments [93]. These results demonstrate that a specific HIV-1 NC zinc finger structure is critical for optimal nucleic acid chaperone activity [92].

Because HIV-1 NC exhibits efficient nucleic acid chaperone activity with finely tuned binding characteristics, it is of interest to determine whether other retroviral nucleocapsid proteins exhibit similar behavior. We examined the DNA binding characteristics of NC proteins from Moloney murine leukemia virus (MLV), Rous sarcoma virus (RSV), and human T-cell lymphotropic virus type 1 (HTLV-1) [94]. A range of binding activities was observed, differing primarily in rates of NC dissociation from ssDNA, as illustrated in Fig. 8.10b–d. For comparison, Fig. 8.10a shows DNA stretching and relaxation in the presence of HIV-1 NC (little hysteresis; rapid kinetics). Figure 8.10b shows DNA stretching and relaxation in the presence of RSV NC (little hysteresis; rapid DNA binding kinetics). In addition, the presence of RSV NC leads to an increase in the apparent DNA melting force, suggesting dsDNA stabilization by RSV NC, similar to that produced by simple multivalent cations [31]. In contrast, MLV and HTLV-1 NC proteins induce moderate to strong hysteresis, similar to that observed for SSBs T4 gp32 and T7 gp2.5. Interestingly, more rapid protein–ssDNA interactions seem to parallel the ability to aggregate nucleic acids, as measured in the bulk aggregation assays [94]. The only exception is MLV NC, which dissociates relatively slowly from ssDNA, but aggregates DNA almost as well as HIV-1 NC. Therefore, rapid NC–ssDNA interactions best correlate with overall chaperone activity, varying from strong for HIV-1 NC to extremely poor for HTLV-1 NC, as measured in in vitro annealing assays [94].

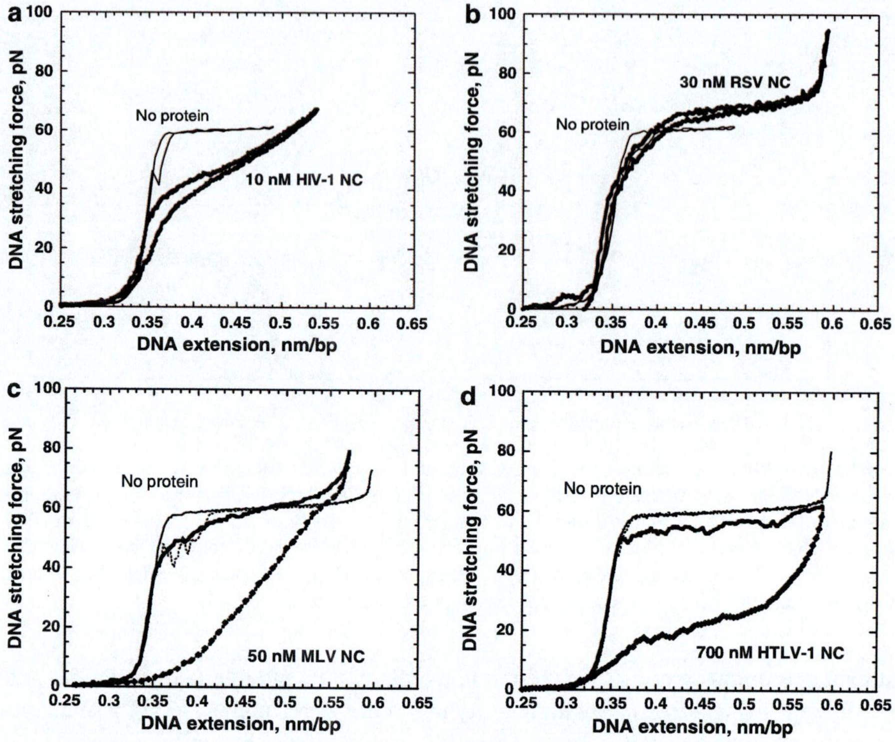

Fig. 8.10 DNA stretching and relaxation curves in the presence of saturating levels of (**a**) HIV-1, (**b**) RSV, (**c**) MLV, or (**d**) HTLV-1 NC. In each *panel*, the *thin lines* represent the curves in the absence of protein. Experiments in the presence of NC are shown by the *thick black lines*. Experiments were performed in 45 mM NaCl and 10 mM HEPES pH 7.5, for a total of 50 mM Na^+. In each *cycle* shown, the *upper curve* represents the stretching and the *lower curve* is the relaxation. Figure is taken from [94]

8.8 HTLV-1 NC Has Properties Intermediate Between SSB and NC Proteins

From a thermodynamic perspective, the HTLV-1 NC protein is, like other NC proteins, a mild NA duplex destabilizer. However, from a kinetic point of view, HTLV-1 NC displays slow interaction kinetics with DNA, more like an SSB protein [94]. The protein off rate can be estimated by measuring the rate of DNA annealing after force-induced melting and relaxation to the midpoint of the transition (Fig. 8.3b, d for T4 gp32 and T7 gp2.5). In the case of T4 gp32, the timescale of the exponential decay of force toward the equilibrium melting force in the presence of protein was on the order of a few minutes. For T7 gp2.5, the timescale was on the order of tens of seconds, or about an order of magnitude faster than T4 gp32. We have performed similar measurements on wild-type HTLV-1 NC, which has an apparent ssDNA off rate of ~10^2/s, similar to the slow dissociation of T4 gp32.

The HTLV-1 NC C-terminal truncate ΔC29 has about a tenfold higher off rate, similar to the off rate of the faster gp2.5 SSB protein. At the same time, the NC proteins from other retroviruses that have so far been measured by DNA stretching have faster off rates that are too high to be measured by this approach. These proteins dissociate from DNA at least an order of magnitude faster than T7 gp2.5 and HTLV-1 NCΔC29 [95].

Based on these results, we propose a model of nucleic acid binding regulation by the CTD of HTLV-1 NC. This model is reminiscent of CTD-regulated binding of T4 gp32 [6–9] and T7 gp2.5 [13, 78] (SSB models shown in Fig. 8.7). For all three proteins, association with nucleic acids requires prior dissociation of an anionic CTD from a cationic nucleic acid binding site. HTLV-1 NC more closely resembles T4 gp32 because its CTD interacts with the DNA binding site of the protein, rather than the DNA binding site of the dimer partner (T7 gp2.5). In addition, HTLV-1 NC appears to bind with at least moderate cooperativity to ssDNA, whereas T7 gp2.5 binds with little or no cooperativity. In contrast to HTLV-1 NC, the CTD of T4 gp32 does not participate in the cooperative interaction between the nucleic acid-bound proteins. For gp32, the cooperative binding interaction is relatively salt independent, and is not eliminated by the removal of CTD (creating T4 gp32 *I) [96, 97]. Therefore, the removal of CTD for gp32 affects the on rate but not the off rate. The off rate is reduced by cooperative binding for both wild-type T4 gp32 and the C-terminal truncate. The case is similar for T7 gp2.5. Therefore, for both T4 gp32 and T7 gp2.5 SSB proteins, on rates (and therefore also the equilibrium binding constants), but not off rates, are increased by the removal of CTD. In contrast, both on and off rates are increased by CTD deletion of HTLV-1 NC [95].

The specific features of HTLV-1 NC binding and regulation by its CTD have presumably evolved for specific biological functions. While those biological functions are currently unknown, there are interesting possibilities. HTLV-1 NC is a small protein, which makes a complex cooperative binding mode such as that of T4 gp32 unlikely. However, HTLV-1 NC has developed a simple cooperative binding mode that limits the protein off rate. In the case of T4 gp32, this low off rate is likely needed for the protection of ssDNA exposed by the helicase during bacteriophage replication. What advantage is conferred by a retroviral NC protein with a low off rate? Such a characteristic seems to contradict the role NC would need to play as a chaperone protein. However, we have shown that not all retroviruses require nucleic acid chaperone proteins to be as powerful as that of HIV-1. Perhaps HTLV-1 has sacrificed nucleic acid chaperone potency for another biological purpose. One possibility is that HTLV-1 NC acts to coat ssRNA to prevent the packaging of APOBEC3G proteins. While HIV-1 NC can coat ssRNA, its rapid kinetics make that ssRNA available for binding by APOBEC3G proteins [98], which are then packaged by HIV-1 [99–101]. In contrast, HTLV-1 does not package APOBEC3G, and HTLV-1 NC has been shown to be responsible for this behavior [102]. Thus, it is possible that the regulation of HTLV-1 NC binding kinetics by its CTD is essential to prevent packaging of APOBEC3G. It is also possible that the slow HTLV-1 NC dissociation kinetics are needed for another viral replication process, or that another protein binds to HTLV-1 NC's CTD, acting as a switch to activate or inactivate HTLV-1 NC's nucleic acid chaperone activity.

8.9 Comparison of SSB Proteins and Nucleic Acid Chaperone Proteins

It is interesting to compare nucleic acid chaperone and single-stranded binding (SSB) proteins because they share common DNA interaction mechanisms, but appear to serve very different biological functions. Both nucleic acid chaperone and SSB proteins are largely sequence nonspecific in their binding to nucleic acids (ssDNA and dsDNA, ssRNA and dsRNA). They display a significant binding preference for single-stranded nucleic acids. Therefore, both classes of proteins destabilize nucleic acid duplexes. However, nucleic acid chaperones such as NC proteins destabilize nucleic acid duplexes only very slightly, while SSB proteins are able to melt dsDNA completely, sometimes creating ssDNA from dsDNA under physiological conditions. These differences reflect the divergent biological roles of these two groups of proteins. Indeed, one of the major functions of the retroviral NC proteins is to assist reverse transcription by facilitation of nucleic acid remodeling [89]. This chaperone activity of NC proteins requires slight destabilization of nucleic acid hairpins and other duplex structures to create folding pathways that can proceed more smoothly through misfolded or less stable conformations toward the final, most stable state. This slight duplex destabilization by NC proteins is reflected by the small free energy of duplex destabilization per base pair (~1 k_BT or less), compared to the average free energy of dsDNA melting per base pair at physiological conditions (~2 k_BT [103]). In contrast, the primary function of the bacteriophage SSB proteins is to support phage replication via stabilization of helicase-unwound ssDNA [104]. This function is consistent with the duplex destabilizing free energy per base pair (~2 k_BT for T4 gp32 [6, 7, 9] and T7 gp2.5 [78]), comparable to the average DNA melting free energy per base pair.

Interestingly, among all SSB and NC proteins studied, the faster kinetics of ssDNA interaction correlates with weaker destabilization of the DNA duplex. Indeed, the SSB and NC proteins discussed here are strongly cationic (or have cationic DNA binding sites), such that their association with dsDNA is primarily nonspecific and electrostatic. Such protein–nucleic acid binding modes are known to lead to high protein mobility (faster sliding) on dsDNA [105, 106]. At the same time, the ability of such proteins to destabilize DNA duplexes depends on their preference for ssDNA, relative to dsDNA, which relies on more specific protein–ssDNA interactions, such as hydrogen bonding or stacking of aromatic residues. These specific interactions are expected to slow the protein's mobility on and dissociation from ssDNA.

We have observed that rapid protein–DNA interaction kinetics appear to positively correlate with the ability of a protein to aggregate nucleic acids. This feature is typical of multivalent cations known to nonspecifically aggregate nucleic acids upon binding [107, 108]. Both rapid interaction kinetics and the ability to aggregate anionic polyelectrolytes, such as nucleic acids, arise from the primarily electrostatic nature of multivalent cation binding [107, 108]. Proteins may also cause nucleic acid aggregation by nonelectrostatic mechanisms, but the resulting aggregates are more likely to be rigid, resisting nucleic acid refolding and annealing. The electrostatic mechanism of aggregation appears to produce protein–nucleic acid complexes characterized by the high mobility needed for refolding of nucleic acid structures.

In summary, the primarily electrostatic nucleic acid binding mode of NCs is responsible for their fast interaction kinetics, strong nucleic acid aggregation, and modest duplex destabilization. These characteristics lead to optimized nucleic acid chaperone function, as observed for HIV-1 NC. Other retroviral NCs have different combinations of these properties, leading to less impressive in vitro chaperone functions. Thus, RSV NC has faster kinetics, but poorer duplex destabilizing properties, while HTLV-1 NC is a better duplex destabilizer, but a poor aggregating agent with slower binding. HTLV-1 NC is thus closer in behavior to SSB proteins than to NC proteins. The two simple SSBs from bacteriophages T4 and T7 belong to the same continuum of nonspecific nucleic acid binding proteins with still slower ssDNA binding kinetics and the pronounced ability to melt DNA duplexes by sequestering single strands as they appear. Interestingly, SSBs appear to employ rapid electrostatic sliding on dsDNA to locate ssDNA sites. This example illustrates how different biological functions can be achieved by evolutionary tuning of protein–DNA binding properties, even when the overall binding occurs through a very similar mechanism.

8.10 Conclusions

We have described the application of new approaches for studying thermodynamics and kinetics of protein binding to ssDNA and dsDNA using single molecule force spectroscopy. These methods have added unprecedented detail to the characterization of two important classes of DNA binding proteins. We find that SSB proteins bind very weakly to dsDNA in order to rapidly diffuse in their search for ssDNA binding sites. Upon binding to such sites, the proteins bind four orders of magnitude more strongly. SSB proteins are thus optimized to rapidly engage rare regions of ssDNA in a sea of genomic dsDNA, as at the bacteriophage replication fork. Unlike nucleic acid chaperone proteins, SSBs bind much more strongly to ssDNA relative to dsDNA, resulting in strong helix-destabilizing properties and a relatively slow ssDNA off rate. In contrast, nucleic acid chaperone proteins, such as retroviral NC proteins, only slightly prefer ssDNA and only weakly destabilize dsDNA. In the case of HIV-1 NC, an excellent nucleic acid chaperone, ssDNA and dsDNA are bound similarly, with slight dsDNA destabilization. When coupled with the ability of HIV-1 NC to strongly aggregate DNA, facilitation of nucleic acid secondary structure rearrangements engenders chaperone activity.

Binding preferences and kinetics differ between SSB and NC proteins. In addition, NC proteins from different retroviruses show varying nucleic acid chaperone capabilities. Proteins less like HIV-1 NC more strongly resemble SSB proteins. Thus, HTLV-1 NC, which displays the least in vitro chaperone activity among the retroviral NC proteins studied, very strongly resembles an SSB protein because of its slow protein dissociation and poor nucleic acid aggregation. Moreover, the nucleic acid interactions of HTLV-1 NC are strongly regulated by its CTD, similar to that observed for T4 gp32 and T7 gp2.5. However, this CTD regulatory mechanism differs from that observed for the SSBs in that both on and off rates are reduced by the CTD, whereas the CTD affects only the on rate for SSBs.

Overall, it appears that simple biophysical mechanisms can regulate all aspects of protein–DNA interactions, from equilibrium binding affinity to binding kinetics. We argue that understanding these biophysical mechanisms is critical for determining how these proteins perform their biological functions. These studies demonstrate that single molecule force spectroscopy can provide an exceptionally powerful approach to study the biophysics of sequence nonspecific protein–DNA interactions.

Acknowledgments The authors would like to thank Margareta Cruceanu, Kiran Pant, Leila Shokri, and Fei Wang for their original work on these systems. This work was funded by the US National Institutes of Health (GM072462) and National Science Foundation (MCB-0744456).

References

1. Haseltine CA, Kowalczykowski SC (2002) A distinctive single-strand DNA-binding protein from the Archaeon Sulfolobus solfataricus. Mol Microbiol 43:1505–1515
2. Richard DJ, Bolderson E, Khanna KK (2009) Multiple human single-stranded DNA binding proteins function in genome maintenance: structural, biochemical and functional analysis. Crit Rev Biochem Mol Biol 44(2–3):98–116
3. Alberts BM (1970) Function of gene 32-protein, a new protein essential for the genetic recombination and replication of T4 bacteriophage DNA. Fed Proc 29:1154–1163
4. Kornberg A, Baker TA (1992) DNA replication. W. H. Freeman and Company, New York
5. Karpel RL (1990) T4 bacteriophage gene 32 protein. In: Revzin A (ed) The biology of non-specific DNA–protein interactions. CRC, Boca Raton, pp 103–130
6. Pant K, Karpel RL, Rouzina I, Williams MC (2004) Mechanical measurement of single-molecule binding rates: kinetics of DNA helix-destabilization by T4 gene 32 protein. J Mol Biol 336:851–858
7. Pant K, Karpel RL, Rouzina I, Williams MC (2005) Salt dependent binding of T4 gene 32 protein to single and double-stranded DNA: single molecule force spectroscopy measurements. J Mol Biol 349:317–330
8. Pant K, Karpel RL, Williams MC (2003) Kinetic regulation of single DNA molecule denaturation by T4 gene 32 protein structural domains. J Mol Biol 327:571–578
9. Rouzina I, Pant K, Karpel RL, Williams MC (2005) Theory of electrostatically regulated binding of T4 gene 32 protein to single- and double-stranded DNA. Biophys J 89:1941–1956
10. Benkovic SJ, Valentine AM, Salinas F (2001) Replisome-mediated DNA replication. Annu Rev Biochem 70:181–208
11. Kim YT, Tabor S, Bortner C, Griffith JD, Richardson CC (1992) Purification and characterization of the bacteriophage T7 gene 2.5 protein. A single-stranded DNA-binding protein. J Biol Chem 267:15022–15031
12. Hollis T, Stattel JM, Walther DS, Richardson CC, Ellenberger T (2001) Structure of the gene 2.5 protein, a single-stranded DNA binding protein encoded by bacteriophage T7. Proc Natl Acad Sci USA 98:9557–9562
13. Shokri L, Marintcheva B, Richardson CC, Rouzina I, Williams MC (2006) Single molecule force spectroscopy of salt-dependent bacteriophage T7 gene 2.5 protein binding to single-stranded DNA. J Biol Chem 281:38689–38696

14. Kim YT, Richardson CC (1993) Bacteriophage T7 gene 2.5 protein: an essential protein for DNA replication. Proc Natl Acad Sci USA 90:10173–10177
15. Kim YT, Tabor S, Churchich JE, Richardson CC (1992) Interactions of gene 2.5 protein and DNA polymerase of bacteriophage T7. J Biol Chem 267:15032–15040
16. Kong D, Griffith JD, Richardson CC (1997) Gene 4 helicase of bacteriophage T7 mediates strand transfer through pyrimidine dimers, mismatches, and nonhomologous regions. Proc Natl Acad Sci USA 94:2987–2992
17. Kong D, Nossal NG, Richardson CC (1997) Role of the bacteriophage T7 and T4 single-stranded DNA-binding proteins in the formation of joint molecules and DNA helicase-catalyzed polar branch migration. J Biol Chem 272:8380–8387
18. Kong D, Richardson CC (1996) Single-stranded DNA binding protein and DNA helicase of bacteriophage T7 mediate homologous DNA strand exchange. EMBO J 15:2010–2019
19. Kong D, Richardson CC (1998) Role of the acidic carboxyl-terminal domain of the single-stranded DNA-binding protein of bacteriophage T7 in specific protein–protein interactions. J Biol Chem 273:6556–6564
20. Scherzinger E, Litfin F, Jost E (1973) Stimulation of T7 DNA polymerase by a new phage-coded protein. Mol Gen Genet 123:247–262
21. Yu M, Masker W (2001) T7 single strand DNA binding protein but not T7 helicase is required for DNA double strand break repair. J Bacteriol 183:1862–1869
22. Pant K, Shokri L, Karpel RL, Morrical SW, Williams MC (2008) Modulation of T4 gene 32 protein DNA binding activity by the recombination mediator protein UvsY. J Mol Biol 380:799–811
23. Berg JM (1986) Potential metal-binding domains in nucleic acid binding proteins. Science 232:485–487
24. Covey SN (1986) Amino acid sequence homology in gag region of reverse transcribing elements and the coat protein gene of cauliflower mosaic virus. Nucleic Acids Res 14:623–633
25. Berkowitz RD, Ohagen A, Hoglund S, Goff SP (1995) Retroviral nucleocapsid domains mediate the specific recognition of genomic viral RNAs by chimeric Gag polyproteins during RNA packaging in vivo. J Virol 69:6445–6456
26. Zhang Y, Barklis E (1995) Nucleocapsid protein effects on the specificity of retrovirus RNA encapsidation [published erratum appears in J Virol 1997 Jul;71(7):5712]. J Virol 69:5716–5722
27. Gorelick R, Gagliardi T, Bosche W, Wiltrout T, Coren L, Chabot D, Lifson J, Henderson L, Arthur A (1999) Strict conservation of the retroviral nucleocapsid protein zinc finger is strongly influenced by its role in viral infection processes: characterization of HIV-1 particles containing mutant nucleocapsid zinc-coordinating sequences. Virology 256:92–104
28. Darlix JL, Gabus C, Nugeyre MT, Clavel F, Barre-Sinoussi F (1990) Cis elements and trans-acting factors involved in the RNA dimerization of the human immunodeficiency virus HIV-1. J Mol Biol 216:689–699
29. Sakaguchi K, Zambrano N, Baldwin ET, Shapiro BA, Erickson JW, Omichinski JG, Clore GM, Gronenborn AM, Appella E (1993) Identification of a binding site for the human immunodeficiency virus type 1 nucleocapsid protein. Proc Natl Acad Sci USA 90:5219–5223
30. Feng YX, Copeland TD, Henderson LE, Gorelick RJ, Bosche WJ, Levin JG, Rein A (1996) HIV-1 nucleocapsid protein induces “maturation” of dimeric retroviral RNA in vitro. Proc Natl Acad Sci USA 93:7577–7581
31. Cruceanu M, Urbaneja MA, Hixson CV, Johnson DG, Datta SA, Fivash MJ, Stephen AG, Fisher RJ, Gorelick RJ, Casas-Finet JR, Rein A, Rouzina I, Williams MC (2006) Nucleic acid binding and chaperone properties of HIV-1 Gag and nucleocapsid proteins. Nucleic Acids Res 34:593–605
32. Mirambeau G, Lyonnais S, Coulaud D, Hameau L, Lafosse S, Jeusset J, Justome A, Delain E, Gorelick RJ, Le Cam E (2006) Transmission electron microscopy reveals an optimal HIV-1 nucleocapsid aggregation with single-stranded nucleic acids and the mature HIV-1 nucleocapsid protein. J Mol Biol 364:496–511
33. Gao K, Gorelick RJ, Johnson DG, Bushman F (2003) Cofactors for human immunodeficiency virus type 1 cDNA integration in vitro. J Virol 77:1598–1603

34. Mirambeau G, Lyonnais S, Coulaud D, Hameau L, Lafosse S, Jeusset J, Borde I, Reboud-Ravaux M, Restle T, Gorelick RJ, Le Cam E (2007) HIV-1 protease and reverse transcriptase control the architecture of their nucleocapsid partner. PLoS One 2:e669
35. Wyma DJ, Kotov A, Aiken C (2000) Evidence for a stable interaction of gp41 with Pr55(Gag) in immature human immunodeficiency virus type 1 particles. J Virol 74:9381–9387
36. Carteau S, Gorelick RJ, Bushman FD (1999) Coupled integration of human immunodeficiency virus type 1 cDNA ends by purified integrase in vitro: stimulation by the viral nucleocapsid protein. J Virol 73:6670–6679
37. Coffin JM, Hughes SH, Varmus HE (1997) Retroviruses. Cold Spring Harbor Laboratory Press, Cold Spring Harbor, NY
38. Barat C, Lullien V, Schatz O, Keith G, Nugeyre MT, Gruninger-Leitch F, Barre-Sinoussi F, LeGrice SF, Darlix JL (1989) HIV-1 reverse transcriptase specifically interacts with the anticodon domain of its cognate primer tRNA. EMBO J 8:3279–3285
39. de Rocquigny H, Gabus C, Vincent A, Fournie-Zaluski MC, Roques B, Darlix JL (1992) Viral RNA annealing activities of human immunodeficiency virus type 1 nucleocapsid protein require only peptide domains outside the zinc fingers. Proc Natl Acad Sci USA 89:6472–6476
40. Li X, Quan Y, Arts EJ, Li Z, Preston BD, de Rocquigny H, Roques BP, Darlix JL, Kleiman L, Parniak MA, Wainberg MA (1996) Human immunodeficiency virus Type 1 nucleocapsid protein (NCp7) directs specific initiation of minus-strand DNA synthesis primed by human tRNA(Lys3) in vitro: studies of viral RNA molecules mutated in regions that flank the primer binding site. J Virol 70:4996–5004
41. Peliska JA, Balasubramanian S, Giedroc DP, Benkovic SJ (1994) Recombinant HIV-1 nucleocapsid protein accelerates HIV-1 reverse transcriptase catalyzed DNA strand transfer reactions and modulates RNase H activity. Biochemistry 33:13817–13823
42. Rodriguez-Rodriguez L, Tsuchihashi Z, Fuentes GM, Bambara RA, Fay PJ (1995) Influence of human immunodeficiency virus nucleocapsid protein on synthesis and strand transfer by the reverse transcriptase in vitro. J Biol Chem 270:15005–15011
43. You JC, McHenry CS (1994) Human immunodeficiency virus nucleocapsid protein accelerates strand transfer of the terminally redundant sequences involved in reverse transcription. J Biol Chem 269:31491–31495
44. Guo J, Wu T, Anderson J, Kane BF, Johnson DG, Gorelick RJ, Henderson LE, Levin JG (2000) Zinc finger structures in the human immunodeficiency virus type 1 nucleocapsid protein facilitate efficient minus- and plus-strand transfer. J Virol 74:8980–8988
45. Johnson PE, Turner RB, Wu ZR, Hairston L, Guo J, Levin JG, Summers MF (2000) A mechanism for plus-strand transfer enhancement by the HIV-1 nucleocapsid protein during reverse transcription. Biochemistry 39:9084–9091
46. Negroni M, Buc H (2001) Mechanisms of retroviral recombination. Ann Rev Genet 35:275–302
47. de Rocquigny H, Shvadchak V, Avilov S, Dong CZ, Dietrich U, Darlix JL, Mely Y (2008) Targeting the viral nucleocapsid protein in anti-HIV-1 therapy. Mini Rev Med Chem 8:24–35
48. Rice WG, Turpin JA (1996) Virus-encoded zinc-finger as targets for antiviral chemotherapy. Rev Med Virol 6:187–196
49. Cruceanu M, Stephen AG, Beuning PJ, Gorelick RJ, Fisher RJ, Williams MC (2006) Single DNA molecule stretching measures the activity of chemicals that target the HIV-1 nucleocapsid protein. Anal Biochem 358:159–170
50. McDonnell NB, De Guzman RN, Rice WG, Turpin JA, Summers MF (1997) Zinc ejection as a new rationale for the use of cystamine and related disulfide-containing antiviral agents in the treatment of AIDS. J Med Chem 40:1969–1976
51. Williams MC, Rouzina I (2002) Force spectroscopy of single DNA and RNA molecules. Curr Opin Struct Biol 12:330–336
52. Marko JF, Siggia ED (1995) Stretching DNA. Macromolecules 28:8759–8770
53. Bustamante C, Smith SB, Liphardt J, Smith D (2000) Single-molecule studies of DNA mechanics. Curr Opin Struct Biol 10:279–285

54. Cluzel P, Lebrun A, Heller C, Lavery R, Viovy J-L, Chatenay D, Caron F (1996) DNA: an extensible molecule. Science 271:792–794
55. Smith SB, Cui Y, Bustamante C (1996) Overstretching B-DNA: the elastic response of individual double-stranded and single-stranded DNA molecules. Science 271:795–799
56. Vladescu ID, McCauley MJ, Rouzina I, Williams MC (2005) Mapping the phase diagram of single DNA molecule force-induced melting in the presence of ethidium. Phys Rev Lett 95:158102
57. van Mameren J, Gross P, Farge G, Hooijman P, Modesti M, Falkenberg M, Wuite GJ, Peterman EJ (2009) Unraveling the structure of DNA during overstretching by using multicolor, single-molecule fluorescence imaging. Proc Natl Acad Sci USA 106:18231–18236
58. Williams MC, Rouzina I, McCauley MJ (2009) Peeling back the mystery of DNA overstretching. Proc Natl Acad Sci USA 106:18047–18048
59. Williams MC, Rouzina I, Bloomfield VA (2002) Thermodynamics of DNA interactions from single molecule stretching experiments. Acc Chem Res 35:159–166
60. Williams MC, Wenner JR, Rouzina I, Bloomfield VA (2001) Entropy and heat capacity of DNA melting from temperature dependence of single molecule stretching. Biophys J 80:1932–1939
61. Williams MC, Wenner JR, Rouzina I, Bloomfield VA (2001) Effect of pH on the overstretching transition of double-stranded DNA: evidence of force-induced DNA melting. Biophys J 80:874–881
62. Shokri L, McCauley MJ, Rouzina I, Williams MC (2008) DNA overstretching in the presence of glyoxal: structural evidence of force-induced DNA melting. Biophys J 95:1248–1255
63. Shokri L, Rouzina I, Williams MC (2009) Interaction of bacteriophage T4 and T7 single-stranded DNA binding proteins with DNA. Phys Biol 6:025002
64. Vladescu ID, McCauley MJ, Nunez ME, Rouzina I, Williams MC (2007) Quantifying force-dependent and zero-force DNA intercalation by single-molecule stretching. Nat Methods 4:517–522
65. McCauley MJ, Williams MC (2007) Mechanisms of DNA binding determined in optical tweezers experiments. Biopolymers 85:154–168
66. McCauley MJ, Williams MC (2009) Review: optical tweezers experiments resolve distinct modes of DNA–protein binding. Biopolymers 91:265–282
67. Shokri L, Rouzina I, Williams MC (2009) Interaction of bacteriophage T4 and T7 single-stranded DNA-binding proteins with DNA. Phys Biol 6:25002
68. Villemain JL, Giedroc DP (1996) Characterization of a cooperativity domain mutant Lys3 right-arrow Ala (K3A) T4 gene 32 protein. J Biol Chem 271:27623–27629
69. Waidner LA, Flynn EK, Wu M, Li X, Karpel RL (2001) Domain effects on the DNA-interactive properties of bacteriophage T4 gene 32 protein. J Biol Chem 276:2509–2516
70. Lonberg N, Kowalczykowski SC, Paul LS, von Hippel PH (1981) Interactions of bacteriophage T4-coded gene 32 protein with nucleic acids. III. Binding properties of two specific proteolytic digestion products of the protein (G32P*I and G32P*III). J Mol Biol 145:123–138
71. Jensen DE, Kelly RC, von Hippel PH (1976) DNA "melting" proteins. II. Effects of bacteriophage T4 gene 32-protein binding on the conformation and stability of nucleic acid structures. J Biol Chem 251:7215–7228
72. von Hippel PH, Delagoutte E (2001) A general model for nucleic acid helicases and their "coupling" within macromolecular machines. Cell 104:177–190
73. Kim YT, Richardson CC (1994) Acidic carboxyl-terminal domain of gene 2.5 protein of bacteriophage T7 is essential for protein–protein interactions. J Biol Chem 269:5270–5278
74. He Z-G, Rezende LF, Willcox S, Griffith JD, Richardson CC (2003) The carboxyl-terminal domain of bacteriophage T7 single-stranded DNA-binding protein modulates DNA binding and interaction with T7 DNA polymerase. J Biol Chem 278:29538–29545
75. He Z-G, Richardson CC (2004) Effect of single-stranded DNA-binding proteins on the helicase and primase activities of the bacteriophage T7 gene 4 protein. J Biol Chem 279:22190–22197
76. Rezende LF, Willcox S, Griffith JD, Richardson CC (2003) A single-stranded DNA-binding protein of bacteriophage T7 defective in DNA annealing. J Biol Chem 278:29098–29105

77. Hyland EM, Rezende LF, Richardson CC (2003) The DNA binding domain of the gene 2.5 single-stranded DNA-binding protein of pacteriophage T7. J Biol Chem 278:7247–7256
78. Shokri L, Marintcheva B, Eldib M, Hanke A, Rouzina I, Williams MC (2008) Kinetics and thermodynamics of salt-dependent T7 gene 2.5 protein binding to single- and double-stranded DNA. Nucleic Acids Res 36:5668–5677
79. Sokolov IM, Metzler R, Pant K, Williams MC (2005) First passage time of N excluded-volume particles on a line. Phys Rev E 72:041102
80. Sokolov IM, Metzler R, Pant K, Williams MC (2005) Target search of N sliding proteins on a DNA. Biophys J 89:895–902
81. Berg OG, Winter RB, von Hippel PH (1981) Diffusion-driven mechanisms of protein translocation on nucleic acids. 1. Models and theory. Biochemistry 20:6929–6948
82. Cherstvy AG, Kolomeisky AB, Kornyshev AA (2008) Protein–DNA interactions: reaching and recognizing the targets. J Phys Chem B 112:4741–4750
83. Halford SE, Marko JF (2004) How do site-specific DNA-binding proteins find their targets? Nucleic Acids Res 32:3040–3052
84. Slutsky M, Mirny LA (2004) Kinetics of protein–DNA interaction: facilitated target location in sequence-dependent potential. Biophys J 87:4021–4035
85. Wang YM, Austin RH, Cox EC (2006) Single molecule measurements of repressor protein 1D diffusion on DNA. Phys Rev Lett 97:048302
86. Winter RB, Berg OG, von Hippel PH (1981) Diffusion-driven mechanisms of protein translocation on nucleic acids. 3. The *Escherichia coli* lac repressor–operator interaction: kinetic measurements and conclusions. Biochemistry 20:6961–6977
87. Hu T, Grosberg AY, Shklovskii BI (2006) How proteins search for their specific sites on DNA: the role of DNA conformation. Biophys J 90:2731–2744
88. Saecker RM, Record MT Jr (2002) Protein surface salt bridges and paths for DNA wrapping. Curr Opin Struct Biol 12:311–319
89. Levin JG, Guo J, Rouzina I, Musier-Forsyth K (2005) Nucleic acid chaperone activity of HIV-1 nucleocapsid protein: critical role in reverse transcription and molecular mechanism. Prog Nucleic Acid Res Mol Biol 80:217–286
90. Williams MC, Gorelick RJ, Musier-Forsyth K (2002) Specific zinc-finger architecture required for HIV-1 nucleocapsid protein's nucleic acid chaperone function. Proc Natl Acad Sci USA 99:8614–8619
91. Williams MC, Rouzina I, Wenner JR, Gorelick RJ, Musier-Forsyth K, Bloomfield VA (2001) Mechanism for nucleic acid chaperone activity of HIV-1 nucleocapsid protein revealed by single molecule stretching. Proc Natl Acad Sci USA 98:6121–6126
92. Cruceanu M, Gorelick RJ, Musier-Forsyth K, Rouzina I, Williams MC (2006) Rapid kinetics of protein-nucleic acid interaction is a major component of HIV-1 nucleocapsid protein's nucleic acid chaperone function. J Mol Biol 363:867–877
93. Guo J, Wu T, Kane BF, Johnson DG, Henderson LE, Gorelick RJ, Levin JG (2002) Subtle alterations of the native zinc finger structures have dramatic effects on the nucleic acid chaperone activity of human immunodeficiency virus type 1 nucleocapsid protein. J Virol 76:4370–4378
94. Stewart-Maynard KM, Cruceanu M, Wang F, Vo MN, Gorelick RJ, Williams MC, Rouzina I, Musier-Forsyth K (2008) Retroviral nucleocapsid proteins display non-equivalent levels of nucleic acid chaperone activity. J Virol 82:10129–10142
95. Qualley DF, Stewart-Maynard KM, Wang F, Mitra M, Gorelick RJ, Rouzina I, Williams MC, Musier-Forsyth K (2009) C-terminal domain modulates the nucleic acid chaperone activity of human T-cell leukemia virus type 1 (HTLV-1) nucleocapsid protein (NC) via an electrostatic mechanism. J Biol Chem 285(1):295–307. http://dx.doi.org/10.1074/jbc.M109.051334.
96. Kowalczykowski SC, Lonberg N, Newport JW, von Hippel PH (1981) Interactions of bacteriophage T4-coded gene 32 protein with nucleic acids. I. Characterization of the binding interactions. J Mol Biol 145:75–104
97. Lohman TM, Kowalczykowski SC (1981) Kinetics and mechanism of the association of the bacteriophage T4 gene 32 (helix destabilizing) protein with single-stranded nucleic acids. Evidence for protein translocation. J Mol Biol 152:67–109

98. Iwatani Y, Chan DS, Wang F, Maynard KS, Sugiura W, Gronenborn AM, Rouzina I, Williams MC, Musier-Forsyth K, Levin JG (2007) Deaminase-independent inhibition of HIV-1 reverse transcription by APOBEC3G. Nucleic Acids Res 35:7096–7108
99. Goila-Gaur R, Strebel K (2008) HIV-1 Vif, APOBEC, and intrinsic immunity. Retrovirology 5:51
100. Holmes RK, Malim MH, Bishop KN (2007) APOBEC-mediated viral restriction: not simply editing? Trends Biochem Sci 32:118–128
101. Malim MH (2009) APOBEC proteins and intrinsic resistance to HIV-1 infection. Philos Trans R Soc Lond B Biol Sci 364:675–687
102. Derse D, Hill SA, Princler G, Lloyd P, Heidecker G (2007) Resistance of human T cell leukemia virus type 1 to APOBEC3G restriction is mediated by elements in nucleocapsid. Proc Natl Acad Sci USA 104:2915–2920
103. Bloomfield VA, Crothers DM, Tinoco I (2000) Nucleic acids structure, properties, and function. University Science Books, Sausalito, CA
104. Chase JW, Williams KR (1986) Single-stranded DNA binding proteins required for DNA replication. Annu Rev Biochem 55:103–136
105. Rouzina I, Bloomfield VA (1996) Competitive electrostatic binding of charged ligands to polyelectrolytes: planar and cylindrical geometries. J Phys Chem 100:4292–4304
106. Rouzina I, Bloomfield VA (1996) Influence of ligand spatial organization on competitive electrostatic binding to DNA. J Phys Chem 100:4305–4313
107. Rouzina I, Bloomfield VA (1996) Macroion attraction due to electrostatic correlation between screening counterions. 1. Mobile surface-adsorbed ions and diffuse ion cloud. J Phys Chem 100:9977–9989
108. Bloomfield VA (1997) DNA condensation by multivalent cations. Biopolymers 44:269–282

Part II
Complex Protein-DNA Interactions

Chapter 9
DNA Looping in Prophage Lambda: New Insight from Single-Molecule Microscopy

Laura Finzi, Carlo Manzo, Chiara Zurla, Haowei Wang, Dale Lewis, Sankar Adhya, and David Dunlap

9.1 Introduction

The lambda (λ) bacteriophage epigenetic switch is a molecular mechanism that permits the quiescent (lysogenic) state of the bacteriophage to irreversibly switch to the virulent (lytic) state. After infection of its host, *E. coli*, λ, a temperate phage, most often grows lysogenically. The phage DNA integrates in the bacterial chromosome and is replicated along with it and transmitted to the bacterial progeny as a prophage. Lysogeny is very stable and yet, the switch to lysis is very efficient. Upon switching to lysis, the viral DNA is excised from the bacterial chromosome and the host machinery is used to produce viral progeny that is then released upon bursting of the host. The pathway to lysis is triggered in response to threats such as starvation, poisoning, or DNA damage.

The lysogenic state of λ prophages is maintained by the λ repressor, or CI protein [35]. During lysogeny, dimers of CI bind to the O_L and O_R control regions, located about 2.3 kbp apart on the phage genome (Fig. 9.1a) and repress P_L and P_R, promoters for the lytic genes. Each control region contains three binding sites for CI, O_L1, O_L2, O_L3 and O_R1, O_R2, O_R3 [11, 26, 33]. CI binds to these operators with an intrinsic affinity $O_L1 > O_R1 > O_L3 > O_L2 > O_R2 > O_R3$ [17, 38]. By studying the O_L and O_R regions separately and in isolation from the rest of the λ chromosome [34], it was found that pairs of dimers interact when bound to adjacent or nearby operators, forming tetramers (Fig. 9.1a). These cooperative interactions improve the specificity and strength of CI binding to O_R1 and O_R2, and O_L1 and O_L2, respectively, so that the order of binding affinity changes to $O_R1 \sim O_L1 \sim O_R2 \sim O_L2 > O_L3 > O_R3$. Biochemical and genetic studies have identified the contacts between amino acids in the C-terminal domain (CTD) that mediate these interactions. These contacts have been confirmed from the crystal structure of the isolated CTD tetramer [7] and are thought to

L. Finzi (✉) and D. Dunlap (✉)
Cell Biology Department, Emory University, 615 Michael St., Atlanta, GA 30322, USA
e-mail: lfinzi@physics.emory.edu

M.C. Williams and L.J. Maher, III (eds.), *Biophysics of DNA-Protein Interactions*, Biological and Medical Physics, Biomedical Engineering,
DOI 10.1007/978-0-387-92808-1_9,

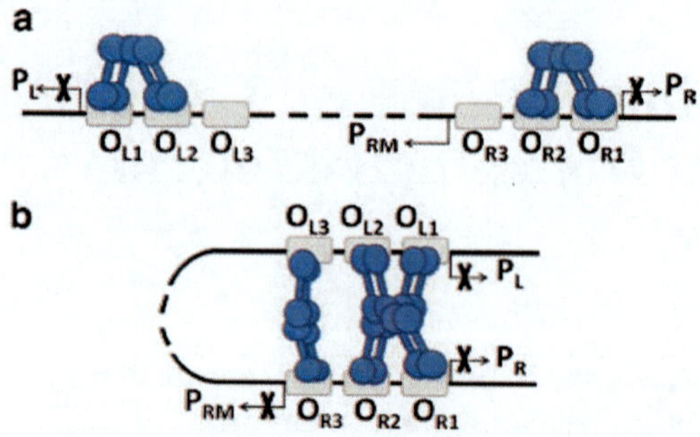

Fig. 9.1 Model of CI regulation by long-range DNA looping proposed by Dodd et al. [11]. (**a**) CI dimers bound cooperatively at O_R1 and O_R2 repress transcription at P_R while the dimer at O_R2 also activates transcription from P_{RM}. The dimers bound cooperatively at O_L1 and O_L2 repress transcription at *pL*. (**b**) Tetramers of CI bound at O_L and O_R interact forming an octameric complex and a 2.3 kbp DNA loop. This higher-order complex facilitates cooperative binding of another pair of CI dimers at O_L3 and O_R3, resulting in the formation of another CI head–head tetramer and repression of transcription from P_{RM}

contribute considerably to the stability of lysogeny. Occupancy of O_R2 by CI also activates transcription of the CI gene from the P_{RM} promoter, giving CI a positive auto-regulatory mechanism [16, 30, 32] (Fig. 9.1a). This increases the level of CI to that required for repression of the lytic genes, as described above. At very high concentrations, CI was observed to also bind to O_R3 and repress its own transcription from P_{RM} [29]. This negative auto-regulation had been suggested to be important to prevent an excessive accumulation of repressor and facilitate efficient switching to lysis when necessary. However, such a negative regulation did not seem possible at physiological concentrations and the role of both O_L3 and O_R3 remained controversial. In particular, the conventional wisdom was that O_L3 was an evolutionary vestige.

A few years ago, it was suggested that, in the context of the intact λ chromosome, CI molecules bind cooperatively not just to adjacent sites, but also to sites separated by thousands of base pairs, inducing a regulatory loop in the viral DNA [11]. Looping is widely used as a gene regulation mechanism [28] but, in most of the prokaryotic cases, the loop length is limited to few hundred base pairs. Therefore, looping is generally viewed as a way to increase the local protein concentration, that is, as a mechanism enhancing the ability of a repressor to bind to a weak operator site through the interaction with a second repressor molecule bound to a stronger operator nearby. Clearly, this is not the case of the λ switch since the large loop size would not produce a significant increase in the local concentration (less than one order of magnitude).

This apparent quandary can be rationalized by the mechanistic hypothesis that the λ loop first forms by the interaction between two tetramers bound at O_R1–O_R2 and O_L1–O_L2, respectively. This octamer-mediated loop brings O_L3 and O_R3 into juxtaposition and favors their occupancy by CI dimers that can interact "head-to-head" in a cooperative fashion, leading to a DNA loop mediated by a CI octamer plus tetramer. According to this hypothesis, the loop provides the correct scaffold for CI binding to weak O_R3 at lysogenic concentrations [9–11] and effective repression of P_{RM} (Fig. 9.1b). *In vivo* experiments also indicate that such a loop would increase repression from lytic P_R fourfold [11, 37], with DNA looping stabilizing the lysogenic state. Thus, stability of the lysogenic state, effective repression of P_{RM}, and efficient switching to lysis in λ may all depend on DNA looping. These considerations piqued the interest of several groups [1–3, 33, 39, 48] led Ptashne to revise the third edition of his book on the λ bacteriophage genetic switch by adding a chapter on the newly proposed CI-mediated long-range interaction and its physiological implications [34].

Here, we summarize new findings on λ repressor interaction with λ DNA. We describe how single-molecule experiments in our laboratories have produced direct evidence of CI-mediated λ DNA looping *in vitro* [47], the critical role of interactions between CI dimers bound at the *O3* sites in loop stabilization, and the correlation between CI-mediated looping and activity of P_{RM} (Sect. 9.2). Furthermore, the complex kinetics of both loop formation and breakdown lead to a postulated kinetic scheme in which nonspecific binding by CI plays a significant role (Sect. 9.3). Strategies for testing this kinetic hypothesis and some preliminary evidence are described at the end of Sect. 9.3, while the conclusions are discussed in Sect. 9.4. The studies provide significant evidence that the multi-operator arrangement found in λ is the minimum necessary to guarantee robust lysogeny (strong repression of lytic genes) and efficient switching to lysis and mechanistic details of how looping confers epigenetic flexibility to cells.

9.2 Loop Stability and Correlation with In Vitro Transcription

9.2.1 CI Mediated Looping

Single molecule techniques have proven to be powerful tools for dissecting protein-induced conformational changes in DNA, such as looping. In particular, using the tethered particle motion (TPM) technique, direct evidence of loop formation and breakdown by CI was provided [47, 48]. In TPM, a submicron-sized bead is tethered by a single DNA molecule to the glass surface of a microscope flow chamber. The bead's lateral displacement with respect to the tether point $\rho(t)=\sqrt{x^2(t)+y^2(t)}$ is recorded as function of time. The running average of the squared displacement over a suitable (4 s) time window $\langle \rho^2 \rangle_{4S}^{1/2}$, is a measurement of the amplitude of the Brownian motion of the bead. In the presence of looping events, this gives rise to a telegraph-like signal [5, 6, 12, 31] and allows loop detection and quantification (Fig. 9.2, right).

In order to reliably analyze the Brownian motion and interpret amplitude fluctuations in TPM measurements, experimental and theoretical methods have been developed and successfully applied in predicting the TPM signal for a broad range of DNA tether lengths [5, 27, 31].

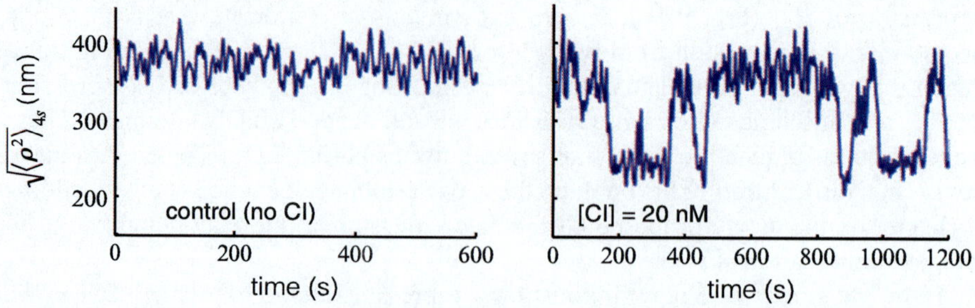

Fig. 9.2 $\langle \rho^2 \rangle_{4S}^{1/2}$ as a function of time for beads tethered by a DNA fragment where the regulatory regions are separated by 2,317 bp. *Left*: In the absence of CI (control experiment); *Right*: In the presence of 20 nM CI (monomer). All measurements are performed in l buffer (10 mM Tris–HCl pH 7.4, 200 mM KCl, 5% DMSO, 0.1 mM EDTA, 0.2 mM DTT, and 0.1 mg/ml a-casein). The length of the DNA tether is 3,477 bp

9.2.2 Dependence of Loop Probability/Stability on CI Concentration and Correlation to P_{RM} Activity

It is now evident that strong stability of the lysogenic state, effective repression of P_{RM}, and efficient switching to lysis in λ all involve DNA looping, which regulates the amount of synthesized CI protein. Therefore, it is important to understand how looping depends on CI concentration, as this will yield insight into the robustness and sensitivity of the λ switch. With this goal, our laboratories have directly detected and thermodynamically characterized. CI-mediated dynamic loop formation [48]. Loop formation was previously found to be correlated with transcriptional regulation of P_L, P_R, and P_{RM} at various CI concentrations in the physiological range, based on the estimated number of CI monomers per lysogen [33]. *In vitro* transcription assays at different CI concentrations (Dale A.E. Lewis, unpublished results; Fig. 9.3a) can be compared with a series of TPM experiments, collectively constituting a "looping titration" (Fig. 9.3b).

For the purpose of this discussion, there are two important features of the transcription results. First, transcription from the lytic promoters P_L and P_R, which is high at low CI concentrations, is reduced at 8 nM CI and is completely repressed at 20 nM CI. Second, CI transcription from P_{RM} is a function of CI concentration, as expected from the model of positive/negative autoregulation described in Sect. 9.1. The assays indicate that transcription from P_{RM}, at a basal level for concentrations lower than 8 nM, increases up to maximal activation at 20 nM CI and is progressively repressed as the concentration of CI is increased further (Fig. 9.3a).

According to the model for P_{RM} autoregulation, repression should result from O_R3 occupation, which in turn results from loop formation and O_L3 occupation. Therefore, the probability of loop formation should be an increasing function of CI concentration (Fig. 9.3a, left panel). In order to test this idea, TPM measurements

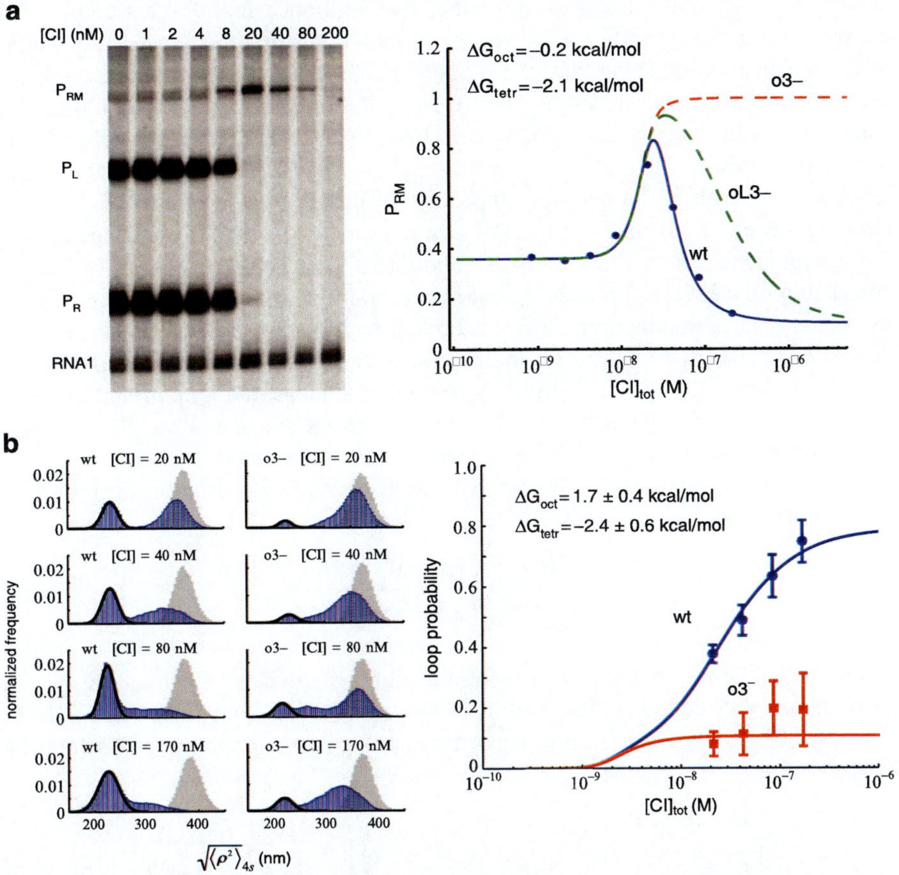

Fig. 9.3 (**a**) P_{RM} transcription concentration dependence. *Left, in vitro* transcription assays showing the amount of transcripts from P_{RM} and the lytic promoters P_L and P_R. *Right*, comparison of the trends of P_{RM} activity. Experimental data for wt λ DNA (*dots*). *Solid line* is best fit for wt λ and dashed lines are simulated repression curves for *O3*⁻ and O_L3^- DNA. All data are normalized to the maximum repression value obtained for *O3*⁻ DNA. (**b**) Dependence of DNA looping on CI concentration. *Left*, probability distribution of the TPM signal observed for tethered beads (about 40 for each condition) in the presence of different CI concentrations. The *left column* refers to wt DNA, while the *right column* refers to *O3*⁻ DNA. The *gray peaks* represent the control measurements without CI. *Right*, loop probability as a function of CI concentration for wt λ (*dots*) and *O3*⁻ DNA (*squares*). The *lines* are the result of the best fit performed as explained in [48]

on several hundred DNA-tethered beads were performed. Cumulative histograms representing the data from all these measurements at CI concentrations of 0, 20, 40, 80, and 170 nM are reported in the left panel of Fig. 9.3b. This figure shows that in the absence of CI, DNA is in the unlooped state only; just one peak appears in the frequency histogram (gray). These control measurements also show the high reproducibility of the TPM experiments. In the presence of CI, the histograms show two peaks: one corresponding to the DNA unlooped state (higher amplitudes of Brownian motion) and the other to the looped conformation (lower amplitudes of Brownian motion). Note that the looped state, although not predominant, does occur at low CI concentrations. At 40 nM CI concentration, DNA molecules spend half of the time in the looped configuration and this correlates well with the partial repression of P_{RM} observed in Fig. 9.3a. In the presence of 80 nM CI, the DNA is prevalently looped, as expected from *in vitro* transcription assays. However, even at 170 nM CI, the DNA is not always looped. This indicates that the dynamic nature of the system is modulated by CI concentration, but never completely disappears at physiological concentrations. Interestingly, the probability of loop formation as derived from the histograms (Fig. 9.3b, right panel) was described very well by an extension of the statistical thermodynamic model developed by Ackers et al. [17, 38]. In order to relate the measured loop probabilities to the microscopic configurations of CI bound to the six operator sites, 81 unlooped and 32 looped configurations were considered according to the scheme proposed by Anderson and Yang [1, 2]. The probability for each DNA–protein configuration was expressed as,

$$fi = \frac{[CI_2]^{s_i} \exp(-\Delta G_i / RT)}{\sum_i [CI_2]^{s_i} \exp(-\Delta G_i / RT)}, \tag{9.1}$$

where ΔG_i is the sum of the free energies for binding, short range cooperativity, and looping of each configuration and s_i is the number of bound CI dimers (CI_2; Fig. 9.4). CI dimer concentration was calculated from the expression for the total concentration of CI:

$$[\mathrm{CI}]_{\mathrm{tot}} = \sqrt{\frac{[\mathrm{CI}_2]}{K_\mathrm{d}}} + 2[\mathrm{CI}_2] + 2K_{\mathrm{NS}} l [\mathrm{DNA}][\mathrm{CI}_2] + [\mathrm{DNA}] \sum_i s_i \cdot f_i, \tag{9.2}$$

where K_d is the dimerization constant for CI [4], K_{NS} is the nonspecific binding constant [4], and l is the DNA length in base pairs (see Table 1 in [48]). The terms of this equation from left to right represent: CI monomers, CI dimers, nonspecifically bound, and specifically bound CI. The concentration-dependent loop probability was then calculated as:

$$\text{Loop probability} = \frac{\sum_{i=82}^{113} f_i}{\sum_{i=1}^{113} f_i} \tag{9.3}$$

in which the sum in the numerator runs over all the 32 looped configurations as in [1, 2] and is normalized by the sum of all the possible (looped and unlooped) configurations.

This thermodynamic analysis assumes that the loop can be secured either by a CI octamer or by an octamer plus a tetramer. The difference in free energy between these two looped species represents the tetramerization free energy ΔG_{tetr}.

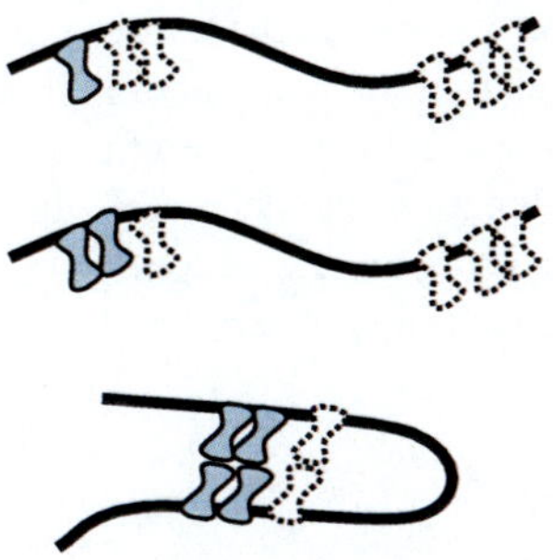

$$\Delta G_{L1} = \Delta G_{L1,\text{binding}}$$

$$\Delta G_{L1,2} = \Delta G_{L1,\text{ binding}} + \Delta G_{L2,\text{ binding}} + \Delta G_{L1,2,\text{ coop}}$$

$$\Delta G_{\text{Loop}1,2} = \Delta G_{L1,\text{ binding}} + \Delta G_{L2,\text{ binding}} + \Delta G_{L1,2,\text{ coop}} + \Delta G_{R1,\text{ binding}} + \Delta G_{R2,\text{ binding}} + \Delta G_{R1,2,\text{ coop}} + \Delta G_{\text{loop octamer}}$$

Fig. 9.4 Illustration of the procedure used to calculate ΔG_i in a few representative cases. The free energy for each unlooped species was expressed as the sum of all the free energies for binding and short-range cooperativity, available from previous work [17, 38]. The free energy expression for the looped species also included the term ΔG_{oct} and the tetramerization term, ΔG_{tetr} was added only for configurations in which two CI dimers not involved in the octamer were juxtaposed [2]. Using this model, the probability of looping was expressed as a function of CI concentration. The experimental data obtained with both wt λ and λ*O3*⁻ DNA were fitted simultaneously to estimate ΔG_{oct} and ΔG_{tetr}

To quantitatively correlate the single molecule experiment with the transcription assay, the same statistical mechanical model described above and used in [48] was employed to test whether it could also describe *in vitro* transcription from P_{RM}. Measurements of mRNA levels were made from the digitized image of the gel shown in the left panel of Fig. 9.3a and are shown in the right panel of Fig. 9.3a. Since the P_{RM} promoter is activated by occupancy of the O_R2 operator and it is repressed by the binding of a CI dimer to O_R3 [29], the expression for transcription from P_{RM} was as:

$$P_{\text{RM}} = p_{\text{repr}} \sum_{i \in O_R 3} f_i + p_{\text{act}} \sum_{\substack{i \in O_R 2 \\ i \in \text{unlooped}}} f_i + \alpha p_{\text{act}} \sum_{\substack{i \in O_R 2 \\ i \in \text{looped}}} f_i + p_{\text{bas}} \sum_{\text{else}} f_i, \tag{9.4}$$

in which a basal transcription value was assigned to all the configurations except those having O_R2 occupied (activated transcription) and those having O_R3 occupied (repressed transcription). Although the effect of looping on the repression of transcription is still a matter of debate [1, 2, 11, 33], we assumed a two- to threefold (α=2–3) increase for the activated transcription in the looped state [1, 2]. In the fitting of this data, the activated transcription p_{act}, ΔG_{oct}, and ΔG_{tetr} values were left as free parameters. The remarkable fit demonstrates that the model also accurately describes how transcription depends on CI concentration.

It must be noted that the *in vitro* transcription experiments were carried out on plasmids having a loop length of only ~400 bp, whereas in the TPM experiment the wt loop region (~2,000 bp) was used. Since the free energy of loop formation also depends on the physical properties of DNA (loop length, supercoiling state, and persistence length), a direct comparison of the looping free energy ΔG_{oct} obtained in the two cases is not possible. However, the lower value found for ΔG_{oct} in the transcription assay presumably reflects the reduced cost of looping due to plasmid supercoiling and shorter loop size. On the other hand, since tetramer formation takes place after loop closure, the value of ΔG_{tetr} is not influenced by the loop length or the degree of supercoiling. Consequently, the values of ΔG_{tetr} determined in fitting the transcription assay (–2.1 kcal/mol) and the TPM-derived ΔG_{tetr} (–2.4 kcal/mol) [48] are very similar.

9.2.3 *Role of the O3 Operators*

If (1) the four sites O_L1, O_L2 and O_R1, O_R2 are occupied first and almost simultaneously because of strong and cooperative binding of CI, and (2) CI octamer-mediated loop formation is further secured by the "head-to-head" interaction of two dimers bound at the O_L3 and O_R3 operators, then mutation of the *O3* operators should interfere with stable CI-mediated looping. This idea was tested with TPM experiments using DNA tethers containing mutated *O3* sites (*O3*⁻ DNA) at four different CI concentrations (Fig. 9.3b). Details regarding these measurements are reported in [48]. First, it was verified that no loop formation was observed in $O_L^-O_R^-$ DNA, where all six λ operators had been mutated (Fig. 9.5, third panel). This indicated that the selected point mutations effectively abrogate specific CI binding to operators. Then the probability of loop formation in *O3*⁻ DNA was found to be approximately 10% for all CI concentrations tested (Fig. 9.3b, middle panel). These results show that intact *O3* operators dramatically shift the equilibrium toward looping. According to the thermodynamic model, concentration-independent *O3*⁻ looping results from cooperative filling of all the remaining sites, even at low CI concentration [30]. Instead, in the wt DNA, concentration-driven occupancy of the *O3* sites drives the formation of octamer-plus-tetramer-mediated loops with lower free energy than CI octamer-mediated loops.

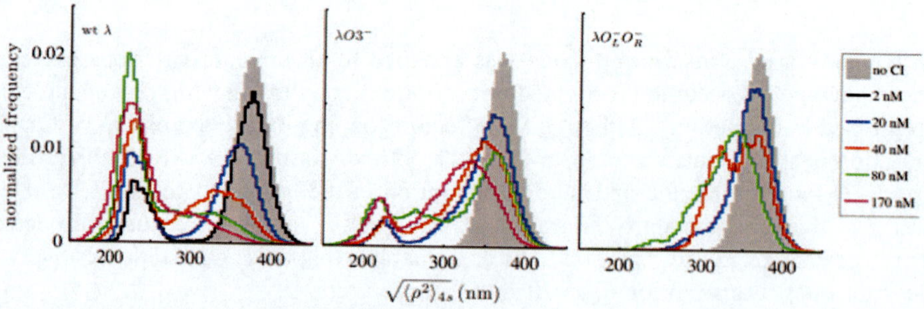

Fig. 9.5 Histograms of the TPM signal measured for wt, *O3*⁻ and null DNA at various nanomolar concentrations of CI reveal that (*left*) looping increases with CI concentration for wild-type DNA, (*middle*) looping dramatically decreases when *O3* operators are mutated, and (*right*) simultaneous mutations to all operators abrogates looping

9.2.4 CI Nonspecific Binding

The peak corresponding to the extended, unlooped state in the frequency distribution histograms of the TPM signal shifts toward smaller values of $\langle \rho^2 \rangle_{4s}^{1/2}$, as CI concentration increases (Figs. 9.3b and 9.5). This could have two different causes: poor temporal resolution preventing resolution of fast loop breakdown events and/or progressive DNA compaction due to bending associated with nonspecific binding of CI to the double helix. Typically, the system is sampled at 50 Hz, but filtering reduces the resolution to 4 s. Even using statistical methods that do not employ filtering (see below), shortening of the unlooped DNA occurs. Thus, the effect of increasing CI concentrations was compared on null (OL^-OR^-), wt and $O3^-$ DNA (Fig. 9.5). The peak corresponding to unlooped DNA shifted leftward in all cases, suggesting that nonspecific CI binding is significant and considerably shortens the DNA tether. Work from the Cox lab lends strong support to the idea that nonspecific CI binding may be significant [44].

Further evidence of nonspecific, CI-induced DNA bending was obtained stretching and relaxing single DNA molecules by means of magnetic tweezers. The *force* versus *extension* curves that were obtained with or without CI were markedly different (Fig. 9.6). In order to produce equivalent DNA extension, more force was necessary in the presence of CI, indicating that CI may induce bends or kinks in the DNA through binding or transient interactions with nearby dimers that bend the intervening DNA. The noteworthy hysteresis between the relaxation and stretching cycles may reveal interesting mechanistic details and is under investigation. Furthermore, quantification of the number of nonspecifically bound proteins is possible using the recent model proposed by Zhang and Marko [46], Liebesny et al. [50].

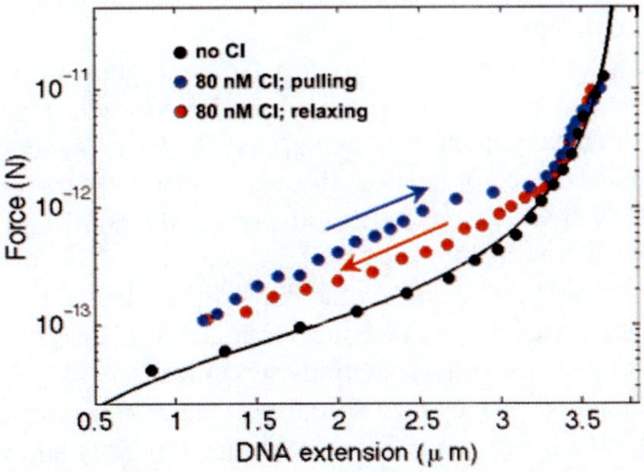

Fig. 9.6 DNA extension versus force for a wt, 11 kbp-long DNA molecule. *Dots* are averages of points from several pulling/relaxation cycles, in the absence of protein (*black*) and in the presence of 80 nM CI (*blue* and *red*). The *solid curve* is a worm-like-chain fit obtained assuming a persistence length for DNA of 52 nm. The *arrows* indicate that the experimental data were collected during stretching (*blue*) or relaxation of the DNA molecule (*red*)

Considering that CI mRNA is transcribed and translated close to P_{RM}, which likely produces a local concentration of CI dimers in excess with respect to the number of operators [33, 34], nonspecific binding may have physiological relevance, just as shown in the case of other proteins [14, 49].

9.3 Kinetics of CI-Mediated DNA Loop Formation and Breakdown

Kinetic analysis of formation and breakdown of the CI-mediated loop yields insight into the mechanism of looping and its regulatory effect on transcription. The kinetics of protein-induced loop formation and breakdown have been deduced from TPM data for a variety of simple systems [12, 40, 41]. In these studies, the TPM signal was analyzed using time filtering (averaging) of the raw data followed by thresholding to determine the state of the system. Then, the measured dwell times of each state were plotted as probability density functions and the average lifetime of the looped or unlooped state was determined. For such systems, characterized by a simple kinetics, a single (or at most a double) exponential function can fit the data satisfactorily. In this way, the rate constants for the looping and unlooping reactions have been obtained along with an estimate for the free energy of loop formation. Note, however, that in all these cases, time filtering of the raw data significantly impacts the time resolution of the measurements and the kinetic constants. Methods have been proposed to either correct for such a drawback [8, 40, 41] or determine the kinetic constants from the raw data [8, 36]. These approaches, however, require the knowledge of the kinetic mechanism of the reaction being considered, and their application is limited to cases in which: (a) there is only a small number of discrete states separated by fixed energy barriers and (b) the kinetic rate constants connecting these states are independent [21].

An ideal method of data analysis should have the highest possible time resolution and should be independent of physical models. In addition, it should avoid user-adjustable parameters, such as filtering, that skew the raw data.

Therefore, an alternative approach to the analysis of TPM traces was devised by Manzo and Finzi [27] based on a method previously published for time traces exhibiting discrete jumps in intensity [45]. A generalized likelihood ratio test is first applied to determine the location of a TPM signal change point (cp). This test is applied recursively to an entire TPM trace, to identify all cps (transitions between different DNA configurations). Expectation–maximization (EM) clustering and the Bayesian information criterion are then used for accurate determination of the number of states accessible to the system. This procedure (cp-EM) allows objective and quantitative determination of TPM cps without the artificial time resolution limitations that arise from filtering and thresholding [45]. The applicability and performance of this analysis was tested on artificial TPM data assembled ad hoc from segments of TPM data acquired in the absence of CI (unlooped) interdispersed between segments of TPM data from a DNA molecule of overall length comparable to that expected for the looped lambda DNA. Subsequently, the cp-EM algorithm was used to analyze TPM data for a lambda DNA molecule in the presence of CI. This analysis confirmed that CI interaction with its operators produces most likely two states, which were commensurate with the looped and unlooped DNA states. The cp-EM approach allowed determination of the looped and unlooped dwell time distributions with a much increased time resolution [27].

9.3.1 Analysis of the λ Loop Kinetics

The cp-EM method described above and in [27] was applied to the analysis of the kinetics of CI-mediated loop formation and breakdown. Interestingly, the probability distribution function (pdf) for the dwell times of the looped and unlooped DNA conformations (states) span several orders of magnitude and show nonexponential tails at long times (Figs. 9.7 and 9.8). In the case of λ DNA, it is reasonable to assume that the tripartite operator organization might lead to complex dynamics for loop formation and breakdown. Nevertheless, the discrete and Markovian nature of the kinetic system still predicts an exponential behavior for the pdfs.

We recall here that nonexponential decays have been reported for other physical systems such as ion channel currents [20, 22–25], quantum dot blinking time [13] and for fluorescence emission by green fluorescent protein [18, 19]. Notably, the dwell time determination by means of the filtering-threshold approach produced qualitatively similar results, excluding the possibility that artifacts might have been

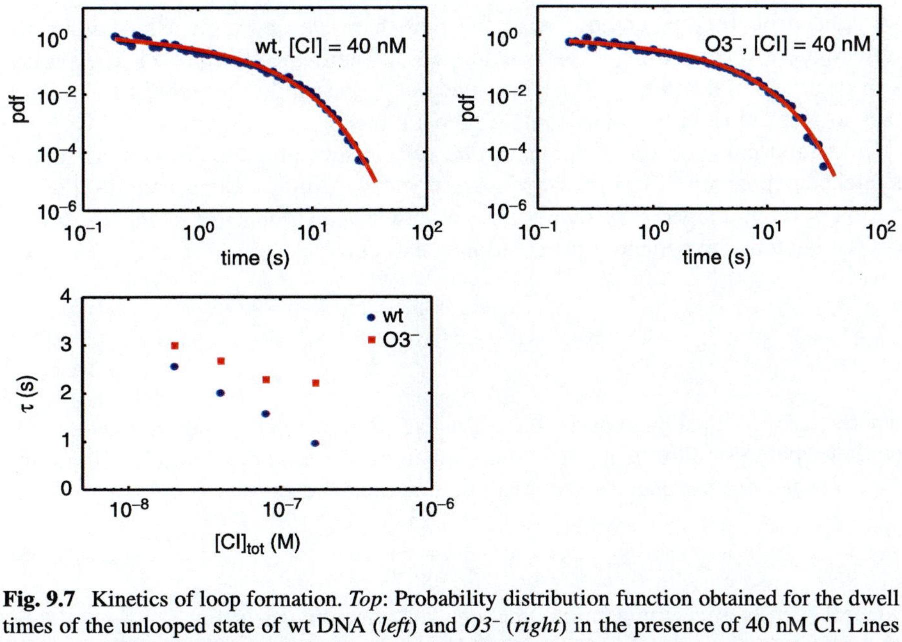

Fig. 9.7 Kinetics of loop formation. *Top*: Probability distribution function obtained for the dwell times of the unlooped state of wt DNA (*left*) and *O3*⁻ (*right*) in the presence of 40 nM CI. Lines are the result of fitting by a stretched exponential pdf. *Bottom*: Average time spent in the unlooped configuration. Wild type and *O3*⁻ data points are *blue* and *red*, respectively

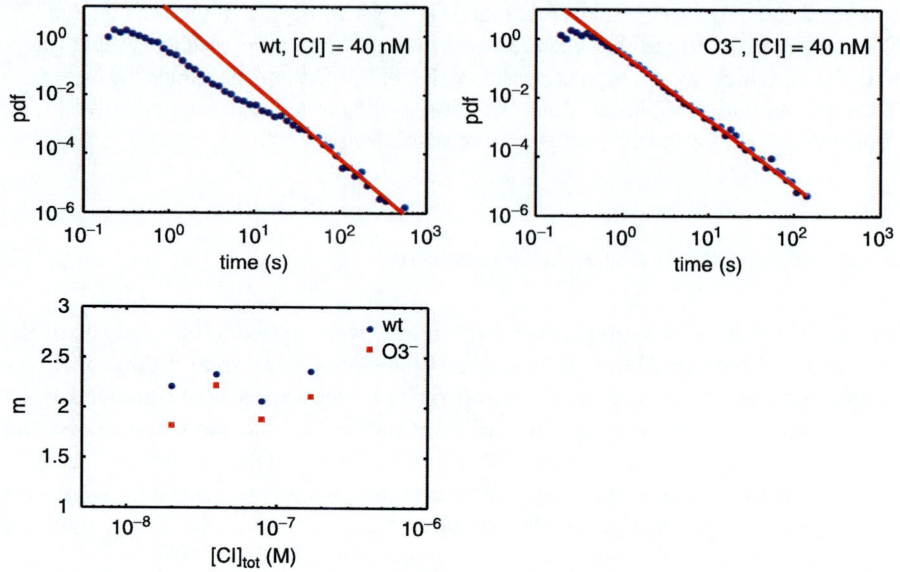

Fig. 9.8 Kinetics of loop breakdown. *Top*: Probability distribution function obtained for the dwell times of the looped state of wt DNA (*left*) and *O3⁻* (*right*) in the presence of 40 nM CI. *Lines* show the power law behavior at large times. *Bottom*: Power law exponent *m* determined from fitting the longer lifetimes. Wild type and *O3⁻* data points are *blue* and *red*, respectively

introduced by the cp method. Also the fact that, even a single DNA molecule observed for a long time, exhibited looped and unlooped dwell times spanning several orders of magnitude (Fig. 9.9) ruled out the possibility that complex kinetics may arise from heterogeneous sample preparations.

To empirically describe the data, we tested several pdfs. We found that a stretched exponential form provided a satisfactory fitting for the probability distribution of the unlooped state dwell times at any tested CI concentration.

The stretched exponential distribution is expressed as:

$$\mathrm{pdf}_{\mathrm{str}} = c\frac{t^{c-1}}{t_0^c}\exp\left\{-\left(\frac{t}{t_0}\right)^c\right\}, \tag{9.5}$$

where the two adjustable parameter are the scale parameter t_0 and the exponent c, the latter being smaller than one and giving an exponential distribution in the limit $c=1$. From these parameters, the mean time is obtained as:

$$\tau = \frac{t_0}{c}\Gamma\left(\frac{1}{c}\right) \tag{9.6}$$

where $\Gamma(x)$ represents the gamma function. The upper panels of Fig. 9.7 show the fit of the data relative to both *wt* and *O3⁻* DNA in the presence of 40 nM CI using the

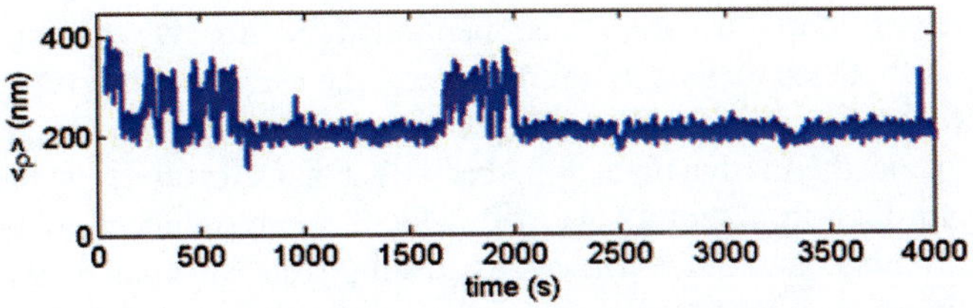

Fig. 9.9 A TPM trace for a *wt* DNA molecule in the presence of 40 nM CI. The DNA is 3,477 bp long

stretched exponential distribution function. Both fitting parameters t_0 and c show dependence on CI concentration and are significantly sensitive to the type of DNA (data not shown). As a consequence, their mean unlooped times, τ, are similar, as shown in the lower panel of Fig. 9.7. For both $wt\lambda$ and $O3^-$ DNA, τ decreases with CI concentration, although it does so more rapidly for the $wt\lambda$ than for the $O3^-$ DNA.

On the other hand, the dwell time distribution for the looped DNA state could not be fitted with a standard pdf; we observe, however, a power law decay at long times, resulting in a straight line on a log–log plot. The power law pdf:

$$\mathrm{pdf}_{\mathrm{pl}} \propto t^{-m} \tag{9.7}$$

is sometimes referred as a "fractal" distribution since it reveals no characteristic size of the system, and therefore makes it impossible to extract a mean lifetime from the data. The log–log plots of the looped time distribution for $wt\lambda$ and $O3^-$ DNA at 40 nM CI (Fig. 9.8, upper panels) show that, while the data from $O3^-$ DNA are well described by a power law function even at very short times, those from *wt* DNA are not. In both cases, however, the data for longer lifetimes are characterized by the exponent $m \approx 2$ in all the experimental conditions tested (lower panel of Fig. 9.8).

9.3.2 *Suggested Mechanism of Loop Formation and Breakdown and Role of Nonspecific Binding*

The single-molecule experiments described in Sects. 9.2 and 9.3.1 highlight four important features of the λ regulatory loop: (1) the pivotal role of the *O3* sites in the thermodynamics of loop formation, (2) the concentration dependence of both loop formation and breakdown, (3) the presence of significant nonspecific CI binding even at low protein concentrations, which seems to bend or soften DNA and, finally, (4) the nonexponential kinetic behavior of both loop formation and breakdown. Taken together, these observations lead us to formulate the kinetic mechanism described below.

First, consider loop formation. The most direct information on this process is provided by the distribution of the lifetimes of the unlooped state, which indicates the average time for loop formation. As mentioned previously, this distribution is described by a stretched exponential function and the average unlooped state lifetime varies with CI concentration and is affected by mutations of the *O3* sites (Fig. 9.7). According to the thermodynamic hypothesis that a CI octamer is the minimum requirement for loop formation [2, 11, 48], which is then stabilized by an additional CI tetramer, the concentration dependence can simply be explained by the increase in the population of the loop-forming substates (i.e., those having at least two pairs of adjacent, specifically bound CI dimers) as the amount of CI increases.

Kinetic complexity may arise from nonspecific binding of CI to DNA. Indeed, our TPM and force spectroscopy measurements show ample evidence of nonspecific CI binding (see Sect. 9.2.4), and there is other evidence from the Cox group [44]. Nonspecifically bound CI dimers may shorten the DNA by bending or softening the double helix upon binding or through interaction with nearby dimers. The ensuing DNA bending could facilitate loop formation by reducing the elastic energy necessary for loop closure which, in turn, increases the encounter probability among the proteins bound at these two regions. The relevance of these effects would depend on the number of nonspecifically bound proteins in the loop region, n_{NS}. The variation of the number of nonspecifically bound dimers would then generate a distribution of rate constants for loop formation (schematically represented by k_L (n_{NS}) in Fig. 9.10) that could explain the observed stretched-exponential pdf. Note that nonspecific CI binding is likely to be relevant *in vivo* as well, since CI is produced in excess with respect to the number of operators [33].

In the case of loop breakdown kinetics, the distribution of the looped lifetimes also depends on CI concentration and on the presence of the *O3* sites, but cannot be described by a standard pdf. Moreover, the distribution of long dwell times shows a power law decay with an exponent which does not depend on the experimental conditions. According to the thermodynamic model described above [2, 48], the many possible looped substates can be grouped in two classes mediated by either four or six CI dimers, respectively. Each class will have a different breakdown rate. Consequently and in qualitative agreement with the experiments, the model predicts concentration-dependent loop breakdown kinetics and a simpler distribution in the case of *O3*$^-$ DNA in which octamer + tetramer loop cannot form. However, as for the distribution of unlooped lifetimes, the observed pdfs do not exhibit simple exponential behavior.

Nonspecific CI binding may contribute to the observed decay at long times. Indeed, nonspecifically bound CI dimers might interact through their CTD residues [7, 15] just as specifically bound dimers do. In particular, nonspecifically bound dimers within the loop may tetramerize [7, 15] to fortify the specific nucleoprotein complex that secures the loop. In Fig. 9.10, k_L (n_{NS}) and k_U represent the rate constants of loop formation and breakdown. If no additional nonspecifically bound proteins are present, k_L $(n_{NS} = 0)$ is single valued and the kinetics of the system are exponential. In the presence of nonspecifically bound CI, the variation of n_{NS} broadens the distribution of k_L (n_{NS}). Any further tetramerization between these nonspecifically bound dimers would be dependent on their number n_{NS} and on their relative separation, ℓ, as schematically diagrammed in Fig. 9.10, where k_a (n_{NS}, ℓ) represents a distribution of rate

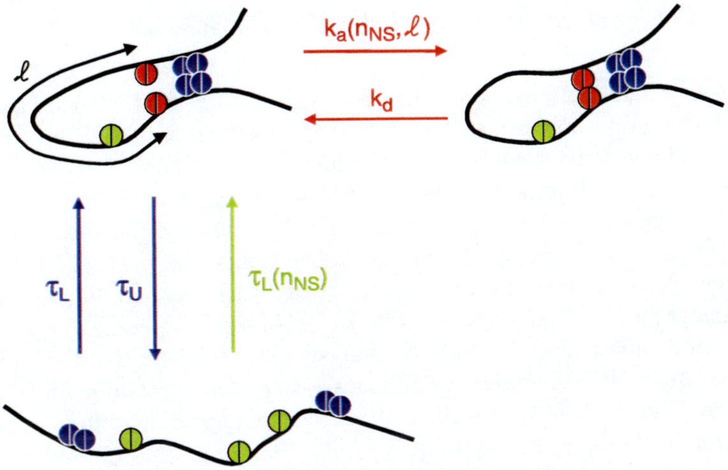

Fig. 9.10 A hypothetical kinetic scheme for loop formation and breakdown. In addition to the CI dimers at the operator sites (*blue*), nonspecifically bound dimers may affect the rate of loop formation (*green*) by DNA bending and may strengthen the loop providing additional loop closure elements (*red*). For simplicity only the case of an octameric loop is sketched

constants for the association between additional, loop-stabilizing CI tetramers (red), and k_d is the rate constant for their dissociation. These manifold looped states would give rise to a continuous distribution of waiting times for the breakdown of the λ-mediated loop, producing a power law-like decay.

The model for loop formation and breakdown hypothesized here is based on two main ideas: the critical role of the *O3* sites and nonspecific CI binding. The latter might facilitate loop formation in this and other physiologically relevant protein-induced loops by decreasing the distance between the interacting sites, maintaining proximity of the specific regions, and by adding stabilizing protein–protein interactions. Indeed, CI has been estimated to be in excess with respect to the number of specific binding sites [33] and extra CI is thought to confer immunity against subsequent infection by additional phages [34].

This proposed mechanism of loop formation and breakdown by the CI repressor seems to be the simplest possible explanation that is consistent with and reconciles: (a) known biochemical data on CI/operator affinity and side-by-side cooperativity [17, 38], (b) *in vivo* transcription data [1], and (c) single-molecule data. This mechanism is also attractive because its key assumptions may easily be tested and the experiments should yield useful guidelines for similar loop-based, regulatory looping systems. Furthermore, we would like to draw the attention of the reader to the fact that, as pointed out by Vilar and Saiz [42], only a narrow range of looping free energies separates high-sensitivity activation and repression of the P_{RM} promoter. In this sense, nonspecific binding might function as a concentration-sensitive mechanism for the fine tuning of the looping free energy.

9.3.3 Evidence in Support of the Kinetic Scheme

The kinetic model outlined above may be tested in at least two ways: First, a theoretical formulation of the distribution of rate constants produced by nonspecific binding could be derived and compared to the experimental pdfs. Second, the various possible looped and unlooped species and their dependence on CI concentration could be characterized using atomic force microscopy. This single-molecule technique can directly reveal the number of nonspecifically bound CI dimers and their interactions, and some evidence has already been obtained. Figure 9.11 shows AFM images of λ DNA without CI, λ DNA bound specifically by CI, and CI-mediated loops in λ DNA. Volume analysis of the CI particle at the loop closure in several such AFM images revealed that the size of the protein particle increased over time (Fig. 9.12) [43]. Together, these data suggest that as time passes additional proteins bind to stabilize loops.

Furthermore, AFM images of DNA molecules in the presence of 40 nM CI included a few DNA loops secured by adjacent protein complexes (Fig. 9.13). These species, reported also in [43], might represent those with additional nonspecific loop closures hypothesized in the kinetic model proposed in Sect. 9.3.2. More experimentation is necessary to confirm these hypotheses. Luckily, experimental and analytical methods are now available to interrogate simple epigenetic switches, such as lambda, and characterize the molecular mechanisms involved.

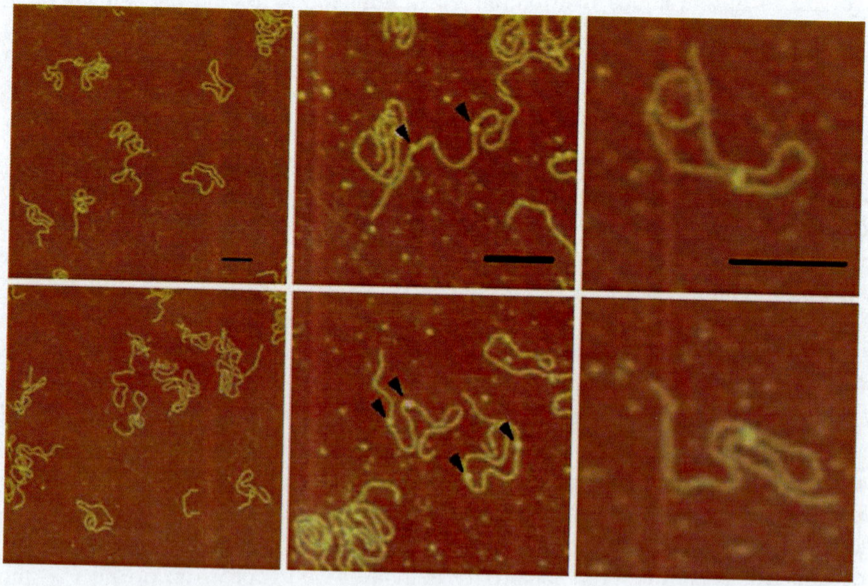

Fig. 9.11 *Left column*: DNA; *center column*: CI bound at operators on unlooped DNA; *right column*: CI-mediated DNA looping. Scale bars represent 100 nm

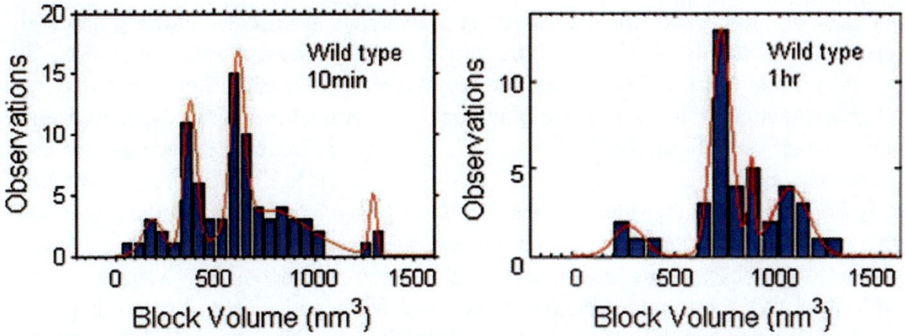

Fig. 9.12 Histograms of particle volumes measured at the loop junction after incubating DNA with 20 nM CI for 10 min (*left*) or 1 h (*right*)

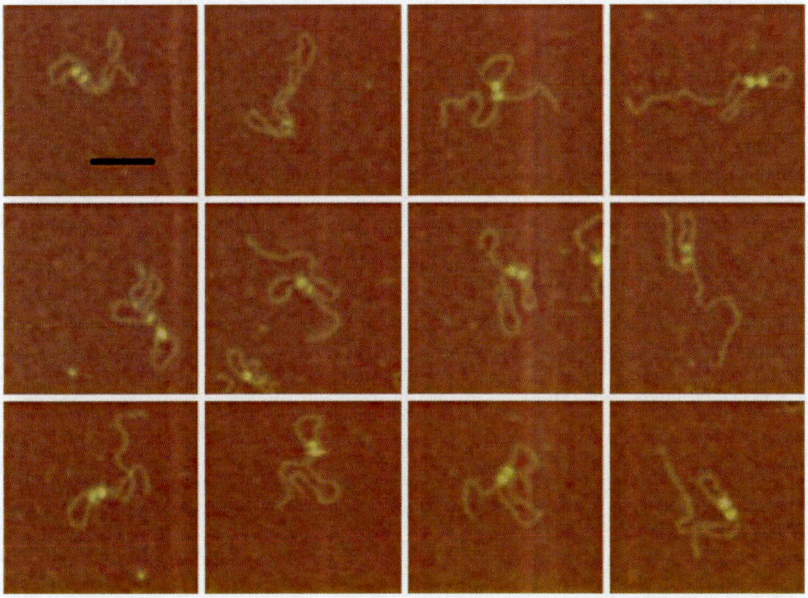

Fig. 9.13 AFM images in which the CI-mediated DNA loop is secured by two distinct protein particles. Scale bar: 100 nm

9.4 Conclusions

From viruses to humans, transcription is regulated by proteins that topologically constrain DNA. In most cases genes are controlled by large, cooperative assemblies of proteins that wrap and loop the DNA. These protein-induced conformational changes in the DNA often constitute real "epigenetic switches" whereby shifting the equilibrium towards one configuration or the other transcriptionally commits the system to one developmental pathway or another. Such is the case of the λ bacteriophage and, perhaps, of most temperate bacteriophages that may proliferate either quiescently (lysogenic mode) or virulently (lytic mode). Most often following infection, the repressor protein, CI, binds to multipartite operators and mediates cooperative, long-range interactions that repress the lytic genes and maintain a stable lysogenic state, until adverse conditions (DNA damage, toxicity, starvation, etc.) induce a cascade of events that leads to the dissociation of CI from the double helix and efficient switching to lysis. This switch that alternatively stabilizes lysogeny or commits the λ bacteriophage to lytic reproduction [16] is a convenient, experimentally tractable epigenetic paradigm of transcriptional regulation. Due to the importance of λ as a model system not only for transcriptional regulation, but also for genetic networks, the underpinnings of this regulatory loop are important for understanding the sensitivity, stability and operation of such networks in epigenetics.

In this chapter, we have discussed new approaches to reveal details of the molecular mechanism of λ repressor-mediated looping. Single molecule observations have, first of all, provided direct evidence of CI-induced looping between the wild-type lambda operators. In addition, a titration of loop closure probability versus CI concentration combined with a thermodynamic model has substantially confirmed the hypothetical octamer + tetramer mechanism of CI-mediated regulation of lysogeny. Finally, kinetic analyses of the looping dynamics suggest an unsuspected, but in retrospect, logical role that non-specific binding of the CI protein likely plays in lysogenic regulation. In a short time single molecule experimentation has provided detailed insight into how long-range interactions may govern epigenetic switching and these mechanisms will be pertinent to a variety of other systems with multipartite operators and multi- or heteromeric protein assemblies [30].

Acknowledgments We would like to thank previous and current members of our groups whose research has facilitated these studies. We are also grateful to Haw Yang, who has provided reagents, analytical tools, and advice. The work described in this chapter was supported by the Italian Funding of Basic Research to LF and DDD, by the HFSP(RGP0050/2002-C) to L.F. and S.A., by the Intramural Research Program of the National Institutes of Health, National Cancer Institute and the Center for Cancer Research to S.A., by the Emory University Research Council and the NIH (RGM084070A) to LF.

References

1. Anderson L, Yang H (2008) A simplified model for lysogenic regulation through DNA looping. Conf Proc IEEE Eng Med Biol Soc 2008:607–610
2. Anderson LM, Yang H (2008) DNA looping can enhance lysogenic CI transcription in phage lambda. Proc Natl Acad Sci U S A 105(15):5827–5832
3. Atsumi S, Little JW (2006) Role of the lytic repressor in prophage induction of phage lambda as analyzed by a module-replacement approach. Proc Natl Acad Sci USA 103(12): 4558–4563
4. Bakk A, Metzler R (2004) Nonspecific binding of the O-R repressors CI and Cro of bacteriophage lambda. J Theor Biol 231(4):525–533
5. Beausang JF, Zurla C, Finzi L, Sullivan L, Nelson PC (2007) Elementary simulation of tethered Brownian motion. Am J Phys 75:520–523

6. Beausang JF, Zurla C, Manzo C, Dunlap D, Finzi L, Nelson PC (2007) DNA looping kinetics analyzed using diffusive hidden Markov model. Biophys J 92(8):L64–L66
7. Bell CE, Frescura P, Hochschild A, Lewis M (2000) Crystal structure of the lambda repressor C-terminal domain provides a model for cooperative operator binding. Cell 101(7):801–811
8. Colquhoun D, Sigworth FJ (1983) Fitting and statistical analysis of single channel recording. Plenum, New York
9. Dodd IB, Perkins AJ, Tsemitsidis D, Egan JB (2001) Octamerization of lambda CI repressor is needed for effective repression of P-RM and efficient switching from lysogeny. Genes Dev 15(22):3013–3022
10. Dodd IB, Shearwin KE, Egan JB (2005) Revisited gene regulation in bacteriophage lambda. Curr Opin Genet Dev 15(2):145–152
11. Dodd IB, Shearwin KE, Perkins AJ, Burr T, Hochschild A, Egan JB (2004) Cooperativity in long-range gene regulation by the lambda CI repressor. Genes Dev 18(3):344–354
12. Finzi L, Gelles J (1995) Measurement of lactose repressor-mediated loop formation and breakdown in single DNA-molecules. Science 267(5196):378–380
13. Frantsuzov P, Kuno M, Janko B, Marcus RA (2008) Universal emission intermittency in quantum dots, nanorods and nanowires. Nat Phys 4(5):519–522 [10.1038/nphys1001]
14. Guerra RF, Imperadori L, Mantovani R, Dunlap DD, Finzi L (2007) DNA compaction by the nuclear factor-Y. Biophys J 93(1):176–182
15. Hochschild A, Ptashne M (1988) Interaction at a distance between lambda-repressors disrupts gene activation. Nature 336(6197):353–357
16. Jain D, Nickels BE, Sun L, Hochschild A, Darst SA (2004) Structure of a ternary transcription activation complex. Mol Cell 13(1):45–53
17. Koblan KS, Ackers GK (1992) Site-specific enthalpic regulation of DNA-transcription at bacteriophage-lambda OR. Biochemistry 31(1):57–65
18. Leiderman P, Huppert D, Agmon N (2006) Transition in the temperature-dependence of gfp fluorescence: from proton wires to proton exit. Biophys J 90(3):1009–1018. doi:10.1529/biophysj.105.069393
19. Leiderman P, Huppert D, Remington SJ, Tolbert LM, Solntsev KM (2008) The effect of pressure on the excited-state proton transfer in the wild-type green fluorescent protein. Chem Phys Lett 455(4–6):303–306. doi:10.1016/j.cplett.2008.02.079
20. Liebovitch LS (1989) Analysis of fractal ion channel gating kinetics – kinetic rates, energy-levels, and activation-energies. Math Biosci 93(1):97–115
21. Liebovitch LS (1989) Testing fractal and Markov-models of ion channel kinetics. Biophys J 55(2):373–377
22. Liebovitch LS, Fischbarg J, Koniarek JP (1987) Ion channel kinetics – a model based on fractal scaling rather than multistate Markov-processes. Math Biosci 84(1):37–68
23. Liebovitch LS, Fischbarg J, Koniarek JP, Todorova I, Wang M (1987) Fractal model of ion-channel kinetics. Biochim Biophys Acta 896(2):173–180
24. Liebovitch LS, Sullivan JM (1987) Fractal analysis of a voltage-dependent potassium channel from cultured mouse hippocampal-neurons. Biophys J 52(6):979–988
25. Liebovitch LS, Toth TI (1991) Distributions of activation-energy barriers that produce stretched exponential probability-distributions for the time spent in each state of the 2 state reaction a-reversible-B. Bull Math Biol 53(3):443–455
26. Maniatis T, Ptashne M (1973) Multiple repressor binding at operators in bacteriophage-lambda – (nuclease protection polynucleotide sizing pyrimidine tracts supercoils E. coli). Proc Natl Acad Sci U S A 70(5):1531–1535
27. Manzo C, Finzi L (2010) Quantitative analysis of DNA looping kinetics from tethered particle motion experiments in *Methods in Enzymology*, volume 475 "Molecule Tools, Part B: Super-Resolution, Particle Tracking, Multiparameter, and Force Based Methods", Ed. Nils G. Walter, pp 199–220.
28. Matthews KS (1992) DNA looping. Microbiol Rev 56(1):123–136
29. Maurer R, Meyer BJ, Ptashne M (1980) Gene-regulation at the right operator (Or) of bacteriophage-lambda.1. Or3 and autogenous negative control by repressor. J Mol Biol 139(2):147–161

30. Meyer BJ, Maurer R, Ptashne M (1980) Gene-regulation at the right operator (Or) of bacteriophage-lambda.2. Or1, Or2, and Or3 – their roles in mediating the effects of repressor and Cro. J Mol Biol 139(2):163–194
31. Nelson PC, Zurla C, Brogioli D, Beausang JF, Finzi L, Dunlap D (2006) Tethered particle motion as a diagnostic of DNA tether length. J Phys Chem B 110(34):17260–17267
32. Nickels BE, Dove SL, Murakami KS, Darst SA, Hochschild A (2002) Protein–protein and protein–DNA interactions of sigma(70) region 4 involved in transcription activation by lambda cl. J Mol Biol 324(1):17–34
33. Oppenheim AB, Kobiler O, Stavans J, Court DL, Adhya S (2005) Switches in bacteriophage lambda development. Annu Rev Genet 39:409–429
34. Ptashne M (2004) A genetic switch: phage lambda revisited, vol 3, 3rd edn. Cold Spring Harbor Laboratory, New York
35. Ptashne M, Gann A (2002) Genes and signals. Cold Spring Harbor Laboratory, New York
36. Qian H (2000) A mathematical analysis for the Brownian dynamics of a DNA tether. J Math Biol 41(4):331–340
37. Revet B, von Wilcken-Bergmann B, Bessert H, Barker A, Muller-Hill B (1999) Four dimers of lambda repressor bound to two suitably spaced pairs of lambda operators form octamers and DNA loops over large distances. Curr Biol 9(3):151–154
38. Senear DF, Brenowitz M, Shea MA, Ackers GK (1986) Energetics of cooperative protein–DNA interactions – comparison between quantitative deoxyribonuclease footprint titration and filter binding. Biochemistry 25(23):7344–7354
39. Stayrook S, Jaru-Ampornpan P, Ni J, Hochschild A, Lewis M (2008) Crystal structure of the lambda repressor and a model for pairwise cooperative operator binding. Nature 452(7190):1022–1025
40. van den Broek B, Vanzi F, Normanno D, Pavone FS, Wuite GJL (2006) Real-time observation of DNA looping dynamics of type IIE restriction enzymes NaeI and NarI. Nucleic Acids Res 34(1):167–174
41. Vanzi F, Broggio C, Sacconi L, Pavone FS (2006) Lac repressor hinge flexibility and DNA looping: single molecule kinetics by tethered particle motion. Nucleic Acids Res 34(12):3409–3420
42. Vilar JMG, Saiz L (2005) DNA looping in gene regulation: from the assembly of macromolecular complexes to the control of transcriptional noise. Curr Opin Genet Dev 15(2):136–144
43. Wang H, Finzi L, Lewis D, Dunlap D (2009) AFM studies of the CI oligomers that secure DNA loops. J Pharm Biotechnol 10:494–501
44. Wang Y, Guo L, Golding I, Cox EC, Ong NP (2009) Quantitative transcription factor binding kinetics at the single-molecule level. Biophys J 96:609–620
45. Watkins LP, Yang H (2005) Detection of intensity change points in time-resolved single-molecule measurements. J Phys Chem B 109(1):617–628
46. Zhang HY, Marko JF (2008) Maxwell relations for single-DNA experiments: monitoring protein binding and double-helix torque with force-extension measurements. Phys Rev E 77(3):031916.1–031916.9
47. Zurla C, Franzini A, Galli G, Dunlap DD, Lewis DEA, Adhya S et al (2006) Novel tethered particle motion analysis of CI protein-mediated DNA looping in the regulation of bacteriophage lambda. J Phys Condens Matter 18(14):S225–S234
48. Zurla C, Manzo C, Dunlap DD, Lewis DEA, Adhya S, Finzi L (2009) Direct demonstration and quantification of long-range DNA looping by the lambda bacteriophage repressor. Nucleic Acids Res 37:2789–2795
49. Zurla C, Samuely T, Bertoni G, Valle F, Dietler G, Finzi L et al (2007) Integration host factor alters LacI-induced DNA looping. Biophys Chem 128(2–3):245–252
50. Liebesny P, Goyal S, Dunlap D, Fereydoon family, Finz L, Fereydoon Family, Determination of the Number of Proteins Bound non-Specifically to DNA. JPCM (in press)

Chapter 10
Dynamics of Protein–ssDNA Interactions in the Bacteriophage T4 Homologous Recombination System

Jie Liu and Scott W. Morrical

10.1 Background

10.1.1 Homologous Recombination

Homologous recombination (HR) is a biological process in which DNA strands are exchanged between DNA molecules of identical or nearly identical sequence. HR helps to maintain genetic diversity in populations of organisms by promoting the gene conversion events that occur during meiosis in eukaryotes and during conjugation in prokaryotes. In mitotic cells, HR is essential for the accurate repair of DNA double-strand breaks (DSBs) (Fig. 10.1), which is important for maintaining genome stability [1, 2]. Defects in homologous recombination pathways are clearly implicated in genome instability syndromes that are associated with human cancer [1, 3].

10.1.2 Generation of ssDNA

The generation of single-stranded DNA is a common early step of HR pathways [4, 5]. Generally, ssDNA production occurs as a result of programmed nucleolytic resection of DSBs (Fig. 10.1), or due to replication fork stalling, breakage, or collapse. The production of ssDNA tails or gaps in otherwise duplex DNA triggers the assembly of core recombination machinery on ssDNA, most notably the *presynaptic filament*. The presynaptic filament is a helical nucleoprotein filament consisting of a recombinase enzyme and its accessory proteins bound cooperatively to ssDNA (Fig. 10.2). The assembly and dynamics of presynaptic filaments are the subjects of this chapter.

S.W. Morrical (✉)
Department of Biochemistry, University of Vermont College of Medicine,
Burlington, VT 05405, USA
e-mail: Scott.Morrical@uvm.edu

M.C. Williams and L.J. Maher, III (eds.), *Biophysics of DNA-Protein Interactions*, Biological and Medical Physics, Biomedical Engineering,
DOI 10.1007/978-0-387-92808-1_10,

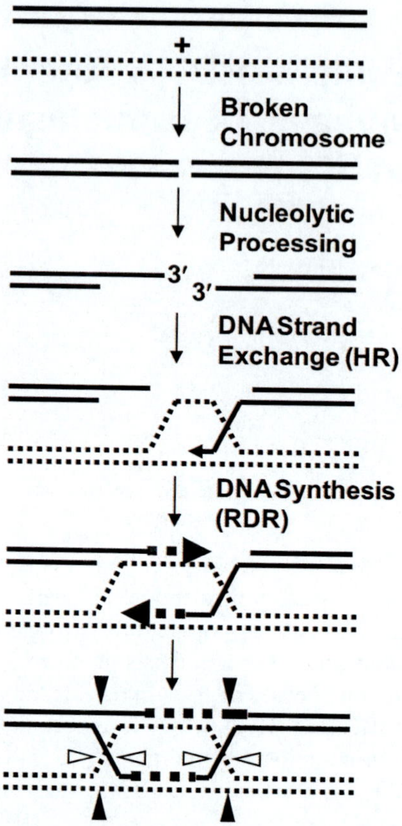

Fig. 10.1 DNA double-strand break repair by homologous recombination (HR). Nucleolytic resection of broken chromosomes *(solid lines)* generates 3′ ssDNA tails on the broken ends. Recombination enzymes assemble into presynaptic filaments (not shown) on the exposed ssDNA and catalyze DNA strand exchange with the homologous chromosome *(dotted lines)*. The invading 3′ ends prime recombination-dependent DNA replication (RDR) *(thick arrows)* using strands of the sister chromosome as templates. Continued DNA synthesis generates a double Holliday junction that is resolved by structure-specific endonucleases (*thin or white arrows*) to generate intact, recombinant chromosomes

10.1.3 Conservation and Biological Function of Presynaptic Filaments

Presynaptic filament formation is a property shared by enzymes of the RecA recombinase family, which is highly conserved in all kingdoms of life [6]. Representatives include the RecA recombinases in prokaryotes, the UvsX recombinases in bacteriophages, the Rad51 and Dmc1 recombinases in eukaryotes, and the RadA recombinases in archaea. These enzymes share a conserved protein fold in their catalytic core domain that includes consensus Walker A/B nucleotide binding motifs, DNA binding sites, and sites for protomer–protomer interactions [6]. Many eukaryotic and archaeal species encode multiple Rad51 or RadA paralogs

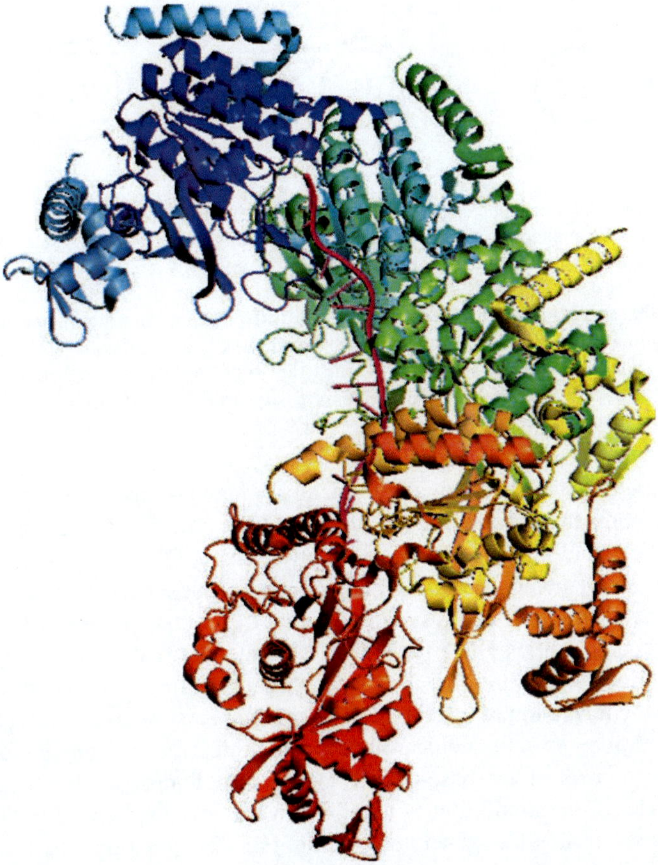

Fig. 10.2 Structure of the RecA recombinase-ssDNA filament. Recombinases of the RecA/Rad51 family form right-handed helical filaments on single-stranded DNA. Shown in the figure is the X-ray crystallographic model for one helical turn of an *E. coli* RecA–ssDNA filament formed in the presence of ADP-AlF_4 (PDB file no. 3CMW) [102]. The six RecA subunits are shown in different colors (*red – blue*). The ssDNA strand (*magenta*) binds to the inner surface of the filament

[7, 8] – that is, proteins related to Rad51 or RadA by gene duplication within the species but with sequences and functions that have diverged somewhat during evolution. RecA-family recombinases form presynaptic filaments by binding cooperatively, in head-to-tail fashion, to ssDNA. The formation of productive presynaptic filaments typically requires the presence of a nucleoside triphosphate cofactor. Filament formation on ssDNA activates the enzymatic activities of a recombinase including ssDNA-stimulated nucleoside triphosphate hydrolysis and homologous DNA strand exchange (Fig. 10.3).

DNA strand exchange reactions occur in three major steps – presynapsis, synapsis, and branch migration, all of which are promoted by the recombinase component of the presynaptic filament [6, 9]. In vivo, the presynapsis phase includes the generation of recombinagenic ssDNA – that is, ssDNA that can be recognized as a substrate for

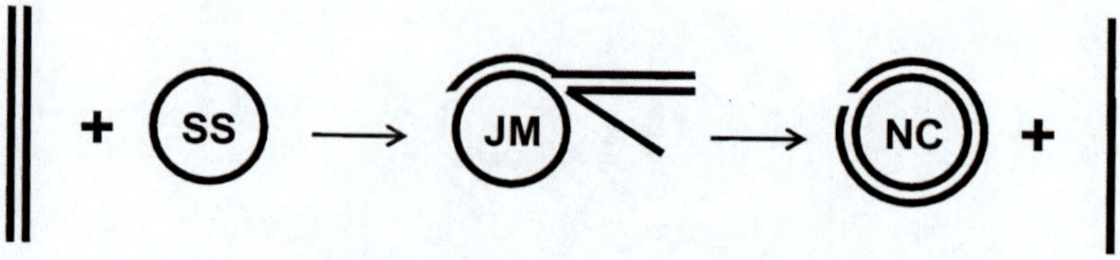

Fig. 10.3 In vitro DNA strand exchange reaction scheme. In vitro assays for DNA strand exchange typically bypass the nucleolytic resection step by providing one DNA substrate that is already single-stranded. Most strand exchange assays employ homologous circular ssDNA (SS) and linear dsDNA substrates derived from bacteriophages such as M13 or PhiX174. Homologous pairing catalyzed by the recombinase in the presence of ATP, ssDNA-binding protein, and recombination mediator proteins convert the substrates into heteroduplex joint molecule (JM) intermediates. Recombinase-catalyzed branch migration extends the joint molecules into heteroduplex nicked circle (NC) and linear ssDNA products, which are easily distinguished from substrates and intermediates by agarose gel electrophoresis

strand exchange by the recombination machinery, and on through the assembly of a catalytically competent presynaptic filament. In vitro strand exchange reactions typically skip the ssDNA generation step by providing a DNA substrate that already contains recombinagenic ssDNA, which shares full sequence homology with the target dsDNA (Fig. 10.3). In the synapsis phase, the competent presynaptic filament searches a second, duplex DNA molecule for sequences that are homologous to the ssDNA contained within the filament. Homologous pairing of sequences is thought to occur via the formation of transient DNA triplex that is relatively slowly converted into a heteroduplex joint molecule structure containing the invading strand from the presynaptic filament, now base-paired to its complement in the target duplex (Fig. 10.3) [10]. The strand that is identical to the invading strand is displaced as ssDNA. The synapsis phase generates a partial Holliday junction that is translocated in the third phase, branch migration, to extend the length of the heteroduplex. Recombinase-driven branch migration, unlike spontaneous branch migration of Holliday junctions, is typically polar. Each recombinase catalyzes branch migration with a characteristic polarity, either 5′ to 3′ or 3′ to 5′ with respect to the invading strand, depending on the enzyme [11, 12]. Recombinase-initiated branch migration may be further stimulated by the action of certain DNA helicase enzymes [13, 14]. In vivo, translocated Holliday junctions are resolved by structure-specific endonucleases to generate intact, recombinant chromosomes (Fig. 10.1), thus completing the HR or DSB repair process.

10.1.4 Accessory Proteins Promote Presynaptic Filament Assembly

The success of recombination and DSB repair depends strongly on the timely assembly, stability, and turnover of presynaptic filaments. Presynaptic filament dynamics are regulated by nucleotide cofactors, and by accessory proteins including ssDNA-binding proteins (SSBs) and recombination mediator proteins (RMPs), which are functionally conserved among diverse species [15]. SSB proteins contain oligonucleotide/oligosaccharide (OB)-fold motifs and bind tightly and specifically to ssDNA. Representatives of the SSB class of proteins include the homooligomeric bacterial SSB proteins, Gp32 protein and its relatives in bacteriophages, and the heterotrimeric RPA proteins in eukaryotes and euryarchaea. The crenarchaeal SSB proteins appear to share characteristics of both bacterial SSB and eukaryotic RPA proteins [16, 17]. The major function of ssDNA binding proteins in presynaptic filament assembly is the removal of inhibitory secondary structure from ssDNA via tight and cooperative binding to the polynucleotide. This provides an optimized ssDNA lattice for the assembly of other recombination proteins. Direct assembly of recombinase-ssDNA filaments on SSB-covered ssDNA is complicated by the high stability of SSB–ssDNA interactions, however. Recombinases typically lack sufficient ssDNA binding energy, under physiological conditions, to directly compete with SSBs for binding sites. Consistent with this notion, biochemical studies have revealed a profound protein order of addition effect on in vitro DNA strand exchange reactions: Recombinase assembly on ssDNA must precede the addition of the SSB component before efficient strand exchange can occur [15, 18]. Presaturation of ssDNA with the SSB component usually strongly inhibits both strand exchange and ssDNA-stimulated ATPase activities of the recombinase component.

The abundance of SSBs in vivo indicates that other factors are necessary to direct the assembly of presynaptic filaments on ssDNA that is likely to already be covered with SSB protein. This role is fulfilled by the recombination mediator proteins, which by definition are proteins capable of overcoming the inhibitory effects of preformed SSB–ssDNA on recombinase catalytic activities [15]. Representatives of the RMP class of proteins include UvsY protein and its relatives in bacteriophages, bacterial RecO, RecR, and RecF proteins, and eukaryotic Rad52 and Brca2 proteins, as well as Rad51 paralogs in both eukaryotes and archaea [18–25]. A general model of presynaptic filament assembly has emerged in which the SSB component sequesters and removes secondary structure from ssDNA, after which an RMP protein mediates loading of recombinase and concomitant displacement of the SSB from ssDNA (Fig. 10.4) [26]. The biochemical mechanism of RMP activity in recombination is addressed in later sections.

10.1.5 The Bacteriophage T4 Homologous Recombination System

Studies of the bacteriophage T4 recombination system have provided important insights on the mechanism of presynaptic filament assembly, on DNA strand exchange mechanisms, and on the coordination of recombination and DNA synthesis activities during recombination-dependent replication and DSB repair pathways [26–28].

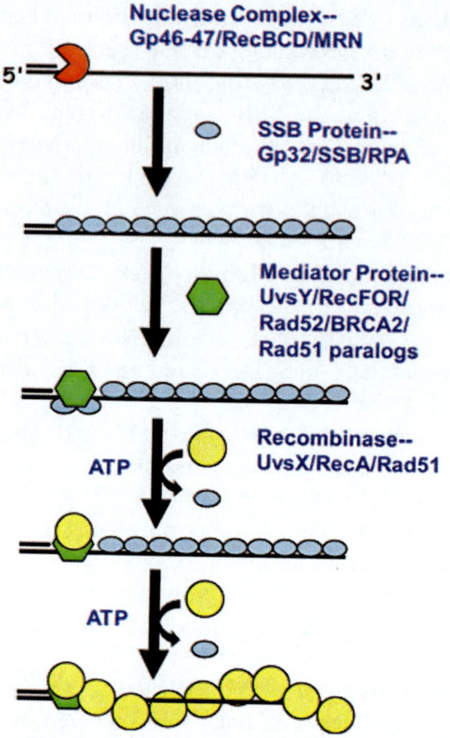

Fig. 10.4 General model for presynaptic filament assembly on a resected DNA double-strand break. DNA ends are resected by nuclease complexes such as T4 Gp46/Gp47, *E. coli* RecBCD, or eukaryotic Mre11/Rad50/Nbs1 (MRN) to expose recombinagenic 3′ ssDNA tails. ssDNA binding proteins such as T4 Gp32, *E. coli* SSB, or eukaryotic RPA rapidly bind and sequester the exposed ssDNA. Recombination mediator proteins/complexes such as T4 UvsY, *E. coli* RecFOR or eukaryotic Rad52, BRCA2, or Rad51 paralogs interact with SSB–ssDNA complexes and/or with dsDNA/ssDNA junctions and recruit recombinase. In the presence of ATP and mediators, recombinases such as T4 UvsX, *E. coli* RecA, or eukaryotic Rad51 nucleate and propagate as filaments on ssDNA while displacing the SSB protein

In addition, the T4 system has provided some of the most detailed information available on the roles of RMP and SSB components in recombination [29–33]. T4 provides a relatively simple set of recombination proteins compared to cellular organisms, yet the functionalities represented are highly conserved, indicating that the molecular principles underlying T4 recombination are relevant to other systems. Therefore, the goal of this chapter is to illustrate the biophysical and biochemical principles elucidated from studies of T4 recombination proteins, in the hope of informing future recombination studies in diverse systems. The emerging story highlights the importance of dynamic protein–ssDNA interactions in fulfilling the DNA recombination and repair operations that are essential for genome stability.

10.2 ssDNA Binding Properties of the T4 Core Recombination Machinery

In the T4 recombination system, three core protein components are required for presynaptic filament assembly under physiological conditions: UvsX, the phage recombinase; Gp32, the prototypical ssDNA-binding protein; and UvsY, the prototypical recombination mediator protein [26, 27]. UvsX is the T4 RecA ortholog, meaning that UvsX and RecA (as well as Rad51, RadA, etc.) evolved from a common ancestral gene by speciation. The ssDNA binding properties of the T4 proteins have been studied in some detail; these are presented below in context with their enzymatic and/or physical properties.

10.2.1 Notes About Salt Effects

The electrostatic interaction between positively charged active binding pockets in the proteins and invariably negatively charged phosphate backbones in DNA polyelectrolytes is the major molecular force driving recombination proteins to their targets. Thus, solution ionic strength has a large impact on the protein–DNA interactions, and anions such as acetate, chloride, and glutamate can compete effectively with DNA to bind proteins. Measuring and comparing the ssDNA binding properties of UvsX, Gp32, and UvsY, therefore, requires an understanding of the differential effects of salt on their ssDNA binding parameters. Thus, many quantitative studies of protein–ssDNA interactions involve salt (usually NaCl) concentration as a variable. For the purposes of this discussion, we define "low-salt conditions" to mean experiments performed in buffered solution containing <0.2 M NaCl, and "high-salt conditions" to mean experiments performed in buffered solution containing ≥ 0.2 M NaCl, unless otherwise noted. 0.2 M NaCl is taken as a reasonable approximation of physiological ionic strength in the T4-infected *E. coli* cell, since the requirements of in vitro recombination assays performed at this salt concentration typically mimic genetic requirements for recombination in vivo [33–35].

10.2.2 UvsX Recombinase

The 44 kDa UvsX protein plays central roles in T4 recombination, recombination-dependent replication, and DSB repair pathways [26]. UvsX is a divergent RecA/Rad51 ortholog, with 28% identity and 51% similarity to the catalytic core domain of *E. coli* RecA protein [6]. The name ortholog defines that UvsX originates from a common ancestor gene with RecA/Rad51 proteins and that they share similar functions in different species. UvsX possesses intrinsic ssDNA and dsDNA binding activities. UvsX has higher intrinsic affinity for dsDNA than for ssDNA (R. Maher, H. Xu, and S. Morrical, unpublished results), but UvsX–ssDNA interactions are selectively stabilized by nucleoside triphosphates (ATP, dATP, or analogs such as ATPγS) and by UvsY protein [36, 37]. Binding of UvsX to ssDNA, not dsDNA, specifically activates the catalytic activities of UvsX including nucleotidase and DNA strand exchange activities.

10.2.2.1 UvsX–ssDNA Interactions

The intrinsic ssDNA binding properties of UvsX have been investigated by several methods including etheno-DNA fluorescence enhancement assays, quantitative DNA-cellulose affinity chromatography experiments, and others [36] (R. Maher, H. Xu, and S. Morrical, unpublished results). UvsX exhibits an average binding site size on ssDNA of four nucleotide residues per protomer. UvsX binds to ssDNA with moderate affinity; the intrinsic affinity parameter, K_{ss}, equals approximately $10^4\ M^{-1}$ at physiological ionic strength [36]. The observed binding affinity of UvsX for ssDNA is higher than the intrinsic affinity due to the effects of cooperativity, given the relationship $K_{obs} = K_{ss}\omega$, where ω is the cooperativity parameter. UvsX–ssDNA interactions have a cooperativity parameter of $\omega = 100$, independent of salt concentration, therefore $K_{ss}\omega \approx 10^6\ M^{-1}$ at physiological ionic strength [36]. This indicates a limited mode of cooperativity for UvsX, consistent with the formation of long filaments at high protein binding densities on ssDNA, but not with all-or-none binding. Studies of ion effects on UvsX–ssDNA interactions revealed a significant electrostatic component for UvsX binding [36]. From salt-dependent binding measurements, it has been determined that UvsX makes as many as three electrostatic contacts with ssDNA phosphate residues per binding site, and at least one anion is displaced per binding site during complex formation. Therefore, UvsX–ssDNA filament stability increases markedly with decreasing salt concentration. The salt-sensitivity of UvsX–ssDNA interactions increases at very low binding densities of UvsX on ssDNA, consistent with a salt-sensitive nucleation step for UvsX filament formation (see below).

10.2.2.2 ssDNA-Dependent ATPase Activity

Binding to ssDNA activates the nucleotidase activity of UvsX. The preferred substrates are ATP and dATP. ATPase activity is strongly ssDNA-dependent under most salt conditions [38]; however, very high salt concentrations (>1 M NaCl) activate a robust ssDNA-independent ATPase reaction, an effect that is also seen in *E. coli* RecA protein [39]. Double-stranded DNA does not activate UvsX ATPase activity. Other noteworthy features of UvsX's ssDNA-dependent ATPase activity include a high rate of catalytic turnover compared to bacterial RecA and eukaryotic Rad51 enzymes [15], and a unique product profile. UvsX ATPase activity generates both ADP and AMP as products [38, 40]. Steady-state kinetics studies revealed that UvsX–ssDNA filaments contain two types of active sites with different K_m and

k_{cat}/K_m values for substrates. One type of active site appears to generate AMP via a sequential mechanism (ATP → ADP → AMP) without releasing the ADP intermediate from the active site, while the other type of active site appears to produce ADP exclusively [40]. Mutational evidence suggests that active site asymmetry is important for UvsX-catalyzed DNA strand exchange [40]. Two classes of active sites are also observed in filaments of *E. coli* RecA and *S. cerevisiae* Rad51 proteins [41, 42], suggesting that active site asymmetry may be an important feature of the mechanisms of all recombinases.

10.2.2.3 DNA Strand Exchange Activity

UvsX–ssDNA filaments promote a rapid search for homology within the dsDNA substrates of DNA strand exchange reactions. Homologous pairing requires ATP binding but apparently not its hydrolysis, whereas propagation of DNA strand exchange via polar branch migration exhibits a stronger requirement for ATP hydrolysis [43–45]. Strand exchange reactions exhibit a strong requirement for Gp32 in the correct order of addition [34, 38, 46]. High-salt conditions and/or incorrect order of Gp32 addition lead to an absolute requirement for UvsY protein for strand exchange reactions in vitro [30, 46]. This condition mimics the in vivo situation wherein T4 *uvsX* and *uvsY* mutants exhibit equivalent recombination deficiencies [47–49]. The products of UvsX-catalyzed DNA strand exchange reactions tend to be branched networks of ssDNA and dsDNA. This observation suggests that UvsX is more efficient at catalyzing homologous pairing than at catalyzing branch migration, so that each molecule of ssDNA substrate participates in many synapsis events before it is completely converted into heteroduplex DNA [14, 38].

10.2.3 Gp32 ssDNA-Binding Protein

Gene 32 protein or Gp32, the prototype ssDNA-binding protein, has been characterized extensively [50–52]. Gp32 is an integral component of the T4 replisome and also plays central roles in homologous recombination and DNA repair. The primary structure of Gp32 (301 residues, 34 kDa) contains three distinct regions including the N-terminal "basic" or B-domain (residues 1–22), the ssDNA-binding "core" domain (residues 22–253), and the C-terminal "acidic" or A-domain (residues 254–301) [53]. The X-ray structure of the core domain was determined upon cocrystallization with oligo$(dT)_6$ [52]. This domain contains an oligonucleotide/oligosaccharide-binding (OB-) fold, and its structure is stabilized by a Zn^{++} atom coordinated to residues His-64, Cys-77, Cys-87, and Cys-90. Molecular studies reveal that the B-domain is the specific region responsible for Gp32 homotypic associations and for the cooperativity of Gp32–ssDNA interactions [54, 55], whereas the A-domain is important for the heterotypic associations of Gp32 with other T4 DNA recombination and replication factors including UvsX, UvsY, DNA polymerase, and associated machinery [56–59].

10.2.3.1 Gp32–ssDNA Interactions

Gp32 exhibits approximately 10^4-fold higher affinity for ssDNA than for dsDNA. Binding is sequence-nonspecific, but the observed affinity ($K_{ss}\omega$) varies with lattice composition according to the general hierarchy of: poly(dT) > ssDNA (random sequence) > ssRNA ≫ dsDNA [60, 61]. The observed affinity, $K_{ss}\omega$, of Gp32 for random-sequence ssDNA approaches 10^9 M^{-1} at salt concentrations that simulate physiological ionic strength [61, 62]. Equilibrium and kinetic parameters of Gp32–ssDNA interactions, including binding site size n, intrinsic affinity K_{ss}, and cooperativity ω, have been determined for a variety of polynucleotide lattices [60]. The average binding site size of Gp32 on ssDNA is approximately seven nucleotide residues per protomer. Gp32–ssDNA interactions are characterized by high cooperativity, with ω generally ≥1,000 for all single-stranded lattices examined. This "unlimited" type of cooperativity means that Gp32 exists almost exclusively in clusters or saturated filaments on ssDNA at binding densities normally encountered in in vitro recombination assays or in vivo.

The affinity of Gp32 for ssDNA shows an unusual dependence on salt concentration. Under high-salt conditions, the affinity constant increases with decreasing salt, but under low-salt conditions (<0.2 M NaCl), the affinity constant becomes essentially independent of salt concentration. This phenomenon has recently been investigated using single-molecule methods [63–65]. An attractive model for the bimodal salt effect has emerged from these studies: Under low-salt conditions, negatively charged residues in the acidic C-terminal domain of Gp32 compete with ssDNA for binding to positively charged residues in the ssDNA binding site. Therefore, binding affinity plateaus as the salt concentration continues to drop below 0.2 M NaCl. On the other hand, high-salt conditions disrupt electrostatic interactions between the C-terminal domain and the ssDNA binding site. Now, small cations (Na^+) interact with negatively charged residues in the C-terminal domain, while small anions (Cl^-) interact with the unmasked positively charged residues in the ssDNA binding site. This sets up a classic binding competition between Cl^- ions and ssDNA phosphate residues under high-salt conditions, resulting in a normal salt effect in which ssDNA binding affinity increases with decreasing salt concentration. This model has been validated by studies of a truncated form of Gp32 called *I or Gp32-A. With *I, which lacks the C-terminal domain, ssDNA binding affinity does not plateau below 0.2 M NaCl, indicating unfettered access of ssDNA to its binding site under low-salt conditions [64]. The bimodal salt effect on Gp32–ssDNA interactions has important implications for the relationship between Gp32, UvsX, and UvsY proteins, which we discuss in a later section.

10.2.3.2 Gp32 in DNA Strand Exchange

Gp32 plays important pre- and postsynaptic roles in DNA strand exchange reactions catalyzed by UvsX protein in concert with UvsY [29, 30, 34, 38, 46, 66].

During presynaptic filament formation, Gp32 denatures ssDNA secondary structure and provides an optimum lattice conformation for assembly of UvsX–ssDNA filaments. The net effect of Gp32 on UvsX–ssDNA filament formation is inhibitory under physiological salt conditions, however, since the higher affinity of Gp32–ssDNA interactions allows Gp32 to effectively compete with UvsX for binding sites on the ssDNA [36]. Proper assembly of presynaptic filaments therefore requires the mediator function of UvsY protein in addition to UvsX and Gp32 (see below) [15, 26].

10.2.4 UvsY Recombination Mediator Protein

UvsY is the prototype recombination mediator protein or RMP [15]. RMP activity is defined as the ability to load an RecA/Rad51 recombinase onto ssDNA that is saturated with prebound cognate ssDNA-binding protein. UvsY is an essential partner for UvsX-catalyzed DNA strand exchange reactions under physiological or high-salt conditions [33–35]. UvsY is also an essential partner for UvsX recombination functions in vivo since mutations knocking out either gene product are equally recombination-deficient [47–49]. UvsY is a small protein (15.8 kDa) but forms stable 95 kDa hexamers that bind to ssDNA as intact units [67]. Thus, UvsY is the only member of the core T4 recombination machinery that forms a discreet oligomeric structure; UvsX and Gp32 both appear to form mixed oligomers in solution [55] (unpublished results).

10.2.4.1 UvsY–ssDNA Interactions

UvsY has a binding site size on ssDNA of four nucleotide residues per protomer, the same as UvsX [68]. A UvsY hexamer therefore occupies 24 nucleotides of ssDNA. Indeed, UvsY hexamers form very stable complexes with 24- or 25-base oligonucleotides, which can be isolated and crystallized [67] (H. Xu and S. Morrical, unpublished results). UvsY binds to ssDNA with high affinity but little or no cooperativity. An intrinsic affinity of $K_{ss} \approx 10^7\ M^{-1}$ has been estimated for UvsY binding to random-sequence ssDNA in 0.2 M NaCl [68]. Therefore, UvsY has higher intrinsic affinity, but lower cooperativity, for ssDNA than both UvsX and Gp32 under salt conditions that approximate physiological ionic strength. The affinity advantage of UvsY is even more striking under low-salt conditions due to a relatively large salt effect on K_{ss} [68, 69], and to the low-salt plateau effect on Gp32–ssDNA interactions described above, which UvsY does not share. High-affinity UvsY–ssDNA interactions are disrupted by mutations in the C-terminus that prevent its hexamerization, and by mutations of basic residues Lys-58 and Arg-60 that are required for normal electrostatic contacts with ssDNA [33, 35, 37, 70, 71].

10.2.4.2 Affinity for Wrapped vs. Stretched DNA

The hexameric structure of UvsY suggests the ability to wrap bound ssDNA. This hypothesis is supported by the observation that a monomeric form of UvsY has 10^4-fold lower affinity for ssDNA than wild-type [72]. DNA stretching studies provide further evidence for ssDNA wrapping by UvsY at the single-molecule level. Force spectroscopy measurements demonstrate that UvsY strongly wraps ssDNA that is created by the exposure of stretched dsDNA molecules to glyoxal [71]. Wrapping occurs at low stretching forces where the DNA is relatively relaxed, whereas wrapping is suppressed when the DNA is under tension. This suppression of wrapping leads to the loss of preferential binding to ssDNA under the conditions of the DNA stretching experiments – in fact, UvsY binds tighter to dsDNA than to ssDNA under high stretching forces [71]. In contrast, under relaxed conditions UvsY exhibits ~1,000-fold higher affinity for ssDNA than for dsDNA [69]. Presumably UvsY cannot wrap dsDNA because its persistence length is much higher than that of ssDNA [73]. The observation that UvsY binds tightly to stretched duplex DNA suggests that it may recognize extended or "stiff" DNA structures such as that created by the binding of Gp32 to ssDNA, then convert them into a wrapped ssDNA structure. This recognition and conversion could play an important role in presynaptic filament assembly as we discuss in a later section.

10.2.4.3 UvsY in DNA Strand Exchange

UvsX exhibits a strict requirement for UvsY in strand exchange assays performed at physiological or high-salt conditions [34, 35, 70], consistent with the codependency of in vivo recombination processes on UvsX and UvsY [47–49]. UvsY also lowers the critical concentration of UvsX for recombination processes including recombination-dependent DNA replication [66, 74]. UvsY stimulates the ssDNA-dependent ATPase activity of UvsX and exhibits synergistic effects with Gp32 on this reaction [35, 46]. UvsY affects these reactions primarily at the stage of presynaptic filament assembly. The mechanism of UvsY action in recombination is explored in detail below.

10.3 Modulation of Protein–ssDNA Interactions During Presynaptic Filament Assembly

10.3.1 Nucleotide Effects

UvsX–ssDNA interactions are regulated by the ATPase cycle. ATPγS, an ATP analog that is bound but hydrolyzed very slowly by UvsX, induces stable, high-affinity binding of ssDNA [36, 37]. The physiological substrate ATP also enhances

UvsX–ssDNA interactions, but it does so transiently because it is rapidly hydrolyzed by UvsX in the presence of ssDNA [33]. The products of UvsX-catalyzed ATP hydrolysis, ADP and AMP, are associated with decreased ssDNA-binding affinity under steady-state conditions [33, 40].

10.3.2 *Effects of UvsY Protein*

While UvsX and Gp32 binding to ssDNA appears to be mutually exclusive [31, 33, 75], there is clear evidence that UvsY can co-occupy ssDNA binding sites simultaneously with either Gp32 or UvsX [30, 37, 43, 75–77]. The presence of UvsY alters the ssDNA binding properties of both Gp32 and UvsX in ways that are critical for presynaptic filament assembly and recombination function.

10.3.2.1 UvsY Destabilizes Gp32–ssDNA Interactions

Ensemble measurements demonstrate that UvsY binds stoichiometrically to Gp32-saturated ssDNA at a ratio of approximately one UvsY hexamer per three Gp32 protomers, which correlates well with their binding site sizes on ssDNA ($n=24$ and 7 nucleotide residues for $UvsY_6$ and Gp32, respectively) [76]. The result is a stable UvsY–Gp32–ssDNA tripartite complex at physiological ionic strength (Fig. 10.5). Within the tripartite complex Gp32–ssDNA interactions are destabilized as shown by their increased sensitivity to disruption by salt compared to Gp32–ssDNA complexes in the absence of UvsY [76]. Destabilization of Gp32–DNA interactions by UvsY has also been observed at the single-molecule level, using DNA stretching methods [71]. It has been proposed that UvsY destabilizes Gp32–ssDNA by lowering Gp32's cooperativity parameter [76]. By destabilizing Gp32–ssDNA, UvsY lowers the energy barrier necessary for UvsX filament formation on and Gp32 displacement from the lattice, under salt conditions that would otherwise favor Gp32- over UvsX–ssDNA interactions. While UvsY does not directly displace Gp32 from ssDNA at physiological salt, it is possible that direct displacement does occur at lower salt concentrations where the intrinsic affinity of UvsY–ssDNA dominates that of Gp32–ssDNA interactions. This suggests that there could be alternative concerted or stepwise mechanisms for Gp32–ssDNA disruption during presynaptic filament assembly, depending on salt conditions [71] (Fig. 10.5). In the concerted mechanism, Gp32 displacement from ssDNA requires the simultaneous action of UvsY and UvsX. This mechanism is observed under high-salt conditions and is most consistent with the genetic requirements of T4 recombination. In the stepwise mechanism, UvsY would directly displace Gp32 from ssDNA *then* recruit UvsX to the lattice. This mechanism is postulated to occur under low-salt conditions, but is difficult to confirm since UvsY-independent presynaptic filament assembly also occurs at low salt [34].

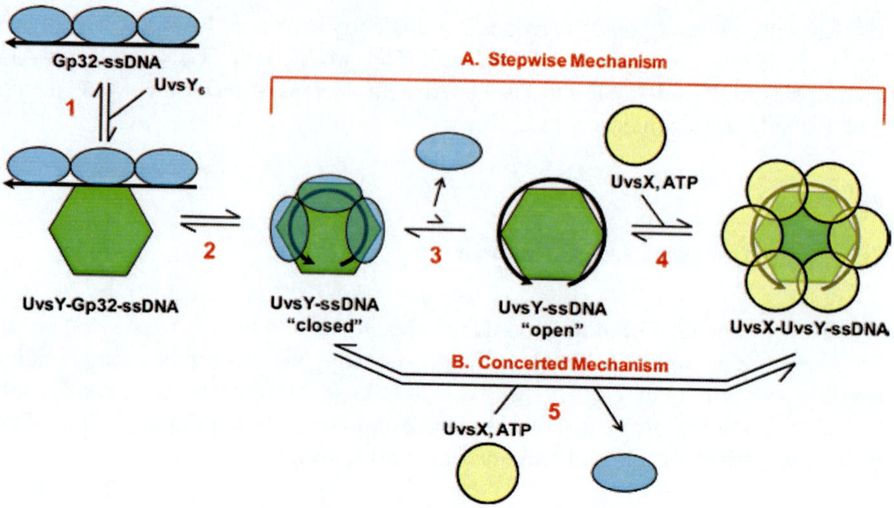

Fig. 10.5 Double hand-off model for the mechanism of mediator protein UvsY in T4 presynaptic filament assembly. UvsY protein facilitates the loading of UvsX recombinase onto ssDNA and the concomitant displacement of Gp32 ssDNA-binding protein from ssDNA. The figure shows UvsX loading and Gp32 displacement from the perspective of a single UvsY hexamer, as if looking down the helical axis of a nascent presynaptic filament. The cooperative binding of Gp32 to ssDNA extends the polynucleotide lattice. The first handoff occurs as hexameric UvsY recognizes and binds to the extended ssDNA (step 1), then converts it into a wrapped conformation(s) (steps 2 and 3), destabilizing Gp32–ssDNA interactions in the process. The UvsY-wrapped ssDNA complex is postulated to be in equilibrium between "closed" and "open" conformations (step 3), the latter of which is recognized by the ATP-bound form of UvsX protein to nucleate presynaptic filament assembly (step 4) while displacing Gp32. (A) Steps 3 and 4 constitute a step-wise mechanism for Gp32 displacement and UvsX loading by UvsY, which may occur under low-salt conditions. (B) Under high-salt conditions UvsY does not displace Gp32 from ssDNA directly, so filament assembly likely occurs by a concerted mechanism in which synergistic action of UvsY and ATP-bound UvsX is required to displace Gp32. See text for details

10.3.2.2 UvsY Stabilizes UvsX–ssDNA Interactions

Ensemble measurements indicate that UvsY and UvsX cooccupy ssDNA at a stoichiometric ratio of approximately one UvsY hexamer per six UvsX protomers, which is in accord with their equivalent protomeric binding site sizes ($n = 4$ nucleotide residues) [37]. UvsY stabilizes UvsX–ssDNA interactions within the UvsY–UvsX–ssDNA tripartite complex, as shown by increased salt-stability of UvsX–ssDNA. Stabilization is observed whether or not UvsX is in *apo* or nucleotide-liganded forms, but the most stable complex is formed in the simultaneous presence of UvsY and ATPγS, indicating that mediator protein and nucleoside triphosphate cofactor act synergistically to stabilize presynaptic filaments [37]. Other evidence suggests that under steady-state conditions for ssDNA-dependent ATP hydrolysis, UvsY acts as a nucleotide exchange factor for UvsX [32]. It is proposed that UvsY's nucleotide exchange factor (NEF) activity increases the lifetime of UvsX in its ATP-bound, high-affinity form, which would promote the stability and recombination activity of presynaptic filaments. Thus, the reciprocal activities of UvsY in stabilizing UvsX–ssDNA and destabilizing Gp32–ssDNA interactions promote the nucleation and growth of UvsX filaments on ssDNA with concomitant displacement of Gp32 (Fig. 10.5).

10.3.2.3 Affinity Changes Are Driven by Changes in ssDNA Structure

All three of the core recombination proteins, UvsX, Gp32, and UvsY, undergo protein–protein interactions with one another, and some of these interactions appear to be important for the overall strand exchange reaction [35, 57, 75]. Both UvsX and UvsY interact with the C-terminal domain of Gp32 [56, 57]. Studies of UvsY, however, have shown that its ability to destabilize Gp32–ssDNA complexes is independent of Gp32's C-terminal domain and of UvsY–Gp32 interactions [71, 76]. Both ensemble and single-molecule data show that UvsY has equal destabilizing effects on the ssDNA-binding activities of wild-type Gp32 and of truncation mutant *I (Gp32-A), which lacks interactions with UvsY. These findings indicate that the ssDNA-binding activity of UvsY is responsible for destabilizing Gp32–ssDNA interactions.

It also appears likely that ssDNA wrapping by UvsY optimizes ssDNA structure for high-affinity binding by UvsX protein (Fig. 10.5). Evidence for this model comes from studies of in vitro complementation of UvsX and UvsY mutations in ssDNA-dependent ATPase and DNA strand exchange reactions [32]. UvsY single and double missense mutants K58A and K58A/R60A have reduced ssDNA binding affinities compared to wild-type [70]; likewise UvsX missense mutants H195Q and H195A have reduced affinities for ssDNA and exhibit differential gain/loss of enzymatic function compared to wild-type [40]. It was shown that unlike wild-type UvsX, the ssDNA-dependent ATPase activities of UvsX-H195Q/A mutants are strongly inhibited by both low and high concentrations of UvsY protein. This inhibition is partially relieved by UvsY mutants with decreased ssDNA-binding affinity [32]. The UvsX-H195Q mutant retains weak DNA strand exchange activity that is inhibited by wild-type UvsY, but stimulated by the ssDNA-binding compromised UvsY mutants [32]. The patterns of complementation and other results support a mechanism in which the formation of competent presynaptic filaments requires a hand-off of ssDNA from UvsY to UvsX, with the efficiency of the hand-off controlled by the relative ssDNA-binding affinities of the two proteins (Fig. 10.5).

10.3.3 A Double Hand-Off Mechanism for Presynaptic Filament Assembly

It has become evident that many nucleic acid pathways channel their DNA or RNA substrates through a series of hand-off transactions in which a nucleic acid structure or intermediate is passed directly from one protein in the pathway to the next without ever being released as free nucleic acid [78]. The advantages of this strategy would appear to be twofold: Potential cytotoxic effects of the nucleic acid structure or intermediate are minimized by keeping it sequestered, and the nucleic acid is protected from inappropriate degradation or side reactions. It seems reasonable that ssDNA in recombination pathways would be similarly channeled. Based on the DNA binding properties of the three protein players, we propose that the loading of UvsX recombinase onto Gp32–ssDNA complexes proceeds through a series of at least two sequential hand-off steps (Fig. 10.5): First, Gp32 binds to ssDNA and converts it into an extended form resembling the mechanically stretched DNA that was sampled in single-molecule experiments. *Hand-off no. 1* occurs as UvsY recognizes and binds to the extended ssDNA, then converts it into a wrapped form that destabilizes Gp32–ssDNA interactions. The UvsY-wrapped ssDNA complex is postulated to be in equilibrium between "closed" and "open" states. The closed form is the one that destabilizes Gp32–ssDNA interactions, and it is inaccessible to UvsX. The open form is optimal for UvsX–ssDNA interactions. *Hand-off no. 2* occurs as the ATP-bound form of UvsX recognizes and binds to the open UvsY–ssDNA structure, nucleating or propagating a UvsX–ssDNA filament while inducing Gp32 to leave. Superimposed on this underlying mechanism are effects involving UvsX cooperativity and the NEF activity of UvsY that have not been fully explored. Also, linking ATP hydrolysis by UvsX to the hand-off scheme creates opportunities for dynamic instability in presynaptic filaments, which we will address in the following section.

10.4 UvsX–Gp32 Exchanges on ssDNA

10.4.1 *Role of Salt, Nucleotides, and UvsY*

A fluorescein-conjugated form of Gp32, Gp32F, has been used as a probe for Gp32 displacement from ssDNA and to study the kinetics of presynaptic filament assembly [33]. The fluorescence intensity of the fluorescein moiety is enhanced by two- to threefold upon binding of Gp32F to ssDNA. Therefore, Gp32F specifically reports on the Gp32–ssDNA association/dissociation status in real time and in the presence of other proteins. Gp32F displays equilibrium and kinetic ssDNA-binding behavior similar to unlabeled Gp32 [33]. Normally, the equilibrium affinity of recombination proteins towards ssDNA increases with decreasing salt concentration, but the extent varies greatly with each individual protein. As described above, K_{ss} of Gp32 increases only slightly when decreasing salt below ~0.2 M NaCl, while K_{ss} of UvsX increases dramatically in the same range [36, 79]. As a consequence, the outcomes of competition towards ssDNA binding between UvsX and Gp32 are quite different at lower and higher salt concentrations. Under low-salt conditions, UvsX possesses high enough intrinsic affinity compared to Gp32 and is able to nucleate and propagate onto Gp32F-covered ssDNA, as measured by the time-dependent decrease in Gp32F fluorescence. ATP- or ATPγS-binding are required to endow UvsX with at least a transient high-affinity form to compete with Gp32F and displace it, since ADP and AMP, or the absence of nucleotides, could not support it [33]. Under high-salt conditions where the ionic strength is more physiologically relevant, ATP- or ATPγS-binding alone is not sufficient for UvsX to compete with and overcome the dominant residence of Gp32 on ssDNA. Under these conditions, UvsY is called for to facilitate UvsX–ssDNA filament formation and Gp32 displacement, as measured by the decrease in Gp32F fluorescence [33]. Even in this UvsY-dependent mode, binding of ATP or ATPγS is an absolute requirement for UvsX to form a presynaptic filament. No Gp32F displacement occurs in the presence of ADP, AMP, and no nucleotide, even with UvsY present [33]. This observation agrees well with the previous finding that UvsY and ATPγS-binding stabilize UvsX on a short 24-mer oligo synergistically, which implies the cooperation of these two factors at the filament nucleation level [37].

10.4.2 Kinetics of Presynaptic Filament Assembly

The special fluorescence properties of Gp32F enabled the first detailed kinetic study on presynaptic filament assembly using a fully-reconstituted in vitro T4 recombination system (UvsX, UvsY, and Gp32), which provides new insights into UvsX–Gp32 exchange and the molecular mechanism of UvsY in recombination mediation [80]. Under low-salt conditions, UvsX filaments assemble on Gp32F-covered ssDNA in an ATP-dependent manner; the nucleation step is highly salt-sensitive. Based on the observation that nucleation rates are faster than propagation rates, it was proposed that UvsX nucleates rapidly at many different sites to initiate filament assembly. This proposal is consistent with observations from direct examination of human Rad51 assembly on dsDNA. Human Rad51 forms multiple nucleation sites rapidly, but the nucleation frequency drops substantially with increasing NaCl concentration as well [81]. Under high-salt conditions, UvsX could not displace Gp32F and no filament assembly could occur without the mediator protein UvsY. Results indicated that UvsY preferentially stabilizes and enhances a prenucleation complex to overcome the salt sensitivity hurdle, permitting UvsX filament formation and concurrent active Gp32F displacement. The observed nucleation rate in the presence of UvsY is several-fold higher than in its absence, and this increase comes from an increased intrinsic binding affinity (K_1) of UvsX at the nucleation phase [80].

In a simplified kinetic model to describe the presynaptic filament assembly process involving UvsX, UvsY, and Gp32 on native ssDNA under no tension, there are two steps in the nucleation phase and two steps in the propagation phase [80] (Fig. 10.6). In both phases, a relatively fast and reversible binding step is followed by a slow and essentially irreversible isomerization step. Unlike the fast nucleation, the propagation of UvsX and concurrent Gp32 displacement are relatively slow. The forward rate constant for the isomerization step in the propagation phase, k_4, is the rate-limiting factor in all tested conditions. It is likely that multiple nucleation events and shorter cluster growth are responsible for the assembly of presynaptic filaments. Consistent with the T4 model, human Rad51 assembles on dsDNA from many rapidly-formed nucleation sites and the cluster growth from these sites is limited in length [81]. The demand for many nucleation events may explain the observation that an apparent 1:1 stoichiometry between UvsX and UvsY has to be maintained for optimal recombination activity [34, 66, 75].

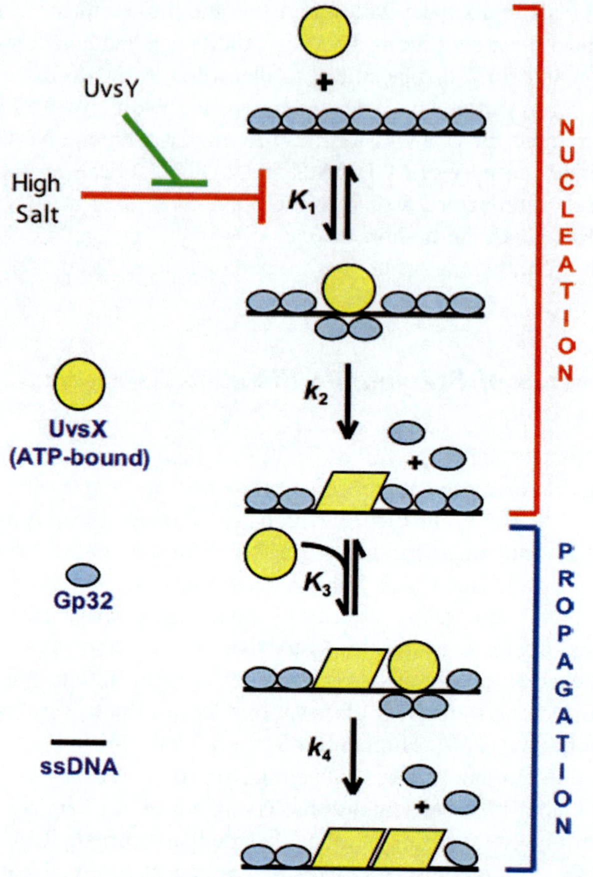

Fig. 10.6 Kinetic model for the nucleation and propagation of UvsX filaments on Gp32-covered ssDNA. Based on observations of the displacement of fluorescein-labeled Gp32 (Gp32F) from ssDNA, UvsX filament assembly occurs in two distinct phases, nucleation and propagation. For both phases, the best-fit model includes a rapid equilibrium step followed by a slow isomerization step that is essentially irreversible under the experimental conditions. K_1, the association constant for the nucleation phase, is highly sensitive to salt concentration; therefore, UvsX by itself cannot displace Gp32 and nucleate a filament under high-salt conditions. A model for UvsY action is that it increases K_1, allowing filament nucleation and Gp32 displacement to occur under high-salt conditions. See text for details

10.4.3 Dynamic Instability of Presynaptic Filaments

Dynamic instability, or vectorial growth and collapse, of presynaptic filaments has been postulated due to the coupling of ATP hydrolysis to ssDNA binding and to the cooperative nature of filament assembly [29, 38, 43, 75]. Studies with the Gp32F probe provide indirect evidence for dynamic instability in UvsX–ssDNA filaments that depends simultaneously on the UvsX ATPase cycle and on competition between

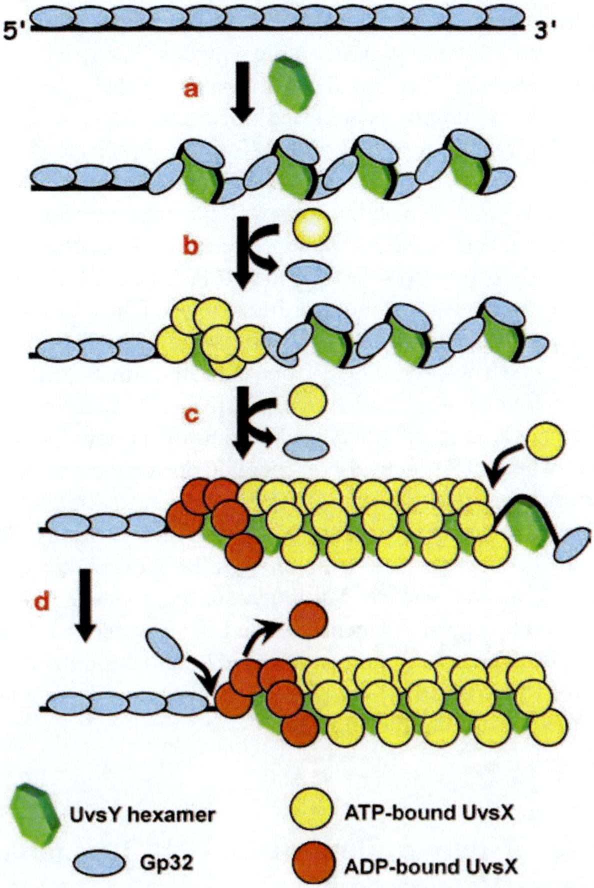

Fig. 10.7 A model for dynamic instability in T4 presynaptic filaments. (**a**) Hexameric UvsY protein weakens Gp32–ssDNA interactions by binding to the complex and wrapping the ssDNA lattice. (**b**) ATP-bound UvsX is recruited to the tripartite UvsY–Gp32–ssDNA intermediate. ATP and UvsY both contribute to a synergistic increase in UvsX–ssDNA binding affinity that allows the recombinase to locally displace Gp32 from the lattice. (**c**) Propagation occurs in the 5′ → 3′ direction as ATP-bound UvsX subunits slowly add to the 3′ filament end, displacing more Gp32 subunits in the process. (**d**) The first UvsX subunits to bind are the first to hydrolyze ATP, generating a relatively aged, ADP-capped 5′ filament end. The ADP-bound UvsX subunits are now vulnerable to displacement by Gp32. Differential competitive effects between Gp32 and the ATP- vs. ADP-capped filament ends create dynamic instability in the complex, which could lead to filament treadmilling

UvsX and Gp32 for binding sites on ssDNA [33] (Fig. 10.7). A presteady-state competition between UvsX and Gp32 for limited ssDNA binding sites reveals that the presynaptic filament formed in the presence of Gp32 undergoes constant assembly/collapse that is tightly linked to the ATP hydrolytic cycle [33]. The experimental setup and solution conditions are designed to resemble in vivo scenarios including physiological ionic strength and the addition of ssDNA to a preexisting mixture of recombination proteins; thus, the observation is probably an accurate reflection of the in vivo assembly/disassembly pathway. The reaction occurs in three temporal phases: First, Gp32 coats all the available ssDNA (Gp32F fluorescence rapidly increases until saturation is achieved). Second, ATP-bound UvsX is loaded by UvsY and gradually displaces Gp32 (Gp32F fluorescence slowly decreases until all or most of it is displaced from the ssDNA). This phase absolutely requires UvsY and either ATP or ATPγS, and is optimal when UvsY concentration is stoichiometric with respect to UvsX and ssDNA binding sites. Third, as the ATP substrate is depleted Gp32 slowly drives UvsX off of the ssDNA (Gp32F fluorescence slowly increases until it re-occupies all or most of the ssDNA). This filament collapse phase requires rapid ATP hydrolysis by UvsX, which converts UvsX from an ATP-bound, high-affinity form that can drive Gp32 from the lattice into an ADP/AMP-bound form that can be driven from the lattice by Gp32 [33, 40]. Therefore, the third phase is sensitive to the nucleotide substrate/product ratio. It is not observed in the presence of ATPγS or when ATP is actively regenerated during the reaction. The addition of ATPγS to an ongoing reaction with ATP is sufficient to stabilize filaments and prevent Gp32F from re-occupying the ssDNA (unpublished results). These observations suggest a model for dynamic instability in T4 presynaptic filaments, which could manifest itself as a classic treadmilling reaction as shown in (Fig. 10.7), with net growth at an ATP-capped filament end and net contraction at an ADP/AMP capped filament end. The key finding is that Gp32 competition and high ATP turnover by UvsX act synergistically to turn the T4 presynaptic filament into a dynamic, metastable intermediate during recombination.

10.5 Roles for Dynamic Filaments in DNA Recombination and Repair Transactions

10.5.1 Other Factors Affecting Filament Dynamics

In addition to ATP binding and hydrolysis by UvsX, competitive effects of Gp32, and mediator functions of UvsY protein, two other known factors are likely to influence the dynamics of the T4 presynaptic filament, and to link filament dynamics more broadly to T4 DNA metabolism: The first factor is the potential coupling of filament assembly to the nucleolytic resection of DNA double-strand breaks. The T4 Gp46/Gp47 complex is a putative recombination nuclease, the subunits of which are homologous to the eukaryotic Rad50 and Mre11 proteins [82, 83]. Both Gp46 and Gp47 have been shown to interact specifically with UvsY [26]. This observation led to the proposal that Gp46/Gp47 could recruit UvsY directly onto the ssDNA that is generated as the complex resects DSBs in the 5′ to 3′ direction. UvsY would then recruit UvsX, circumventing the competition with Gp32, at least during the filament nucleation stage [26]. This model is similar in principle to the resection-coupled nucleation of presynaptic filaments that is observed in the *E. coli* recombination system, wherein RecA protein is directly recruited onto ssDNA generated by the RecBCD helicase/nuclease complex after *chi* site activation [84].

The second factor with the potential to affect filament dynamics is the DNA helicase and translocase enzymes, which could facilitate the turnover of recombinase during the postsynapsis stage. *E. coli* UvrD and yeast Srs2 proteins are two good examples of helicases functioning to remove recombinases from ssDNA and to prevent improper presynaptic filament formation [85–87]. In T4, no helicase has been found to be the functional homolog of UvrD or Srs2. However, T4 Dda protein interacts with and facilitates recombination functions of UvsX protein, although little is known about the underlying mechanism. Thus, Dda might be a candidate factor that contributes to UvsX turnover on the lattice. Dda is an ATP-driven DNA helicase that loads onto the 5′ ssDNA tail adjacent to a duplex region and unwinds the dsDNA with a $5' \rightarrow 3'$ polarity [88]. It translocates on ssDNA and unwinds simple dsDNA substrates while acting as a monomeric helicase [89, 90]. Dda is involved in T4 DNA replication and is required to move the replication fork past DNA-bound proteins on the template in vitro [91–93]. The DNA replication activities of Dda require protein–protein interactions between Dda and the C-terminal domain of Gp32 [58]. This helicase also physically interacts with UvsX protein and modulates its recombination activities [94, 95]. Equally important, UvsX and Dda act synergistically in template switching to allow DNA lesion bypass and to rescue stalled replication forks [96]. The molecular mechanism by which Dda cooperates with UvsX and facilitates template switching in T4 remains largely unknown, but presumably it could affect presynaptic filament assembly and turnover through direct interactions with UvsX or Gp32 or by recognizing DNA structures generated during recombination.

In budding yeast, Rad54, an ATP-powered motor protein, disassembles Rad51 from dsDNA and facilitates Rad51 turnover during the postsynapsis phase [97]. Recent single molecule studies of human Rad51 filament disassembly under tension revealed a pattern of bursts with long pauses, which justified the need for an accessory protein such as Rad54 to remove the ADP-bound Rad51 from filament ends and to speed up the complete disassembly of filaments [98]. Thus, there is a strong possibility that the translocation of Dda on DNA might act similarly to guarantee the complete dissociation of UvsX and to provide the free template access for the incoming polymerase during recombination-dependent replication and DSB repair processes. Consistent with this hypothesis, inclusion of UvsY actually inhibits the two sequential template switching events [96], since UvsY greatly stabilizes the UvsX filament and thus decreases the dynamics of UvsX on the lattice [37].

10.5.2 Biological Significance of Presynaptic Filament Dynamics

The high- and low-affinity transitions caused by ATP hydrolysis, constant competition with Gp32 for available DNA binding sites, and tentative disassembly by certain helicases cooperate to impose a dynamic filament and to prevent rigidity.

Thus, the T4 presynaptic filament is a metastable intermediate under tight surveillance. The significance of this biological process is illustrated by the fact that disruption of human presynaptic filament assembly is associated with severe cancer predisposition diseases [99, 100]. The dynamic feature is critical not to interfere with the next enzymatic step temporally or spatially. For example, the presence of ATPγS endows UvsX with a transient acceleration in the initial rate of branch migration, as well as an initial increase in strand exchange product formation. However, the final yield is greatly decreased and the reaction efficiency is significantly compromised, comparing ATPγS to ATP as the nucleotide cofactor [43].

The dynamic instability of the presynaptic filament also defines the key element of the treadmilling model, which is developed to explain unidirectional branch migration during initial synapsis and the subsequent bubble-migration synthesis stages of recombination-dependent DNA synthesis in the phage [38, 43, 101]. Two actions provide the driving force for UvsX treadmilling on ssDNA (Fig. 10.7): (1) a polar displacement of Gp32 by ATP-bound UvsX at the 3′ growing end of the filament; (2) another concurrent polar displacement of ADP-bound UvsX by Gp32 at the opposite, aged 5′ end. The molecular force originates from not only the ATP binding/hydrolysis cycle of UvsX but also from the constant competition between Gp32 and interchangeable nucleotide-bound forms of UvsX for available ssDNA binding sites [33]. The metastability of presynaptic filaments is a special property that enables reorganization of UvsX on a DNA lattice and eliminates accumulation of dead-end recombination intermediates. The constant assembly/collapse process simplifies the homology search at the initial stage and prevents UvsX from interfering with downstream events, guiding a timely and correct assembly of the presynaptic filament only when absolutely needed.

References

1. Jasin M (2002) Homologous repair of DNA damage and tumorigenesis: the BRCA connection. Oncogene 21(58):8981–8993
2. Symington LS (2002) Role of RAD52 epistasis group genes in homologous recombination and double-strand break repair. Microbiol Mol Biol Rev 66(4):630–670, table of contents
3. Pierce AJ, Stark JM, Araujo FD, Moynahan ME, Berwick M, Jasin M (2001) Double-strand breaks and tumorigenesis. Trends Cell Biol 11(11):S52–S59
4. Sun H, Treco D, Szostak JW (1991) Extensive 3′-overhanging, single-stranded DNA associated with the meiosis-specific double-strand breaks at the ARG4 recombination initiation site. Cell 64(6):1155–1161
5. Haber JE (1995) In vivo biochemistry: physical monitoring of recombination induced by site-specific endonucleases. Bioessays 17(7):609–620
6. Bianco PR, Tracy RB, Kowalczykowski SC (1998) DNA strand exchange proteins: a biochemical and physical comparison. Front Biosci 3:570–603
7. Thacker J (1999) A surfeit of RAD51-like genes? Trends Genet 15(5):166–168
8. McRobbie AM, Carter LG, Kerou M, Liu H, McMahon SA, Johnson KA, Oke M, Naismith JH, White MF (2009) Structural and functional characterisation of a conserved archaeal RadA paralog with antirecombinase activity. J Mol Biol 389(4):661–673

9. Cox MM (2007) Motoring along with the bacterial RecA protein. Nat Rev Mol Cell Biol 8(2):127–138
10. Haber JE (1997) A super new twist on the initiation of meiotic recombination. Cell 89(2): 163–166
11. Konforti BB, Davis RW (1990) The preference for a 3′ homologous end is intrinsic to RecA-promoted strand exchange. J Biol Chem 265(12):6916–6920
12. Sung P, Robberson DL (1995) DNA strand exchange mediated by a RAD51-ssDNA nucleoprotein filament with polarity opposite to that of RecA. Cell 82(3):453–461
13. Oh SD, Lao JP, Hwang PY, Taylor AF, Smith GR, Hunter N (2007) BLM ortholog, Sgs1, prevents aberrant crossing-over by suppressing formation of multichromatid joint molecules. Cell 130(2):259–272
14. Salinas F, Kodadek T (1995) Phage T4 homologous strand exchange: a DNA helicase, not the strand transferase, drives polar branch migration. Cell 82(1):111–119
15. Beernink HT, Morrical SW (1999) RMPs: recombination/replication mediator proteins. Trends Biochem Sci 24(10):385–389
16. Wadsworth RI, White MF (2001) Identification and properties of the crenarchaeal single-stranded DNA binding protein from *Sulfolobus solfataricus*. Nucleic Acids Res 29(4):914–920
17. Kerr ID, Wadsworth RI, Cubeddu L, Blankenfeldt W, Naismith JH, White MF (2003) Insights into ssDNA recognition by the OB fold from a structural and thermodynamic study of Sulfolobus SSB protein. EMBO J 22(11):2561–2570
18. Sung P, Krejci L, Van Komen S, Sehorn MG (2003) Rad51 recombinase and recombination mediators. J Biol Chem 278(44):42729–42732
19. Morimatsu K, Kowalczykowski SC (2003) RecFOR proteins load RecA protein onto gapped DNA to accelerate DNA strand exchange: a universal step of recombinational repair. Mol Cell 11(5):1337–1347
20. New JH, Sugiyama T, Zaitseva E, Kowalczykowski SC (1998) Rad52 protein stimulates DNA strand exchange by Rad51 and replication protein A. Nature 391(6665):407–410
21. Shinohara A, Ogawa T (1998) Stimulation by Rad52 of yeast Rad51-mediated recombination. Nature 391(6665):404–407
22. Yang H, Li Q, Fan J, Holloman WK, Pavletich NP (2005) The BRCA2 homologue Brh2 nucleates RAD51 filament formation at a dsDNA-ssDNA junction. Nature 433(7026):653–657
23. Sung P (1997) Yeast Rad55 and Rad57 proteins form a heterodimer that functions with replication protein A to promote DNA strand exchange by Rad51 recombinase. Genes Dev 11(9):1111–1121
24. Komori K, Miyata T, DiRuggiero J, Holley-Shanks R, Hayashi I, Cann IK, Mayanagi K, Shinagawa H, Ishino Y (2000) Both RadA and RadB are involved in homologous recombination in *Pyrococcus furiosus*. J Biol Chem 275(43):33782–33790
25. DiRuggiero J, Brown JR, Bogert AP, Robb FT (1999) DNA repair systems in archaea: mementos from the last universal common ancestor? J Mol Evol 49(4):474–484
26. Bleuit JS, Xu H, Ma Y, Wang T, Liu J, Morrical SW (2001) Mediator proteins orchestrate enzyme-ssDNA assembly during T4 recombination-dependent DNA replication and repair. Proc Natl Acad Sci USA 98(15):8298–8305
27. Mosig G (1994) Molecular biology of bacteriophage T4. American Society for Microbiology, Washington, DC, pp 54–82
28. Kreuzer KN (2000) Recombination-dependent DNA replication in phage T4. Trends Biochem Sci 25(4):165–173
29. Kodadek T (1990) The role of the bacteriophage T4 gene 32 protein in homologous pairing. J Biol Chem 265(34):20966–20969
30. Kodadek T, Gan DC, Stemke-Hale K (1989) The phage T4 uvsY recombination protein stabilizes presynaptic filaments. J Biol Chem 264(28):16451–16457
31. Griffith J, Formosa T (1985) The uvsX protein of bacteriophage T4 arranges single-stranded and double-stranded DNA into similar helical nucleoprotein filaments. J Biol Chem 260(7): 4484–4491

32. Farb JN, Morrical SW (2009) Functional complementation of UvsX and UvsY mutations in the mediation of T4 homologous recombination. Nucleic Acids Res 37(7):2336–2345
33. Liu J, Qian N, Morrical SW (2006) Dynamics of bacteriophage T4 presynaptic filament assembly from extrinsic fluorescence measurements of Gp32-single-stranded DNA interactions. J Biol Chem 281(36):26308–26319
34. Harris LD, Griffith JD (1989) UvsY protein of bacteriophage T4 is an accessory protein for in vitro catalysis of strand exchange. J Mol Biol 206(1):19–27
35. Yassa DS, Chou KM, Morrical SW (1997) Characterization of an amino-terminal fragment of the bacteriophage T4 uvsY recombination protein. Biochimie 79(5):275–285
36. Ando RA, Morrical SW (1998) Single-stranded DNA binding properties of the UvsX recombinase of bacteriophage T4: binding parameters and effects of nucleotides. J Mol Biol 283(4):785–796
37. Liu J, Bond JP, Morrical SW (2006) Mechanism of presynaptic filament stabilization by the bacteriophage T4 UvsY recombination mediator protein. Biochemistry 45(17):5493–5502
38. Formosa T, Alberts BM (1986) Purification and characterization of the T4 bacteriophage uvsX protein. J Biol Chem 261(13):6107–6118
39. Pugh BF, Cox MM (1988) High salt activation of recA protein ATPase in the absence of DNA. J Biol Chem 263(1):76–83
40. Farb JN, Morrical SW (2009) Role of allosteric switch residue histidine 195 in maintaining active-site asymmetry in presynaptic filaments of bacteriophage T4 UvsX recombinase. J Mol Biol 385(2):393–404
41. Conway AB, Lynch TW, Zhang Y, Fortin GS, Fung CW, Symington LS, Rice PA (2004) Crystal structure of a Rad51 filament. Nat Struct Mol Biol 11(8):791–796
42. Lauder SD, Kowalczykowski SC (1991) Asymmetry in the recA protein-DNA filament. J Biol Chem 266(9):5450–5458
43. Kodadek T, Wong ML, Alberts BM (1988) The mechanism of homologous DNA strand exchange catalyzed by the bacteriophage T4 uvsX and gene 32 proteins. J Biol Chem 263(19):9427–9436
44. Riddles PW, Lehman IR (1985) The formation of plectonemic joints by the recA protein of *Escherichia coli*. Requirement for ATP hydrolysis. J Biol Chem 260(1):170–173
45. Kowalczykowski SC, Krupp RA (1995) DNA-strand exchange promoted by RecA protein in the absence of ATP: implications for the mechanism of energy transduction in protein-promoted nucleic acid transactions. Proc Natl Acad Sci USA 92(8):3478–3482
46. Yonesaki T, Minagawa T (1989) Synergistic action of three recombination gene products of bacteriophage T4, uvsX, uvsY, and gene 32 proteins. J Biol Chem 264(14):7814–7820
47. Melamede RJ, Wallace SS (1977) Properties of the nonlethal recombinational repair x and y mutants of bacteriophage T4. II. DNA synthesis. J Virol 24(1):28–40
48. Melamede RJ, Wallace SS (1980) Properties of the nonlethal recombinational repair deficient mutants of bacteriophage T4. III. DNA replicative intermediates and T4w. Mol Gen Genet 177(3):501–509
49. Kreuzer KN, Morrical SM (1994) Molecular biology of bacteriophage T4. American Society for Microbiology, Washington, DC, pp 28–42
50. Chase JW, Williams KR (1986) Single-stranded DNA binding proteins required for DNA replication. Annu Rev Biochem 55:103–136
51. Karpel R, Karpel RL (1990) The biology of nonspecific DNA-protein interactions. CRC, Boca Raton, FL, pp 103–130
52. Shamoo Y, Friedman AM, Parsons MR, Konigsberg WH, Steitz TA (1995) Crystal structure of a replication fork single-stranded DNA binding protein (T4 gp32) complexed to DNA. Nature 376(6538):362–366
53. Williams KR, Shamoo Y, Spicer EK, Coleman JE, Konigsberg WH (1994) Molecular biology of bacteriophage T4. American Society for Microbiology, Washington, DC, pp 301–304
54. Giedroc DP, Khan R, Barnhart K (1990) Overexpression, purification, and characterization of recombinant T4 gene 32 protein22-301 (g32P-B). J Biol Chem 265(20):11444–11455

55. Xu H, Wang Y, Bleuit JS, Morrical SW (2001) Helicase assembly protein Gp59 of bacteriophage T4: fluorescence anisotropy and sedimentation studies of complexes formed with derivatives of Gp32, the phage ssDNA binding protein. Biochemistry 40(25):7651–7661
56. Hurley JM, Chervitz SA, Jarvis TC, Singer BS, Gold L (1993) Assembly of the bacteriophage T4 replication machine requires the acidic carboxy terminus of gene 32 protein. J Mol Biol 229(2):398–418
57. Jiang H, Giedroc D, Kodadek T (1993) The role of protein-protein interactions in the assembly of the presynaptic filament for T4 homologous recombination. J Biol Chem 268(11): 7904–7911
58. Ma Y, Wang T, Villemain JL, Giedroc DP, Morrical SW (2004) Dual functions of single-stranded DNA-binding protein in helicase loading at the bacteriophage T4 DNA replication fork. J Biol Chem 279(18):19035–19045
59. Morrical SW, Beernink HT, Dash A, Hempstead K (1996) The gene 59 protein of bacteriophage T4. Characterization of protein-protein interactions with gene 32 protein, the T4 single-stranded DNA binding protein. J Biol Chem 271(33):20198–20207
60. Kowalczykowski SC (1990) Thermodynamic data for protein-nucleic acid interactions. In: Saenger W (ed) Landolt-Bornstein: numerical data and functional relationships in science and technology. Springer, Berlin, pp 244–263
61. Newport JW, Lonberg N, Kowalczykowski SC, von Hippel PH (1981) Interactions of bacteriophage T4-coded gene 32 protein with nucleic acids. II. Specificity of binding to DNA and RNA. J Mol Biol 145(1):105–121
62. Kowalczykowski SC, Lonberg N, Newport JW, von Hippel PH (1981) Interactions of bacteriophage T4-coded gene 32 protein with nucleic acids. I. Characterization of the binding interactions. J Mol Biol 145(1):75–104
63. Rouzina I, Pant K, Karpel RL, Williams MC (2005) Theory of electrostatically regulated binding of T4 gene 32 protein to single- and double-stranded DNA. Biophys J 89(3): 1941–1956
64. Pant K, Karpel RL, Rouzina I, Williams MC (2005) Salt dependent binding of T4 gene 32 protein to single and double-stranded DNA: single molecule force spectroscopy measurements. J Mol Biol 349(2):317–330
65. Shokri L, Rouzina I, Williams MC (2009) Interaction of bacteriophage T4 and T7 single-stranded DNA-binding proteins with DNA. Phys Biol 6(2):25002
66. Morrical SW, Alberts BM (1990) The UvsY protein of bacteriophage T4 modulates recombination-dependent DNA synthesis in vitro. J Biol Chem 265(25):15096–15103
67. Beernink HT, Morrical SW (1998) The uvsY recombination protein of bacteriophage T4 forms hexamers in the presence and absence of single-stranded DNA. Biochemistry 37(16):5673–5681
68. Sweezy MA, Morrical SW (1997) Single-stranded DNA binding properties of the uvsY recombination protein of bacteriophage T4. J Mol Biol 266(5):927–938
69. Xu H, Beernink HT, Morrical SW (2010) DNA binding properties of T4 UvsY recombination mediator protein: polynudeotide wrapping promotes high-affinity binding to ssDNA. Nucleic Acids Res 38(14):4821–4833
70. Bleuit JS, Ma Y, Munro J, Morrical SW (2004) Mutations in a conserved motif inhibit single-stranded DNA binding and recombination mediator activities of bacteriophage T4 UvsY protein. J Biol Chem 279(7):6077–6086
71. Pant K, Shokri L, Karpel RL, Morrical SW, Williams MC (2008) Modulation of T4 gene 32 protein DNA binding activity by the recombination mediator protein UvsY. J Mol Biol 380(5):799–811
72. Ando RA, Morrical SW (1999) Relationship between hexamerization and ssDNA binding affinity in the uvsY recombination protein of bacteriophage T4. Biochemistry 38(50):16589–16598
73. McGhee JD (1976) Theoretical calculations of the helix-coil transition of DNA in the presence of large, cooperatively binding ligands. Biopolymers 15(7):1345–1375
74. Morrical SW, Wong ML, Alberts BM (1991) Amplification of snap-back DNA synthesis reactions by the uvsX recombinase of bacteriophage T4. J Biol Chem 266(21):14031–14038

75. Kodadek T (1990) Functional interactions between phage T4 and *E. coli* DNA-binding proteins during the presynapsis phase of homologous recombination. Biochem Biophys Res Commun 172(2):804–810
76. Sweezy MA, Morrical SW (1999) Biochemical interactions within a ternary complex of the bacteriophage T4 recombination proteins uvsY and gp32 bound to single-stranded DNA. Biochemistry 38(3):936–944
77. Hashimoto K, Yonesaki T (1991) The characterization of a complex of three bacteriophage T4 recombination proteins, uvsX protein, uvsY protein, and gene 32 protein, on single-stranded DNA. J Biol Chem 266(8):4883–4888
78. Echols H (1986) Multiple DNA-protein interactions governing high-precision DNA transactions. Science 233(4768):1050–1056
79. Lohman TM, Kowalczykowski SC (1981) Kinetics and mechanism of the association of the bacteriophage T4 gene 32 (helix destabilizing) protein with single-stranded nucleic acids. Evidence for protein translocation. J Mol Biol 152(1):67–109
80. Liu J, Berger C, Morrical SW Kinetics of presynaptic filament assembly in the presence of SSB and mediator proteins. Manuscript under revision
81. Hilario J, Amitani I, Baskin RJ, Kowalczykowski SC (2009) Direct imaging of human Rad51 nucleoprotein dynamics on individual DNA molecules. Proc Natl Acad Sci USA 106(2):361–368
82. Mickelson C, Wiberg JS (1981) Membrane-associated DNase activity controlled by genes 46 and 47 of bacteriophage T4D and elevated DNase activity associated with the T4 das mutation. J Virol 40(1):65–77
83. Cromie GA, Connelly JC, Leach DR (2001) Recombination at double-strand breaks and DNA ends: conserved mechanisms from phage to humans. Mol Cell 8(6):1163–1174
84. Anderson DG, Kowalczykowski SC (1997) The translocating RecBCD enzyme stimulates recombination by directing RecA protein onto ssDNA in a chi-regulated manner. Cell 90(1):77–86
85. Krejci L, Van Komen S, Li Y, Villemain J, Reddy MS, Klein H, Ellenberger T, Sung P (2003) DNA helicase Srs2 disrupts the Rad51 presynaptic filament. Nature 423(6937):305–309
86. Veaute X, Jeusset J, Soustelle C, Kowalczykowski SC, Le Cam E, Fabre F (2003) The Srs2 helicase prevents recombination by disrupting Rad51 nucleoprotein filaments. Nature 423(6937):309–312
87. Veaute X, Delmas S, Selva M, Jeusset J, Le Cam E, Matic I, Fabre F, Petit MA (2005) UvrD helicase, unlike Rep helicase, dismantles RecA nucleoprotein filaments in *Escherichia coli*. EMBO J 24(1):180–189
88. Jongeneel CV, Formosa T, Alberts BM (1984) Purification and characterization of the bacteriophage T4 dda protein. A DNA helicase that associates with the viral helix-destabilizing protein. J Biol Chem 259(20):12925–12932
89. Nanduri B, Byrd AK, Eoff RL, Tackett AJ, Raney KD (2002) Pre-steady-state DNA unwinding by bacteriophage T4 Dda helicase reveals a monomeric molecular motor. Proc Natl Acad Sci USA 99(23):14722–14727
90. Raney KD, Benkovic SJ (1995) Bacteriophage T4 Dda helicase translocates in a unidirectional fashion on single-stranded DNA. J Biol Chem 270(38):22236–22242
91. Gauss P, Park K, Spencer TE, Hacker KJ (1994) DNA helicase requirements for DNA replication during bacteriophage T4 infection. J Bacteriol 176(6):1667–1672
92. Barry J, Alberts B (1994) A role for two DNA helicases in the replication of T4 bacteriophage DNA. J Biol Chem 269(52):33063–33068
93. Bedinger P, Hochstrasser M, Jongeneel CV, Alberts BM (1983) Properties of the T4 bacteriophage DNA replication apparatus: the T4 dda DNA helicase is required to pass a bound RNA polymerase molecule. Cell 34(1):115–123
94. Formosa T, Alberts BM (1984) The use of affinity chromatography to study proteins involved in bacteriophage T4 genetic recombination. Cold Spring Harb Symp Quant Biol 49:363–370

95. Kodadek T, Alberts BM (1987) Stimulation of protein-directed strand exchange by a DNA helicase. Nature 326(6110):312–314
96. Kadyrov FA, Drake JW (2004) UvsX recombinase and Dda helicase rescue stalled bacteriophage T4 DNA replication forks in vitro. J Biol Chem 279(34):35735–35740
97. Solinger JA, Kiianitsa K, Heyer WD (2002) Rad54, a Swi2/Snf2-like recombinational repair protein, disassembles Rad51:dsDNA filaments. Mol Cell 10(5):1175–1188
98. van Mameren J, Modesti M, Kanaar R, Wyman C, Peterman EJ, Wuite GJ (2009) Counting RAD51 proteins disassembling from nucleoprotein filaments under tension. Nature 457(7230):745–748
99. Scully R (2000) Role of BRCA gene dysfunction in breast and ovarian cancer predisposition. Breast Cancer Res 2(5):324–330
100. Yu VP, Koehler M, Steinlein C, Schmid M, Hanakahi LA, van Gool AJ, West SC, Venkitaraman AR (2000) Gross chromosomal rearrangements and genetic exchange between nonhomologous chromosomes following BRCA2 inactivation. Genes Dev 14(11):1400–1406
101. Formosa T, Alberts BM (1986) DNA synthesis dependent on genetic recombination: characterization of a reaction catalyzed by purified bacteriophage T4 proteins. Cell 47(5):793–806
102. Chen Z, Yang H, Pavletich NP (2008) Mechanism of homologous recombination from the RecA-ssDNA/dsDNA structures. Nature 453(7194):489–494

Chapter 11
Polymerase Switching in Response to DNA Damage

Jaylene N. Ollivierre, Michelle C. Silva, Jana Sefcikova, and Penny J. Beuning

11.1 Introduction

DNA polymerases are highly efficient and accurate macromolecular machines. They are capable of replicating DNA at up to 1,000 nucleotides per second while making less than one error in 100,000 additions. However, DNA is constantly subjected to damage from myriad sources. DNA damage disrupts normal cellular DNA replication by interfering with the accuracy and efficiency of replicative DNA polymerases. Specialized Y family DNA polymerases exist that can copy damaged DNA, although that ability often has a mutagenic cost. Therefore, Y family DNA polymerase activity is highly regulated in the cell. This chapter presents the functions of both replicative and Y family DNA polymerases and the cellular mechanisms of polymerase management. The focus is on *Escherichia coli* systems but also briefly discusses eukaryotic Y family polymerases. We first present DNA replication carried out by prokaryotic DNA polymerase III and describe its subunits and the coordination of leading and lagging strand replication. We then discuss DNA damage and specialized Y family DNA polymerases. Different models for the management of replicative and Y family DNA polymerases are presented. Finally, we briefly compare the eukaryotic systems with their prokaryotic counterparts.

11.2 *Escherichia coli* DNA Polymerase III

DNA replication requires the coordination of many different proteins to accomplish the goal of simultaneous replication of the two antiparallel stands of DNA. This process is tightly regulated so that DNA is replicated in a timely and accurate

P.J. Beuning (✉)
Department of Chemistry & Chemical Biology, Center for Interdisciplinary,
Research on Complex Systems, Northeastern University, 360 Huntington Ave 102 Hurtig Hall,
Boston, MA 02115, USA
e-mail: beuning@neu.edu

M.C. Williams and L.J. Maher, III (eds.), *Biophysics of DNA-Protein Interactions*,
Biological and Medical Physics, Biomedical Engineering,
DOI 10.1007/978-0-387-92808-1_11,

manner. Moreover, DNA replication is highly processive, which allows efficient replication of over three million base pairs in every *E. coli* cell cycle.

In *E. coli,* DNA polymerase III (DNA pol III) is responsible for the majority of DNA replication (Fig. 11.1). Each cell expresses approximately ten copies of DNA pol III core (see below) [1]. Along with auxiliary proteins, DNA pol III semidiscontinuously replicates DNA at a speed of approximately 1 kb per second [2], making less than one error in approximately 10^5 nucleotide additions [1, 3]. In the presence of proofreading and mismatch repair, the error frequency is approximately 10^{-10} per base pair [1, 4]. At each replication fork, DNA pol III acts as an asymmetric dimer [5–9]: one monomer acts in the continuous replication of the leading strand and another acts in the discontinuous replication of the lagging strand (Fig. 11.1). Even though only two polymerases are needed to replicate DNA, it has been shown that the replisome may contain three DNA polymerases. The third polymerase may function on the lagging strand or serve as a spare polymerase, able to replace either polymerase when needed [10].

Escherichia coli DNA pol III consists of ten subunits [2, 6, 9, 11] that can be classified into three subassemblies: the core, the clamp, and the clamp loader complex [9, 12] (Table 11.1). The core is composed of three subunits: α, ε, and θ. The α subunit contains the polymerase activity and is responsible for synthesizing DNA. The processivity of the isolated α subunit, defined as the number of nucleotides

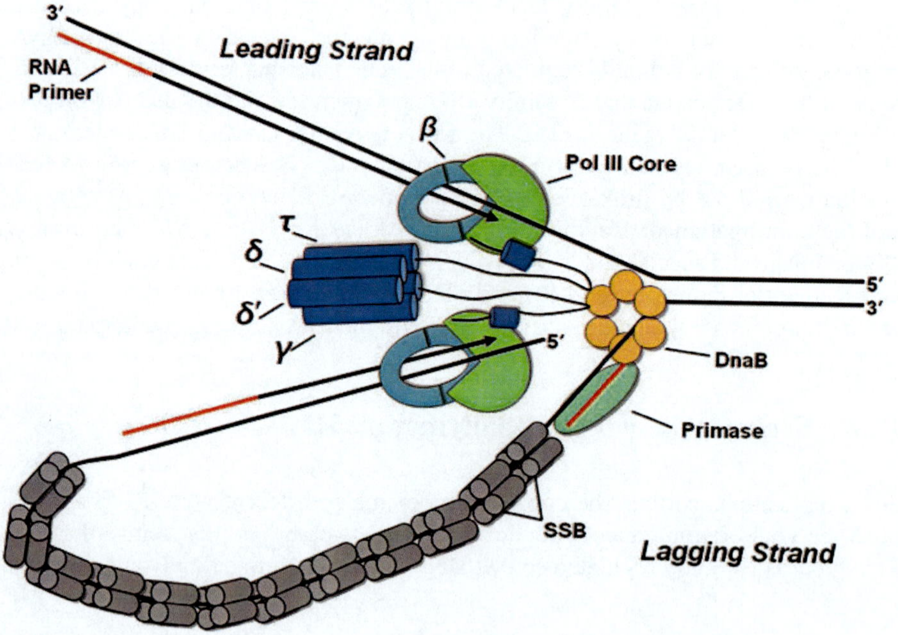

Fig. 11.1 DNA polymerase III holoenzyme (dimer form) at a replication fork. The two polymerase cores (*green*), which are tethered to the β clamps (*light blue*), contain the three subunits α, ε, and θ. The γ complex (*blue*), assembled with two τ subunits, couples the polymerization of both the leading strand and the lagging strand. The single-stranded lagging strand is threaded through DnaB helicase (*orange*) and is coated with SSB (*gray*). Also shown is the primase (*light green*) that synthesizes the RNA primers (*red*) on the lagging strand. The ψ and χ subunits are not shown but would connect the γ complex to SSB

Table 11.1 Components of DNA polymerase III [1, 9, 11]

Mass (kDa)		Gene	Function	References
Polymerase core				
α	130	*dnaE*	Polymerase	[25, 33, 37, 40]
ε	27.5	*dnaQ (mutD)*	3′-5′ Exonuclease	[18, 43, 45–48, 52, 368, 369]
θ	10	*holE*	Stabilizes the core	[48, 50–53]
Clamp loader [20, 59, 80]				
τ	71	*dnaX*	Coordinates replication	[9, 21, 34, 36, 61, 100, 106, 107, 109–112]
γ	47.5	*dnaX*	ATPase	[61, 64, 65, 77]
δ	35	*holA*	"Wrench"; opens clamp	[60, 64]
δ′	33	*holB*	Mediator between γ and δ	[64, 66]
χ	15	*holC*	Binds SSB	[22, 23, 87, 88]
ψ	12	*holD*	Bridges γ complex and χ	[22, 87]
Clamp				
β	40.6	*dnaN*	Clamp	[11, 37, 59, 71, 89–91, 93]

incorporated into the nascent DNA per association event, is low (1–10 nucleotides) compared with >50 kb for the holoenzyme [1, 13–17]. The ε subunit is a 3′–5′ exonuclease that is responsible for the proofreading capability of the core. In the absence of ε, the frequency of mutations due to misincorporations during replication increases by approximately 40-fold [18]. When coupled, the polymerase and exonuclease activities of the α and ε subunits increase significantly [18]. The third component of the core, θ, is not required for high processivity [19].

The β clamp, also known as the processivity clamp, is the major contributor to the processivity of DNA pol III [1]. The β clamp encircles DNA and tethers the α subunit to its DNA substrate (Fig. 11.1). The β clamp is loaded onto DNA by a complex known as the clamp loader, which is composed of six subunits: τ, γ, δ, δ′, χ, and ψ (Table 11.1) [20]. The τ subunits coordinate replication on both strands by coupling the polymerase cores to the clamp loader complex [9, 21]. When τ is coupled to the core, processivity increases approximately six-fold [1]. The γ subunit is an ATPase. Along with δ and δ′, γ is responsible for loading the clamp onto DNA [2, 11]. The two other subunits, ψ and χ, bind single-stranded DNA-binding protein (SSB) and help regulate replication on the lagging strand [22, 23]. The ψ subunit also has a role in clamp loading [24].

11.3 The Polymerase Core

The DNA pol III core includes the polymerase and proofreading exonuclease activity of the replisome. Alone, the core can replicate DNA at a rate of approximately 20 nucleotides per second (nt/s) with a processivity of 11 nucleotides, values much lower than the entire replisome [2, 11]. The following is an in-depth description of each of the subunits of the core.

11.3.1 The Polymerase Subunit (α)

The α subunit, encoded by the *dnaE* gene and a member of the polymerase C family, is a protein of approximately 130 kDa, containing several distinct domains. Although this polymerase has been studied for decades, the crystal structure of the α subunit of *E. coli* was only solved recently [25]. Like other polymerases, the structure resembles a right hand with three characteristic domains: the palm, the fingers, and the thumb domains [26–28]. The palm domain contains the active site of the polymerase consisting of three aspartic acid residues: Asp401, Asp403, and Asp555 [29]. This domain is similar to the palm domains of polymerases in the X family, especially to that of DNA pol β [25]. Modeling DNA onto the *E. coli* DNA pol III α structure using the human DNA pol β cocrystal structure with DNA [30] as the modeling template showed that the DNA strand collides with a short α-helix in the palm domain in a sterically unfavorable interaction [25]. This suggests that although DNA pol III and pol β have similar active sites, there may be differences in how they bind DNA.

The finger domain includes four subdomains: the index finger, the middle finger, the ring finger, and the little finger. These subdomains are responsible for binding the incoming nucleotide. The thumb domain guides the newly formed DNA duplex as it leaves the active site. The polymerase and histidinol phosphatase (PHP) domain is located in the "wrist" position, relative to the hand of the polymerase domain. The exact role of this domain is unknown but because of the domain's sequence similarity to histidinol phosphatases, it was proposed to possess pyrophosphatase activity for the pyrophosphate produced during replication [31]. Based on the crystal structure of DNA pol III α, this domain is unlikely to harbor such activity [25]. However, it has been demonstrated that the *Thermus thermophilus* DNA pol III α subunit contains a Zn^{2+}-dependent 3′–5′ exonuclease activity [32].

The C-terminal domain is not present in the crystal structure of *E. coli* DNA pol III α. This domain, which is located C-terminal to the tip of the little finger domain, includes an oligonucleotide/oligosaccharide binding (OB) fold [25, 33] and the binding sites for both the τ subunit [34, 35] and the β clamp [36, 37]. The recently solved structure of DNA pol III α from *Thermus aquaticus* [33] includes this C-terminal domain, consisting of an αβ fold. The OB fold domain consists of five β-strands arranged in a β barrel, similar to that of other OB folds [38]. The OB fold domain of α has been shown to bind single-stranded DNA (ssDNA) specifically [39]. Compared with other ssDNA-binding proteins, the OB fold domain of the α subunit is somewhat unusual in that it does not actively melt DNA, but rather binds to ssDNA that is preformed, in this case by force-induced melting [39]. Along with the rest of the polymerase, this function provides insight into the regulation of DNA pol III α, discussed below.

A cocrystal structure of *T. aquaticus* DNA pol III α bound to primer-template DNA and an incoming deoxynucleoside 5′-triphosphate has been determined with a resolution of 4.6 Å [40]. When compared with the structure of *T. aquaticus* DNA pol III α without DNA, it is possible to see significant movements of the thumb, finger, and β-binding domains. These movements position the protein on the DNA, allowing for the interaction with the DNA backbone at the minor groove. This structure also indicates that the DNA and incoming nucleotide bind in a similar fashion to that of DNA pol β. The C-terminal domain undergoes an approximately 30° rotation putting the OB fold in position to bind the single-stranded template DNA. The internal β-binding motif also seems to be correctly positioned in the structure with DNA in order to bind to the hydrophobic pocket on the β clamp [37, 40].

The structure of a ternary complex of PolC, the replicative polymerase of the gram-positive bacterium *Geobacillus kaustophilus*, with primed DNA and an incoming dideoxynucleoside substrate has been solved to 2.4 Å resolution [41]. Unlike the DNA pol III α polymerase, PolC contains an intrinsic 3′–5′ exonuclease domain as an insertion within the PHP domain. Instead of being located in the C-terminal domain, the OB fold of PolC is located N-terminal to the palm domain and is positioned so it could bind the single-stranded template strand approximately 15–20 nucleotides away from the polymerase active site. The thumb domain contains β-strands that bind DNA in the minor groove, possibly allowing for the detection of mismatched base pairs after incorporation [41]. Flexibility in the palm domain suggests a large conformational change upon DNA binding [41, 42].

11.3.2 The ε Subunit

The ε subunit, a 27.5-kDa protein encoded by the *dnaQ* (also known as *mutD*) gene, is responsible for the 3′–5′ exonuclease activity of the polymerase core [1] and forms a tight complex with DNA pol III α. Whereas in *E. coli*, the polymerase and exonuclease reside on two different polypeptides, in other cases, the exonuclease activity and the polymerase activity are part of the same polypeptide, as in gram-positive PolC and in DNA pol I [26, 28, 41], another eubacterial polymerase. Such an interaction between α and ε enhances the overall activity of each protein. In fact, it has been shown that α has greater polymerase activity in complex with ε than alone [18]. The exonuclease activity of the ε subunit is also substantially stimulated within the complex [18].

Although the structure of the N-terminal domain of ε has been determined by X-ray crystallography [43] and by nuclear magnetic resonance spectroscopy (NMR) [44], no structures for the full-length ε subunit have been determined due to the difficulty in obtaining large amounts of pure protein. The structures include the 186 N-terminal residues of ε responsible for its exonuclease activity. The 57 residues that are not present in the structure include the C-terminal domain, which contains a flexible linker that has been shown to bind to the α subunit [45–48]. The ε subunit binds α in the PHP domain located in the α N-terminal domain [49].

11.3.3 The θ Subunit

The third subunit of the polymerase core is θ, the 10-kDa product of the *holE* gene [1]. The θ subunit binds to ε close to the ε active site [50, 51], although it is unlikely to play a direct role in exonuclease activity [52]. Extensive hydrophobic surfaces define the interactions between θ and ε [52–54]. The θ subunit has not been shown to bind directly to α [51, 55], but it may have a stabilizing effect on the α:ε complex. Such a function is supported by the results of a series of yeast two-hybrid experiments indicating that the interaction between α and ε is strengthened in the presence of θ [56]. The θ subunit also seems to enhance the exonuclease activity of the ε subunit by stabilizing ε [51, 57]. Deletion of θ results in a slight increase in the spontaneous mutation frequency of *E. coli* [56]. In biochemical experiments, the exonuclease activity of ε I170T/V215A double mutant was not substantially stimulated in the presence of either α or θ [58]. Upon addition of both α and θ, however, activity of the ε I170T/V215A variant was stimulated [58].

11.4 The Clamp Loader Complex and the β Clamp Subunit

The clamp loader complex consists of at least two τ subunits, up to three γ subunits, and one each of the δ, δ′, χ, and ψ subunits [2, 9, 11]. As mentioned above, DNA pol III can be assembled as a trimer [10]. In this case, three τ subunits are present in the absence of γ subunits. Although not required for the clamp loading process, the τ subunits play a critical role in managing replication at the fork [9, 11]. The γ subunits, closely related to τ, bind ATP and facilitate loading the β clamp onto DNA [11, 59]. The δ and δ′ subunits are directly involved in the loading of the clamp, δ as the "wrench" and δ′ as the mediator between τ and δ [59, 60].

11.4.1 The γ Complex: δ':γ_1:γ_2:γ_3:δ; Loading the Clamp

The *dnaX* gene encodes both τ and γ. The τ subunit is the full-length product of the gene and γ is produced due to a –1 frameshift, which causes a stop codon to be inserted prematurely, forming the shorter product, γ [61–63]. This frameshift is caused by two factors: a heptanucleotide sequence that induces frameshifts and a downstream RNA stem-loop structure (Fig. 11.2a). Therefore, γ consists of only the first three of the five τ domains (Fig. 11.2b). These three domains contain the ATPase site and so both τ and γ are ATPases. In the crystal structure of the γ complex, containing $\gamma_3\delta\delta'$ [64], it is possible to distinguish these domains. Domains I and II contain the nucleotide-binding site in which ATP binds at the interface between the subunits, and Domain III forms a circular collar with the other subunits of the γ complex.

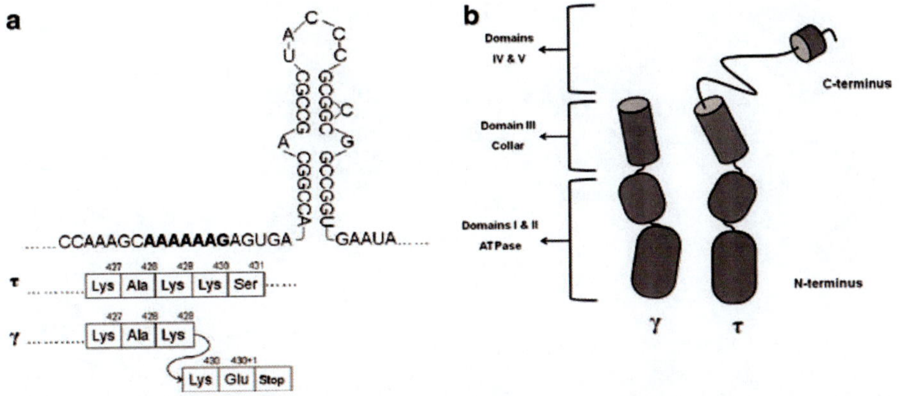

Fig. 11.2 (**a**) The segment of the *dnaX* mRNA, responsible for coding the γ and τ subunits. The two factors that cause the −1 frameshift are a heptanucleotide sequence (*bold*) and a downstream stem-loop. As a result, the γ subunit consists of the first 430 residues of the *dnaX* gene product shared with the τ subunit, followed by a glutamic acid residue and a stop codon. (**b**) A side-by-side comparison of the domains of the τ and γ subunits. The γ subunit, the shorter protein, contains only the first three domains of the entire gene product. These domains include the ATPase active site and the collar domain. Domains IV and V of the τ subunit contain the binding sites for both the pol III α subunit and DnaB

A crystal structure of the first 243 residues of γ (Domains I and II) solved with and without nucleotides [65] shows a conformational change upon nucleotide binding, suggesting a mechanism for binding to the clamp and ATP hydrolysis (Fig. 11.3). In the crystal structure of the γ complex [64], the γ subunits, together with δ and δ′, are arranged in a heptameric complex resembling an opened ring in the order $\delta':\gamma_1:\gamma_2:\gamma_3:\delta$ (Fig. 11.3). At each γ interface, there is an ATP-binding site [64]. Such a configuration suggests that upon ATP binding, the γ complex undergoes a conformational change from a closed state (without ATP) to an open state (with ATP) [65]. The δ subunit is then free to interact with the β clamp.

The δ′ subunit, the 33-kDa member of the γ complex encoded by the *holB* gene, acts as a mediator between γ and δ [1]. Although sequence and structure alignments of δ′ and γ suggest that these two subunits are homologous, δ′ does not have a functional nucleotide-binding domain, as shown in the crystal structure of the δ′ subunit [66]. This crystal structure shows that δ′, like γ, consists of three consecutive domains organized in a C-shaped architecture. The first domain consists of a β sheet with five parallel strands surrounded by six α-helices similar to the nucleotide-binding domain of RecA [66, 67]. The δ′ subunit contains a zinc-binding module whose function is unknown but because it is found on what resembles a phosphate-binding loop, it may help couple DNA binding with ATP hydrolysis by the clamp loader [66].

The δ subunit, a 35-kDa product of the *holA* gene, is considered the "wrench" of the clamp loader complex because its binding to the β clamp causes a spring-like conformational change [60]. This allows the dimeric ring of the β clamp to transition from its default closed state, where both dimeric interfaces are intact, to an

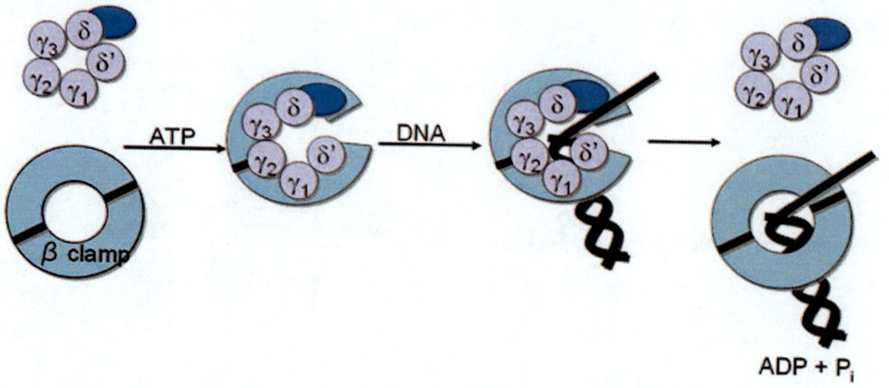

Fig. 11.3 The process of loading the β clamp onto the primer-template DNA duplex. ATP binds the γ complex, causing a conformational change from a closed state to an open state, which allows the N-terminal domain of the δ subunit (*dark blue*) to bind to the β clamp. This interaction disrupts the dimer interface of the β clamp, creating an opening for the primer-template DNA duplex to enter. The bound ATP is then hydrolyzed, causing the γ complex to relax back to its closed state. The δ subunit then releases the β clamp, re-establishing the dimer interface

open state, where only one dimeric interface exists (Fig. 11.3) [60, 68–70]. This spring-like mechanism facilitates loading of the β clamp onto the primer-template DNA duplex [2, 9, 11, 60]. The "wrench"-like interaction between δ and the β clamp is shown in a crystal structure involving the N-terminal 140 residues of δ and a variant form of the β clamp that cannot dimerize [60]. Attempts to crystallize δ with the full-length β clamp dimer have been unsuccessful [60]. In fact, δ binds to the monomer form approximately 50-fold more tightly than to the dimer. The binding interface between these two subunits is contained in a hydrophobic tip of δ and a hydrophobic pocket on the surface of the β clamp that is also known for binding other components of the replisome [60, 71, 72] and DNA polymerases [73–76]. When the structure of the monomer form of the β clamp without δ was compared to that with δ, a distortion of the curvature of the β clamp that results in a ~15 Å opening can be seen [60]. Such an opening is enough to allow ssDNA into the center of the clamp [59, 60].

The crystal structures of the clamp loader subunits suggest a multistep process for loading the β clamp onto a primed DNA strand [9, 11, 59], powered by the binding of two ATP molecules [77] (Fig. 11.3). When ATP binds to the closed ring-shaped γ complex, a conformational change takes place disrupting the interaction between δ and δ′. The δ subunit is then free to bind to the hydrophobic pocket of the β clamp dimer. This binding event distorts the β dimer interface allowing DNA to enter the clamp, creating an opened ring-like structure consisting of the opened γ complex and clamp. Site-directed mutagenesis [78], electron microscopy [79], and X-ray crystallography [80] experiments have shown that DNA can bind to the center chamber of the complex. Once ATP is hydrolyzed, the dimer interface of the β clamp is restored, allowing the γ complex to relax back to its closed state. This causes the clamp loader to release its hold on the β clamp, the rate-limiting step of the clamp loading process [81]. The β clamp and primer-template DNA duplex are then competent for polymerase loading [82].

The recent crystal structure that shows the clamp loader complex coupled to DNA [80] supports the notched screw-cap model for clamp loading as described above. The structure shows that the complex forms a right-handed spiral-like structure and is loaded onto dsDNA like a cap, allowing ssDNA to exit through a slit formed by the complex. In this structure, the γ complex does not recognize DNA via both strands of the primer–template complex as previously thought [79, 83]. Rather, recognition occurs on the phosphate backbone of the template strand alone and at just the 3′ nucleotide of the primer strand [80]. This allows for both DNA and RNA primers to be recognized, the mechanism of which was previously unclear [80, 84].

11.4.2 The χ and ψ Subunits

The χ and ψ subunits, products of the *holC* and *holD* genes, respectively, are subunits of the clamp loader complex [2, 9, 11, 85]. These two subunits function during replication on the lagging strand [22, 23]. Because DNA polymerases replicate DNA only in the 5′ to 3′ direction, the lagging strand must be replicated in a direction opposite to the movement of the replication fork. This is accomplished by replicating DNA in ~1 kb fragments, called Okazaki fragments [86]. As a result, an abundance of ssDNA is present and coated with SSB (Fig. 11.1). The χ and ψ subunits help to coordinate replication on the lagging strand [22, 23]. The ψ subunit acts as a mediator between the γ complex and χ by binding the collar domains (Domain III) of the τ and γ subunits [80, 87]. The χ subunit binds to the C-terminal domain of SSB [23], thereby coupling it to the replisome and allowing the clamp loader complex to be in close proximity to the primer-template DNA on the lagging strand [88]. Together, the χ and ψ subunits constitute a tightly held complex that increases the affinity of τ and γ for δ and δ′ [87]. The ψ subunit also serves to increase affinity of the clamp loader for the β clamp in the presence of ATPγS [24].

11.4.3 The β Clamp Subunit

Once it is loaded onto the primer-template DNA duplex, the β clamp (*dnaN*) has two specific roles. It tethers the polymerase to the DNA and contributes to the mobility of the polymerase on the DNA strand. The β clamp, a ring-shaped homodimer (Fig. 11.4) [89, 90], is the major contributor to processivity [1], by allowing the polymerase to maintain close contact with the DNA. In order to facilitate processive DNA synthesis, the clamp must remain bound to the DNA with or

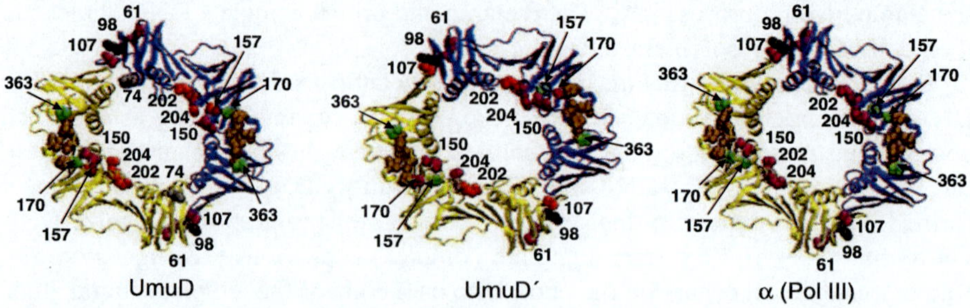

Fig. 11.4 Residue substitutions in the β clamp that are implicated in interactions with UmuD (*left*), UmuD′ (*middle*), and α subunit of Pol III (*right*) [98]. Positions in *green* are important to the interaction between the β clamp and all three proteins listed above. Positions in *purple* exhibit only a modest effect. Substitutions that result in an increase or decrease in the affinity of UmuD and UmuD′ for the β clamp by formaldehyde or glutaraldehyde cross-linking are shown in *red*. Residue Lys74 shown in *gray* (left) cross-links to UmuD using formaldehyde. The hydrophobic channel is shown in *brown* (residues Leu177, Pro242, Val247, Val360, and Met362) [60, 103], while the rim interaction residue Leu98 is shown in *black*. Structures were generated using VMD [379] and coordinates for β (2POL) from the PDB [89]

without the presence of polymerase, which has been shown [91, 92]. Other studies show that the clamp can remain on a circular plasmid two to three times longer than the time it takes for cells to divide [93]. With the use of single molecule fluorescence spectroscopy, it was found that the diffusion constant for the β clamp is at least three orders of magnitude lower than for diffusion through water. The β clamp seems to be held at the 3′-end of the primer in the presence of SSB [94]. The relatively slow motion of the clamp may be due to the attractive interactions of positively charged residues on the inside of the clamp that come into contact with the negatively charged phosphate backbone of the DNA [94, 95].

Binding experiments of the clamp with DNA pol III α and other components suggest that the same hydrophobic region on the surface of β to which δ binds is also responsible for binding DNA pol III α and other DNA polymerases [60, 71, 73–76, 96–98]. It has also been suggested that the DNA pol III α C-terminus binds to the β clamp (see below) and binds τ. These observations suggest that the polymerases, δ, and τ compete with one another for binding to the β clamp, creating a mechanism for polymerase loading and switching on the β clamp [36, 71, 99, 100].

An internal binding site (α residues 920–924), rather than the 20 C-terminal residues, was shown to be responsible for the interaction between the β clamp and α [37]. Replacement of all residues in this internal binding site eliminated binding to the β clamp, but binding between the τ subunit and α was not affected. When an analogous set of mutations was made in the C-terminal binding site of α, the β clamp still showed affinity for the α subunit [37]. In fact, α participated in processive replication even when the entire C-terminal binding site was removed. The absence of the C-terminal peptide, however, specifically effected the interaction of τ with α. These findings suggest that the clamp does not bind to the C-terminus of α, but instead to an internal site. This discrepancy might be due to the use of an α variant with a relatively large C-terminal truncation in the previous study [36, 37], rather than site-directed mutant variants or more modest deletions [37].

There is also evidence that two different polymerases can simultaneously bind to the clamp. The β subunit, a homodimer, has two hydrophobic pockets per functional protein, thus allowing it to bind two DNA polymerases [11, 98, 101, 102]. Such a situation allows replication to alternate between the two DNA polymerases without the need for dissociation from the β clamp. This "tool-belt" hypothesis allows for high processivity, even under conditions where multiple polymerases are used. The ability of the clamp to bind both DNA pol III and pol IV (a Y family polymerase) was investigated using fluorescence resonance energy transfer (FRET). Proximity of the pol III α subunit and pol IV was detected in a β clamp-dependent manner, providing experimental evidence that two different DNA polymerase molecules can simultaneously bind to the β clamp [102]. The crystal structure of β with the C-terminal little finger domain of pol IV [103] showed that the pol IV polymerase domain is angled off to the side providing enough room for another polymerase, such as DNA pol III α, to bind. Modeling the full-length structure of Dpo4 [104] (a pol IV homolog from *Sulfolobus solfataricus*), containing a primer-template duplex and incoming nucleotide, onto that of the little finger structure of *E. coli* pol IV with the β clamp, showed that when bound in this position the polymerase likely does not have access to the DNA strand [103]. The little finger domain can likely undergo a conformational change [103], positioning the polymerase onto the DNA substrate.

Another model for polymerase switching is "dynamic processivity" [105]. This model involves polymerase replacement without affecting apparent overall processivity. Such a scheme was observed during bacteriophage T4 DNA replication [105]. The addition of a catalytically inactive variant D408N of gp43, the T4 DNA polymerase, to an active replication fork arrested replication while still retaining wild-type-like affinity for DNA and the clamp (gp45). This observation suggested that the active polymerase is quickly (<1 min) replaced by the inactive variant [105]. This dynamic processivity of polymerases implies that multiple replicative polymerases may be required to replicate normal, undamaged DNA.

11.4.4 The τ Subunit

Although similar to the γ subunit, the role of the τ subunit is distinct from the remainder of the γ complex. As a central component of the DNA pol III replisome, τ coordinates replication on both strands by connecting the subassemblies of the core and β clamp as τ binds to the polymerase [106] and the helicase DnaB [107, 108], and is part of the clamp loader complex [87]. Numerous distinct roles have been assigned to τ, as summarized in a review [9]. The roles of the τ subunit in replication are to: (1) coordinate replication on both strands;

(2) bind to DnaB; (3) prevent premature removal of the β clamp; and (4) function in the processivity switch.

When DNA polymerase III′ (core + τ) was first isolated [21], it was observed that two polymerases were coupled by two τ subunits, suggesting that τ may be a key factor in coordinating replication (Fig. 11.1). This hypothesis was tested by varying the concentration of τ [109]. At low concentrations of τ, shorter DNA fragments were observed upon agarose gel electrophoresis analysis of replication reaction samples [109]. These results suggest that the two polymerase cores must be coupled through τ in order to effectively replicate both the lagging and leading strands.

DNA pol III α binds τ through the α C-terminal domain [36, 37] with an equilibrium dissociation constant (K_d) of 4 nM [106]. The α binding site on τ was determined to be at the τ C-terminal domain [107], which is not part of γ (Fig. 11.2b). Along with determining the NMR solution structure for this domain [110], combinatorial binding studies were conducted to determine which residues bind to α [34]. It was concluded that the 18 C-terminal residues of τ are required for binding to α [34].

Four different reconstituted clamp loader complexes, $\gamma_3\delta\delta'\chi\psi$, $\tau_1\gamma_2\delta\delta'\chi\psi$, $\tau_2\gamma_1\delta\delta'\chi\psi$, and $\tau_3\delta\delta'\chi\psi$, have similar rates of loading the β clamp onto DNA [10]. The complex containing τ_3 may coordinate three DNA polymerases at the replication fork. A "triple-polymerase" model has been proposed in which two of the pol III cores function on the lagging strand to synthesize Okazaki fragments. Alternatively, one pol III core is utilized on the lagging strand with the third pol III core held "in reserve" off of the DNA [10]. The latter model suggests a possible switching mechanism between high- and low-fidelity DNA polymerases [10].

The affinity of the τ subunit for DnaB [107] also seems to affect the rate at which the replication fork proceeds [8, 111]. DnaB, a hexameric helicase, unwinds DNA ahead of the fork while encircling only the lagging strand (Fig. 11.1) [8]. Without the replisome, the helicase unwinds DNA at a rate of approximately 35 nt/s, similar to that in the presence of the complex without τ. When τ is added, the rate increases to at least 400 nt/s, approaching that of DNA pol III [107, 111].

The τ subunit may also indirectly prevent the premature removal of the β clamp by the γ complex, thereby maintaining processivity. The length of the DNA produced in the absence of τ is directly proportional to the concentration of β and inversely proportional to the concentration of the γ clamp loader complex [112]. This suggests that the clamp removal function is inhibited during normal replication in the presence of the γ complex, allowing the β clamp to stay on the DNA.

Discontinuous replication of the lagging strand requires that DNA pol III α must constantly dissociate from and reassociate with different β clamps. Such a cycle is known as the processivity switch and is thought to involve τ [100]. It has been shown that the polymerase core/τ complex must complete the newly formed strand, leaving only a nick, before the switch can be activated [100]. In such a scenario, τ loses affinity for α when primed DNA is present. Then when replication is completed, τ gains affinity for α, releasing α from the clamp [100]. The ssDNA-binding function of α also appears to play a role in this process [99].

11.5 DNA Damage Disrupts DNA Replication

DNA pol III function requires finely tuned interactions of multiple proteins for efficient and accurate DNA replication. When this complex encounters noncanonical DNA structures, including DNA damage, it is not equipped to replicate them. DNA damage is ubiquitous, arising from numerous exogenous and endogenous sources (Fig. 11.5). For example, it is estimated that 10,000 abasic sites are formed per human cell per day [113]. The outcome of the replisome encountering DNA damage in the template may depend on whether the damage is encountered during leading strand or lagging strand synthesis, but typically the rate of progress of replication is decreased [114, 115]. A lesion in the leading strand has been observed to slow progression of

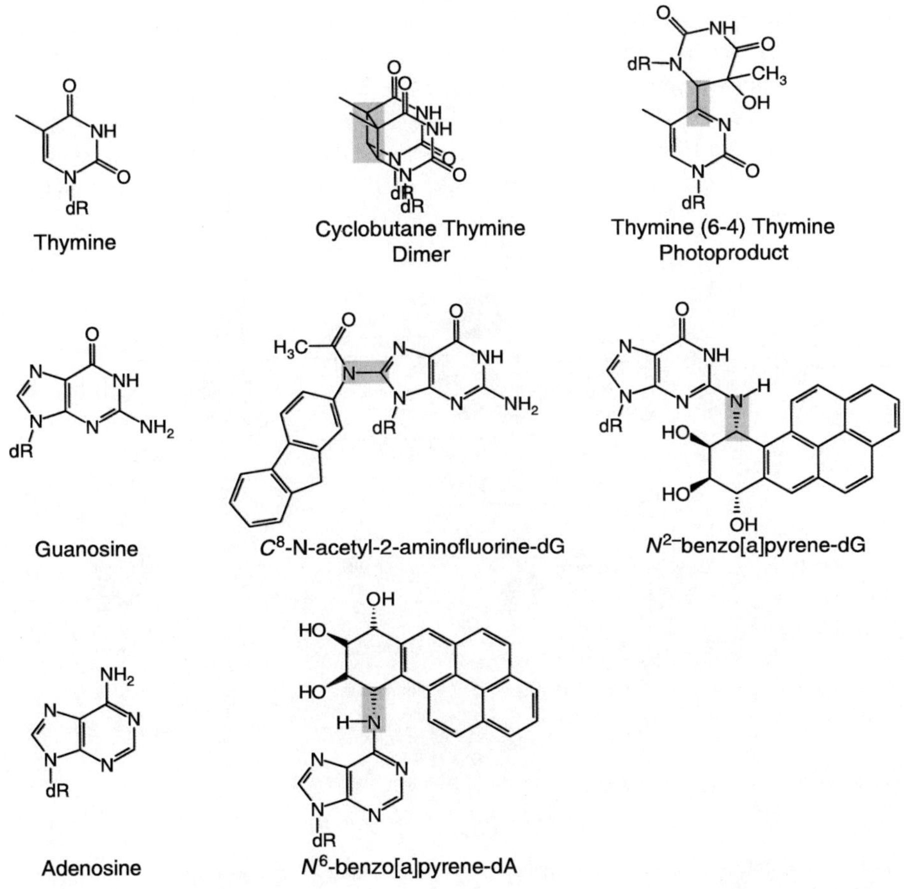

Fig. 11.5 Examples of DNA damage. Undamaged bases are shown on the *left*. Both types of thymine dimers shown result from exposure to UV light. N-2-acetylaminofluorine is a carcinogenic aromatic amine that forms C^8-dG and N^2-dG adducts (not shown). Benzo[*a*]pyrene is a carcinogenic hydrocarbon that is metabolized to form a 7,8-diol-9,10-epoxide derivative, which forms adducts with both dA and dG

the replication fork, but was not observed to interfere with lagging strand replication [116]. On the other hand, a lesion in the lagging strand does not block overall progression of the replication fork. Instead, the lagging strand DNA polymerase appears to reinitiate downstream of the lesion at the next Okazaki fragment, leaving a gap [116–119]. Indeed, it has been shown that replication can restart downstream of obstacles, even in the case of leading strand synthesis [120].

11.6 Specialized DNA Polymerases Facilitate DNA Damage Tolerance

In *E. coli* and some other bacteria, DNA damage and other stresses lead to induction of the SOS response [113]. Stalling of DNA replication at damaged sites results in the accumulation of single-stranded DNA (ssDNA). RecA polymerizes on the ssDNA, forming a nucleoprotein filament that serves as the inducing signal for the SOS response. As a result, the expression of at least 57 genes is induced [121, 122]. SOS-regulated genes code for proteins involved in the regulation of cell division, nonmutagenic repair of chemically modified DNA, or in damage tolerance mechanisms, which can be mutagenic or error prone (Fig. 11.6) [113, 121, 123, 124].

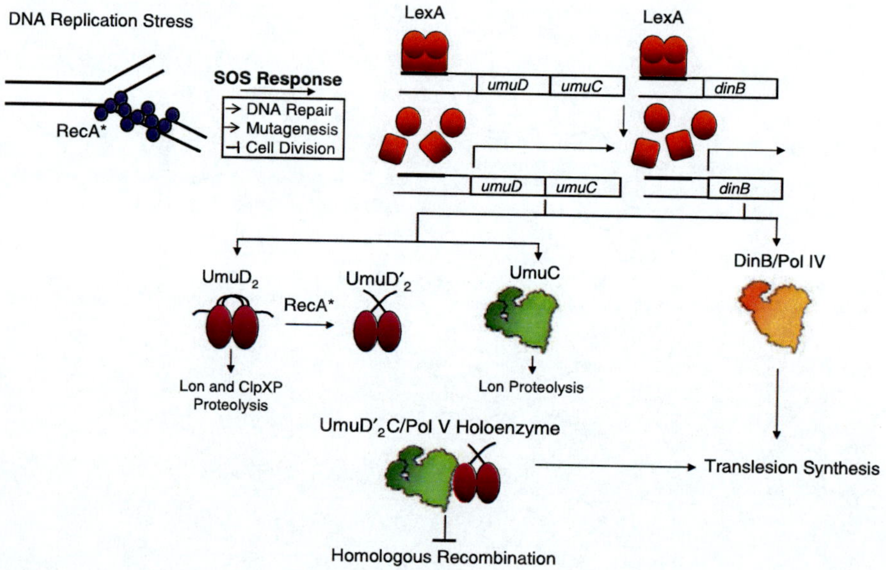

Fig. 11.6 Regulation of SOS-induced genes after DNA damage. The SOS response is induced by the formation of a RecA nucleoprotein filament on single-stranded DNA (RecA*). This stimulates the autoproteolysis of the LexA repressor which leads to the induction of at least 57 genes. Among these genes are Y family polymerases pol IV (DinB) and pol V (UmuD′$_2$C). UmuD$_2$ undergoes RecA* facilitated cleavage of its N-terminal 24 amino acids to yield UmuD′$_2$, the form that is active in translesion synthesis. Lon and ClpXP proteases play a role in regulating the levels of UmuD$_2$, UmuD′$_2$, and UmuC in the cell. ClpXP also specifically targets UmuD′ in UmuDD′ heterodimers (not shown)

Table 11.2 Error frequencies of selected TLS polymerases [370]

			Error frequency (×10^{-5})		
Origin	Polymerase[a]	Family	Single base sub	Single base in/del	References
Ec	DinB[a]	Y	5.1	21	[371]
Ec	$UmuD'_2C$[a]	Y	21	–	[372]
Sc	ζ[a]	B	130	4.4	[314]
Sc	η[a]	Y	950	87/93	[373]
Hs	η[a]	Y	3,500	130/240	[374]
Hs	κ[a]	Y	580	180	[296]
Hs	ι[b]	Y	72,000 (T.dGTP)	–	[375]

Ec *Escherichia coli*, Sc *Saccharomyces cerevisiae*, Hs *Homo sapiens*
[a]Values are average error rates obtained using gap-filling DNA synthesis assays [370]
[b]Determined using a 5-nt gap-filling assay, since pol ι does not fill long gaps [370, 375]

Three of the five known *E. coli* DNA polymerases are under SOS inducible regulation [123, 125]. These SOS inducible DNA polymerases are pol II (*polB*), pol IV (*dinB*), and pol V (*umuDC*; $UmuD'_2C$). The latter two belong to the Y family of DNA polymerases, which are characterized by their ability to perform translesion synthesis (TLS) on damaged DNA templates, as well as their relatively low fidelity on undamaged DNA (Table 11.2) [126]. Y family DNA polymerases also lack intrinsic 3′–5′ exonucleolytic proofreading and exhibit low processivity [123, 126, 127].

Although high-resolution structures of *E. coli* Y family polymerases have not yet been experimentally determined, the structures of Y family polymerases from *S. solfataricus* and *Sulfolobus acidocaldaricus* homologs provide insights into their function [104, 128, 129]. While there is no obvious sequence homology between replicative and Y family DNA polymerases, the crystal structures of the latter reveal a similar right-hand structure of the catalytic domain consisting of thumb, palm, and finger domains, common to other DNA polymerases. Another domain, the little finger domain, is present in the Y family polymerases, providing additional DNA-binding contacts in the major groove [104, 126]. This domain may be responsible for both substrate specificity and processivity [130]. Moreover, the O helix that is responsible for high fidelity in the replicative polymerases [131–133] is not present in Y family polymerases, suggesting a structural basis for the low fidelity of Y family DNA polymerases while replicating undamaged DNA. The specialized ability of Y family polymerases to replicate damaged DNA has been attributed to their loose, flexible active sites that accommodate aberrant DNA structures [104, 126, 129]. In addition, Y family polymerases have fewer contacts with their DNA substrates than replicative DNA polymerases [104, 129]. Although the crystal structures show that the catalytic domains of Y family polymerases have similar overall folds, these polymerases exert different efficiencies and fidelity in bypassing various DNA lesions [126, 134].

11.6.1 Specificity of *E. coli* Y Family DNA Polymerases

Escherichia coli Y family polymerases appear to possess specificity for certain lesions. Pol V ($UmuD'_2C$) is responsible for most of the mutagenesis induced by

UV radiation [113, 123, 124, 135–139]. Pol V bypasses the major products of UV radiation, including replication across *cis-syn* thymine-thymine cyclobutane pyrimidine dimers (Fig. 11.5). Accuracy is achieved by incorporating two dA nucleotides opposite the lesion. At thymine-thymine (6-4) photoproducts, insertion of dG opposite the 3′-thymine and dA opposite the 5′-thymine causes transition mutations [140]. Further, pol V can bypass abasic (AP) sites by inserting dA opposite the lesion and can bypass the *N*-2-acetylaminofluorene adduct at the C^8 position of guanine (C^8-dG-AAF) [101, 140–143]. Pol V copies N^2-benzo[a]pyrene-dG adducts with remarkable inaccuracy by favoring dA and dT incorporation over dC, and N^6-benzo[a]pyrene-dA adducts with relative accuracy and efficiency [144–146]. Pol V also copies a number of different mutagenic guanine oxidation products [147]. The biochemical requirements for pol V activity are still a matter of some debate. However, it is clear that RecA is required for activity and it seems that the β clamp and SSB stimulate pol V activity [148–156].

Pol IV (DinB) copies DNA containing bulky N^2-dG adducts, including N^2-furfuryl-dG and N^2-benzo[a]pyrene-dG [157, 158]. In fact, pol IV is approximately 15-fold more active in incorporation of dC opposite N^2-furfuryl-dG than incorporation of dC opposite unmodified dG [158]. Other N^2-dG adducts are substrates for pol IV, including those formed from methylglyoxal as well as DNA–DNA and DNA–peptide cross-links involving the N^2 position of dG [159–161]. C^8-dG-AAF is weakly bypassed by pol IV [162]. Pol IV can also bypass abasic sites *in vitro* [163]. Thus, while there is some overlap in the lesions that can be bypassed by the two *E. coli* Y family DNA polymerases, in general, their repertoires are distinct.

11.6.2 Regulation of Y Family DNA Polymerases

Expression of the *umuDC* and *dinB* gene products is negatively regulated by the LexA repressor as part of the SOS transcriptional response. LexA binds to a sequence in the operator region of the genes [113, 164, 165]. Derepression of the *umuDC* and *dinB* operons occurs when the RecA protein binds to single-stranded regions of DNA that develop at replication forks that are stalled by DNA damage [166]. The RecA/ssDNA nucleoprotein filament serves as a coprotease to facilitate cleavage of the LexA repressor (Fig. 11.6). As the cellular concentration of LexA diminishes, the genes whose expression is normally repressed by LexA are transcribed [113].

The *umuDC* genes are among the most tightly regulated SOS genes; the equilibrium dissociation constant (K_d) is 0.2 nM for LexA binding to the "SOS-box" in the promoter region [164]. In comparison, the K_d values for LexA binding to the "SOS-boxes" of the *recA* and *lexA* genes are estimated to be 2 nM and 20 nM, respectively [167]. Immunoblotting assays have shown the cellular steady-state levels of UmuD to be ~180 copies per uninduced cell and ~2,400 copies per cell under SOS induction [168]. A single protein in a compartment with the volume of a typical *E. coli* cell is present at a concentration of ~1 nM. The level of UmuC is approximately

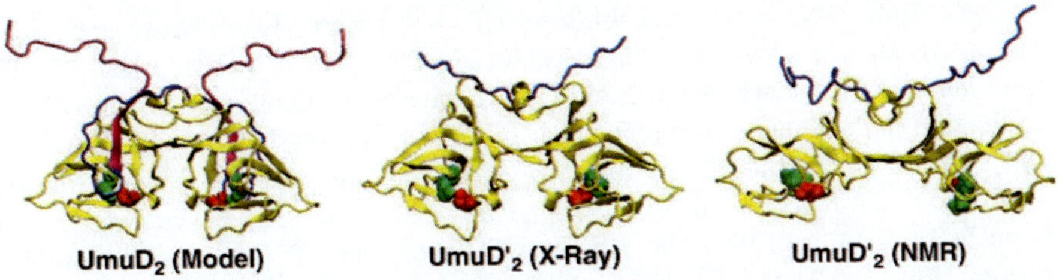

Fig. 11.7 Model of full-length UmuD [195], crystal [175], and NMR [174] structures of UmuD′. UmuD model shown in the *trans*, elbows down conformation (*left*). Crystal structure (*middle*) and NMR structure (*right*) of UmuD′ in *trans* conformation. The N-terminal arms of UmuD′ are cleaved between Cys24 and Gly25. Residues 1–24 are shown in *magenta*; residues 25–40 are shown in *blue*. Active site Ser60 and Lys97 are highlighted in *red* and *green*, respectively

12-fold lower than UmuD with about 15 molecules per cell in the absence of induction and ~200 molecules of UmuC per cell under SOS-induced conditions [168]. There are ~250 molecules of DinB per cell in the absence of induction and ~2,500 molecules of DinB per cell after treatment with the DNA damaging agent mitomycin C [169]. It should be noted that the *dinB* gene is also present on the *E. coli* F′ episome, and expression levels of DinB from the episome under uninduced and induced conditions are approximately threefold higher than from the chromosome [169]. Expression of the *umuDC* genes initially produces UmuD (139 amino acids) which undergoes a RecA/ssDNA-stimulated autodigestion reaction after induction resulting in UmuD′ (115 amino acids) (Figs. 11.6 and 11.7) [143, 170]. UmuD is the predominant species for the first approximately 20–40 min after SOS induction, after which UmuD′ is the predominant species [171]. UmuD proteins exist in solution as $UmuD_2$ and $UmuD'_2$ homodimers as well as the UmuD–UmuD′ heterodimer, which is more stable than either of the homodimers [172–175]. The K_d for $UmuD_2$ dimerization is estimated to be in the low-pM range, so UmuD is likely to be present in the cell as a dimer under most conditions [176].

SOS-induced mutagenesis is also regulated at the posttranslational level. UmuD is functionally inactive for facilitating TLS until it undergoes RecA/ssDNA-mediated cleavage to generate UmuD′ [170, 177, 178]. Full-length UmuD also inhibits −1 frameshift mutagenesis by DinB [179]. Efficient cleavage of UmuD *in vitro* and *in vivo* was observed at elevated levels of activated RecA, suggesting that TLS likely occurs when cells are under more severe environmental stress [168, 177]. The removal of the UmuD N-terminal 24 amino acids through the cleavage of the Cys24-Gly25 bond occurs via an intermolecular pathway, that is, one protomer of the dimer acts as an enzyme while the other is the substrate [180]. The crystal structure of $UmuD'_2$ revealed that the active site, consisting of conserved serine (Ser60) and lysine (Lys97) residues, is found at the end of a cleft within the C-terminal globular domain of the protein, and these residues are poised for cleavage (Fig. 11.7) [175]. In the NMR structure of $UmuD'_2$, Ser60 and Lys97 are further apart and not correctly oriented for catalysis. It has been suggested that the crystal structure of $UmuD_2$ mimics the effect of the RecA/ssDNA nucleoprotein filament as it serves to realign these residues, thereby activating $UmuD_2$ for self-cleavage [174, 181].

The two different forms of UmuD provide a temporal switch between accurate and mutagenic phases of the cellular response to DNA damage [171, 182, 183]. The combination of uncleaved UmuD and UmuC specifically decreases the rate of DNA replication and increases the resistance of cells to killing by UV radiation [171, 184]. Uncleaved $UmuD_2C$ improves DNA damage survival by allowing time for error-free repair mechanisms to act before the combination of cleaved $UmuD'_2$, and UmuC ($UmuD'_2C$, pol V) initiates potentially error-prone TLS [171]. The combination of UmuC and noncleavable UmuD(S60A) significantly delayed the recovery of cell growth after UV radiation [171, 183]. Therefore, a model for a *umuDC*-dependent DNA damage checkpoint in *E. coli* was proposed wherein a delay in DNA synthesis provides time for error-free nucleotide excision repair to remove DNA lesions. TLS is then enabled by the presence of UmuD′ [141, 150, 185]. This model suggests that the different *umuD* gene products, in combination with UmuC, are involved in distinct survival pathways after UV damage.

Increasing $UmuD'_2C$ protein complex concentration was found to antagonize RecA-mediated recombination of a UV-damaged gene [186]; this effect was also observed *in vitro* [187]. In the proposed model, high concentrations of the $UmuD'_2C$ proteins induce replisome switching from recombination to SOS mutagenesis. Indeed, the $UmuD'_2C$ complex has been shown to bind directly to the RecA/ssDNA filament and could disrupt the DNA pairing activity of RecA [188, 189]. Notably, in the presence of homologous DNA sequences, homologous recombination repair is more prevalent than TLS in responding to DNA damage [190].

The mutagenic potential of Y family polymerases may be further regulated by preferential formation of heterodimers between UmuD and UmuD′, thereby depleting the cell of mutagenically active UmuD′ homodimers [172]. UmuD′ is degraded by the ATP-dependent protease ClpXP while in a heterodimeric complex with UmuD [191, 192]. Formation of UmuDD′ heterodimers in preference to mutagenically active UmuD′ homodimers therefore specifically targets UmuD′ for proteolysis. UmuD also targets its UmuD homodimer partner for proteolytic degradation by ClpXP [193]. The ATP-dependent serine protease Lon is responsible for the degradation of both UmuD and UmuC proteins in vivo [194]. Targeted proteolysis of the *umuD* and *umuC* gene products is one mechanism for returning protein levels to their uninduced state.

11.6.3 Structural Dynamics of UmuD and UmuD′

The *umuD* gene products interact with multiple replication factors such as polymerases UmuC (as $UmuD'_2C$, pol V), DinB (pol IV), and components of the pol III holoenzyme. The latter include the polymerase subunit α, proofreading subunit ε, and the processivity clamp β [168, 176, 179, 195–197]. These interactions are due in part to the relative flexibility of full-length UmuD and its UmuD′ cleavage product as shown biochemically and by X-ray crystallography, NMR spectroscopy, and circular dichroism (CD) [174–176]. The cleaved form UmuD′ contains disordered

N-terminal arms that expose the C-terminal globular domain to solvent upon cleavage, while in full-length UmuD, the arms are more stably bound to the globular domain [174, 175]. Therefore, UmuD and UmuD′ make specific contacts that facilitate a variety of protein–protein interactions [174, 176, 196, 198].

The crystal structure of UmuD′ reveals extended N-terminal arms (residues 25–39) and a globular C-terminal body (residues 40–139) that contains the catalytic dyad Ser60 and Lys97 [175]. Although two dimer interfaces (designated as molecular and filament) were observed in the crystal structure, NMR and cross-linking experiments support the conclusion that the so-called filament dimer interface is the form present in solution (Fig. 11.7) [174, 175, 198–203]. However, there is evidence that the filament structure may be biologically relevant [199]. To date, there is no high-resolution structure of the $UmuD_2$ homodimer. Cross-linking studies of a series of single-cysteine derivatives of UmuD are consistent with the $UmuD_2$ homodimer interface resembling the interface of the $UmuD'_2$ homodimer, involving contacts between the C-termini of the monomers and intermolecular interactions between Asn41 and Leu44 of α-helix 1 [174, 203]. NMR experiments also suggest that the UmuDD′ heterodimer most closely resembles the $UmuD_2$ homodimer [174]. Dimerization appears to be important for biological activity, as UmuD′ variants that resulted in decreased UV-induced mutagenesis also have severe deficiencies in their abilities to form homodimers *in vivo* [200, 204, 205].

Four models of the UmuD homodimer have been generated from NMR, electron paramagnetic resonance (EPR), and cross-linking studies, and by homology to LexA [174, 195, 206, 207]. One model shows UmuD with the N-terminal arms in *trans* with the elbows down, where the N-terminal arm of one monomer folds down across the C-terminal body of the adjacent monomer and crosses the catalytic site (Fig. 11.7). Each UmuD monomer cleaves the N-terminal arm of its partner at Cys24-Gly25 [195]. A *trans*, elbows up model positions the arms along the outer edge of the globular domains. Two *cis* versions with elbows up and elbows down suggest that each N-terminal arm could bind over its own globular domain [195]. The N-terminal region (residues 1–14) is likely to be in a random extended conformation. Cross-linking and chemical modification experiments suggest that the *trans*, elbows down conformation of the N-terminal domain is the most prevalent in solution [195, 206]. However, given the dynamic nature of UmuD, all four conformations may be physiologically relevant [176, 195].

Circular dichroism (CD) spectroscopy simulating physiological conditions detected random coil conformations for both UmuD dimers [176], rather than the β-sheet-rich structure determined by X-ray crystallography and NMR spectroscopy [174, 175, 198]. At higher salt concentration both UmuD and UmuD′ dimers have more typical β-sheet appearance. Thus, the *umuD* gene products belong to the group of intrinsically disordered proteins (IDPs) [208]. Like their IDP counterparts, UmuD dimers are capable of making a remarkable number of specific protein–protein contacts.

It has been shown that UmuD and UmuD′ interact with the α, β, and ε subunits of DNA polymerase III [196]. UmuD interacts less strongly with the α subunit *in vitro* than cleaved UmuD′ [196]. Moreover, uncleaved UmuD interacts more strongly with the β clamp than cleaved UmuD′ as observed by affinity chromatography [196]. Different interactions of UmuD and UmuD′ with the β clamp suggest that these interactions regulate how the *umuD* gene products access the replication fork. Overexpression of the α subunit or the β clamp inhibits cleavage of UmuD to UmuD′ *in vivo*, further supporting the specific interactions between the pairs of proteins [196]. It has also been found that overexpression of the β clamp reduces UV-induced mutagenesis [209].

11.6.4 UmuD–β Clamp Interactions

To date, interactions between the *umuD* gene products and the β clamp have been studied in much more detail than other interactions involving the *umuD* gene products. Proteins that interact with the β clamp, with the exception of UmuD and UmuD′, contain the eubacterial clamp-binding motif (QL[S/D]LF) [210]. UmuD contains a ^{14}TFPLF^{18} sequence within its N-terminal arm [210]. Although the motif lies in a region of UmuD that is important for its interaction with the β clamp [197], the interaction does not depend on the sequence identity of the motif [195]. That is, a UmuD variant containing mutations in this motif binds to the β clamp with similar affinity as that of wild-type UmuD (see below) [195].

UmuD and UmuD′ affinity chromatography and *in vitro* cross-linking studies confirm that the β clamp has a higher affinity for UmuD than UmuD′ [197]. However, it has been shown that both the N-terminal arms and C-terminal globular domains of UmuD are important for interaction with the β clamp [197]. UmuD lacking its N-terminal nine residues is proficient for interactions with the β clamp, while UmuD lacking the N-terminal 19 residues results in reduced formaldehyde cross-linking to β [197]. The UmuD variant UmuD-3A (T14A L17A F18A), a noncleavable variant with mutations of the most conserved residues of the TFPLF motif, possesses some of the biological functions of the cleaved form UmuD′. Although the K_d values for interaction of the β clamp with UmuD (5.5 ± 0.8 μM) and UmuD-3A (6.1 ± 0.5 μM) are similar, their interactions with the β clamp may be different, as observed by intrinsic tryptophan fluorescence of the β clamp [195]. Tryptophan fluorescence is a relatively sensitive probe of the microenvironment. The single tryptophan of the β clamp is located on a flexible loop between Domains I and II of β, and thus is a sensitive reporter of conformational changes [89, 195]. The β clamp tryptophan fluorescence differed dramatically upon UmuD vs. UmuD-3A binding, suggesting they have different binding modes [195].

UmuD, UmuD′, the α catalytic subunit, UmuC, DinB, and clamp loader all interact with the β clamp around the β clamp hydrophobic pocket, approximately defined by residues Leu177, Pro242, Val247, Val360, and Met362 (Fig. 11.4) [64, 73, 76, 98, 103]. UmuD, UmuD′, and the α subunit interact with overlapping regions of β, suggesting that there may be competition for binding (Fig. 11.4) [196]. This implies that UmuD plays a regulatory role following the SOS response, interacting with the components of pol III, interfering with α binding to β, slowing replication, and allowing time for error-free repair mechanisms to act [171, 196]. It is also possible that α and UmuD or UmuD′ bind the homodimeric β clamp simultaneously. A model has been proposed in which the cleavage of UmuD to form UmuD′ reduces binding to the β clamp, thereby releasing the DNA damage checkpoint and enabling TLS [211].

11.7 DNA Polymerase Switching

Replicative DNA polymerases are unable to copy damaged DNA under most circumstances [212, 213]. In the polymerase-switching model [101, 102, 214, 215], a lesion blocks the progress of the replicative DNA polymerase at the replication fork, and translesion polymerases act to allow replication to proceed (Fig. 11.8). Since the SOS-inducible polymerases are characterized by low processivity and low-fidelity replication, specialized polymerases must subsequently be replaced by replicative polymerases to restore efficient and accurate DNA replication.

11.7.1 Replisome Dynamics

Elevated levels of UmuD and UmuC inhibit DNA replication, a phenomenon suggested to represent a primitive DNA damage checkpoint [171, 184]. DinB (pol IV) also interferes with replication fork progression both *in vitro* and *in vivo*, and its overexpression inhibits cell growth [215, 216]. Neither DinB catalytic activity nor its canonical β clamp-binding motif is required for the inhibition of replication [216]. Both pol II and DinB inhibit replication fork progression by altering the speed of the DnaB helicase [217]. In this case, the β clamp-binding motif is required for DinB to occupy the replisome and slow the helicase [217]. This decrease in helicase velocity allows time for accurate DNA repair processes to take place while maintaining the overall structure of the replication fork [217]. Taken together, these observations suggest a high degree of plasticity in the replisome as well as parallels between the physiological effects of pol IV and pol V on the replication fork.

The β clamp binds to, and increases the processivity of, all five known *E. coli* DNA polymerases [1, 96, 218] and is important for the lesion-bypass activity of $UmuD'_2C$ [73, 74, 76]. Both UmuD and UmuD′ interact with the β clamp and the α and ε subunits of pol III [196]. In general, the β clamp-binding proteins possess a canonical peptide motif that binds to the hydrophobic channel on the β clamp [97, 210], suggesting that they compete for the same binding site. Extensive analysis of DNA polymerase binding to site-directed mutants of the β clamp shows that the different polymerases have partly, but not completely, overlapping sites of interaction on the β clamp [219–221]. Based on the co-crystal structure of the little finger domain of DinB with the β clamp, a physical basis for recruitment of Y family polymerases to the replisome was proposed [103]. Two interfaces between the DinB little finger domain and the β clamp

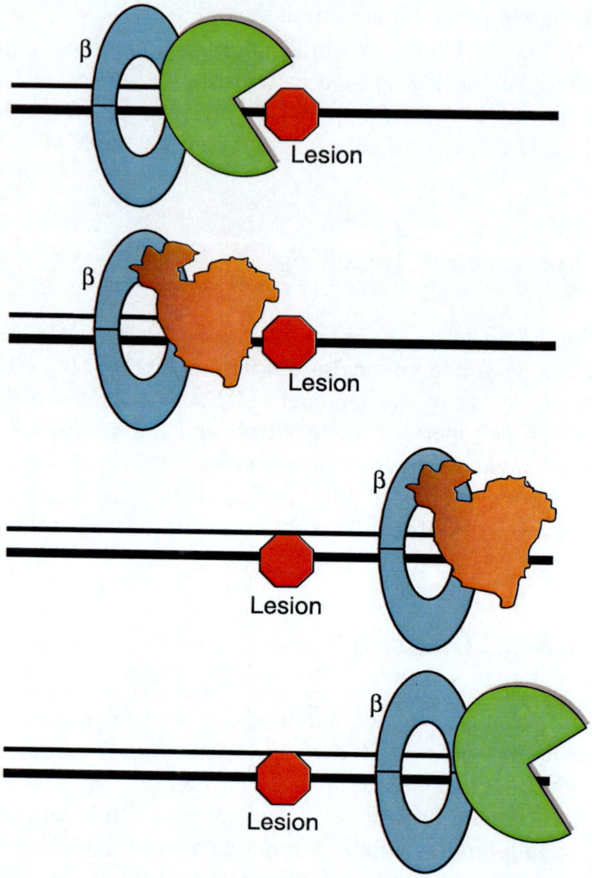

Fig. 11.8 Polymerase switching in response to DNA damage. Replicative DNA polymerases (*green*) are generally unable to copy damaged DNA. A polymerase switch occurs allowing a TLS polymerase (*orange*) access to DNA. The TLS polymerase synthesizes DNA opposite the lesion and far enough beyond it that the replicative polymerase can resume synthesis without disruption due to the lesion. In eukaryotes, there is evidence that two different TLS polymerases act in a stepwise fashion to insert nucleotides opposite the lesion and then extend the primer beyond the lesion [345]

were identified. One is a peptide–protein interaction involving the extended C-terminal tail of the DinB little finger domain and hydrophobic channel, or "cleft," on the surface of the β clamp [103]. A second interaction was observed between surface loops of the DinB little finger (residues 303–305) and the outer rim of the β clamp at the dimer interface (residue 98) [103]. Modeling of the entire *S. solfataricus* DinB-ortholog Dpo4 protein onto the structure of the DinB little finger domain showed that, when bound in this position, DinB would likely not have access to a linear DNA duplex [103]. Therefore, DinB could have two modes of binding to the β clamp: a "locked-down" conformation with no access to DNA, and a "tethered" complex attached to the β clamp via the C-terminal canonical β clamp-binding motif, which would allow interactions with DNA [103]. By analogy, mutations that disrupt the proposed β clamp outer rim interaction of UmuC result in increased UV-induced mutagenesis, presumably because the "lock" is released, giving UmuC more access to DNA [73]. Recent crystal structures of Dpo4 show that it also adopts multiple conformations whether in the apo form, bound to DNA or bound to the sliding clamp [222]. Specifically, the little finger rotates around the linker region and adopts different conformations depending on the presence of different binding partners of Dpo4. While the Dpo4 little finger domain makes additional contacts with the clamp beyond the canonical interaction motif (PIP-box), unlike DinB the Dpo4 little finger domain lacks interactions with the outer rim of the clamp.

11.7.2 Polymerase Switching: Tool-Belt and Active Exchange Models

The multiple conformations and flexibility of polymerase binding to the β clamp support the "tool-belt" model of polymerase switching [73, 76, 102, 103, 214], in which several polymerases simultaneously attach to the clamp. Depending on the DNA substrate, either a replicative or a translesion polymerase is given access to the DNA. The proposed triple-polymerase replisome provides another possible mechanism for efficient switching between high- and low-fidelity polymerases [10]. The observed tilt angle of the β clamp on DNA may also facilitate polymerase switching [95]. It was recently reported that only one of the two possible hydrophobic channels that are sites of polymerase interaction on the β clamp is required for switching between pol III and DinB [223]. This result challenges the "tool-belt" model [223]. Other studies suggest an active exchange model in which one polymerase readily displaces another at the replication fork [105].

DNA replication assays reveal the exchange of the replicative pol III and TLS pol V on DNA in the presence of DNA damage [101, 224]. The pol III holoenzyme replicates up to the nucleotide position preceding the damaged base, but not beyond it. Through contacts between their minor groove recognition domain and the DNA, replicative DNA polymerases sense distortions in the DNA within the last four to five nucleotides replicated [225–227]. In agreement with these structural studies of other replicative polymerases, it has been experimentally shown that pol III requires the primer to be four to five nucleotides beyond the lesion for resumption of efficient replication [101]. If the primer is not extended sufficiently beyond the lesion and pol III has access to the DNA, the proofreading subunit ε will degrade the primer [101]. In the presence of DNA damage, the SOS response is induced and RecA coats ssDNA. Pol V binds to RecA and the β clamp, and inserts nucleotides opposite the lesion. The less-constrained active site of pol V permits insertion and extension of the 3′ terminus of the primer beyond the lesion. Pol V synthesizes a "TLS patch" ranging from 1 to 60 nucleotides, averaging 20 nucleotides in a single binding event [101, 224]. Once pol V synthesizes a TLS patch of at least four nucleotides, the pol III holoenzyme is able to efficiently resume DNA replication even in the presence of a RecA/ssDNA nucleoprotein filament [101].

There is growing genetic evidence that multiple polymerases act during replication and in lesion bypass [142, 228–231]. It was hypothesized that SOS-inducible polymerases II, IV, and V alter the fidelity of DNA replication by interfering with the cellular functions of pol I and III. This was tested by exploiting the observation that each polymerase leaves a unique mutagenic "fingerprint" when copying DNA [228]. The spectrum of mutations in the *rpoB* gene arising in various Δ*mutL* strains was measured [228]. Deletion of *mutL* provides a genetic background deficient in postreplicative mismatch repair to allow determination of actual polymerase-specific misincorporation events, rather than simply assaying mutations that escape repair. Furthermore, polymerase expression levels were modulated using strains with mutations in *recA* and *lexA* [228]. Analysis of the *rpoB* genes in strains with deletions of different combinations of polymerases revealed changes in the spectrum of mutations. This result was interpreted as evidence that *E. coli* DNA polymerases compete for access to genomic DNA, thereby promoting or suppressing mutagenesis [228]. Specifically, multiple spontaneous transversion mutations are due to pol V or to the combined action of pol IV and pol V [228]. Moreover, the chromosomally encoded levels of pol V are the limiting factor for the production of the mutations, and even modest low-level expression causes an increase in transversion at specific hot spots. Facile switching between the five DNA polymerases of *E. coli* during genome replication suggests that spontaneous mutations arise due to the interactions of multiple polymerases [228].

Additional evidence for switching between pol III and pol V on UV-damaged DNA came from a study revealing that strains harboring variants of *E. coli* pol V with mutations of F10 or Y11 (the steric gate residue) display hypersensitivity to UV light and dominant negative genetics [231]. The steric gate residue has been shown to prevent incorporation of ribonucleotides by sterically occluding the 2′-OH moiety in the Y family polymerases [232, 233]. The dominant negative character of *umuDC Y11A* is suppressed by disruption of *dnaQ*, the gene encoding the ε proofreading subunit (Table 11.1). It seems likely that pol V Y11A is recruited to damaged DNA and prevents other DNA polymerases from accessing the replication fork. The steric gate variants of pol V (Y11A) and pol IV (F13V) conferred on a wild-type *E. coli* strain extreme sensitivity to both nitrofurazone and UV radiation [158, 231]. Considering that pol IV copies N^2-furfuryl-dG lesions efficiently and accurately [158], this finding suggest that the pol V steric gate variant prevents another Y family polymerase from accessing damaged DNA.

11.7.3 Gap-Filling Model

Several studies have suggested that lesion bypass does not occur at the replication fork, but at gaps that are left by the replication machinery [119]. This model is supported by the observation that the rate of replication recovery after DNA damage is independent of the presence of pol IV or pol V [234, 235]. Moreover, as discussed above, lesions in the lagging strand do not slow the overall rate of replication, but lead to re-initiation downstream of the lesion, leaving a gap [116–118]. Finally, it has been observed that β clamps can be left behind when the polymerase dissociates from DNA [92], possibly as an aid to recruitment of other replication factors, including Y family polymerases. The observation that Rev1, a Y family polymerase in yeast, is upregulated in the G_2-M transition of the cell cycle, rather than in S phase when most DNA replication occurs, is also consistent with this model [236]. Ultimately, the two models of coordinated polymerase switching and gap filling are not mutually exclusive. The specific mechanism a particular cell utilizes likely depends on the type of lesion encountered and the conditions in the cell at that time.

11.8 Eukaryotic DNA Polymerases at the Replication Fork

DNA replication in eukaryotes is substantially more complicated than prokaryotic replication. This may be related to DNA packaging into chromatin, the larger size of eukaryotic genomes, the requirement for numerous origins of replication, and the need to coordinate replication with the cell cycle. It is estimated that there are up to 100 proteins present at a single replication fork in eukaryotic cells [237]. However, many of the basic mechanisms and processes of replication, and even the overall structures of some of the proteins involved, are widely conserved. For example, processivity clamps from throughout evolution adopt similar structures, and prokaryotic and eukaryotic clamp loading share similarities. One of the most striking and fundamental differences in replication between prokaryotes and eukaryotes is the presence in eukaryotes of two different replicative DNA polymerases and a primase-polymerase that synthesizes both DNA and RNA at the replication fork. Moreover, each of these DNA polymerases is a multisubunit complex (Table 11.3) [238–240]. Eukaryotic chromosomal DNA replication requires the coordinated efforts of three B family DNA polymerases: pol α, pol δ, and pol ε (Fig. 11.9) [240, 241]. The proposed model of the replication fork assigns the synthesis of the leading strand to the highly processive pol ε [238, 239]. Pol α, the only polymerase with primase activity, initiates DNA synthesis at the origin of replication and synthesizes short RNA–DNA fragments later extended by pol δ on the lagging strand [238, 239]. Pol δ also completes lagging strand synthesis after the removal of RNA primers [238, 242, 243].

In budding yeast, DNA pol α, ε, and δ contain catalytic subunits denoted Pol 1, Pol 2, and Pol 3, respectively (Table 11.3) [240]. The size of the catalytic subunit varies because of additional N-terminal or C-terminal extensions whose structure and function remain largely unknown [244]. Polymerases ε and δ possess the conserved 3′–5′ exonuclease domains, while pol α lacks exonuclease activity [238, 240, 244]. B family polymerases contain nonfunctional uracil-recognition domains as well as two Zn-finger domains located at the C-terminal ends of the polymerases that are essential for the assembly of the holoenzyme [245–248].

Table 11.3 DNA polymerases at the *S. cerevisiae* replication fork

DNA polymerase	Genes and subunit sizes	Activity	Fidelity	Function	References
Pol α-primase	Pol1-p167	Polymerase	10^{-4} to 10^{-5}	Replication initiation	[376]
	Pol12-p79				
	Pri1-p48	Primase		Okazaki fragment initiation	
	Pri2-p62				
Pol δ	Pol3-p125	Polymerase 3′-exonuclease	10^{-6} to 10^{-7}	Elongation and maturation of Okazaki fragment	[377]
	Pol31-p55				
	Pol32 p40				
Pol ε	Pol2-p256	Polymerase 3′-exonuclease	10^{-6} to 10^{-7}	Leading strand synthesis and replisome assembly	[378]
	Dpb2-p78				
	Dpb3-p23	DNA binding			
	Dpb4-p22				

The large subunits contain polymerase and 3′-exonuclease activity with the exception of Pol α. For details, see [251]

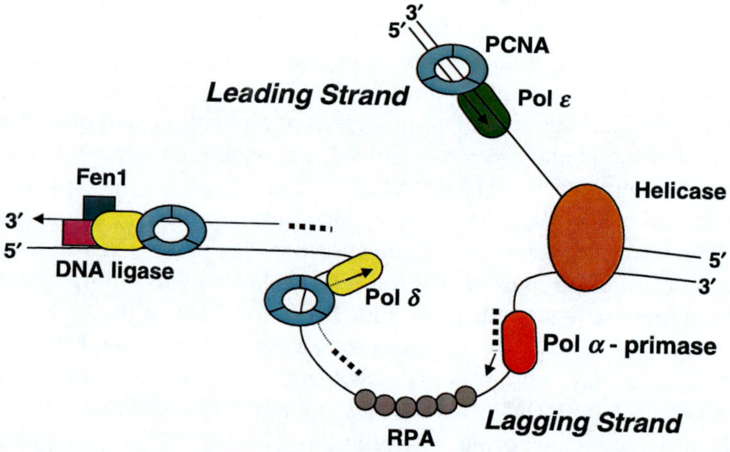

Fig. 11.9 Simplified model of the eukaryotic replication fork. In this three-polymerase model, Pol α-primase (*red*) synthesizes RNA–DNA fragments (*dots*) on the lagging strand. Pol ε (*green*) synthesizes DNA on the leading stand and Pol δ (*yellow*) extends the DNA–RNA fragments on the lagging strand. Also shown is the helicase (*orange*), replication protein A (RPA; *gray*), PCNA (*light blue*), Fen1 (*teal*), and DNA ligase (*pink*). Fen1 and DNA ligase are involved in maturation of Okazaki fragments

Pol α is a four subunit complex, the largest being the catalytic subunit that displays low processivity and lower fidelity in comparison to pol ε and pol δ (Table 11.3) [249–251]. The crystal structure of yeast pol α fragments along with electron microscopy images suggest that the catalytic subunit is connected to the C-terminal Zn-finger domains by a flexible linker [247]. The smallest of the four subunits is the catalytic primase subunit. This catalytic primase subunit is tightly associated with an accessory primase subunit. The second largest subunit has an iron-sulfur domain that is also essential for the priming reaction and for association with the origin recognition complex [252, 253]. Because pol α lacks exonucleolytic proofreading activity, the short DNA fragments that are synthesized by this polymerase are mutational hotspots if the errors are not corrected. Genetic studies suggest that pol δ can proofread and correct errors made by pol α [238, 254]. This evidence also supports the three-polymerase model of the replication fork that assigns pol δ to the lagging strand and pol ε to the leading strand [238].

Human and *Schizosaccharomyces pombe* pol δ are composed of four subunits, whereas *Saccharomyces cerevisiae* pol δ has three subunits [240]. The catalytic subunit is the largest and contains 3′–5′ exonuclease activity, a PCNA-binding motif and other protein–protein interaction sites [248, 251, 255]. The catalytic subunit is stabilized by the second largest subunit [255]. The third largest subunit encoded by *Pol32* contains a conserved PCNA-binding motif and a motif that interacts with pol α [256–258]. *Pol32* is essential for UV-induced mutagenesis in yeast, which is also the case for the TLS polymerases encoded by the *Rev1* and *Rev3* genes [259–261]. Pol 32 interacts with Rev1 and subsequently recruits pol ζ to DNA damage by means of this interaction [261]. This evidence suggests that Pol 32 may play a role in regulating TLS [240, 260, 262]. Although pol δ is primarily responsible for lagging strand synthesis, there is evidence that it can replicate the leading strand in the absence of pol ε. The N-terminal region of the Pol 2 subunit of pol ε, containing its polymerase and exonuclease functions, is nonessential for DNA replication, repair, and cell viability [263].

The catalytic subunit is the largest of the four subunits that comprise the pol ε complex [238, 240]. This subunit displays high accuracy because it also contains proofreading exonuclease activity [240]. Leading strand synthesis requires that the polymerase be highly processive, although processivity of pol ε is similar to that of pol δ [243]. The PCNA-binding site, which is dispensable for replication but contributes to DNA damage tolerance, is located in the large catalytic subunit [264]. The second largest subunit is essential and mediates interactions within the holoenzyme [265]. Mutations in the Dpb2 subunit weaken the affinity of Dpb2 for the Pol2 subunit resulting in a reduction of high-fidelity synthesis in yeast [266]. The third and fourth subunits are nonessential for growth in yeast and have been shown to bind double-stranded DNA [240]. The fourth subunit has also been implicated in chromatin remodeling [267]. Overall, genetic and biochemical analysis places pol ε at origins of replication where it is additionally regulated by a host of accessory factors [265, 268–270].

11.8.1 Damage Bypass by TLS Polymerases

Five TLS polymerases have been identified in eukaryotes including pol η, pol ι, pol κ, and Rev1 (Y family) and Pol ζ (B family) [262, 271, 272]. As in the cases of their bacterial counterparts, these enzymes lack intrinsic 3′–5′ proofreading exonuclease activity and contain relatively nonrestrictive active sites that contribute to their lesion bypass ability and low fidelity (Table 11.2). The Y family polymerases also possess low processivity [273].

Mutations in the human *RAD30A/POLH* gene, which encodes pol η, cause a variant form of Xeroderma pigmentosum (XP-V) that leads to increased susceptibility to UV-induced DNA damage and consequently skin cancer [274, 275]. Pol η was the first DNA polymerase shown to act as a tumor suppressor in humans [276]. Human pol η has low processivity on undamaged DNA, but primer extension assays indicate that it displays higher processivity in replicating across *cis-syn* thymine-thymine (T-T) dimers (Fig. 11.5) than undamaged DNA [277]. Pol η then switches to less processive synthesis 1–2 nucleotides after damage bypass, perhaps facilitating the transition to pol δ or another polymerase [277]. In spite of the distortions caused by the cyclobutane ring, Pol η incorporates nucleotides in an error-free manner opposite *cis-syn* T-T dimers, but it misincorporates nucleotides at a frequency of 10^{-2} to10^{-3} on undamaged templates (Table 11.2) [274, 278]. The low fidelity and low processivity of pol η on undamaged DNA and the unusual affinity for accurately bypassing T-T dimers suggest that pol η is specialized for this type of damage [275, 277–280].

Pol ι is encoded by the *RAD30B/POLI* gene in humans [281, 282]. Both pol η and pol ι have been shown to interact as they localize in replication factories [283]. Pol ι is the only polymerase known to incorporate nucleotides opposite minor groove adducts through Hoogsteen base pairing [284–286]. Pol ι is highly error-prone when replicating undamaged DNA (Table 11.2) [284]. It also has very low processivity and cannot extend past the single nucleotide incorporated opposite a DNA lesion [287]. Pol ι is particularly inefficient at adding the correct base opposite T; it preferentially inserts G rather than the correct base A [288].

Pol κ, a DinB homolog, is present in the yeast *S. pombe* and vertebrates, but appears to be absent from *S. cerevisiae* [272]. Unlike pol ι and pol η that colocalize with PCNA in replication factories during S phase, pol κ is found in replication foci in only a small fraction of untreated cells or in cells treated with hydroxyurea, UV irradiation, or benzo[*a*]pyrene (B[*a*]P) (see Fig. 11.5) [289]. Mouse embryonic stem (ES) cells that are deficient in pol κ are only minimally affected by UV and X-ray radiation, but are extremely sensitive to B[*a*]P [290]. Pol κ bypasses bulky adducts, such as N^2-B[*a*]P-dG, with higher efficiency than pol ι or pol η by correctly inserting C opposite adducted G [158, 290–292]. Pol κ aids in DNA lesion bypass events by efficiently extending from nucleotides inserted by another DNA polymerase opposite the lesion site [293, 294]. Sequential action of pol δ and pol κ is necessary for efficient replication through O^6-methyl-dG lesions with pol δ efficiently inserting a T or C residue opposite the lesion and pol κ acting at the extension step [293]. In addition, pol κ bypasses abasic sites and acetylaminofluorene-adducted guanine in an error-prone manner, and is unable to copy DNA containing a thymine-thymine (6-4) dimer or a cisplatin-adduct [295, 296].

Rev1 from budding yeast is a deoxycytidyl transferase that inserts dCMP efficiently opposite template G and A as well as opposite abasic sites *in vivo* and *in vitro* [297]. Interestingly, Rev1 incorporates the incoming dCMP by using a protein "template" Arg residue instead of template G [298]. The G is expelled from the DNA helix ensuring nucleotide selection in a DNA template-independent manner [298]. Rev1 is essential for mutagenesis resulting from TLS through abasic sites or other DNA damage [299–301]. However, Rev1 catalytic activity is not required for many lesion bypass events, suggesting that Rev1 plays a structural role in this process [302]. This structural role is facilitated by the BRCA1 C-terminal (BRCT) domain, which is necessary for the targeting of Rev1 to replication foci, and the little finger domain (also called the polymerase-associated domain, or PAD) which is known to be essential in these polymerases for DNA synthetic activity [303, 304]. Rev1 interacts with Pol η, Pol κ, and Pol ι, which is also suggestive of its function as a scaffold during TLS [305–308].

Pol ζ is a B family polymerase composed of the Rev3 catalytic subunit and the Rev7 regulatory subunit [309]. Pol ζ efficiently extends from nucleotides inserted opposite an abasic site (AP), particularly A. It is the combined action of pol δ which inserts A opposite the AP site, and extension by pol ζ that facilitates the bypass of this lesion [310]. Insertion of A maintains the structural stability of the double helix as A·AP, which is geometrically and thermodynamically similar to an A·T base pair [311, 312]. Pol ζ is critical for damage-induced mutagenesis in yeast, but disruption of murine Rev3 is embryonic lethal suggesting that it may have more complex functions in higher eukaryotes [313]. Unlike other B family polymerases, pol ζ lacks intrinsic proofreading exonuclease activity but is more accurate in comparison to Y family polymerases (Table 11.2) [287, 314].

11.8.2 Regulation of TLS Polymerases by PCNA Modification

Given that TLS polymerases are potentially mutagenic, their access to DNA must be regulated, and the processivity clamp PCNA plays a major role in this process. PCNA is a ring-shaped homotrimeric complex that encircles DNA and operates as a scaffold. PCNA belongs to a family of sliding clamps, including the *E. coli* β clamp, that are structurally and functionally conserved in all domains of life [315, 316]. During replication, PCNA slides along DNA, tethering high-fidelity polymerases δ and ε to the DNA. PCNA monomers have two similar globular domains linked by a flexible interdomain connecting loop (IDCL, Fig. 11.10) [317]. The global ring structure is formed by the head-to-tail arrangement of the three monomers to form a pseudo-six-fold axis of symmetry [317]. The inner positively charged surface formed by α-helices interacts with DNA while the outer surface is composed of β sheets [316, 317]. Although the *E. coli* β clamp is a homodimer that shares little sequence conservation with homotrimeric PCNA, they display a high degree of overall structural similarity (Figs. 11.4 and 11.10) [89, 317].

PCNA is loaded onto DNA by the replication factor C (RFC), which, like the *E. coli* clamp loader, is an arc-shaped complex of structurally similar proteins [20, 315]. The RFC ATPase contains four similarly sized subunits p36, p37, p38, and p40 (Rfc2-5), and a large subunit pl40 (Rfc1). The RFC subunits are arranged so as to associate with PCNA like a screw cap [315], similar to the interaction of the *E. coli*

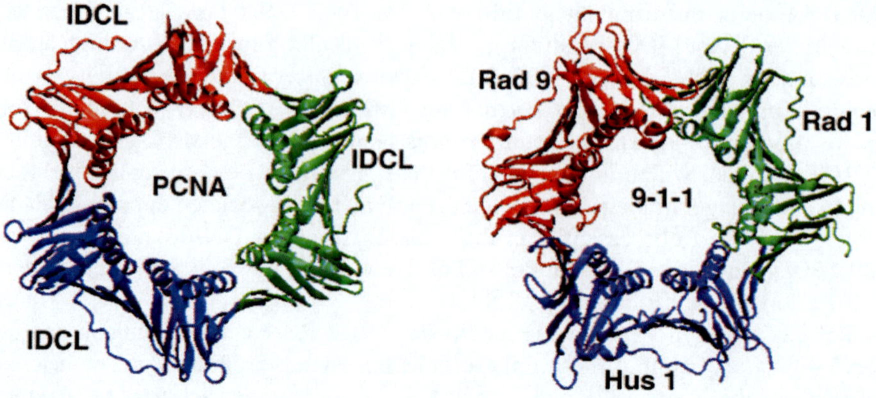

Fig. 11.10 Structural comparison of human homotrimeric PCNA and the 9-1-1 complex. Crystal structure of homotrimeric PCNA (PDB: 3IFV) and the Rad9-Hus1-Rad1 (PDB: 3A1J) complex. Both clamps contain a basic central channel through which DNA is passed. Each subunit is shown in *red*, *blue*, and *green*, respectively [352, 380, 381]. The interdomain connecting loops (IDCL) of PCNA are indicated

clamp loader with the β clamp [20, 60, 64, 80]. In an ATP-dependent process, the PCNA-RFC complex is loaded onto the 3′-end of the template-primer before DNA binding activates ATP hydrolysis leading to dissociation of RFC from the loaded clamp [315]. PCNA is loaded directionally, allowing the C side (the face from which the C-termini of the monomers project) and the polymerases that bind to it, to face the 3′-end of the nascent DNA [316]. The directionality also allows for differentiation between the template and the newly synthesized strand during DNA repair processes [318]. PCNA interactors bind the C side by means of the conserved PCNA-interacting protein motif (PIP box). The PIP box consensus sequence, reminiscent of the β-binding motif (QL[S/D]LF) [210], is Qxx[L/I/M]xxF[F/Y], where x is any amino acid; the PIP box may be N-terminally flanked by the sequence KAx [319]. The PIP box folds into a 3_{10} helix that plugs into a hydrophobic pocket of PCNA formed in part by the interdomain connecting loop [320, 321]. Binding of factors containing PIP boxes to PCNA may be competitive, but because PCNA is a homotrimer, it is possible that multiple binding events can occur simultaneously [316]. The Y family polymerases pol η and pol κ possess canonical PIP boxes while pol ι possesses a noncanonical PIP box and the three polymerases display differential interactions with PCNA [273, 322]. However, a hierarchical model of Y family DNA polymerase binding to PCNA based on the relative affinities of their respective PIP boxes appears to be too simplistic, especially considering the proposed scaffolding role of Rev1, the two-step lesion bypass model in eukaryotes, and the roles of posttranslational modifications of PCNA [272, 273, 323]. All four eukaryotic Y family DNA polymerases interact with PCNA, but the interaction is modulated by posttranslational modification of PCNA [273, 323]. Posttranslational modification of PCNA with ubiquitin, in particular, enhances the affinity of Y family polymerases for PCNA [273, 323], described below.

11.8.2.1 Ubiquitylation of PCNA

As part of the regulation of DNA damage responses, PCNA is posttranslationally modified by conjugation to the proteins ubiquitin and SUMO (small ubiquitin-related modifier) [324–326]. Ubiquitin (Ub) is a highly conserved 76 amino-acid globular protein that is reversibly attached to Lys residues of target proteins [327]. Ub modifications influence a wide range of biological activities but perhaps the most characterized function is proteosomal degradation of poly-Ub-modified substrates with Lys48-linked Ub chains [327–330]. Lys63-linked Ub chains have a range of functions including roles in DNA repair, cell cycle progression, translation, IκB kinase activation, and endocytosis and transport [327]. Ub is conjugated to its substrates in three steps: (1) activation of Ub by an E1 activating enzyme, (2) transfer to an E2 Ub conjugating enzyme, and (3) conjugation to the target protein by an E3 Ub ligase [327]. E3 enzymes can also transfer Ub to growing Ub chains in the case of polyubiquitylation [327]. Other ubiquitin-like modifiers (UBLs) including SUMO adopt structures that are similar to that of Ub [331–333].

PCNA is monoubiquitylated on Lys164 by a Rad6-Rad18 (E2-E3, respectively)-dependent process (Fig. 11.11), in response to stalled replication forks. Monoubiquitylated PCNA has also been shown to enhance the affinity for Y family DNA polymerases to facilitate error-prone pathways via TLS. Y family polymerases Pol η, Pol ι, Pol κ, and Rev1 all contain ubiquitin-binding domains (designated UBM in Pol ι and Rev1 or UBZ in Pol η and Pol κ) and PCNA-binding motifs that together recruit these polymerases to monoubiquitylated PCNA at stalled replication sites [276, 323].

As a result of DNA damage, PCNA can also be polyubiquitylated at Lys164 by the E2-E3 complex Mms2-Ubc13-Rad5 to form Lys63-linked chains (Fig. 11.11) [323, 325, 326]. Polyubiquitylated PCNA promotes error-free DNA damage tolerance [326, 334, 335]. PCNA is de-ubiquitylated by ubiquitin-specific protease 1 (USP1). Upon UV irradiation, USP1 is inactivated allowing the accumulation of monoubiquitylated PCNA and the subsequent activation of TLS [336]. In the absence of DNA damage, it appears that USP1 keeps the cellular levels of ubiquitylated PCNA low so as to prevent unnecessary usage of TLS polymerases [323]. Therefore, PCNA ubiquitylation plays an important role in determining whether replication bypass will occur in an error-free or error-prone fashion [337].

11.8.2.2 Sumoylation of PCNA

SUMO is one of the more extensively studied UBL modifiers that alter the activity of target proteins [338]. The machinery responsible for the attachment of SUMO to its substrates is similar to the systems that result in ubiquitylation [331–333]. First, a thioester conjugate is formed between mature SUMO and E1 activating enzyme. Next, SUMO is transferred to the E2 conjugating enzyme Ubc9, and then transferred

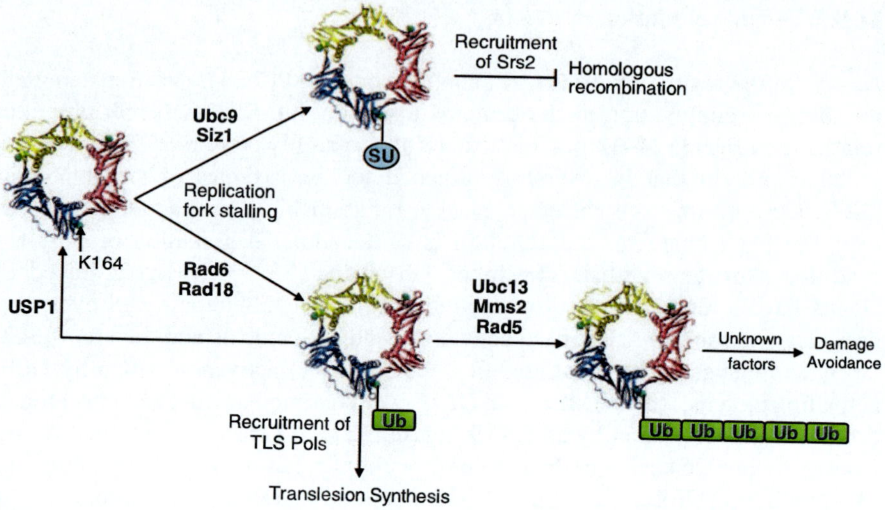

Fig. 11.11 PCNA modifications determine DNA damage response pathways. PCNA (structure shown) can be either monoubiquitylated (Ub, *green*) (Rad6-Rad18-dependent) or sumoylated (SU, *light blue*) (Ubc9-Siz1-dependent) at Lys164 in response to DNA damage. In a Ubc13-Mms2-Rad5-dependent process, polyubiquitylation can also occur on the Lys63 residue of Ub. PCNA monoubiquitylation promotes translesion synthesis by recruitment of TLS polymerases. USP1 regulates the level of ubiquitylated PCNA in the cell by removing ubiquitin from monoubiquitylated PCNA. PCNA polyubiquitylation promotes an error-free damage avoidance pathway that is not well understood. Sumoylation of PCNA prevents inappropriate homologous recombination

from Ubc9 to the substrate protein by one of several E3 ligases. Analogous to ubiquitylation, SUMO modification is reversible [330, 331].

PCNA can be sumoylated at Lys164 by the Ubc9-Siz1 E2-E3 complex and at Lys127 in a Siz1-independent manner [326]. In yeast, PCNA sumoylated at Lys164 recruits Srs2 helicase, which prevents inappropriate or "unscheduled" homologous recombination (Fig. 11.11) [326, 337]. The details of the regulation of the switch between ubiquitylation and sumoylation remain to be determined, but one mechanism may involve USP1, which deubiquitylates PCNA [336]. Because PCNA is ubiquitylated and sumoylated at the same residue, it has been suggested that PCNA sumoylation may be antagonistic to ubiquitylation and that these modifications influence which DNA damage response pathways are used by a cell [324, 335, 339].

These specific types of posttranslational modifications are not known to occur in *E. coli*. However, the posttranslational modification of $UmuD_2$ may play an analogous role [98]. The interactions of $UmuD_2$ and $UmuD'_2$ with the β clamp, and the temporal control of the conversion of $UmuD_2$ to $UmuD'_2$, regulate the transition from primitive DNA damage checkpoint control to UmuC-dependent DNA-damage-induced mutagenesis [98, 150, 171, 182, 185, 211]. Thus, though the specific mechanisms vary tremendously, some of the overarching themes of the DNA damage responses in prokaryotes and eukaryotes appear to be conserved.

11.8.3 Eukaryotic Polymerase Switch Mechanisms

Several ubiquitin-mediated bypass mechanisms have been proposed for polymerase switching and bypass replication. Monoubiquitylation of PCNA facilitates TLS polymerase binding to PCNA [334, 340]. A "trial and error" mechanism, akin to the active exchange model proposed above, may be used to select the optimal TLS polymerase for lesion bypass [126]. There is also evidence that a second TLS polymerase is required to extend the primer from the position opposite the lesion for several additional nucleotides before the replicative polymerase is restored [287, 341–345]. Such a task could generally be accomplished by Pol ζ, which proficiently extends distorted primer termini [310, 346]. Thus, two polymerase switches involving two different TLS polymerases may be required for lesion bypass. On the other hand, monoubiquitylation of Pol η and Pol ι disrupts their ubiquitin-interaction domains and may prevent the polymerases from binding monoubiquitylated PCNA [347]. Given the suggested scaffolding role of Rev1, it is also likely to be important in polymerase switching [224, 323].

11.8.4 DNA Damage and the Eukaryotic 9-1-1 Complex

In addition to posttranslational modifications of PCNA, eukaryotic cells also possess an alternative clamp-like protein involved in DNA damage responses. The human 9-1-1 complex, composed of the Rad9-Hus1-Rad1 proteins (in *S. cerevisiae*: Ddc1-Rad17-Mec3, respectively) [348], is a critical participant in the cellular responses to DNA damage [349]. As predicted [350], Rad9-Hus1-Rad1 has a toroidal structure resembling homotrimeric PCNA with a secondary structure topology similar to other DNA clamp proteins (Figs. 11.4 and 11.10) [351, 352]. In response to DNA damage, the heterotrimeric complex is loaded onto DNA by a distinct clamp loader complex Rad17-RFC, in which Rad17 is substituted for the large pl40 subunit [315, 349, 353–356]. *In vivo* experiments show that Rad17 is critical to the DNA damage response as its disruption blocks damage-induced 9-1-1 chromatin binding [349]. Studies in yeast have shown that the 9-1-1 complex, particularly Rad9, is essential for the survival of cells treated with ionizing radiation [357]. Recent results suggest that the 9-1-1 clamp is involved in mediating an SOS-like transcriptional response to DNA damage via its ubiquitylation by Rad6-Rad18 [358].

Like PCNA, the 9-1-1 complex has direct roles in DNA damage responses including potentially mutagenic TLS by Y family polymerases [359–363]. The 9-1-1 complex also functions in the recruitment of a variety of enzymes associated with long patch base excision repair [364]. Overall, 9-1-1 efficiently directs responses to DNA damage by regulating the appropriate checkpoint response, DNA repair, or by recruiting TLS polymerases for damage bypass [349, 364]. Thus, eukaryotes contain multiple mechanisms for polymerase switching that utilize posttranslational modifications as well as alternative pathways to tightly control DNA damage responses. Perhaps analogously, *E. coli* also has an alternative clamp, called β*, that is involved in DNA damage responses [365–367]. Its function is not yet as well-characterized as 9-1-1.

11.9 Conclusions

DNA damage is ubiquitous. Cells have developed numerous pathways to respond to potentially mutagenic or lethal DNA damage. Specialized translesion DNA polymerases represent one such mechanism, albeit a mechanism that comes at the possible cost of mutagenesis. Due to the potentially mutagenic nature of translesion polymerases, their activities are tightly regulated, in part by protein–protein interactions that may be modulated by posttranslational modifications. Many questions remain regarding the exact nature of polymerase selection for the multiple types of DNA damage encountered in cells. In addition, how is damage identified on the relatively rapid time scale of DNA replication and the cell cycle? Another area of active investigation involves the relative contributions to DNA damage tolerance of coordinated polymerase switching (tool-belt vs. dynamic exchange model) vs. gap filling. How is the access of DNA polymerases to damaged DNA regulated biophysically? What are the molecular details of polymerase exchange within the respective models? In the 10 years since the Y family polymerases were recognized as a distinct family, remarkable progress has been made in understanding their regulation. With wider application of biophysical methods to address future questions, progress will be made in understanding the multiple layers of regulation that control the activity of translesion polymerases.

Acknowledgments Research in the Beuning laboratory is supported by a New Faculty Award from the Camille & Henry Dreyfus Foundation and a CAREER Award from the NSF (MCB-0845033 to PJB). PJB is a Cottrell Scholar of the Research Corporation for Science Advancement. We thank Sebastian Zawiślak for assistance with the preparation of Figs. 11.6 and 11.8.

References

1. Kornberg A, Baker TA (1992) DNA replication, 2nd edn. W.H. Freeman and Company, New York, NY
2. Kelman Z, O'Donnell M (1995) DNA polymerase III holoenzyme: structure and function of a chromosomal replicating machine. Annu Rev Biochem 64:171–200
3. Bloom LB, Chen X, Fygenson DK, Turner J, O'Donnell M, Goodman MF (1992) Fidelity of *Escherichia coli* DNA polymerase III holoenzyme. The effects of beta, gamma complex processivity proteins and epsilon proofreading exonuclease on nucleotide misincorporation efficiencies. J Biol Chem 272:27919–27930
4. Topal MD, Fresco JR (1976) Complementary base pairing and the origin of substitution mutations. Nature 263:285–289
5. Glover BP, McHenry CS (2001) The DNA polymerase III holoenzyme: an asymmetric dimeric replicative complex with leading and lagging strand polymerases. Cell 105:925–934

6. Maki H, Maki S, Kornberg A (1988) DNA polymerase III holoenzyme of *Escherichia coli*. IV. The holoenzyme is an asymmetric dimer with twin active sites. J Biol Chem 263:6570–6578
7. O'Donnell M, Studwell PS (1990) Total reconstitution of DNA polymerase III holoenzyme reveals dual accessory protein clamps. J Biol Chem 265:1179–1187
8. Yuzhakov A, Turner J, O'Donnell M (1996) Replisome assembly reveals the basis for asymmetric function in leading and lagging strand replication. Cell 86:877–886
9. McHenry CS (2003) Chromosomal replicases as asymmetric dimers: studies of subunit arrangement and functional consequences. Mol Microbiol 49:1157–1165
10. McInerney P, Johnson A, Katz F, O'Donnell M (2007) Characterization of a triple DNA polymerase replisome. Mol Cell 27:527–538
11. Johnson A, O'Donnell M (2005) Cellular DNA replicases: components and dynamics at the replication fork. Annu Rev Biochem 74:283–315
12. Onrust R, Finkelstein J, Turner J, Naktinis V, O'Donnell M (1995) Assembly of a chromosomal replication machine: two DNA polymerases, a clamp loader, and sliding clamps in one holoenzyme particle. III. Interface between two polymerases and the clamp loader. J Biol Chem 270:13366–13377
13. Fay PJ, Johanson KO, McHenry CS, Bambara RA (1981) Size classes of products synthesized processively by DNA polymerase III and DNA polymerase III holoenzyme of *Escherichia coli*. J Biol Chem 256:976–983
14. Fay PJ, Johanson KO, McHenry CS, Bambara RA (1982) Size classes of products synthesized processively by two subassemblies of *Escherichia coli* DNA polymerase III holoenzyme. J Biol Chem 257:5692–5699
15. LaDuca RJ, Crute JJ, McHenry CS, Bambara RA (1986) The beta subunit of the *Escherichia coli* DNA polymerase III holoenzyme interacts functionally with the catalytic core in the absence of other subunits. J Biol Chem 261:7550–7557
16. Maki H, Kornberg A (1985) The polymerase subunit of DNA polymerase III of *Escherichia coli*. II. Purification of the alpha subunit, devoid of nuclease activities. J Biol Chem 260:12987–12992
17. Mok M, Marians KJ (1987) Formation of rolling-circle molecules during phi X174 complementary strand DNA replication. J Biol Chem 262:2304–2309
18. Maki H, Kornberg A (1987) Proofreading by DNA polymerase III of *Escherichia coli* depends on cooperative interaction of the polymerase and exonuclease subunits. Proc Natl Acad Sci USA 84:4389–4392
19. Studwell PS, O'Donnell M (1990) Processive replication is contingent on the exonuclease subunit of DNA polymerase III holoenzyme. J Biol Chem 265:1171–1178
20. Bloom LB (2009) Loading clamps for DNA replication and repair. DNA Repair 8:570–578
21. McHenry CS (1982) Purification and characterization of DNA polymerase III'. Identification of tau as a subunit of the DNA polymerase III holoenzyme. J Biol Chem 257:2657–2663
22. Glover BP, McHenry CS (1998) The chi psi subunits of DNA polymerase III holoenzyme bind to single-stranded DNA-binding protein (SSB) and facilitate replication of an SSB-coated template. J Biol Chem 273:23476–23484
23. Kelman Z, Yuzhakov A, Andjelkovic J, O'Donnell M (1998) Devoted to the lagging strand – the chi subunit of DNA polymerase III holoenzyme contacts SSB to promote processive elongation and sliding clamp assembly. EMBO J 17:2436–2449
24. Anderson SG, Williams CR, O'Donnell M, Bloom LB (2007) A function for the psi subunit in loading the *Escherichia coli* DNA polymerase sliding clamp. J Biol Chem 282:7035–7045
25. Lamers MH, Georgescu RE, Lee SG, O'Donnell M, Kuriyan J (2006) Crystal structure of the catalytic alpha subunit of *E. coli* replicative DNA polymerase III. Cell 126:881–892
26. Brautigam CA, Steitz TA (1998) Structural and functional insights provided by crystal structures of DNA polymerases and their substrate complexes. Curr Opin Struct Biol 8:54–63
27. Rothwell PJ, Waksman G (2005) Structure and mechanism of DNA polymerases. Adv Protein Chem 71:401–440
28. Steitz TA (1999) DNA polymerases: structural diversity and common mechanisms. J Biol Chem 274:17395–17398

29. Pritchard AE, McHenry CS (1999) Identification of the acidic residues in the active site of DNA polymerase III. J Mol Biol 285:1067–1080
30. Sawaya MR, Prasad R, Wilson SH, Kraut J, Pelletier H (1997) Crystal structures of human DNA polymerase beta complexed with gapped and nicked DNA: evidence for an induced fit mechanism. Biochemistry 36:11205–11215
31. Aravind L, Koonin EV (1998) Phosphoesterase domains associated with DNA polymerases of diverse origins. Nucleic Acids Res 26:3746–3752
32. Stano NM, Chen J, McHenry CS (2006) A coproofreading Zn(2+)-dependent exonuclease within a bacterial replicase. Nat Struct Mol Biol 13:458–459
33. Bailey S, Wing RA, Steitz TA (2006) The structure of *T. aquaticus* DNA polymerase III is distinct from eukaryotic replicative DNA polymerases. Cell 126:893–904
34. Jergic S, Ozawa K, Williams NK, Su XC, Scott DD, Hamdan SM, Crowther JA, Otting G, Dixon NE (2007) The unstructured C-terminus of the tau subunit of *Escherichia coli* DNA polymerase III holoenzyme is the site of interaction with the alpha subunit. Nucleic Acids Res 35:2813–2824
35. Kim DR, McHenry CS (1996) Biotin tagging deletion analysis of domain limits involved in protein–macromolecular interactions. Mapping the tau binding domain of the DNA polymerase III alpha subunit. J Biol Chem 271:20690–20698
36. Lopez de Saro FJ, Georgescu RE, O'Donnell M (2003) A peptide switch regulates DNA polymerase processivity. Proc Natl Acad Sci USA 100:14689–14694
37. Dohrmann PR, McHenry CS (2005) A bipartite polymerase-processivity factor interaction: only the internal beta binding site of the alpha subunit is required for processive replication by the DNA polymerase III holoenzyme. J Mol Biol 350:228–239
38. Theobald DL, Mitton-Fry RM, Wuttke DS (2003) Nucleic acid recognition by OB-fold proteins. Annu Rev Biophys Biomol Struct 32:115–133
39. McCauley MJ, Shokri L, Sefcikova J, Venclovas C, Beuning PJ, Williams MC (2008) Distinct double- and single-stranded DNA binding of *E. coli* replicative DNA polymerase III alpha subunit. ACS Chem Biol 3:577–587
40. Wing RA, Bailey S, Steitz TA (2008) Insights into the replisome from the structure of a ternary complex of the DNA polymerase III alpha-subunit. J Mol Biol 382:859–869
41. Evans RJ, Davies DR, Bullard JM, Christensen J, Green LS, Guiles JW, Pata JD, Ribble WK, Janjic N, Jarvis TC (2008) Structure of PolC reveals unique DNA binding and fidelity determinants. Proc Natl Acad Sci USA 105:20695–20700
42. Lamers MH, O'Donnell M (2008) A consensus view of DNA binding by the C family of replicative DNA polymerases. Proc Natl Acad Sci USA 105:20565–20566
43. Hamdan S, Carr PD, Brown SE, Ollis DL, Dixon NE (2002) Structural basis for proofreading during replication of the *Escherichia coli* chromosome. Structure 10:535–546
44. DeRose EF, Li D, Darden T, Harvey S, Perrino FW, Schaaper RM, London RE (2002) Model for the catalytic domain of the proofreading epsilon subunit of *Escherichia coli* DNA polymerase III based on NMR structural data. Biochemistry 41:94–110
45. Perrino FW, Harvey S, McNeill SM (1999) Two functional domains of the epsilon subunit of DNA polymerase III. Biochemistry 38:16001–16009
46. Taft-Benz SA, Schaaper RM (1999) The C-terminal domain of DnaQ contains the polymerase binding site. J Bacteriol 181:2963–2965
47. Taft-Benz SA, Schaaper RM (1998) Mutational analysis of the 3′–>5′ proofreading exonuclease of *Escherichia coli* DNA polymerase III. Nucleic Acids Res 26:4005–4011
48. Ozawa K, Jergic S, Park AY, Dixon NE, Otting G (2008) The proofreading exonuclease subunit epsilon of *Escherichia coli* DNA polymerase III is tethered to the polymerase subunit alpha via a flexible linker. Nucleic Acids Res 36:5074–5082
49. Wieczorek A, McHenry CS (2006) The NH_2-terminal PHP domain of the alpha subunit of the *Escherichia coli* replicase binds the epsilon proofreading subunit. J Biol Chem 281:12561–12567
50. DeRose EF, Darden T, Harvey S, Gabel S, Perrino FW, Schaaper RM, London RE (2003) Elucidation of the epsilon-theta subunit interface of *Escherichia coli* DNA polymerase III by NMR spectroscopy. Biochemistry 42:3635–3644

51. Studwell-Vaughan PS, O'Donnell M (1993) DNA polymerase III accessory proteins. V. Theta encoded by holE. J Biol Chem 268:11785–11791
52. Keniry MA, Park AY, Owen EA, Hamdan SM, Pintacuda G, Otting G, Dixon NE (2006) Structure of the theta subunit of *Escherichia coli* DNA polymerase III in complex with the epsilon subunit. J Bacteriol 188:4464–4473
53. Mueller GA, Kirby TW, DeRose EF, Li D, Schaaper RM, London RE (2005) Nuclear magnetic resonance solution structure of the *Escherichia coli* DNA polymerase III theta subunit. J Bacteriol 187:7081–7089
54. Gupta R, Hamdan SM, Dixon NE, Sheil MM, Beck JL (2004) Application of electrospray ionization mass spectrometry to study the hydrophobic interaction between the epsilon and theta subunits of DNA polymerase III. Protein Sci 13:2878–2887
55. Jonczyk P, Nowicka A, Fijalkowska IJ, Schaaper RM, Ciesla Z (1998) In vivo protein interactions within the *Escherichia coli* DNA polymerase III core. J Bacteriol 180:1563–1566
56. Taft-Benz SA, Schaaper RM (2004) The theta subunit of *Escherichia coli* DNA polymerase III: a role in stabilizing the epsilon proofreading subunit. J Bacteriol 186:2774–2780
57. Hamdan S, Bulloch EM, Thompson PR, Beck JL, Yang JY, Crowther JA, Lilley PE, Carr PD, Ollis DL, Brown SE, Dixon NE (2002) Hydrolysis of the 5′-p-nitrophenyl ester of TMP by the proofreading exonuclease (epsilon) subunit of *Escherichia coli* DNA polymerase III. Biochemistry 41:5266–5275
58. Lehtinen DA, Perrino FW (2004) Dysfunctional proofreading in the *Escherichia coli* DNA polymerase III core. Biochem J 384:337–348
59. Ellison V, Stillman B (2001) Opening of the clamp: an intimate view of an ATP-driven biological machine. Cell 106:655–660
60. Jeruzalmi D, Yurieva O, Zhao Y, Young M, Stewart J, Hingorani M, O'Donnell M, Kuriyan J (2001) Mechanism of processivity clamp opening by the delta subunit wrench of the clamp loader complex of *E. coli* DNA polymerase III. Cell 106:417–428
61. Tsuchihashi Z, Kornberg A (1990) Translational frameshifting generates the gamma subunit of DNA polymerase III holoenzyme. Proc Natl Acad Sci USA 87:2516–2520
62. Blinkowa AL, Walker JR (1990) Programmed ribosomal frameshifting generates the *Escherichia coli* DNA polymerase III gamma subunit from within the tau subunit reading frame. Nucleic Acids Res 18:1725–1729
63. Flower AM, McHenry CS (1990) The gamma subunit of DNA polymerase III holoenzyme of *Escherichia coli* is produced by ribosomal frameshifting. Proc Natl Acad Sci USA 87:3713–3717
64. Jeruzalmi D, O'Donnell M, Kuriyan J (2001) Crystal structure of the processivity clamp loader gamma complex of *E. coli* DNA polymerase III. Cell 106:429–441
65. Podobnik M, Weitze TF, O'Donnell M, Kuriyan J (2003) Nucleotide-induced conformational changes in an isolated *Escherichia coli* DNA polymerase III clamp loader subunit. Structure 11:253–263
66. Guenther B, Onrust R, Sali A, O'Donnell M, Kuriyan J (1997) Crystal structure of the delta′ subunit of the clamp-loader complex of *E. coli* DNA polymerase III. Cell 91:335–345
67. Story RM, Steitz TA (1992) Structure of the RecA protein-ADP complex. Nature 355:374–376
68. Millar D, Trakselis MA, Benkovic SJ (2004) On the solution structure of the T4 sliding clamp (gp45). Biochemistry 43:12723–12727
69. Stewart J, Hingorani MM, Kelman Z, O'Donnell M (2001) Mechanism of beta clamp opening by the delta subunit of *Escherichia coli* DNA polymerase III holoenzyme. J Biol Chem 276:19182–19189
70. Turner J, Hingorani MM, Kelman Z, O'Donnell M (1999) The internal workings of a DNA polymerase clamp-loading machine. EMBO J 18:771–783
71. Lopez de Saro FJ, Georgescu RE, Goodman MF, O'Donnell M (2003) Competitive processivity-clamp usage by DNA polymerases during DNA replication and repair. EMBO J 22:6408–6418
72. Burnouf DY, Olieric V, Wagner J, Fujii S, Reinbolt J, Fuchs RP, Dumas P (2004) Structural and biochemical analysis of sliding clamp/ligand interactions suggest a competition between replicative and translesion DNA polymerases. J Mol Biol 335:1187–1197

73. Beuning PJ, Sawicka D, Barsky D, Walker GC (2006) Two processivity clamp interactions differentially alter the dual activities of UmuC. Mol Microbiol 59:460–474
74. Sutton MD, Duzen JM, Maul RW (2005) Mutant forms of the *Escherichia coli* beta sliding clamp that distinguish between its roles in replication and DNA polymerase V-dependent translesion DNA synthesis. Mol Microbiol 55:1751–1766
75. Scouten Ponticelli SK, Duzen JM, Sutton MD (2009) Contributions of the individual hydrophobic clefts of the *Escherichia coli* beta sliding clamp to clamp loading, DNA replication and clamp recycling. Nucleic Acids Res 37:2796–2809
76. Becherel OJ, Fuchs RP, Wagner J (2002) Pivotal role of the beta clamp in translesion DNA synthesis and mutagenesis in *E. coli* cells. DNA Repair 1:703–708
77. Hingorani MM, O'Donnell M (1998) ATP binding to the *Escherichia coli* clamp loader powers opening of the ring-shaped clamp of DNA polymerase III holoenzyme. J Biol Chem 273:24550–24563
78. Goedken ER, Kazmirski SL, Bowman GD, O'Donnell M, Kuriyan J (2005) Mapping the interaction of DNA with the *Escherichia coli* DNA polymerase clamp loader complex. Nat Struct Mol Biol 12:183–190
79. Miyata T, Suzuki H, Oyama T, Mayanagi K, Ishino Y, Morikawa K (2005) Open clamp structure in the clamp-loading complex visualized by electron microscopic image analysis. Proc Natl Acad Sci USA 102:13795–13800
80. Simonetta KR, Kazmirski SL, Goedken ER, Cantor AJ, Kelch BA, McNally R, Seyedin SN, Makino DL, O'Donnell M, Kuriyan J (2009) The mechanism of ATP-dependent primer-template recognition by a clamp loader complex. Cell 137:659–671
81. Anderson SG, Thompson JA, Paschall CO, O'Donnell M, Bloom LB (2009) Temporal correlation of DNA binding, ATP hydrolysis, and clamp release in the clamp loading reaction catalyzed by the *Escherichia coli* gamma complex. Biochemistry 48:8516–8527
82. O'Donnell ME (1987) Accessory proteins bind a primed template and mediate rapid cycling of DNA polymerase III holoenzyme from *Escherichia coli*. J Biol Chem 262:16558–16565
83. Bowman GD, O'Donnell M, Kuriyan J (2004) Structural analysis of a eukaryotic sliding DNA clamp-clamp loader complex. Nature 429:724–730
84. Lahue RS, Au KG, Modrich P (1989) DNA mismatch correction in a defined system. Science 245:160–164
85. Gulbis JM, Kazmirski SL, Finkelstein J, Kelman Z, O'Donnell M, Kuriyan J (2004) Crystal structure of the chi:psi sub-assembly of the *Escherichia coli* DNA polymerase clamp-loader complex. Eur J Biochem 271:439–449
86. Okazaki R, Okazaki T, Sakabe K, Sugimoto K, Sugino A (1968) Mechanism of DNA chain growth. I. Possible discontinuity and unusual secondary structure of newly synthesized chains. Proc Natl Acad Sci USA 59:598–605
87. Olson MW, Dallmann HG, McHenry CS (1995) DnaX complex of *Escherichia coli* DNA polymerase III holoenzyme. The chi psi complex functions by increasing the affinity of tau and gamma for delta.delta' to a physiologically relevant range. J Biol Chem 270:29570–29577
88. Witte G, Urbanke C, Curth U (2003) DNA polymerase III chi subunit ties single-stranded DNA binding protein to the bacterial replication machinery. Nucleic Acids Res 31:4434–4440
89. Kong XP, Onrust R, O'Donnell M, Kuriyan J (1992) Three-dimensional structure of the beta subunit of *E. coli* DNA polymerase III holoenzyme: a sliding DNA clamp. Cell 69:425–437
90. Oakley AJ, Prosselkov P, Wijffels G, Beck JL, Wilce MCJ, Dixon NE (2003) Flexibility revealed by the 1.85. A crystal structure of the beta sliding-clamp subunit of *Escherichia coli* DNA polymerase III. Acta Crystallogr D Biol Crystallogr 59:1192–1199
91. Stukenberg PT, Studwell-Vaughan PS, O'Donnell M (1991) Mechanism of the sliding beta-clamp of DNA polymerase III holoenzyme. J Biol Chem 266:11328–11334
92. Stukenberg PT, Turner J, O'Donnell M (1994) An explanation for lagging strand replication: polymerase hopping among DNA sliding clamps. Cell 78:877–887
93. Yao N, Turner J, Kelman Z, Stukenberg PT, Dean F, Shechter D, Pan ZQ, Hurwitz J, O'Donnell M (1996) Clamp loading, unloading and intrinsic stability of the PCNA, beta and gp45 sliding clamps of human, *E. coli* and T4 replicases. Genes Cells 1:101–113

94. Laurence TA, Kwon Y, Johnson A, Hollars CW, O'Donnell M, Camarero JA, Barsky D (2008) Motion of a DNA sliding clamp observed by single molecule fluorescence spectroscopy. J Biol Chem 283:22895–22906
95. Georgescu RE, Kim S-S, Yurieva O, Kuriyan J, Kong X-P, O'Donnell M (2008) Structure of a sliding clamp on DNA. Cell 132:43–54
96. Wagner J, Fujii S, Gruz P, Nohmi T, Fuchs RP (2000) The beta clamp targets DNA polymerase IV to DNA and strongly increases its processivity. EMBO Rep 1:484–488
97. Georgescu RE, Yurieva O, Kim SS, Kuriyan J, Kong XP, O'Donnell M (2008) Structure of a small-molecule inhibitor of a DNA polymerase sliding clamp. Proc Natl Acad Sci USA 105:11116–11121
98. Duzen JM, Walker GC, Sutton MD (2004) Identification of specific amino acid residues in the *E. coli* beta processivity clamp involved in interactions with DNA polymerase III, UmuD and UmuD′. DNA Repair 3:301–312
99. Georgescu RE, Kurth I, Yao NY, Stewart J, Yurieva O, O'Donnell M (2009) Mechanism of polymerase collision release from sliding clamps on the lagging strand. EMBO J 28:2981–2991
100. Leu FP, Georgescu R, O'Donnell M (2003) Mechanism of the *E. coli* tau processivity switch during lagging-strand synthesis. Mol Cell 11:315–327
101. Fujii S, Fuchs RP (2004) Defining the position of the switches between replicative and bypass DNA polymerases. EMBO J 23:4342–4352
102. Indiani C, McInerney P, Georgescu R, Goodman MF, O'Donnell M (2005) A sliding-clamp toolbelt binds high- and low-fidelity DNA polymerases simultaneously. Mol Cell 19:805–815
103. Bunting KA, Roe SM, Pearl LH (2003) Structural basis for recruitment of translesion DNA polymerase pol IV/DinB to the beta-clamp. EMBO J 22:5883–5892
104. Ling H, Boudsocq F, Woodgate R, Yang W (2001) Crystal structure of a Y-family DNA polymerase in action: a mechanism for error-prone and lesion-bypass replication. Cell 107:91–102
105. Yang J, Zhuang Z, Roccasecca RM, Trakselis MA, Benkovic SJ (2004) The dynamic processivity of the T4 DNA polymerase during replication. Proc Natl Acad Sci USA 101:8289–8294
106. Gao D, McHenry CS (2001) Tau binds and organizes *Escherichia coli* replication through distinct domains. Partial proteolysis of terminally tagged tau to determine candidate domains and to assign domain V as the alpha binding domain. J Biol Chem 276:4433–4440
107. Dallmann HG, Kim S, Pritchard AE, Marians KJ, McHenry CS (2000) Characterization of the unique C terminus of the *Escherichia coli* tau DnaX protein. Monomeric C-tau binds alpha AND DnaB and can partially replace tau in reconstituted replication forks. J Biol Chem 275:15512–15519
108. Gao D, McHenry CS (2001) Tau binds and organizes *Escherichia coli* replication proteins through distinct domains. Domain IV, located within the unique C terminus of tau, binds the replication fork, helicase, DnaB. J Biol Chem 276:4441–4446
109. Kim S, Dallmann HG, McHenry CS, Marians KJ (1996) Tau couples the leading- and lagging-strand polymerases at the *Escherichia coli* DNA replication fork. J Biol Chem 271:21406–21412
110. Su XC, Jergic S, Keniry MA, Dixon NE, Otting G (2007) Solution structure of domains IVa and V of the tau subunit of *Escherichia coli* DNA polymerase III and interaction with the alpha subunit. Nucleic Acids Res 35:2825–2832
111. Kim S, Dallmann HG, McHenry CS, Marians KJ (1996) Coupling of a replicative polymerase and helicase: a tau–DnaB interaction mediates rapid replication fork movement. Cell 84:643–650
112. Kim S, Dallmann HG, McHenry CS, Marians KJ (1996) Tau protects beta in the leading-strand polymerase complex at the replication fork. J Biol Chem 271:4315–4318
113. Friedberg EC, Walker GC, Siede W, Wood RD, Schultz RA, Ellenberger T (2006) DNA repair and mutagenesis, 2nd edn. American Society for Microbiology, Washington, DC
114. Rupp WD (1996) DNA repair mechanisms. In: Neidhardt FC (ed) *Escherichia coli* and *Salmonella*: cellular and molecular biology. ASM Press, Washington, DC, pp 2277–2294

115. Rudolph CJ, Upton AL, Lloyd RG (2007) Replication fork stalling and cell cycle arrest in UV-irradiated *Escherichia coli.* Genes Dev 21:668–681
116. Pages V, Fuchs RP (2003) Uncoupling of leading- and lagging-strand DNA replication during lesion bypass in vivo. Science 300:1300–1303
117. Higuchi K, Katayama T, Iwai S, Hidaka M, Horiuchi T, Maki H (2003) Fate of DNA replication fork encountering a single DNA lesion during oriC plasmid DNA replication in vitro. Genes Cells 8:437–449
118. McInerney P, O'Donnell M (2004) Functional uncoupling of twin poymerases. J Biol Chem 279:21543–21551
119. Rupp WD, Howard-Flanders P (1968) Discontinuities in the DNA synthesized in an excision-defective strain of *Escherichia coli* following ultraviolet irradiation. J Mol Biol 31:291–304
120. Heller RC, Marians KJ (2006) Replication fork reactivation downstream of a blocked nascent leading strand. Nature 439:557–562
121. Simmons LA, Foti JJ, Cohen SE, Walker GC (2008) The SOS regulatory network. In: Bock A et al (eds) EcoSal – *Escherichia coli* and *Salmonella*: cellular and molecular biology. ASM Press, Washington, DC
122. Foti JJ, Simmons LA, Beuning PJ, Walker GC (2009) Signal transduction in the *Escherichia coli* SOS response. In: Bradshaw RA, Dennis EA (eds) Handbook of cell signaling. Elsevier Science, San Diego, CA, pp 2127–2136
123. Jarosz DF, Beuning PJ, Cohen SE, Walker GC (2007) Y-family DNA polymerases in *Escherichia coli.* Trends Microbiol 15:70–77
124. Schlacher K, Goodman MF (2007) Lessons from 50 years of SOS DNA-damage-induced mutagenesis. Nat Rev Mol Cell Biol 8:587–594
125. Nohmi T (2006) Environmental stress and lesion-bypass DNA polymerases. Annu Rev Microbiol 60:231–253
126. Yang W (2003) Damage repair DNA polymerases Y. Curr Opin Struct Biol 13:23–30
127. Beard WA, Wilson SH (2001) DNA lesion bypass polymerases open up. Structure 9:759–764
128. Silvian LF, Toth EA, Pham P, Goodman MF, Ellenberger T (2001) Crystal structure of a DinB family error-prone DNA polymerase from *Sulfolobus solfataricus.* Nature 8:984–989
129. Zhou BL, Pata JD, Steitz TA (2001) Crystal structure of a DinB lesion bypass DNA polymerase catalytic fragment reveals a classic polymerase catalytic domain. Mol Cell 8:427–437
130. Boudsocq F, Kokoska RJ, Plosky BS, Vaisman A, Ling H, Kunkel TA, Yang W, Woodgate R (2004) Investigating the role of the little finger domain of Y-family DNA polymerases in low fidelity synthesis and translesion replication. J Biol Chem 279:32932–32940
131. Beard WA, Wilson SH (2003) Structural insights into the origins of DNA polymerase fidelity. Structure 11:489–496
132. Bell JB, Eckert KA, Joyce CM, Kunkel TA (1997) Base miscoding and strand misalignment errors by mutator Klenow polymerases with amino acid substitutions at tyrosine 766 in the O helix of the fingers subdomain. J Biol Chem 272:7345–7351
133. Zhang H, Rhee C, Bebenek A, Drake JW, Wang J, Konigsberg W (2006) The L561A substitution in the nascent base-pair binding pocket of RB69 DNA polymerase reduces base discrimination. Biochemistry 45:2211–2220
134. Boudsocq F, Ling H, Yang W, Woodgate R (2002) Structure-based interpretation of missense mutations in Y-family DNA polymerases and their implications for polymerase function and lesion bypass. DNA Repair 1:343–358
135. Elledge SJ, Walker GC (1983) Proteins required for ultraviolet light and chemical mutagenesis. Identification of the products of the umuC locus of *Escherichia coli.* J Mol Biol 164:175–192
136. Kato T (1977) Effects of chloramphenicol and caffeine on postreplication repair in uvr A-umuC- und uvrA-recF-strains of *Escherichia coli* K-12. Mol Gen Genet 156:115–120
137. Kato T, Shinoura Y (1977) Isolation and characterization of mutants of *Escherichia coli* deficient in induction of mutagenesis by ultraviolet light. Mol Gen Genet 156:121–131
138. Shinagawa H, Kato T, Ise T, Makino K, Nakata A (1983) Cloning and characterization of the umu operon responsible for inducible mutagenesis in *Escherichia coli.* Gene 23:167–174

139. Steinborn G (1978) Uvm mutants of *Escherichia coli* K12 deficient in UV mutagenesis. I. Isolation of uvm mutants and their phenotypical characterization in DNA repair and mutagenesis. Mol Gen Genet 165:87–93
140. Tang M, Pham P, Shen X, Taylor J-S, ODonnell M, Woodgate R, Goodman MF (2000) Roles of *E. coli* DNA polymerases IV and V in lesion-targeted and untargeted SOS mutagenesis. Nature 404:1014–1018
141. Tang M, Bruck I, Eritja R, Turner J, Frank EG, Woodgate R, O'Donnell M, Goodman MF (1998) Biochemical basis of SOS-induced mutagenesis in *Escherichia coli*: reconstitution of in vitro lesion bypass dependent on the UmuD′2C mutagenic complex and RecA protein. Proc Natl Acad Sci USA 95:9755–9760
142. Fuchs RP, Fujii S (2007) Translesion synthesis in *Escherichia coli*: lessons from the NarI mutation hot spot. DNA Repair 6:1032–1041
143. Rajagopalan M, Lu C, Woodgate R, ODonnell M, Goodman MF, Echols H (1992) Activity of the purified mutagenesis proteins UmuC, UmuD′, and RecA in replicative bypass of an abasic DNA lesion by DNA polymerase III. Proc Natl Acad Sci USA 89:10777–10781
144. Shen X, Sayer JM, Kroth H, Ponten I, ODonnell M, Woodgate R, Jerina DM, Goodman MF (2002) Efficiency and accuracy of SOS-induced DNA polymerases replicating benso[a] pyrene-7, 8-diol 9, 10-epoxide A and G adducts. J Biol Chem 277:5265–5274
145. Lenne-Samuel N, Janel-Bintz R, Kolbanovsky A, Geacintov NE, Fuchs RP (2000) The processing of benzo(a)pyrene adduct into a frameshift or a base substitution mutation requires a different set of genes in *Escherichia coli*. Mol Microbiol 38:299–307
146. Yin J, Seo KY, Loechler EL (2004) A role for DNA polymerase V in G->T mutations from the major benzo[a]pyrene N2-dG adduct when studied in a 5′-TGT sequence in *E. coli*. DNA Repair 3:323–334
147. Neeley WL, Delaney S, Alekseyev YO, Jarosz DF, Delaney JC, Walker GC, Essigmann JM (2007) DNA polymerase V allows bypass of toxic guanine oxidation products in vivo. J Biol Chem 282:12741–12748
148. Fujii S, Gasser V, Fuchs RP (2004) The biochemical requirements of DNA polymerase V-mediated translesion synthesis revisited. J Mol Biol 341:405–417
149. Pham P, Seitz EM, Saveliev S, Shen X, Woodgate R, Cox MM, Goodman MF (2002) Two distinct modes of RecA action are required for DNA polymerase V-catalyzed translesion synthesis. Proc Natl Acad Sci USA 99:11061–11066
150. Reuven NB, Arad G, Maor-Shoshani A, Livneh Z (1999) The mutagenesis protein UmuC is a DNA polymerase activated by UmuD′, RecA, and SSB and is specialized for translesion replication. J Biol Chem 274:31763–31766
151. Schlacher K, Cox MM, Woodgate R, Goodman MF (2006) RecA acts in trans to allow replication of damaged DNA by DNA polymerase V. Nature 442:883–887
152. Jiang Q, Karata K, Woodgate R, Cox MM, Goodman MF (2009) The active form of DNA polymerase V is UmuD′2C-RecA-ATP. Nature 460:359–363
153. Schlacher K, Leslie K, Wyman C, Woodgate R, Cox M, Goodman M (2005) DNA polymerase V and RecA protein, a minimal mutasome. Mol Cell 17:561–572
154. Arad G, Hendel A, Urbanke C, Curth U, Livneh Z (2008) Single-stranded DNA-binding protein recruits DNA polymerase V to primer termini on RecA-coated DNA. J Biol Chem 283:8274–8282
155. Fujii S, Fuchs RP (2009) Biochemical basis for the essential genetic requirements of RecA and the β-clamp in Pol V activation. Proc Natl Acad Sci USA 106:14825–14830
156. Reuven NB, Arad G, Stasiak AZ, Stasiak A, Livneh Z (2001) Lesion bypass by the *Escherichia coli* DNA polymerase V requires assembly of a RecA nucleoprotein filament. J Biol Chem 276:5511–5517
157. Seo KY, Nagalingam A, Miri S, Yin J, Chandani S, Kolbanovskiy A, Shastry A, Loechler EL (2006) Mirror image stereoisomers of the major benzo[a]pyrene N2-dG adduct are bypassed by different lesion-bypass DNA polymerases in *E. coli*. DNA Repair 5:515–522
158. Jarosz DF, Godoy VG, DeLaney JC, Essigmann JM, Walker GC (2006) A single amino acid governs enhanced activity of DinB DNA polymerases on damaged templates. Nature 439:225–228

159. Kumari A, Minko IG, Harbut MB, Finkel SE, Goodman MF, Lloyd RS (2008) Replication bypass of interstrand cross-link intermediates by *Escherichia coli* DNA polymerase IV. J Biol Chem 283:27433–27437
160. Minko IG, Yamanaka K, Kozekov ID, Kozekova A, Indiani C, O'Donnell ME, Jiang Q, Goodman MF, Rizzo CJ, Lloyd RS (2008) Replication bypass of the acrolein-mediated deoxyguanine DNA-peptide cross-links by DNA polymerases of the DinB family. Chem Res Toxicol 21:1983–1990
161. Yuan B, Cao H, Jiang Y, Hong H, Wang Y (2008) Efficient and accurate bypass of N2-(1-carboxyethyl)-2′-deoxyguanosine by DinB DNA polymerase in vitro and in vivo. Proc Natl Acad Sci USA 105:8679–8684
162. Suzuki N, Okashi E, Hayashi K, Ohmori H, Grollman AP, Shibutani S (2001) Translesional synthesis past acetylaminofluorene-derived DNA adducts catalyzed by human DNA polymerase κ and *Escherichia coli* DNA polymerase IV. Biochemistry 40:15176–15183
163. Maor-Shoshani A, Hayashi K, Ohmori H, Livneh Z (2003) Analysis of translesion replication across an abasic site by DNA polymerase IV of *Escherichia coli*. DNA Repair 2:1227–1238
164. Kitagawa Y, Akaboshi E, Shinagawa H, Horii T, Ogawa H, Kato T (1985) Structural analysis of the umu operon required for inducible mutagenesis in *Escherichia coli*. Proc Natl Acad Sci USA 82:4336–4340
165. Courcelle J, Khodursky A, Peter B, Brown PO, Hanawalt PC (2001) Comparative gene expression profiles following UV exposure in wild-type and SOS-deficient *Escherichia coli*. Genetics 158:41–64
166. Sassanfar M, Roberts JW (1990) Nature of the SOS-inducing signal in *Escherichia coli*. The involvement of DNA replication. J Mol Biol 212:79–96
167. Brent R, Ptashne M (1981) Mechanism of action of the lexA gene product. Proc Natl Acad Sci USA 78:4204–4208
168. Woodgate R, Ennis DG (1991) Levels of chromosomally encoded Umu proteins and requirements for in vivo UmuD cleavage. Mol Gen Genet 229:10–16
169. Kim S-R, Matsui K, Yamada M, Gruz P, Nohmi T (2001) Roles of chromosomal and episomal dinB genes encoding DNA pol IV in targeted and untargeted mutagenesis in *Escherichia coli*. Mol Genet Genomics 266:207–215
170. Nohmi T, Battista JR, Dodson LA, Walker GC (1988) RecA-mediated cleavage activates UmuD for mutagenesis: mechanistic relationship between transcriptional derepression and posttranslational activation. Proc Natl Acad Sci USA 85:1816–1820
171. Opperman T, Murli S, Smith BT, Walker GC (1999) A model for a umuDC-dependent prokaryotic DNA damage checkpoint. Proc Natl Acad Sci USA 96:9218–9223
172. Battista JR, Ohta T, Nohmi T, Sun W, Walker GC (1990) Dominant negative umuD mutations decreasing RecA-mediated cleavage suggest roles for intact UmuD in modulation of SOS mutagenesis. Proc Natl Acad Sci USA 87:7190–7194
173. Woodgate R, Rajagopalan M, Lu C, Echols H (1989) UmuC mutagenesis protein of *Escherichia coli*: purification and interaction with UmuD and UmuD′. Proc Natl Acad Sci USA 86:7301–7305
174. Ferentz AE, Walker GC, Wagner G (2001) Converting a DNA damage checkpoint effector (UmuD2C) into a lesion bypass polymerase (UmuD′2C). EMBO J 20:4287–4298
175. Peat TS, Frank EG, McDonald JP, Levine AS, Woodgate R, Hendrickson WA (1996) Structure of the UmuD′ protein and its regulation in response to DNA damage. Nature 380:727–730
176. Simon SM, Sousa FJ, Mohana-Borges R, Walker GC (2008) Regulation of *Escherichia coli* SOS mutagenesis by dimeric intrinsically disordered umuD gene products. Proc Natl Acad Sci USA 105:1152–1157
177. Burckhardt SE, Woodgate R, Scheuermann RH, Echols H (1988) UmuD mutagenesis protein of *Escherichia coli*: overproduction, purification and cleavage by RecA. Proc Natl Acad Sci USA 85:1811–1815

178. Shinagawa H, Iwasaki H, Kato T, Nakata A (1988) RecA protein-dependent cleavage of UmuD protein and SOS mutagenesis. Proc Natl Acad Sci USA 85:1806–1810
179. Godoy VG, Jarosz DF, Simon SM, Abyzov A, Ilyin V, Walker GC (2007) UmuD and RecA directly modulate the mutagenic potential of the Y family DNA polymerase DinB. Mol Cell 28:1058–1070
180. McDonald JP, Frank EG, Levine AS, Woodgate R (1998) Intermolecular cleavage by UmuD-like mutagenesis proteins. Proc Natl Acad Sci USA 95:1478–1483
181. Sutton MD, Kim M, Walker GC (2001) Genetic and biochemical characterization of a novel umud mutation: insights into a mechanism for UmuD self-cleavage. J Bacteriol 183:347–357
182. Sutton MD, Walker GC (2001) UmuDC-mediated cold sensitivity is a manifestation of functions of the $UmuD_2C$ complex involved in a DNA damage checkpoint control. J Bacteriol 183:1215–1224
183. Opperman T, Murli S, Walker GC (1996) The genetic requirements for UmuDC-mediated cold sensitivity are distinct from those for SOS mutagenesis. J Bacteriol 178:4400–4411
184. Marsh L, Walker GC (1985) Cold sensitivity induced by overproduction of UmuDC in *Escherichia coli*. J Bacteriol 162:155–161
185. Tang M, Shen X, Frank EG, O'Donnell M, Woodgate R, Goodman MF (1999) UmuD'2C is an error-prone DNA polymerase, *Escherichia coli* pol V. Proc Natl Acad Sci USA 96:8919–8924
186. Sommer S, Bailone A, Devoret R (1993) The appearance of the UmuD'C protein complex in *Escherichia coli* switches repair from homologous recombination to SOS mutagenesis. Mol Microbiol 10:963–971
187. Szpilewska H, Bertrand P, Bailone A, Dutreix M (1995) In vitro inhibition of RecA-mediated homologous pairing by UmuD'C proteins. Biochimie 77:848–853
188. Frank EG, Cheng N, Do CC, Cerritelli ME, Bruck I, Goodman MF, Egelman EH, Woodgate R, Steven AC (2000) Visualization of two binding sites for the *Escherichia coli* UmuD'(2)C complex (DNA pol V) on RecA-ssDNA filaments. J Mol Biol 297:585–597
189. Rehrauer WM, Bruck I, Woodgate R, Goodman MF, Kowalczykowski SC (1998) Modulation of RecA nucleoprotein function by the mutagenic UmuD'C protein complex. J Biol Chem 273:32384–32387
190. Berdichevsky A, Izhar L, Livneh Z (2002) Error-free recombinational repair predominates over mutagenic translesion replication in *E. coli*. Mol Cell 10:917–924
191. Gonzalez M, Frank EG, Levine AS, Woodgate R (1998) Lon-mediated proteolysis of the *Escherichia coli* UmuD mutagenesis protein: in vitro degradation and identification of residues required for proteolysis. Genes Dev 12:3889–3899
192. Gonzalez M, Rasulova F, Maurizi MR, Woodgate R (2000) Subunit-specific degradation of the UmuD/UmuD′ heterodimer by the ClpXp protease: the role of trans recognition in UmuD′ stability. EMBO J 19:5251–5258
193. Neher SB, Sauer RT, Baker TA (2003) Distinct peptide signals in the UmuD and UmuD′ subunits of UmuD/D′ mediate tethering and substrate processing by the ClpXP protease. Proc Natl Acad Sci USA 100:13219–13224
194. Frank EG, Ennis DG, Gonzalez M, Levine AS, Woodgate R (1996) Regulation of SOS mutagenesis by proteolysis. Proc Natl Acad Sci USA 93:10291–10296
195. Beuning PJ, Simon SM, Zemla A, Barsky D, Walker GC (2006) A non-cleavable UmuD variant that acts as a UmuD′ mimic. J Biol Chem 281:9633–9640
196. Sutton MD, Opperman T, Walker GC (1999) The *Escherichia coli* SOS mutagenesis proteins UmuD and UmuD′ interact physically with the replicative DNA polymerase. Proc Natl Acad Sci USA 96:12373–12378
197. Sutton MD, Narumi I, Walker GC (2002) Posttranslational modification of the umuD-encoded subunit of *Escherichia coli* DNA polymerase V regulates its interactions with the beta processivity clamp. Proc Natl Acad Sci USA 99:5307–5312
198. Ferentz AE, Opperman T, Walker GC, Wagner G (1997) Dimerization of the UmuD′ protein in solution and its implications for regulation of SOS mutagenesis. Nat Struct Biol 4:979–983

199. Peat TS, Frank EG, McDonald JP, Levine AS, Woodgate R, Hendrickson WA (1996) The UmuD′ protein filament and its potential role in damage induced mutagenesis. Structure 4:1401–1412
200. Ohta T, Sutton MD, Guzzo A, Cole S, Ferentz AE, Walker GC (1999) Mutations affecting the ability of the *Escherichia coli* UmuD′ protein to participate in SOS mutagenesis. J Bacteriol 181:177–185
201. Guzzo A, Lee MH, Oda K, Walker GC (1996) Analysis of the region between amino acids 30 and 42 of intact UmuD by a monocysteine approach. J Bacteriol 178:7295–7303
202. Lee MH, Guzzo A, Walker GC (1996) Inhibition of RecA-mediated cleavage in covalent dimers of UmuD. J Bacteriol 178:7304–7307
203. Lee MH, Ohta T, Walker GC (1994) A monocysteine approach for probing the structure and interactions of the UmuD protein. J Bacteriol 176:4825–4837
204. McLenigan M, Peat TS, Frank EG, McDonald JP, Gonzalez M, Levine AS, Hendrickson WA, Woodgate R (1998) Novel *Escherichia coli* UmuD′ mutants: structure function insights into SOS mutagenesis. J Bacteriol 180:4658–4666
205. Beuning PJ, Chan S, Waters LS, Addepalli H, Ollivierre JN, Walker GC (2009) Characterization of novel alleles of the *Escherichia coli* umuDC genes identifies additional interaction sites of UmuC with the beta clamp. J Bacteriol 191:5910–5920
206. Sutton MD, Guzzo A, Narumi I, Costanzo M, Altenbach C, Ferentz AE, Hubbell WL, Walker GC (2002) A model for the structure of the *Escherichia coli* SOS-regulated UmuD(2) protein. DNA Repair 1:77–93
207. Luo Y, Pfuetzner RA, Mosimann S, Paetzel M, Frey EA, Cherney M, Kim B, Little JW, Strynadka NC (2001) Crystal structure of LexA: a conformational switch for regulation of self-cleavage. Cell 106:585–594
208. Radivojac P, Iakoucheva LM, Oldfield CJ, Obradovic Z, Uversky VN, Dunker AK (2007) Intrinsic disorder and functional proteomics. Biophys J 92:1439–1456
209. Tadmor Y, Ascarelli-Goell R, Skaliter R, Livneh Z (1992) Overproduction of the beta subunit of DNA polymerase III holoenzyme reduces UV mutagenesis in *Escherichia coli.* J Bacteriol 174:2517–2524
210. Dalrymple BP, Kongsuwan K, Wijffels G, Dixon NE, Jennings PA (2001) A universal protein–protein interaction motif in the eubacterial DNA replication and repair systems. Proc Natl Acad Sci USA 98:11627–11632
211. Sutton MD (2006) Damage signals triggering the *Escherichia coli* SOS response. In: Wolfram Seide YWK, Doetsch PW (eds) DNA damage and recognition. Taylor & Francis, New York, NY, pp 781–802
212. Hogg M, Wallace SS, Doublie S (2005) Bumps in the road: how replicative DNA polymerases see DNA damage. Curr Opin Struct Biol 15:86–93
213. Yang W (2008) Structure and mechanism for DNA lesion recognition. Cell Res 18:184–197
214. Sutton MD (2009) Coordinating DNA polymerase traffic during high and low fidelity synthesis. Biochim Biophys Acta. doi: 10.1016/j.bbapap.2009.06.010
215. Furukohri A, Goodman MF, Maki H (2008) A dynamic polymerase exchange with *Escherichia coli* DNA polymerase IV replacing DNA polymerase III on the sliding clamp. J Biol Chem 283:11260–11269
216. Uchida K, Furukohri A, Shinozaki Y, Mori T, Ogawara D, Kanaya S, Nohmi T, Maki H, Akiyama M (2008) Overproduction of *Escherichia coli* DNA polymerase DinB (Pol IV) inhibits replication fork progression and is lethal. Mol Microbiol 70:608–622
217. Indiani C, Langston LD, Yurieva O, Goodman MF, O'Donnell M (2009) Translesion DNA polymerases remodel the replisome and alter the speed of the replicative helicase. Proc Natl Acad Sci USA 106:6031–6038
218. Maor-Shoshani A, Livneh Z (2002) Analysis of the stimulation of DNA polymerase V of *Escherichia coli* by processivity proteins. Biochemistry 41:14438–14446
219. Maul RW, Ponticelli SKS, Duzen JM, Sutton MD (2007) Differential binding of *Escherichia coli* DNA polymerases to the beta-sliding clamp. Mol Microbiol 65:811–827
220. Sutton MD (2004) The *Escherichia coli* dnaN159 mutant displays altered DNA polymerase usage and chronic SOS induction. J Bacteriol 186:6738–6748

221. Sutton MD, Duzen JM (2006) Specific amino acid residues in the beta sliding clamp establish a DNA polymerase usage hierarchy in *Escherichia coli*. DNA Repair 5:312–323
222. Xing G, Kirouac K, Shin YJ, Bell SD, Ling H (2009) Structural insight into recruitment of translesion DNA polymerase Dpo4 to sliding clamp PCNA. Mol Microbiol 71:678–691
223. Heltzel JM, Maul RW, Scouten Ponticelli SK, Sutton MD (2009) A model for DNA polymerase switching involving a single cleft and the rim of the sliding clamp. Proc Natl Acad Sci USA 106:12664–12669
224. Friedberg EC, Lehmann AR, Fuchs RP (2005) Trading places: how do DNA polymerases switch during translesion DNA synthesis? Mol Cell 18:499–505
225. Hsu GW, Kiefer JR, Burnouf D, Becherel OJ, Fuchs RP, Beese LS (2004) Observing translesion synthesis of an aromatic amine DNA adduct by a high-fidelity DNA polymerase. J Biol Chem 279:50280–50285
226. Johnson SJ, Taylor JS, Beese LS (2003) Processive DNA synthesis observed in a polymerase crystal suggests a mechanism for the prevention of frameshift mutations. Proc Natl Acad Sci USA 100:3895–3900
227. Kiefer JR, Mao C, Braman JC, Beese LS (1998) Visualizing DNA replication in a catalytically active Bacillus DNA polymerase crystal. Nature 391:304–307
228. Curti E, McDonald JP, Mead S, Woodgate R (2009) DNA polymerase switching: effects on spontaneous mutagenesis in *Escherichia coli*. Mol Microbiol 71(2):315–331
229. Delmas S, Matic I (2006) Interplay between replication and recombination in *Escherichia coli*: impact of the alternative DNA polymerases. Proc Natl Acad Sci USA 103:4564–4569
230. Napolitano R, Janel-Bintz R, Wagner J, Fuchs RP (2000) All three SOS-inducible DNA polymerases (Pol II, Pol IV and Pol V) are involved in induced mutagenesis. EMBO J 19:6259–6265
231. Shurtleff BW, Ollivierre JN, Tehrani M, Walker GC, Beuning PJ (2009) Steric gate variants of UmuC confer UV hypersensitivity on *Escherichia coli*. J Bacteriol 191:4815–4823
232. DeLucia AM, Chaudhuri S, Potapova O, Grindley NDF, Joyce CM (2006) The properties of steric gate mutants reveal different constraints within the active sites of Y-family and A-family DNA polymerases. J Biol Chem 281:27286–27291
233. DeLucia AM, Grindley NDF, Joyce CM (2003) An error-prone family Y DNA polymerase (DinB homolog from *Sulfolobus solfataricus*) uses a "steric gate" residue for discrimination against ribonucleotides. Nucleic Acids Res 31:4129–4137
234. Courcelle CT, Belle JJ, Courcelle J (2005) Nucleotide excision repair or polymerase V-mediated lesion bypass can act to restore UV-arrested replication forks in *Escherichia coli*. J Bacteriol 187:6953–6961
235. Courcelle CT, Chow KH, Casey A, Courcelle J (2006) Nascent DNA processing by RecJ favors lesion repair over translesion synthesis at arrested replication forks in *Escherichia coli*. Proc Natl Acad Sci USA 103:9154–9159
236. Waters LS, Walker GC (2006) The critical mutagenic translesion synthesis DNA polymerase Rev1 is highly expressed during G2/M phase rather than S phase. Proc Natl Acad Sci USA 103:8971–8976
237. Kelly TJ, Stillman B (2006) Duplication of DNA in eukaryotic cells. In: DePamphilis ML (ed) DNA replication and human disease. Cold Spring Harbor Laboratory Press, Cold Spring Harbor, NY
238. Burgers PM (2009) Polymerase dynamics at the eukaryotic DNA replication fork. J Biol Chem 284:4041–4045
239. Kunkel TA, Burgers PM (2008) Dividing the workload at a eukaryotic replication fork. Trends Cell Biol 18:521–527
240. Pavlov YI, Shcherbakova PV, Rogozin IB (2006) Roles of DNA polymerases in replication, repair, and recombination in eukaryotes. Int Rev Cytol 255:41–132
241. Syvaoja J, Suomensaari S, Nishida C, Goldsmith JS, Chui GS, Jain S, Linn S (1990) DNA polymerases alpha, delta, and epsilon: three distinct enzymes from HeLa cells. Proc Natl Acad Sci USA 87:6664–6668
242. Morrison A, Araki H, Clark AB, Hamatake RK, Sugino A (1990) A third essential DNA polymerase in *S. cerevisiae*. Cell 62:1143–1151

243. Chilkova O, Stenlund P, Isoz I, Stith CM, Grabowski P, Lundstrom EB, Burgers PM, Johansson E (2007) The eukaryotic leading and lagging strand DNA polymerases are loaded onto primer-ends via separate mechanisms but have comparable processivity in the presence of PCNA. Nucleic Acids Res 35:6588–6597
244. Tahirov TH, Makarova KS, Rogozin IB, Pavlov YI, Koonin EV (2009) Evolution of DNA polymerases: an inactivated polymerase-exonuclease module in Pol epsilon and a chimeric origin of eukaryotic polymerases from two classes of archaeal ancestors. Biol Direct 4:11
245. Wardle J, Burgers PM, Cann IK, Darley K, Heslop P, Johansson E, Lin LJ, McGlynn P, Sanvoisin J, Stith CM, Connolly BA (2008) Uracil recognition by replicative DNA polymerases is limited to the archaea, not occurring with bacteria and eukarya. Nucleic Acids Res 36:705–711
246. Dua R, Levy DL, Campbell JL (1998) Role of the putative zinc finger domain of *Saccharomyces cerevisiae* DNA polymerase epsilon in DNA replication and the S/M checkpoint pathway. J Biol Chem 273:30046–30055
247. Klinge S, Nunez-Ramirez R, Llorca O, Pellegrini L (2009) 3D architecture of DNA Pol alpha reveals the functional core of multi-subunit replicative polymerases. EMBO J 28:1978–1987
248. Sanchez Garcia J, Ciufo LF, Yang X, Kearsey SE, MacNeill SA (2004) The C-terminal zinc finger of the catalytic subunit of DNA polymerase delta is responsible for direct interaction with the B-subunit. Nucleic Acids Res 32:3005–3016
249. Copeland WC, Wang TS (1991) Catalytic subunit of human DNA polymerase alpha overproduced from baculovirus-infected insect cells. Structural and enzymological characterization. J Biol Chem 266:22739–22748
250. Takada-Takayama R, Suzuki M, Enomoto T, Hanaoka F, Ui M (1990) Purification and characterization of mouse DNA polymerase alpha devoid of primase activity. FEBS Lett 273:27–30
251. Garg P, Burgers PM (2005) DNA polymerases that propagate the eukaryotic DNA replication fork. Crit Rev Biochem Mol Biol 40:115–128
252. Klinge S, Hirst J, Maman JD, Krude T, Pellegrini L (2007) An iron-sulfur domain of the eukaryotic primase is essential for RNA primer synthesis. Nat Struct Mol Biol 14:875–877
253. Uchiyama M, Wang TS (2004) The B-subunit of DNA polymerase alpha-primase associates with the origin recognition complex for initiation of DNA replication. Mol Cell Biol 24:7419–7434
254. Pavlov YI, Frahm C, Nick McElhinny SA, Niimi A, Suzuki M, Kunkel TA (2006) Evidence that errors made by DNA polymerase alpha are corrected by DNA polymerase delta. Curr Biol 16:202–207
255. Baranovskiy AG, Babayeva ND, Liston VG, Rogozin IB, Koonin EV, Pavlov YI, Vassylyev DG, Tahirov TH (2008) X-ray structure of the complex of regulatory subunits of human DNA polymerase delta. Cell Cycle 7:3026–3036
256. Gomes XV, Burgers PM (2000) Two modes of FEN1 binding to PCNA regulated by DNA. EMBO J 19:3811–3821
257. Gray FC, Pohler JR, Warbrick E, MacNeill SA (2004) Mapping and mutation of the conserved DNA polymerase interaction motif (DPIM) located in the C-terminal domain of fission yeast DNA polymerase delta subunit Cdc27. BMC Mol Biol 5:21
258. Johansson E, Garg P, Burgers PM (2004) The Pol32 subunit of DNA polymerase delta contains separable domains for processive replication and proliferating cell nuclear antigen (PCNA) binding. J Biol Chem 279:1907–1915
259. Gerik KJ, Li X, Pautz A, Burgers PM (1998) Characterization of the two small subunits of *Saccharomyces cerevisiae* DNA polymerase delta. J Biol Chem 273:19747–19755
260. Lawrence CW (2002) Cellular roles of DNA polymerase zeta and Rev1 protein. DNA Repair 1:425–435
261. Acharya N, Johnson RE, Pages V, Prakash L, Prakash S (2009) Yeast Rev1 protein promotes complex formation of DNA polymerase zeta with Pol32 subunit of DNA polymerase delta. Proc Natl Acad Sci USA 106:9631–9636

262. Prakash S, Johnson RE, Prakash L (2005) Eukaryotic translesion synthesis DNA polymerases: specificity of structure and function. Annu Rev Biochem 74:317–353
263. Kesti T, Flick K, Keranen S, Syvaoja JE, Wittenberg C (1999) DNA polymerase epsilon catalytic domains are dispensable for DNA replication, DNA repair, and cell viability. Mol Cell 3:679–685
264. Dua R, Levy DL, Li CM, Snow PM, Campbell JL (2002) In vivo reconstitution of *Saccharomyces cerevisiae* DNA polymerase epsilon in insect cells. Purification and characterization. J Biol Chem 277:7889–7896
265. Pursell ZF, Kunkel TA (2008) DNA polymerase epsilon: a polymerase of unusual size (and complexity). Prog Nucleic Acid Res Mol Biol 82:101–145
266. Jaszczur M, Rudzka J, Kraszewska J, Flis K, Polaczek P, Campbell JL, Fijalkowska IJ, Jonczyk P (2009) Defective interaction between Pol2p and Dpb2p, subunits of DNA polymerase epsilon, contributes to a mutator phenotype in *Saccharomyces cerevisiae*. Mutat Res 669:27–35
267. Iida T, Araki H (2004) Noncompetitive counteractions of DNA polymerase epsilon and ISW2/yCHRAC for epigenetic inheritance of telomere position effect in *Saccharomyces cerevisiae*. Mol Cell Biol 24:217–227
268. Takayama Y, Kamimura Y, Okawa M, Muramatsu S, Sugino A, Araki H (2003) GINS, a novel multiprotein complex required for chromosomal DNA replication in budding yeast. Genes Dev 17:1153–1165
269. Feng W, Rodriguez-Menocal L, Tolun G, D'Urso G (2003) *Schizosacchromyces pombe* Dpb2 binds to origin DNA early in S phase and is required for chromosomal DNA replication. Mol Biol Cell 14:3427–3436
270. Masumoto H, Sugino A, Araki H (2000) Dpb11 controls the association between DNA polymerases alpha and epsilon and the autonomously replicating sequence region of budding yeast. Mol Cell Biol 20:2809–2817
271. Washington MT, Carlson KD, Freudenthal BD, Pryor JM (2009) Variations on a theme: eukaryotic Y-family DNA polymerases. Biochim Biophys Acta. doi: 10.1016/j.bbapap.2009.07.004
272. Waters LS, Minesinger BK, Wiltrout ME, D'Souza S, Woodruff RV, Walker GC (2009) Eukaryotic translesion polymerases and their roles and regulation in DNA damage tolerance. Microbiol Mol Biol Rev 73:134–154
273. Yang W, Woodgate R (2007) What a difference a decade makes: insights into translesion DNA synthesis. Proc Natl Acad Sci USA 104:15591–15598
274. Johnson RE, Kondratick CM, Prakash S, Prakash L (1999) hRAD30 mutations in the variant form of xeroderma pigmentosum. Science 285:263–265
275. Masutani C, Kusumoto R, Yamada A, Dohmae N, Yokoi M, Yuasa M, Araki M, Iwai S, Takio K, Hanaoka F (1999) The XPV (xeroderma pigmentosum variant) gene encodes human DNA polymerase eta. Nature 399:700–704
276. Lee KY, Myung K (2008) PCNA modifications for regulation of post-replication repair pathways. Mol Cells 26:5–11
277. McCulloch SD, Kokoska RJ, Masutani C, Iwai S, Hanaoka F, Kunkel TA (2004) Preferential cis-syn thymine dimer bypass by DNA polymerase eta occurs with biased fidelity. Nature 428:97–100
278. Washington MT, Johnson RE, Prakash S, Prakash L (2001) Mismatch extension ability of yeast and human DNA polymerase eta. J Biol Chem 276:2263–2266
279. Vaisman A, Lehmann AR, Woodgate R (2004) DNA polymerases eta and iota. Adv Protein Chem 69:205–228
280. Trincao J, Johnson RE, Escalante CR, Prakash S, Prakash L, Aggarwal AK (2001) Structure of the catalytic core of *S. cerevisiae* DNA polymerase eta: implications for translesion DNA synthesis. Mol Cell 8:417–426
281. Friedberg EC, Feaver WJ, Gerlach VL (2000) The many faces of DNA polymerases: strategies for mutagenesis and for mutational avoidance. Proc Natl Acad Sci USA 97:5681–5683
282. McDonald JP, Rapic-Otrin V, Epstein JA, Broughton BC, Wang X, Lehmann AR, Wolgemuth DJ, Woodgate R (1999) Novel human and mouse homologs of *Saccharomyces cerevisiae* DNA polymerase eta. Genomics 60:20–30

283. Kannouche P, Fernandez de Henestrosa AR, Coull B, Vidal AE, Gray C, Zicha D, Woodgate R, Lehmann AR (2003) Localization of DNA polymerases eta and iota to the replication machinery is tightly co-ordinated in human cells. EMBO J 22:1223–1233
284. Tissier A, McDonald JP, Frank EG, Woodgate R (2000) Poliota, a remarkably error-prone human DNA polymerase. Genes Dev 14:1642–1650
285. Choi JY, Guengerich FP (2006) Kinetic evidence for inefficient and error-prone bypass across bulky N2-guanine DNA adducts by human DNA polymerase iota. J Biol Chem 281:12315–12324
286. Nair DT, Johnson RE, Prakash S, Prakash L, Aggarwal AK (2004) Replication by human DNA polymerase-iota occurs by Hoogsteen base-pairing. Nature 430:377–380
287. Johnson RE, Washington MT, Haracska L, Prakash S, Prakash L (2000) Eukaryotic polymerases iota and zeta act sequentially to bypass DNA lesions. Nature 406:1015–1019
288. Tissier A, Frank EG, McDonald JP, Vaisman A, Fernandez de Henestrosa AR, Boudsocq F, McLenigan MP, Woodgate R (2001) Biochemical characterization of human DNA polymerase iota provides clues to its biological function. Biochem Soc Trans 29:183–187
289. Ogi T, Kannouche P, Lehmann AR (2005) Localisation of human Y-family DNA polymerase kappa: relationship to PCNA foci. J Cell Sci 118:129–136
290. Ogi T, Shinkai Y, Tanaka K, Ohmori H (2002) Polkappa protects mammalian cells against the lethal and mutagenic effects of benzo[a]pyrene. Proc Natl Acad Sci USA 99:15548–15553
291. Choi JY, Angel KC, Guengerich FP (2006) Translesion synthesis across bulky N2-alkyl guanine DNA adducts by human DNA polymerase kappa. J Biol Chem 281:21062–21072
292. Suzuki N, Ohashi E, Kolbanovskiy A, Geacintov NE, Grollman AP, Ohmori H, Shibutani S (2002) Translesion synthesis by human DNA polymerase kappa on a DNA template containing a single stereoisomer of dG-(+)- or dG-(-)-anti-N(2)-BPDE (7, 8-dihydroxy-anti-9, 10-epoxy-7, 8, 9, 10-tetrahydrobenzo[a]pyrene). Biochemistry 41:6100–6106
293. Haracska L, Prakash L, Prakash S (2002) Role of human DNA polymerase kappa as an extender in translesion synthesis. Proc Natl Acad Sci USA 99:16000–16005
294. Lone S, Townson SA, Uljon SN, Johnson RE, Brahma A, Nair DT, Prakash S, Prakash L, Aggarwal AK (2007) Human DNA polymerase kappa encircles DNA: implications for mismatch extension and lesion bypass. Mol Cell 25:601–614
295. Ohashi E, Ogi T, Kusumoto R, Iwai S, Masutani C, Hanaoka F, Ohmori H (2000) Error-prone bypass of certain DNA lesions by the human DNA polymerase kappa. Genes Dev 14:1589–1594
296. Ohashi E, Bebenek K, Matsuda T, Feaver WJ, Gerlach VL, Friedberg EC, Ohmori H, Kunkel TA (2000) Fidelity and processivity of DNA synthesis by DNA polymerase kappa, the product of the human DINB1 gene. J Biol Chem 275:39678–39684
297. Nelson JR, Lawrence CW, Hinkle DC (1996) Deoxycytidyl transferase activity of yeast REV1 protein. Nature 382:729–731
298. Nair DT, Johnson RE, Prakash L, Prakash S, Aggarwal AK (2005) Rev1 employs a novel mechanism of DNA synthesis using a protein template. Science 309:2219–2222
299. Lawrence CW, Christensen R (1976) UV mutagenesis in radiation-sensitive strains of yeast. Genetics 82:207–232
300. Johnson RE, Torres-Ramos CA, Izumi T, Mitra S, Prakash S, Prakash L (1998) Identification of APN2, the *Saccharomyces cerevisiae* homolog of the major human AP endonuclease HAP1, and its role in the repair of abasic sites. Genes Dev 12:3137–3143
301. Baynton K, Bresson-Roy A, Fuchs RP (1999) Distinct roles for Rev1p and Rev7p during translesion synthesis in *Saccharomyces cerevisiae*. Mol Microbiol 34:124–133
302. Lawrence CW (2004) Cellular functions of DNA polymerase zeta and Rev1 protein. Adv Protein Chem 69:167–203
303. Guo C, Sonoda E, Tang TS, Parker JL, Bielen AB, Takeda S, Ulrich HD, Friedberg EC (2006) REV1 protein interacts with PCNA: significance of the REV1 BRCT domain in vitro and in vivo. Mol Cell 23:265–271
304. Acharya N, Haracska L, Johnson RE, Unk I, Prakash S, Prakash L (2005) Complex formation of yeast Rev1 and Rev7 proteins: a novel role for the polymerase-associated domain. Mol Cell Biol 25:9734–9740

305. Ohashi E, Murakumo Y, Kanjo N, Akagi J, Masutani C, Hanaoka F, Ohmori H (2004) Interaction of hREV1 with three human Y-family DNA polymerases. Genes Cells 9:523–531
306. Tissier A, Kannouche P, Reck MP, Lehmann AR, Fuchs RP, Cordonnier A (2004) Co-localization in replication foci and interaction of human Y-family members. DNA polymerase pol eta and REVl protein. DNA Repair 3:1503–1514
307. Guo C, Fischhaber PL, Luk-Paszyc MJ, Masuda Y, Zhou J, Kamiya K, Kisker C, Friedberg EC (2003) Mouse Rev1 protein interacts with multiple DNA polymerases involved in translesion DNA synthesis. EMBO J 22:6621–6630
308. Murakumo Y, Ogura Y, Ishii H, Numata S, Ichihara M, Croce CM, Fishel R, Takahashi M (2001) Interactions in the error-prone postreplication repair proteins hREV1, hREV3, and hREV7. J Biol Chem 276:35644–35651
309. Nelson JR, Lawrence CW, Hinkle DC (1996) Thymine-thymine dimer bypass by yeast DNA polymerase zeta. Science 272:1646–1649
310. Haracska L, Unk I, Johnson RE, Johansson E, Burgers PM, Prakash S, Prakash L (2001) Roles of yeast DNA polymerases delta and zeta and of Rev1 in the bypass of abasic sites. Genes Dev 15:945–954
311. Cuniasse P, Sowers LC, Eritja R, Kaplan B, Goodman MF, Cognet JA, LeBret M, Guschlbauer W, Fazakerley GV (1987) An abasic site in DNA. Solution conformation determined by proton NMR and molecular mechanics calculations. Nucleic Acids Res 15:8003–8022
312. Kalnik MW, Chang CN, Grollman AP, Patel DJ (1988) NMR studies of abasic sites in DNA duplexes: deoxyadenosine stacks into the helix opposite the cyclic analogue of 2-deoxyribose. Biochemistry 27:924–931
313. Bemark M, Khamlichi AA, Davies SL, Neuberger MS (2000) Disruption of mouse polymerase zeta (Rev3) leads to embryonic lethality and impairs blastocyst development in vitro. Curr Biol 10:1213–1216
314. Zhong X, Garg P, Stith CM, Nick McElhinny SA, Kissling GE, Burgers PM, Kunkel TA (2006) The fidelity of DNA synthesis by yeast DNA polymerase zeta alone and with accessory proteins. Nucleic Acids Res 34:4731–4742
315. Majka J, Burgers PM (2004) The PCNA-RFC families of DNA clamps and clamp loaders. Prog Nucleic Acid Res Mol Biol 78:227–260
316. Moldovan GL, Pfander B, Jentsch S (2007) PCNA, the maestro of the replication fork. Cell 129:665–679
317. Krishna TS, Kong XP, Gary S, Burgers PM, Kuriyan J (1994) Crystal structure of the eukaryotic DNA polymerase processivity factor PCNA. Cell 79:1233–1243
318. Lau PJ, Kolodner RD (2003) Transfer of the MSH2.MSH6 complex from proliferating cell nuclear antigen to mispaired bases in DNA. J Biol Chem 278:14–17
319. Xu H, Zhang P, Liu L, Lee MY (2001) A novel PCNA-binding motif identified by the panning of a random peptide display library. Biochemistry 40:4512–4520
320. Gulbis JM, Kelman Z, Hurwitz J, O'Donnell M, Kuriyan J (1996) Structure of the C-terminal region of p21(WAF1/CIP1) complexed with human PCNA. Cell 87:297–306
321. Bruning JB, Shamoo Y (2004) Structural and thermodynamic analysis of human PCNA with peptides derived from DNA polymerase-delta p66 subunit and flap endonuclease-1. Structure 12:2209–2219
322. Hishiki A, Hashimoto H, Hanafusa T, Kamei K, Ohashi E, Shimizu T, Ohmori H, Sato M (2009) Structural basis for novel interactions between human translesion synthesis polymerases and proliferating cell nuclear antigen. J Biol Chem 284:10552–10560
323. Lehmann AR, Niimi A, Ogi T, Brown S, Sabbioneda S, Wing JF, Kannouche PL, Green CM (2007) Translesion synthesis: Y-family polymerases and the polymerase switch. DNA Repair 6:891–899
324. Stelter P, Ulrich HD (2003) Control of spontaneous and damage-induced mutagenesis by SUMO and ubiquitin conjugation. Nature 425:188–191
325. Bergink S, Jentsch S (2009) Principles of ubiquitin and SUMO modifications in DNA repair. Nature 458:461–467
326. Ulrich HD (2009) Regulating post-translational modifications of the eukaryotic replication clamp PCNA. DNA Repair 8:461–469

327. Weissman AM (2001) Themes and variations on ubiquitylation. Nat Rev Mol Cell Biol 2:169–178
328. Jentsch S, Pyrowolakis G (2000) Ubiquitin and its kin: how close are the family ties? Trends Cell Biol 10:335–342
329. Pickart CM (2001) Mechanisms underlying ubiquitination. Annu Rev Biochem 70:503–533
330. Schwartz DC, Hochstrasser M (2003) A superfamily of protein tags: ubiquitin. SUMO and related modifiers. Trends Biochem Sci 28:321–328
331. Gill G (2004) SUMO and ubiquitin in the nucleus: different functions, similar mechanisms? Genes Dev 18:2046–2059
332. Kim KI, Baek SH, Chung CH (2002) Versatile protein tag. SUMO: its enzymology and biological function. J Cell Physiol 191:257–268
333. Melchior F (2000) SUMO – nonclassical ubiquitin. Annu Rev Cell Dev Biol 16:591–626
334. Budzowska M, Kanaar R (2009) Mechanisms of dealing with DNA damage-induced replication problems. Cell Biochem Biophys 53:17–31
335. Hoege C, Pfander B, Moldovan GL, Pyrowolakis G, Jentsch S (2002) RAD6-dependent DNA repair is linked to modification of PCNA by ubiquitin and SUMO. Nature 419:135–141
336. Huang TT, Nijman SM, Mirchandani KD, Galardy PJ, Cohn MA, Haas W, Gygi SP, Ploegh HL, Bernards R, D'Andrea AD (2006) Regulation of monoubiquitinated PCNA by DUB autocleavage. Nat Cell Biol 8:339–347
337. Andersen PL, Xu F, Xiao W (2008) Eukaryotic DNA damage tolerance and translesion synthesis through covalent modifications of PCNA. Cell Res 18:162–173
338. Sheng W, Liao X (2002) Solution structure of a yeast ubiquitin-like protein Smt3: the role of structurally less defined sequences in protein-protein recognitions. Protein Sci 11:1482–1491
339. Haracska L, Torres-Ramos CA, Johnson RE, Prakash S, Prakash L (2004) Opposing effects of ubiquitin conjugation and SUMO modification of PCNA on replicational bypass of DNA lesions in *Saccharomyces cerevisiae*. Mol Cell Biol 24:4267–4274
340. Haracska L, Unk I, Prakash L, Prakash S (2006) Ubiquitylation of yeast proliferating cell nuclear antigen and its implications for translesion DNA synthesis. Proc Natl Acad Sci USA 103:6477–6482
341. Fernandes PH, Lloyd RS (2007) Mutagenic bypass of the butadiene-derived 2′-deoxyuridine adducts by polymerases eta and zeta. Mutat Res 625:40–49
342. Shachar S, Ziv O, Avkin S, Adar S, Wittschieben J, Reissner T, Chaney S, Friedberg EC, Wang Z, Carell T, Geacintov N, Livneh Z (2009) Two-polymerase mechanisms dictate error-free and error-prone translesion DNA synthesis in mammals. EMBO J 28:383–393
343. Ziv O, Geacintov N, Nakajima S, Yasui A, Livneh Z (2009) DNA polymerase zeta cooperates with polymerases kappa and iota in translesion DNA synthesis across pyrimidine photodimers in cells from XPV patients. Proc Natl Acad Sci USA 106:11552–11557
344. Zhang Y, Wu X, Guo D, Rechkoblit O, Geacintov NE, Wang Z (2002) Two-step error-prone bypass of the (+)- and (-)-trans-anti-BPDE-N2-dG adducts by human DNA polymerases eta and kappa. Mutat Res 510:23–35
345. Prakash S, Prakash L (2002) Translesion DNA synthesis in eukaryotes: a one- or two-polymerase affair. Genes Dev 16:1872–1883
346. Haracska L, Prakash S, Prakash L (2003) Yeast DNA polymerase zeta is an efficient extender of primer ends opposite from 7, 8-dihydro-8-Oxoguanine and O6-methylguanine. Mol Cell Biol 23:1453–1459
347. Bienko M, Green CM, Crosetto N, Rudolf F, Zapart G, Coull B, Kannouche P, Wider G, Peter M, Lehmann AR, Hofmann K, Dikic I (2005) Ubiquitin-binding domains in Y-family polymerases regulate translesion synthesis. Science 310:1821–1824
348. Majka J, Burgers PM (2003) Yeast Rad17/Mec3/Ddc1: a sliding clamp for the DNA damage checkpoint. Proc Natl Acad Sci USA 100:2249–2254

349. Parrilla-Castellar ER, Arlander SJ, Karnitz L (2004) Dial 9-1-1 for DNA damage: the Rad9-Hus1-Rad1 (9-1-1) clamp complex. DNA Repair 3:1009–1014
350. Venclovas C, Thelen MP (2000) Structure-based predictions of Rad1, Rad9, Hus1 and Rad17 participation in sliding clamp and clamp-loading complexes. Nucleic Acids Res 28:2481–2493
351. Xu M, Bai L, Gong Y, Xie W, Hang H, Jiang T (2009) Structure and functional implications of the human rad9-hus1-rad1 cell cycle checkpoint complex. J Biol Chem 284:20457–20461
352. Dore AS, Kilkenny ML, Rzechorzek NJ, Pearl LH (2009) Crystal structure of the rad9-rad1-hus1 DNA damage checkpoint complex–implications for clamp loading and regulation. Mol Cell 34:735–745
353. Shiomi Y, Shinozaki A, Nakada D, Sugimoto K, Usukura J, Obuse C, Tsurimoto T (2002) Clamp and clamp loader structures of the human checkpoint protein complexes, Rad9-1-1 and Rad17-RFC. Genes Cells 7:861–868
354. Griffith JD, Lindsey-Boltz LA, Sancar A (2002) Structures of the human Rad17-replication factor C and checkpoint Rad 9-1-1 complexes visualized by glycerol spray/low voltage microscopy. J Biol Chem 277:15233–15236
355. Bermudez VP, Lindsey-Boltz LA, Cesare AJ, Maniwa Y, Griffith JD, Hurwitz J, Sancar A (2003) Loading of the human 9-1-1 checkpoint complex onto DNA by the checkpoint clamp loader hRad17-replication factor C complex in vitro. Proc Natl Acad Sci USA 100:1633–1638
356. Ellison V, Stillman B (2003) Biochemical characterization of DNA damage checkpoint complexes: clamp loader and clamp complexes with specificity for 5′ recessed DNA. PLoS Biol 1:E33
357. Roos-Mattjus P, Hopkins KM, Oestreich AJ, Vroman BT, Johnson KL, Naylor S, Lieberman HB, Karnitz LM (2003) Phosphorylation of human Rad9 is required for genotoxin-activated checkpoint signaling. J Biol Chem 278:24428–24437
358. Fu Y, Zhu Y, Zhang K, Yeung M, Durocher D, Xiao W (2008) Rad6-Rad18 mediates a eukaryotic SOS response by ubiquitinating the 9-1-1 checkpoint clamp. Cell 133:601–611
359. Paulovich AG, Armour CD, Hartwell LH (1998) The *Saccharomyces cerevisiae* RAD9, RAD17, RAD24 and MEC3 genes are required for tolerating irreparable, ultraviolet-induced DNA damage. Genetics 150:75–93
360. Kai M, Wang TS (2003) Checkpoint activation regulates mutagenic translesion synthesis. Genes Dev 17:64–76
361. Kai M, Wang TS (2003) Checkpoint responses to replication stalling: inducing tolerance and preventing mutagenesis. Mutat Res 532:59–73
362. Sabbioneda S, Minesinger BK, Giannattasio M, Plevani P, Muzi-Falconi M, Jinks-Robertson S (2005) The 9-1-1 checkpoint clamp physically interacts with polzeta and is partially required for spontaneous polzeta-dependent mutagenesis in *Saccharomyces cerevisiae*. J Biol Chem 280:38657–38665
363. Kai M, Furuya K, Paderi F, Carr AM, Wang TS (2007) Rad3-dependent phosphorylation of the checkpoint clamp regulates repair-pathway choice. Nat Cell Biol 9:691–697
364. Helt CE, Wang W, Keng PC, Bambara RA (2005) Evidence that DNA damage detection machinery participates in DNA repair. Cell Cycle 4:529–532
365. Paz-Elizur T, Skaliter R, Blumenstein S, Livneh Z (1996) Beta*, a UV-inducible smaller form of the beta subunit sliding clamp of DNA polymerase III of *Escherichia coli*. I. Gene expression and regulation. J Biol Chem 271:2482–2490
366. Skaliter R, Bergstein M, Livneh Z (1996) Beta*, a UV-inducible shorter form of the beta subunit of DNA polymerase III of *Escherichia coli*. II. Overproduction, purification, and activity as a polymerase processivity clamp. J Biol Chem 271:2491–2496
367. Skaliter R, Paz-Elizur T, Livneh Z (1996) A smaller form of the sliding clamp subunit of DNA polymerase III is induced by UV irradiation in *Escherichia coli*. J Biol Chem 271:2478–2481
368. Kirby TW, Harvey S, DeRose EF, Chalov S, Chikova AK, Perrino FW, Schaaper RM, London RE, Pedersen LC (2006) Structure of the *Escherichia coli* DNA polymerase III epsilon-HOT proofreading complex. J Biol Chem 281:38466–38471

369. Scheuermann RH, Echols H (1984) A separate editing exonuclease for DNA replication: the epsilon subunit of *Escherichia coli* DNA polymerase III holoenzyme. Proc Natl Acad Sci USA 81:7747–7751
370. McCulloch SD, Kunkel TA (2008) The fidelity of DNA synthesis by eukaryotic replicative and translesion synthesis polymerases. Cell Res 18:148–161
371. Kobayashi S, Valentine MR, Pham P, O'Donnell M, Goodman MF (2002) Fidelity of *Escherichia coli* DNA polymerase IV. Preferential generation of small deletion mutations by dNTP-stabilized misalignment. J Biol Chem 277:34198–34207
372. Maor-Shoshani A, Reuven NB, Tomer G, Livneh Z (2000) Highly mutagenic replication by DNA polymerase V (UmuC) provides a mechanistic basis for SOS untargeted mutagenesis. Proc Natl Acad Sci USA 97:565–570
373. McCulloch SD, Wood A, Garg P, Burgers PM, Kunkel TA (2007) Effects of accessory proteins on the bypass of a cis-syn thymine-thymine dimer by *Saccharomyces cerevisiae* DNA polymerase eta. Biochemistry 46:8888–8896
374. Matsuda T, Bebenek K, Masutani C, Rogozin IB, Hanaoka F, Kunkel TA (2001) Error rate and specificity of human and murine DNA polymerase eta. J Mol Biol 312:335–346
375. Bebenek K, Tissier A, Frank EG, McDonald JP, Prasad R, Wilson SH, Woodgate R, Kunkel TA (2001) 5′-Deoxyribose phosphate lyase activity of human DNA polymerase iota in vitro. Science 291:2156–2159
376. Kunkel TA, Hamatake RK, Motto-Fox J, Fitzgerald MP, Sugino A (1989) Fidelity of DNA polymerase I and the DNA polymerase I-DNA primase complex from *Saccharomyces cerevisiae*. Mol Cell Biol 9:4447–4458
377. Fortune JM, Pavlov YI, Welch CM, Johansson E, Burgers PM, Kunkel TA (2005) *Saccharomyces cerevisiae* DNA polymerase delta: high fidelity for base substitutions but lower fidelity for single- and multi-base deletions. J Biol Chem 280:29980–29987
378. Shimizu K, Hashimoto K, Kirchner JM, Nakai W, Nishikawa H, Resnick MA, Sugino A (2002) Fidelity of DNA polymerase epsilon holoenzyme from budding yeast *Saccharomyces cerevisiae*. J Biol Chem 277:37422–37429
379. Humphrey W, Dalke A, Schulten K (1996) VMD – visual molecular dynamics. J Mol Graph 14:33–38
380. Sohn SY, Cho Y (2009) Crystal structure of the human rad9-hus1-rad1 clamp. J Mol Biol 390:490–502
381. Winter JA, Christofi P, Morroll S, Bunting KA (2009) The crystal structure of *Haloferax volcanii* proliferating cell nuclear antigen reveals unique surface charge characteristics due to halophilic adaptation. BMC Struct Biol 9:55

Chapter 12
Topological Probes of a Cooperative, Nonspecific Protein–DNA Complex

Claire A. Adams, Manana Melikishvili, and Michael G. Fried

12.1 Introduction

Many proteins bind DNA with moderate cooperativity and low sequence discrimination. Important among these are bacterial and eukaryotic chromosome-structuring proteins [2, 22] and single-stranded DNA-binding proteins [14]. In addition, many sequence-specific DNA-binding proteins interact with nontarget sequences cooperatively and with low sequence discrimination as part of their target-search mechanisms. Examples can be found among the transcription-regulatory proteins (human glucocorticoid receptor [10] and NFκB [23]; *E. coli* CAP [27, 30] and lambda repressor [24]), among the bacterial restriction endonucleases [31], and among the DNA-repair enzymes [18]. The ubiquity of such complexes calls for a better understanding of the interactions that stabilize them and the structures that result. Here we describe DNA-directed cross-linking and topoisomer analyses that are useful for characterizing topologies of protein–DNA complexes. Each is a variation on a classical approach, adapting it to features that are present in cooperative, nonspecific assemblies.

The system that we use to demonstrate these approaches is the interaction of human O^6-alkylguanine-DNA alkyltransferase (AGT) with duplex DNA molecules. AGT is a well-characterized DNA-repair protein [5, 6, 16, 20] that is easily purified to near homogeneity [5]. It is a small, monomeric protein (M_r=21,519; [9]) that binds DNAs containing O^6-alkylguanines (and in some cases O^4-alkylthymines), flips the damaged base out of the duplex stack, and promotes the transfer of the alkyl group to a cysteine residue in its nucleotide binding cleft [5, 32]. Crystal structures of 1:1 complexes formed on short DNAs indicate that individual molecules of AGT occupy ~8 bp on the minor groove face of B-form DNA; this isolated binding is accompanied by modest DNA bending (~15° away from the protein)

M.G. Fried (✉)
Center for Structural Biology, Department of Molecular and Cellular Biochemistry, University of Kentucky, 741 South Limestone, Lexington, KY 40536-0509, USA
e-mail: michael.fried@uky.edu

M.C. Williams and L.J. Maher, III (eds.), *Biophysics of DNA-Protein Interactions*, Biological and Medical Physics, Biomedical Engineering,
DOI 10.1007/978-0-387-92808-1_12,

[5, 6, 32]. In free solution, AGT binds undamaged DNA with modest cooperativity at densities approaching 1 protein/4 bp (or nt for single stranded substrates) [18, 25]. This binding has little base composition- or sequence-specificity [9, 18, 25]. Studies with alkyl transfer-inactive forms suggest that the enzyme binds O^6-alkylguanine lesion-containing DNA with densities and affinities similar to those observed with lesion-free substrates [25], while at comparable binding densities, wild-type protein mediates rapid alkyl transfer from DNA to protein [26]. These results are consistent with the ideas that cooperative, nonspecific binding is part of AGT's mechanism of lesion search and that the binding to lesion-containing DNA sites also takes place in the context of cooperative protein assemblies.

Together, the available crystal structures of 1:1 AGT–DNA complexes [5, 6] and data on binding densities of cooperative complexes [18, 25] allow construction of plausible models of cooperative AGT–DNA assemblies [1]. These models predict a range of structural and functional properties that are amenable to experimental test, among which are the identities of the protein–protein interaction surfaces and the presence of torsionally relaxed, B-form DNA in the complex (Fig. 12.1). The experiments described below test these predictions.

12.2 DNA-Directed Protein Cross-Linking

Within a cooperative assembly, the local concentrations of residues across a protein-interaction interface can be very high, possibly even exceeding 1 M. In contrast, the local concentrations of the same residues in the solution of free protein at equilibrium with the complex may be micromolar or lower. All else being equal, this difference should have a dramatic effect on the relative rates of the bimolecular reactions that contribute to cross-linking (Fig. 12.2a). In addition, the structure of the cooperative complex brings together a subset of reactive groups while others are kept apart. Thus, the sequence distribution of intermolecular cross-links has the potential to identify groups that are in close proximity across a contact interface. Short reagents offer the best spatial resolution; note that the relatively long reagent DMS (spacer length ~11 Å) cross-links as many as four proteins present in a 4:1 AGT–DNA complex, while formaldehyde (spacer length 2–3 Å [19]) cross-links only two (Fig. 12.2b). For this reason, the reagents we prefer for mapping are formaldehyde and Tris (2,2′-Bipyridyl) ruthenium(II) chloride (Ru(II)), a reagent that produces direct cross-links between reactive amino-acid side chains [7].

AGT–DNA complexes were formed under conditions of DNA-excess and cross-linked with either formaldehyde or Ru(II) reagent, according to published methods [7, 21]. Monomeric and covalently dimeric proteins were resolved by SDS-PAGE [11], subjected to in-gel digestion with sequencing-grade trypsin and/or chymotrypsin [28], and the resulting soluble polypeptides were resolved by MALDI-TOF mass spectrometry (Fig. 12.2c). Comparison of the spectra from cross-linked and uncross-linked protein samples allowed identification of cross-linked species, while comparison of cross-linked species from monomeric and dimeric protein samples

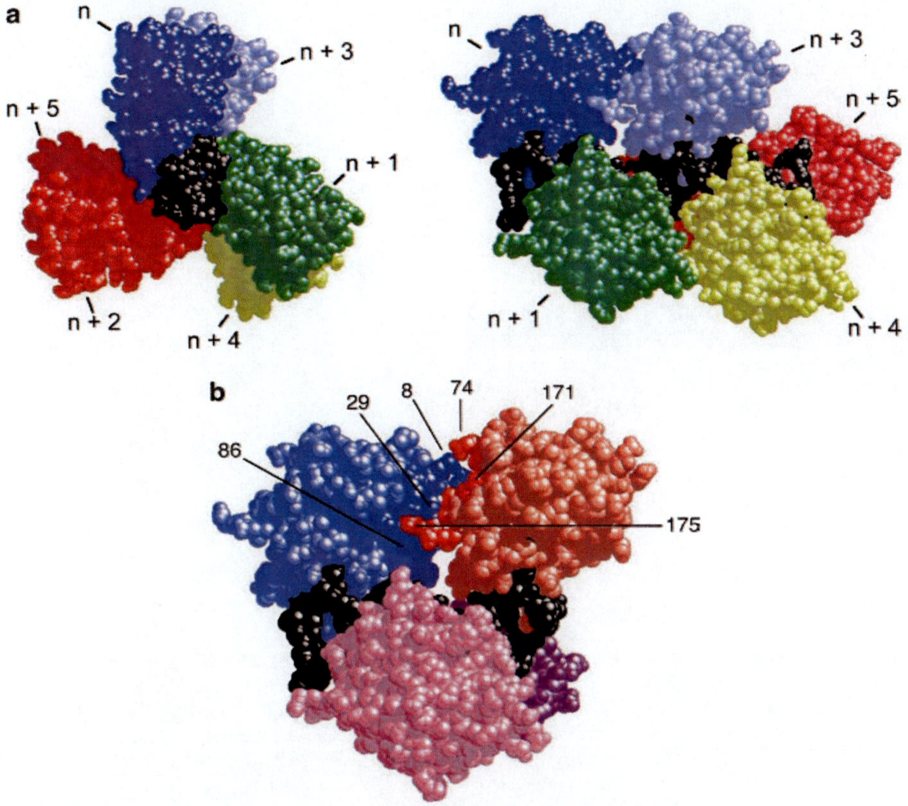

Fig. 12.1 Models of cooperative AGT–DNA complexes containing double-stranded DNA. Panel (**a**). Model of a cooperative complex formed with 24 bp of duplex DNA. End view (*left*) and side view (*right*). The repeating unit of this model is one molecule of AGT (*colors*) plus four base-pairs of DNA (*black*); coordinates were derived from the crystal structure of Daniels et al. [5]. Juxtaposition of repeating units preserved B-DNA helical parameters (twist = 34.6°, rise = 3.4 Å) between base-pairs of adjacent units. This results in a helical distribution of proteins about the DNA axis, with significant contacts between proteins n and n + 3. In the side view, proteins are oriented with C-terminal surfaces toward the left. This model predicts strong contacts between proteins n and n + 3 and little or no contact between proteins n and n + 1 or n + 2. Panel (**b**). Major protein–protein contacts predicted by the model. This version of the model contains four protein molecules and 16 bp of duplex DNA. Proteins are oriented with C-terminal surfaces toward the left. Protein monomers colored *blue* and *pink* are in contact; contact residues on the N-terminal face (residues 6–9; 25, 26, 29, 30, 32, 85–87 and 91) are colored *deep blue*, contact residues on the C-terminal face (residues 73, 74, 108, 109 and 171–176) are colored *red*. The DNA is shown in *black*. Reproduced with modification from [1], with permission

allowed identification of cross-links that were unique to the covalent dimer. The most abundant species that met these criteria are listed in Table 12.1. These candidates may represent intermolecular cross-links or intramolecular cross-links that have been made possible by a conformational change accompanying binding. However, cross-links between polypeptides that are nonadjacent in the structure of

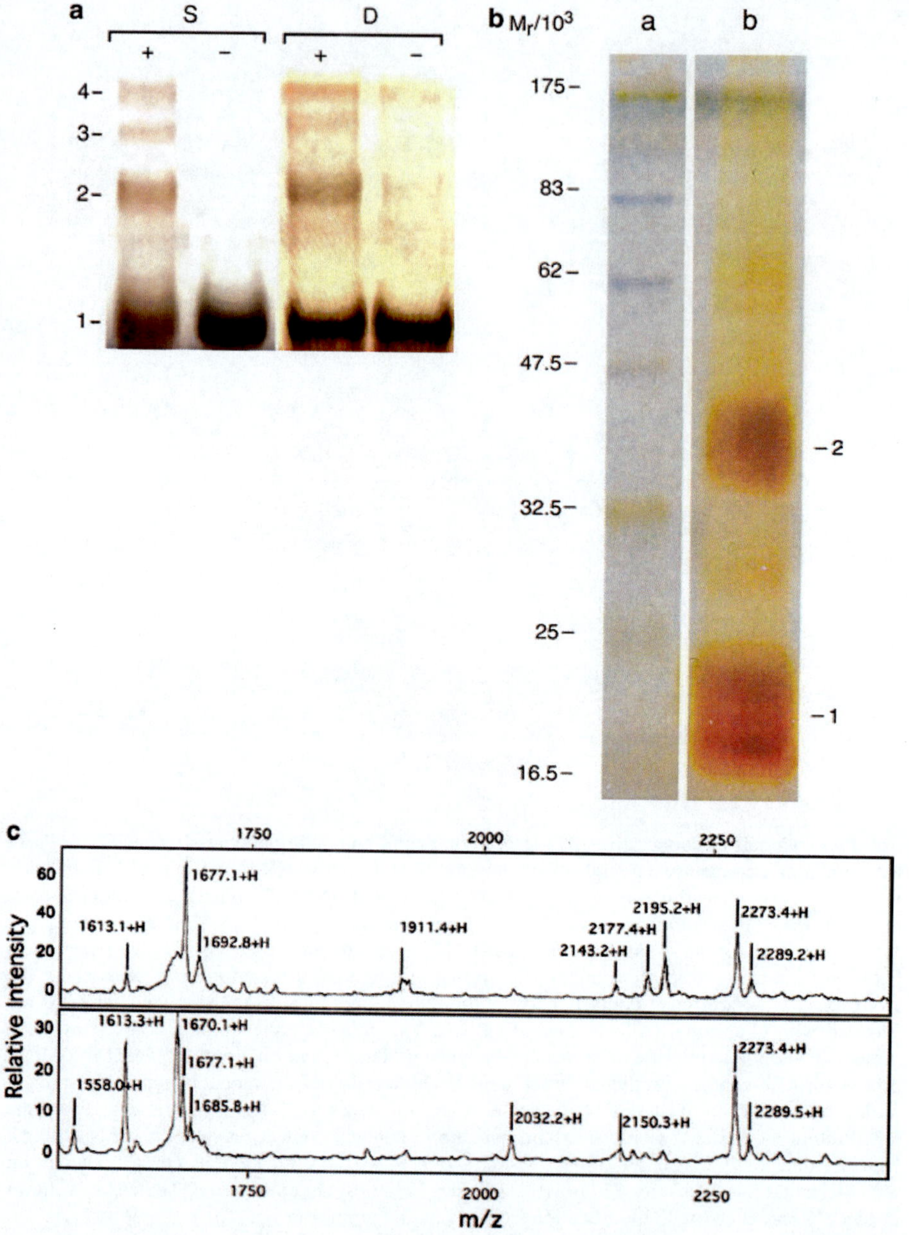

Fig. 12.2 Chemical cross-linking as a probe of protein topology. Panel (**a**). DNA-directed cross-linking of AGT protein. AGT protein (~8.3 μM) was cross-linked with 2.5 mM sulfosuccinimidyl 6-(4′ azido-2′-nitro-phenylamino) hexanoate (sulfo-SANPAH; S) or with 4 mM dimethyl-suberimidate (DMS; D) for 30 min at 4°C in buffer containing 20 mM potassium phosphate (pH 7.4 at 20°C), 100 mM KCl, 1 mM EDTA, in the presence (+) or absence (−) of 16 bp duplex DNA. Samples were resolved by SDS-PAGE on a 12% gel and visualized by silver staining. The dominant AGT complex formed with this DNA contains four protein molecules [18]. Reaction

Table 12.1 Cross-linked species present in covalent dimer but absent in monomer

Cross-linking reagent[a]	Proteolytic enzyme[b]	Mass observed	Mass predicted[c]	Residue numbers	Peptides adjacent across protein–protein interface in model[d]
F	C	1,786	1,788	1–7 and 163–169	+
F	C	2,272	2,276	8–19 and 159–167	+
F	C	2,877	2,874	17–28 and 104–117	+
F	C+T	1,673	1,672	1–7 and 170–176	+
F	C+T	2,001	1,993	1–7 and 105–114	+
F	C+T	2,246	2,243	8–19 and 167–176	+
F	C+T	2,427	2,431	22–34 and 159–167	+
Ru(II)	T	2,169	2,167	61–65 and 193–207	[e]
Ru(II)	C+T	2,372	2,375	19–32 and 169–175	+
Ru(II)	C+T	2,552	2,549	22–34 and 105–114	+
Ru(II)	C+T	2,695	2,698	19–32 and 105–114	+

Table modified from [1], with permission

[a] *F* formaldehyde, *Ru(II)* Tris (2,2′-Bipyridyl) ruthenium(II) chloride

[b] *C* chymotrypsin, *T* trypsin, *C + T* double digest with chymotrypsin and trypsin

[c] Predicted using the ExPASy Compute pI/MW tool (http://www.expasy.org/tools/pi_tool.html) using average isotopic masses for amino acids

[d] Symbol: + denotes juxtaposition in structure

[e] Contact between polypeptides could not be evaluated because the protein structure used in the model does not extend beyond residue 176

the protein monomer are not readily explained by modest conformational transitions. Representative pairs of nonadjacent polypeptides are shown in Fig. 12.3. Our interpretation of the corresponding cross-links is that they span protein–protein interfaces within the cooperative assembly. It is significant that these pairs of polypeptides have one member that is located close to the protein's N-terminus in the protein–DNA structure, while the other is located near the C-terminus. This pairing is consistent with the head-to-tail arrangement of proteins predicted by the models shown in Fig. 12.1.

Fig. 12.2 (continued) products that migrate as monomers through tetramers are indicated by numbers 1–4. Panel (**b**). Formaldehyde cross-linking of AGT-complexes detected by SDS-PAGE. *Lane a*, molecular weight standards (scale shown at *left*). *Lane b*, AGT plus 16mer DNA cross-linked for 120 min with formaldehyde (5 mM). Samples were resolved in a 15% SDS-polyacrylamide gel and visualized by silver staining. Modified from [1], with permission. Panel (**c**). Representative MALDI-mass spectra of trypsin fragments obtained from cross-linked AGT monomer (*upper panel*) and dimer (*lower panel*). Gel slices corresponding to monomer and dimer gel bands were digested with sequencing-grade trypsin and/or chymotrypsin. Soluble polypeptides were extracted into aqueous buffer, adjusted to 2.5 mg/mL alpha-cyanohydroxycinnamic acid, 0.05% trifluoroacetic acid, 50% acetonitrile, and spotted on a gold target. Dried spots were washed five times with 0.1% trifluoroacetic acid and redried. Mass spectrometry was performed using a Ciphergen (Fremont, CA) PBSIIc MALDI-TOF mass spectrometer operated at a digitizer rate of 250 MHz, pulse voltage of 3,000 V, and source voltage 20,000 V. Figure modified from [1], with permission

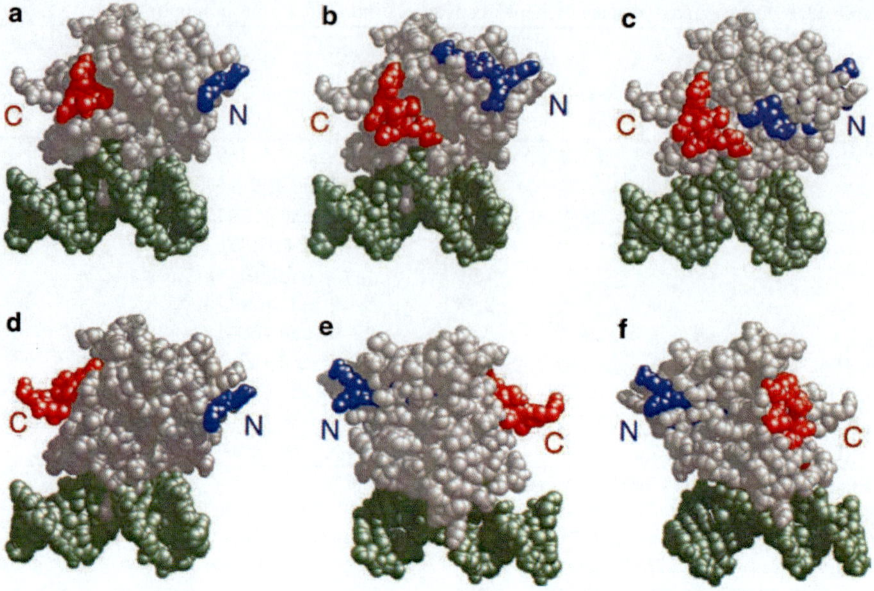

Fig. 12.3 Proteolytic fragments of AGT assigned to intermolecular cross-links. Representative fragment-pairs from the set listed in Table 12.1, mapped on a crystal structure of the monomeric AGT–DNA complex [5]. In each pair, the fragment with sequence nearest the N-terminal end of the protein is colored *blue*, the one nearest the C-terminal end, *red*. Panel (**a**): Fragments 1–7 and 163–169. Panel (**b**): Fragments 8–19 and 159–167. Panel (**c**): Fragments 22–34 and 159–167. Panel (**d**): Fragments 1–7 and 169–176. Panel (**e**): Fragments 19–32 and 169–176. Panel (**f**): Fragments 19–32 and 105–114. Amino-terminal surfaces are labeled with a *blue* N, carboxyl-terminal surfaces with a *red* C. Reproduced from [1], with permission

12.3 Analysis of DNA Topology

Our models of AGT–DNA complexes were built with DNA that deviates little from the average helical parameters of relaxed B-form DNA. However, because these models are based on the structure of a 1:1 AGT–DNA complex, they may not capture features of the cooperative assembly that could result in changes in DNA twist or writhe. Since several hundred AGT molecules can bind to a large DNA molecule (see below), even a small perturbation of DNA structure by each protein may result in very significant net winding or unwinding. For these reasons, it was essential to test the effects of cooperative AGT binding on DNA topology.

The supercoil relaxation (or topoisomerase) experiment [15] is a classical method for determining whether protein binding causes net DNA winding or unwinding. This assay involves formation of complexes on covalently closed circular DNA, torsional relaxation of the DNA by topoisomerase, removal of the proteins, and measurement of any change in linking number evaluated by gel electrophoresis. The data analysis requires correlation of protein binding stoichiometry with the measured change in DNA linking number. In the past, difficulty in determining the stoichiometries of large protein–DNA complexes hampered use of this method for study of cooperative, nonspecific DNA interactions. However, as described below, sedimentation equilibrium analysis provides a direct and rigorous solution to this problem.

AGT–DNA mixtures were formed using closed-circular pUC19 plasmid DNA (2,686 bp) as the binding substrate. Topoisomer relaxation was carried out with topoisomerase I enzymes from *E. coli* or from *Vaccinia* virus.[1] As shown in Fig. 12.4a, increasing AGT concentration shifted topoisomer distributions in favor of ones with greater electrophoretic mobility. In these gels, each topoisomer band differs by one linking number from the ones migrating immediately above and below it. Two-dimensional electrophoresis, with the second dimension run in buffer containing ethidium bromide established that AGT binding caused net unwinding with respect to relaxed circular DNA (not shown). Integrated band intensities for representative topoisomer populations are graphed in Fig. 12.4b. These intensities reflect the Boltzmann distribution of torsional free energies of the DNAs; the mean linking difference, $\Delta \mathrm{Lk}_m$, was calculated by fitting band intensity as a function of ΔLk using (12.1).

$$\mathrm{I} = \mathrm{I}_M \exp\left[-w\left(\Delta\,\mathrm{Lk} - \Delta\,\mathrm{Lk}_m\right)^2\right] \quad (12.1)$$

Here, I is the integrated intensity of a given band and I_M the maximum intensity of the envelope of bands; w is the width of the distribution; ΔLk is the difference in linking number of the band in question from that of the fully relaxed topoisomer; and $\Delta \mathrm{Lk}_m$ is the mean linking difference for the DNA population [33]. This analysis shows that the mean linking difference ($\Delta \mathrm{Lk}_m$) decreases from 0 to ~–7 over the range of AGT concentrations tested.

To obtain binding stoichiometry information, aliquots from each topoisomerase reaction were analyzed at sedimentation equilibrium (Fig. 12.5a). For large DNAs such as pUC19, the concentration of protein-free DNA molecules becomes negligible long before all available binding sites are saturated. These solutions contain mixtures of free protein and protein–DNA complex, and the weight-average molecular weight of the complex increases with [protein] until saturation is reached. For such a solution, the radial dependence of absorbance at sedimentation equilibrium is given by (12.2).

$$A(r) = \alpha_P \exp\left[\sigma_P\left(r^2 - r_0^2\right)\right] + \alpha_{P_nD} \exp\left[\sigma_{P_nD}\left(r^2 - r_0^2\right)\right] + \varepsilon \quad (12.2)$$

[1] *E. coli* topoisomerase I acts preferentially on negatively supercoiled DNAs, while the *vaccinia* enzyme relaxes both negatively and positively supercoiled molecules [8, 13, 29]. Topoisomer distributions obtained with the *vaccinia* enzyme are offset in the direction of positive supercoiling when compared to those obtained with the *E. coli* enzyme; this is an expected consequence of the preferential binding of *vaccinia* topoisomerase with positively supercoiled DNA. When this offset is taken into account, the linking differences observed with *vaccinia* and *E. coli* topoisomerases were mutually consistent.

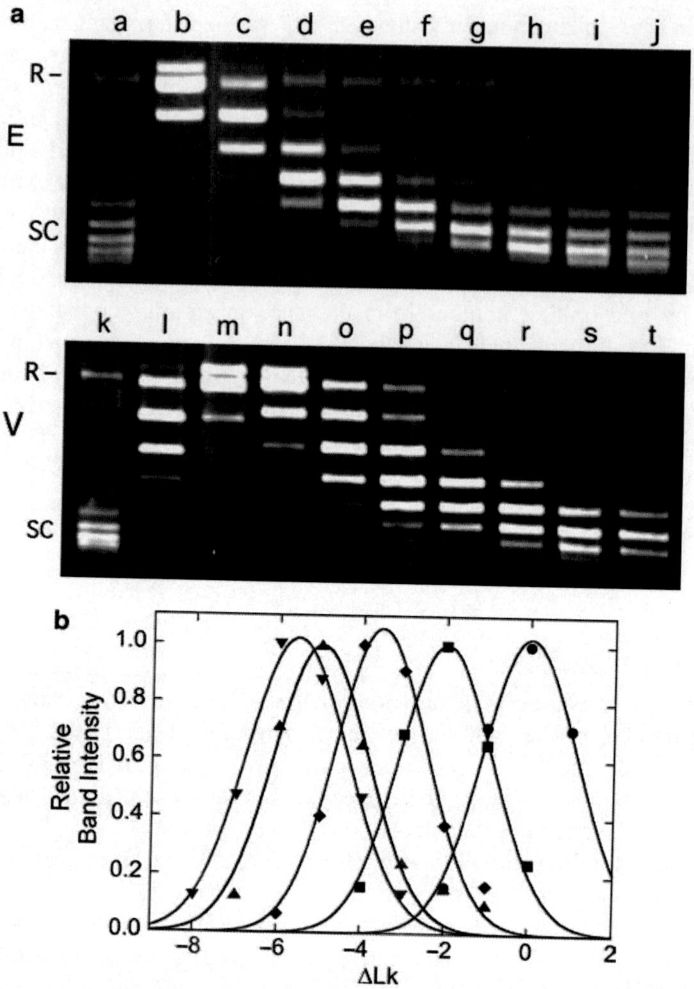

Fig. 12.4 Topoisomerase analysis of AGT complexes formed with pUC19 DNA. Panel (**a**). Electrophoretic profiles following topoisomerase treatment. *Top*: experiment using *E. coli* topoisomerase I (E). *Bottom*: experiment using *Vaccinia* topoisomerase I (V). All samples contained 12 nM closed circular pUC19 DNA. Samples *c–j* contained AGT protein at final concentrations of 4.2, 8.5, 10.9, 14.6, 18.2, 22.5, 26.7, and 36.4 μM, respectively. Samples *m–t* contained AGT protein at final concentrations of 4.2, 8.3, 15, 22, 28, 36, 46 and 60 μM, respectively. Samples *b–j* were treated with *E. coli* topoisomerase I (2 units) for 1.5 h at 20°C; samples *l–t* were treated with *Vaccinia* topoisomerase I (2 units) for 2.5 h at 20°C. Panel (**b**). Distribution of ΔLk for samples *c–g*. Integrated band intensities were normalized to that of the most intense band in each supercoiled set. The smooth curves are fits of (Equation 12.1) to this data, allowing the most probable value of the linking difference to be determined for each sample. Reproduced with modification from [1], with permission

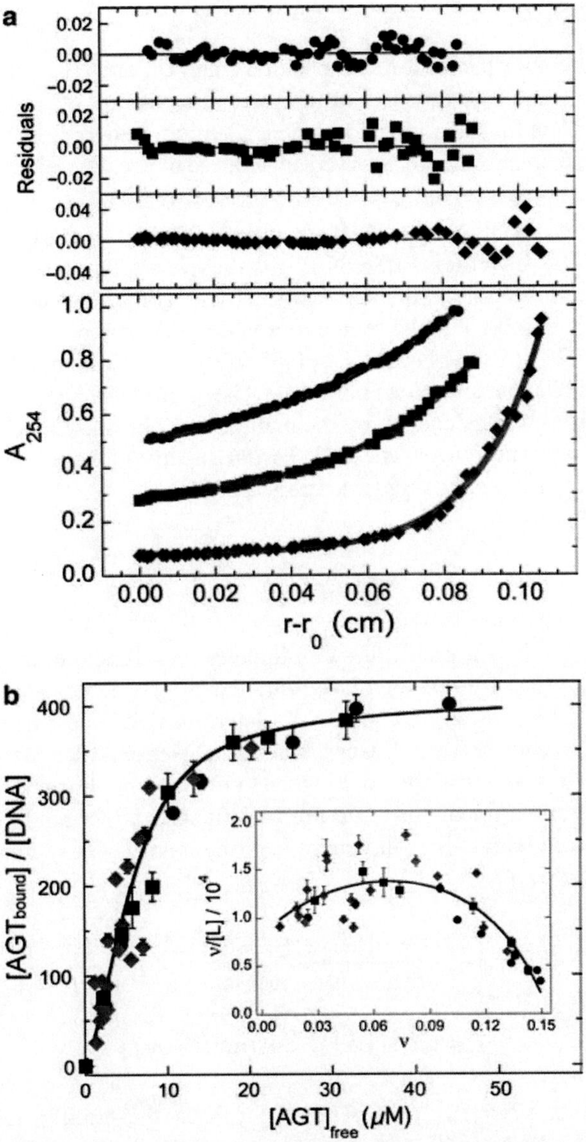

Fig. 12.5 Sedimentation equilibrium analysis of AGT complexes formed with pUC19 DNA. Panel (**a**). Representative data taken at 254 nm, 3,000 rpm and 20±0.1°C. Samples contained DNA (12 nM) and AGT in buffer consisting of 10 mM Tris (pH 7.6), 1 mM DTT, 1 mM EDTA, 100 mM NaCl. AGT concentrations were: 0 M (*filled circle*), 4.2 μM (*filled square*), 22 μM (*filled circle with filled diamond*). The smooth curves shown in the *lower panel* are fits of (Equation 12.2) to each data set. Small, symmetric residuals show that the model represented by this equation is consistent with the data. Panel (**b**). Binding stoichiometry as a function of free AGT concentration. Stoichiometries were calculated from the reduced molecular weights of complexes. *Black symbols* (*filled square, filled circle*) are values of samples taken from topoisomerase reactions. *Gray diamonds* are values from experiments carried out with linear pUC19. The error bars show the 95% confidence limits for values. *Inset*: Scatchard plot of the data. The *solid curve* is a fit of (Equation 12.4) to the data, returning K=8266±1018, ω=43.9±3.2 and s=6.8±0.2. From [1], with permission

Here, A(r) is the absorbance at radial position r and α_P, and α_{P_nD} are absorbances of protein and protein–DNA complex at the reference position, r_o, and ε is a baseline offset that accounts for radial position-independent differences in the absorbances of different cell assemblies. The weight-average, reduced molecular weight of the ensemble of AGT–DNA complexes is given by $\sigma_{P_nD} = (nM_P + M_D)(1-\bar{v}_{P_nD}\rho)\omega^2/(2RT)$, while that of free AGT is $\sigma_P = M_P(1-\bar{v}_P\rho)\omega^2/(2RT)$. Here, M_P and M_D are the molecular weights of protein and DNA, n is the average protein:DNA ratio of the ensemble of complexes; ρ is the solvent density, ω is the rotor angular velocity, R is the gas constant, and T is the temperature (Kelvin). The partial specific volume of duplex NaDNA at 0.1 M NaCl ($\bar{v}_D$ =0.55 mL/g) was estimated by interpolation of the data of Cohen and Eisenberg [3]. The partial specific volume of AGT ($\bar{v}_P$ =0.744 mL/g) was calculated by the method of Cohn and Edsall [4], using data tabulated by Laue and coworkers [12]. Partial specific volumes of protein–DNA complexes were estimated using (Equation 12.3).

$$\bar{v}_{P_nD} = \frac{\left(nM_P\,\bar{v}_P + M_D\,\bar{v}_D\right)}{\left(nM_P + M_D\right)} \tag{12.3}$$

Shown in Fig. 12.5b is a graph of stoichiometry as a function of free [AGT] for binding to topoisomerase-relaxed closed-circular pUC19 DNA; also shown is data for AGT binding to linear pUC19 DNA, determined by the same method. The overlap of these data sets establishes that the closed-circular DNAs were fully relaxed and that structures that distinguish circular and linear DNAs (such as molecular ends in the linear form) do not significantly affect binding. The inset is a Scatchard plot in which the solid curve is a fit of the McGhee–von Hippel equation (12.4) to the data.

$$\frac{v}{[P]} = K(1-sv)\left(\frac{(2\omega-1)(1-sv)+v-R}{(2\omega-1)(1-sv)}\right)^{n-1} \times \left(\frac{1-(s+1)v+R}{2(1-sv)}\right)^2 \tag{12.4}$$
$$R = \left((1-(s+1)v)^2 + 4\omega v(1-sv)\right)^{1/2}$$

Here, v=n/2,686, $[P]=[P]_{input} - n[pUC19]_{input}$, K is the equilibrium association constant for binding a single site, ω is the cooperativity parameter, and s is the size of the site (in base pairs) from which a protein molecule excludes others [17]. For this data set, K=8,266±1,018, ω=43.9±3.2, and s=6.8±0.2. These values of K and ω are similar to ones found for short duplex oligodeoxyribonucleotides [18], while the occluded binding site size (6.8±0.2) is larger. It is likely that the larger value of s reflects inhomogeneous protein packing that is expected for binding to large DNA, but that is largely absent when short templates are used.

The dependence of the mean linking difference ΔLk_m on binding stoichiometry is shown in Fig. 12.6. Over a wide stoichiometry range, AGT binding is accompanied by a linking difference of −0.020±0.001/protein molecule, equivalent to a net unwinding of 7.2±0.3°/protein. Consistent values were obtained with *E. coli* and

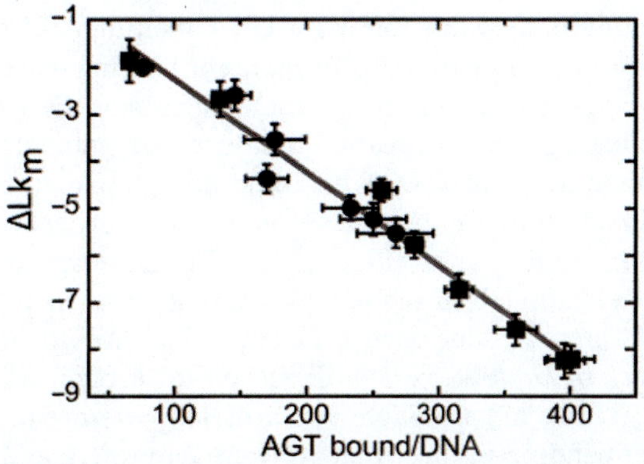

Fig. 12.6 Dependence of the linking difference ΔLk on binding stoichiometry, n. Data from the experiments shown in Figs. 12.4 and 12.5. The error bars represent standard errors of measurement distributions. The *line* is a linear fit; the slope $d\Delta Lk/dn = -0.020 \pm 0.001$ corresponds to net unwinding of $7.2 \pm 0.3°$/protein bound. Modified from [1], with permission

Vaccinia topoisomerases, indicating that the linking difference is unlikely to be an artifact of topoisomerase mechanism. This small linking difference is consistent with the initial prediction that AGT binds B-form DNA and causes little net torsional deformation.

12.4 Discussion

Topological and binding data of the kinds described here can provide useful insights into the structures of cooperative complexes, their mechanisms of assembly, and possibly even their cell biology. For example, the cross-linking data given above tests and confirms predictions about the identities of the surfaces of protein–protein contact within AGT complexes. The polypeptides that have been identified can now be subjected to mutagenesis and the mutants tested for effects on binding cooperativity in vitro and DNA repair *in vivo*. Thus, cross-linking can make possible a "reverse genetics" approach to analysis of interacting systems. In addition, the cross-linking approach described here is model-independent; it can be carried out successfully without detailed structural information like that available for the AGT system. A further and important advantage is that this approach can be generalized to complex mixtures of proteins organized around a nucleic acid or some other structure. In this case, it can be a valuable tool for mapping the juxtaposition of proteins in assemblies that are too large or heterogeneous to be easily approached by high-resolution structural biology techniques.

Topoisomer analysis provides a second and independent way to view the features of protein–DNA complexes. In the example shown above, AGT binding was accompanied by a small (~7°/protein) net unwinding of the double helix. This amount of unwinding is incompatible with DNA melting (full-strand separation would require a net unwinding of ~138°/protein for binding that can reach a limiting density of 4 bp/protein). This result establishes that the isolated protein is incapable of mediating strand separation under our standard conditions, an important conclusion given that AGT can bind both single-stranded and duplex DNAs [18, 25]. In addition, the juxtaposition of proteins depends on the helical periodicity of their binding sites (Fig. 12.1). The measured linking difference suggests that the helical pitch of the complex might be as small as 131°/protein, instead of the 138°/protein value used in our models. This would reduce the rotational displacement of protein n+3 with respect to protein n and increase the overlap of surfaces that are available for protein–protein contact. Finally, the small degree of unwinding implies that AGT binds preferentially to torsionally relaxed substrates. Because many cellular processes generate torsional stress in DNA, this result suggests that AGT might partition into regions containing relaxed DNA. In turn, this suggests that AGT and topoisomerase enzymes may colocalize within nuclei. This prediction can be tested in whole cells, using standard immunohistochemical methods.

Acknowledgments Mass spectrometric analyses were performed at the University of Kentucky Center for Structural Biology Protein Core Facility. This facility is supported, in part, by funds from NIH National Center for Research Resources (NCRR) grant P20 RR020171. We gratefully acknowledge the help of Dr. Carol Beach in acquiring these data. This research was supported by NIH grant GM-070662 (to M.G.F.).

References

1. Adams CA, Melikishvili M, Rodgers DW, Rasimas JJ, Pegg AE, Fried MG (2009) Topologies of complexes containing O^6-alkylguanine-DNA alkyltransferase and DNA. J Mol Biol 389:248–263
2. Bouffartigues E, Buckle M, Badaut C, Travers A, Rimsky S (2007) H-NS cooperative binding to high-affinity sites in a regulatory element results in transcriptional silencing. Nat Struct Mol Biol 14:441–448
3. Cohen G, Eisenberg H (1968) Deoxyribonucleate solutions: sedimentation in a density gradient, partial specific volumes, density and refractive density increments and preferential interactions. Biopolymers 6:1077–1100
4. Cohn EJ, Edsall JT (1943) Proteins, amino acids and peptides as ions and dipolar ions. In: Cohn EJ, Edsall JT (eds) Proteins, amino acids and peptides as ions and dipolar ions. Reinhold, New York, pp 370–381, and 428–431
5. Daniels DS, Woo TT, Luu KX, Noll DM, Clarke ND, Pegg AE, Tainer JA (2004) DNA binding and nucleotide flipping by the human DNA repair protein AGT. Nat Struct Mol Biol 11:714–720
6. Duguid EM, Rice PA, He C (2005) The structure of the human AGT protein bound to DNA and its implications for damage detection. J Mol Biol 350:657–666
7. Fancy DA, Kodadek T (1999) Chemistry for the analysis of protein-protein interactions: rapid and efficient cross-linking triggered by long wavelength light. Proc Natl Acad Sci USA 96:6020–6024
8. Fernandez-Beros M-E, Tse-Dinh Y-C (1996) Vaccinia virus DNA topoisomerase i preferentially removes positive supercoils from DNA. FEBS Lett 384:265–268

9. Fried MG, Kanugula S, Bromberg JL, Pegg AE (1996) DNA binding mechanisms of O^6-alkylguanine-DNA alkyltransferase: stoichiometry and effects of DNA base composition and secondary structures on complex stability. Biochemistry 35:15295–15301
10. Hard T, Dahlman K, Carlstedt-Duke J, Gustafsson J-A, Rigler R (1990) Cooperativity and specificity in the interactions between DNA and the glucocorticoid receptor DNA-binding domain. Biochemistry 29:5358–5364
11. Laemmli UK (1970) Cleavage of structural proteins during the assembly of the head of bacteriophage T4. Nature 227:680–685
12. Laue TM, Shah BD, Ridgeway TM, Pelletier SL (1992) Computer-aided inter-pretation of analytical sedimentation data for proteins. In: Harding SE, Rowe AJ, Harding JC (eds) Analytical ultracentrifugation in biochemistry and polymer science. The Royal Society of Chemistry, Cambridge, England, pp 90–125
13. Lima CD, Wang JC, Mondragon A (1994) Three-dimensional structure of the 67K N-terminal fragment of *E. coli* topoisomerase I. Nature 367:138–146
14. Lohman TM, Ferrari ME (1994) *Escherichia coli* single-stranded DNA-binding protein: multiple DNA-binding modes and cooperativities. Annu Rev Biochem 63:527–570
15. Lutter LC, Drabik CE, Halvorson HR (2000) Use of topology to measure protein-induced DNA bend and unwinding angles. In: Travers A, Buckle M (eds) DNA-protein interactions: a practical approach. Oxford University Press, Oxford, UK, pp 47–64
16. Margison GP, Santibáñez-Koref MF (2002) *O*^{6}-alkylguanine-DNA alkyltransferase: role in carcinogenesis and chemotherapy. Bioessays 24:255–266
17. McGhee J, von Hippel PH (1974) Theoretical aspects of DNA-protein interactions: co-operative and non-co-operative binding of large ligands to a one-dimensional homo-geneous lattice. J Mol Biol 86:469–489
18. Melikishvili M, Rasimas JJ, Pegg AE, Fried MG (2008) Interactions of human O^6-alkylguanine-DNA alkyltransferase (AGT) with short double-stranded DNAs. Biochemistry 47:13754–13763
19. Nadeau OW, Carlson GM (2005) Protein interactions captured by chemical cross-linking. In: Golemis E, Adams PD (eds) Protein-protein interactions. Cold Spring Harbor Laboratory Press, Cold Spring Harbor, NY, pp 75–92
20. Pegg AE (2000) Repair of O^6-alkylguanine by alkyltransferases. Mutat Res 462:83–100
21. Peters K, Richards FM (1977) Chemical cross-linking: reagents and problems in studies of membrane structure. Annu Rev Biochem 46:523–551
22. Peterson SN, Dahlquist FW, Reich NO (2007) The role of high affinity non-specific DNA binding by LP in transcriptional regulation and DNA organization. J Mol Biol 369:1307–1317
23. Phelps CB, Sengchanthalangsy LL, Shiva Malek S, Ghosh G (2000) Mechanism of B-DNA binding by Rel/NFk-B dimers. J Biol Chem 275:24392–24399
24. Pray TR, Burz DS, Ackers GK (1998) Cooperative non-specific DNA binding by octamerizing lambda cI repressors: a site-specific thermodynamic analysis. J Mol Biol 282:947–958
25. Rasimas JJ, Kar SR, Pegg AE, Fried MG (2007) Interactions of human O^6-alkylguanine-DNA alkyltransferase (AGT) with short single-stranded dnas. J Biol Chem 282:3357–3366
26. Rasimas JJ, Pegg AE, Fried MG (2003) DNA-binding mechanism of O^6-alkylguanine-DNA alkyltransferase. Effects of protein and DNA alkylation on complex stability. J Biol Chem 278:7973–7980
27. Saxe SA, Revzin A (1979) Cooperative binding to DNA of catabolite activator protein of *Escherichia coli*. Biochemistry 18:255–263
28. Shevchenko A, Tomas H, Havli J, Olsen JV, Mann M (2007) In-gel digestion for mass spectrometric characterization of proteins and proteomes. Nat Protoc 1:2856–2860
29. Stivers JT, Harris TK, Mildvan AS (1997) Vaccinia DNA topoisomerase I: evidence supporting a free rotation mechanism for DNA supercoil relaxation. Biochemistry 36:5212–5222
30. Takahashi M, Blazy B, Baudras A (1979) Non-specific interactions of CRP from *E. coli* with native and denatured DNAs: control of binding by cAMP and cGMP and by cation concentration. Nucleic Acids Res 7:1699–1712

31. Taylor JD, Badcoe IG, Clarke AR, Halford SE (1991) EcoRV restriction endonuclease binds all DNA sequences with equal affinity. Biochemistry 30:8743–8753
32. Tubbs JL, Pegg AE, Tainer JA (2007) DNA binding, nucleotide flipping, and the helix-turn-helix motif in base repair by O^6-alkylguanine-DNA alkyltransferase and its implications for cancer chemotherapy. DNA Repair (Amst) 6(8):1100–1115
33. Zeman SM, Crothers DM (2001) Simultaneous measurement of binding constants and unwinding angles by gel electrophoresis. Methods Enzymol 340:51–68

Chapter 13
How to Think Like a Single Molecule: Obtaining Quantitative Measurements on Single DNA Molecules and Chromatin Fibers

Sanford H. Leuba and Richard A. Steinman

13.1 Introduction

There is a dichotomy between bulk, population methods and single-molecule studies to understand biological systems although both approaches generally seek to learn mechanistic information. Using single-molecule methods requires a different kind of thinking in order to achieve interpretable results. With a limited set of examples from our laboratories, we will demonstrate some successful single-molecule studies of protein–DNA interactions as well as to explain some caveats to improve future experiments. Our hope is that the reader of our chapter may learn to conceive, design, and successfully perform new single-molecule experiments by careful consideration of the desired outputs.

One often encounters molecular and cellular biologists who are intrigued with single-molecule approaches and would like to apply them to their own research directions. An advantage of incorporating such an approach is that single-molecule methods have the potential for quantitatively elucidating fundamental biophysical interactions, and unmasking key interactions that may be undetectable in bulk analyses. However, although cell biologists are intrigued by single-molecule methods, they may not have the knowledge about optimal design of a single-molecule biology experiment to generate meaningful and quantifiable results. For example, in atomic force microscopy (AFM) experiments, it is necessary to underscore the importance of using purified protein complexes so that the protein of interest can be specifically identified in DNA–protein complexes. Further, single-pair fluorescence resonance energy transfer (spFRET) can only be used when fluorescence events are detectable above fluorescence noise, which also requires pure and properly labeled protein samples.

S.H. Leuba (✉)
Departments of Cell Biology and Physiology and Bioengineering, Hillman Cancer Center, Petersen Institute of NanoScience and Engineering and University of Pittsburgh Cancer Institute, University of Pittsburgh School of Medicine and Swanson School of Engineering, 5117 Centre Avenue, 2.26a, Pittsburgh, PA 15213, USA
e-mail: leuba@pitt.edu

M.C. Williams and L.J. Maher, III (eds.), *Biophysics of DNA-Protein Interactions*, Biological and Medical Physics, Biomedical Engineering,
DOI 10.1007/978-0-387-92808-1_13, © Springer Science+Business Media, LLC 2011

Although single-molecule data are generally relatively easy to interpret in well-designed experiments, it is important to note that the information generated may fit multiple models. For example, in their elegant seminal work involving optical tweezers stretching of single DNA molecule experiments, Smith et al. [1] (see also [2]) posited that a plateau in the stretch curve that occurred about 65 pN was of the formation of a new DNA structure that was called "S-DNA" or stretched DNA. However, the experiments did not actually provide structural information, and S-DNA was a hypothetical model. Later, using the same kinds of optical tweezer technologies, Williams and collaborators were able to show using thermodynamic arguments that the plateau is created from a combination mixture of single-stranded (ss) and double-stranded (ds) DNA [3] (for a review, see [4]). Recently, van Mameren et al. used a combination of fluorescence and force spectroscopy to structurally demonstrate DNA melting during overstretching [5] (also, see commentary: [6]).

As illustrated above, while single-molecule biology techniques are sensitive approaches, they must be applied with care. It helps to envision what the data may look like, and to design experiments that will generate quantifiable data. Here, we will discuss a number of approaches that we have successfully used to make quantitative and repeatable measurements on DNA/protein complexes. Additionally, we will discuss some of the pitfalls that we have experienced and how we have or will overcome these problems. We will begin by discussing measurements of DNA alone. We will then describe how AFM, spFRET, and magnetic tweezers (MT) can be used to study chromatin structure and dynamics. We end with a description of how these methods can be used to evaluate drugs that may interact with chromatin

13.2 Atomic Force Microscope

The AFM instrument allows for nanometer scale imaging of macromolecules such as DNA and DNA–protein complexes. There are two methods used: rinsing the sample and then fluxing with air prior to imaging in air or direct imaging in buffer solution. The atomically flat surface (freshly cleaved mica or flamed and rinsed glass) on which macromolecules are imaged in AFM provides both advantages and disadvantages. This flat surface allows a planar surface for DNA–protein interactions to be probed with a fine tip by scanning the surface and mapping out the height as a function of position. The major disadvantage is that there must be some sort of interaction between the macromolecule(s) and the surface so that the molecule will adhere throughout the deposition process and as the sample is scanned. Typically during deposition for imaging in air, the deposited molecules are rinsed with purified water and then fluxed with nitrogen gas; after this typical deposition process only stable complexes are left intact on the surface. For AFM imaging studies where the sample and AFM probe are immersed in buffer, there must be sufficient interactions between the DNA or DNA–protein complexes and the surface so that these macromolecules adhere to the surface and are not dislodged by the raster scans of the AFM tip.

13.2.1 Height Measurements in AFM

Typically, the radius of curvature of the AFM tip is on the order of 5–10 nm. Macromolecules with features of 1 nm heights or less can be imaged, although a larger range of features is optimal. Imaging is generally successful in experiments in which the molecule of interest projects, i.e., it is double the height of other nearby objects. For imaging in air, DNA is often measured to be about 0.5 nm high off the surface of the underlying atomically flat mica or glass. Therefore, proteins bound to a molecule of DNA that are double or triple that standard (0.5 nm) apparent height are easily recognizable.

13.2.2 DNA Length Measurements in AFM

DNA has a minimal length that can be readily discerned in AFM images. About 200 nm of dsDNA (~600 bp) is a typical minimum length for analysis.

13.2.3 Chromatin and AFM

The repeating nature of nucleosomes in chromatin fibers renders these fibers an excellent system for study using the imaging capabilities of the AFM (Fig. 13.1). In this instance, the overall 11 nm dimension of a nucleosome is slightly larger than the typical radius of curvature of an AFM tip, making nucleosomes easily recognizable in AFM images. Furthermore, the typical 30 nm spacing of individual nucleosomes and the repeated pattern of nucleosomal structure makes them readily identifiable in AFM images, akin to a bird's eye view of a range of mountain peaks. Even individual nucleosomes can be imaged because their apparent heights (~2 nm in tapping mode AFM images in air) compared to the apparent heights of naked DNA molecules (~0.5 nm in tapping mode images in air), and the hemispherical shape makes them readily identifiable. The compatibility of AFM capabilities and chromatin fiber structure will be advantageous in analyses of the effects of posttranscriptional histone modifications on chromatin structure.

When we started AFM imaging of chromatin fibers, we first considered using the same conditions that were used for imaging naked DNA molecules [7], i.e., to perform the deposition in 10 mM $MgCl_2$ buffers. However, we were concerned that such high magnesium conditions would disrupt the protein/DNA structure in chromatin, so magnesium was absent in our approach. Ultimately, we were able to determine that depositing chromatin over time in tri-ethanolamine buffers under low ionic strength conditions was sufficient for the adherence of chromatin fibers to the surface. The addition of 10 mM NaCl caused disruption of the chromatin fibers leading to an image analogous to DNA "spaghetti" with a few histone "meatballs"

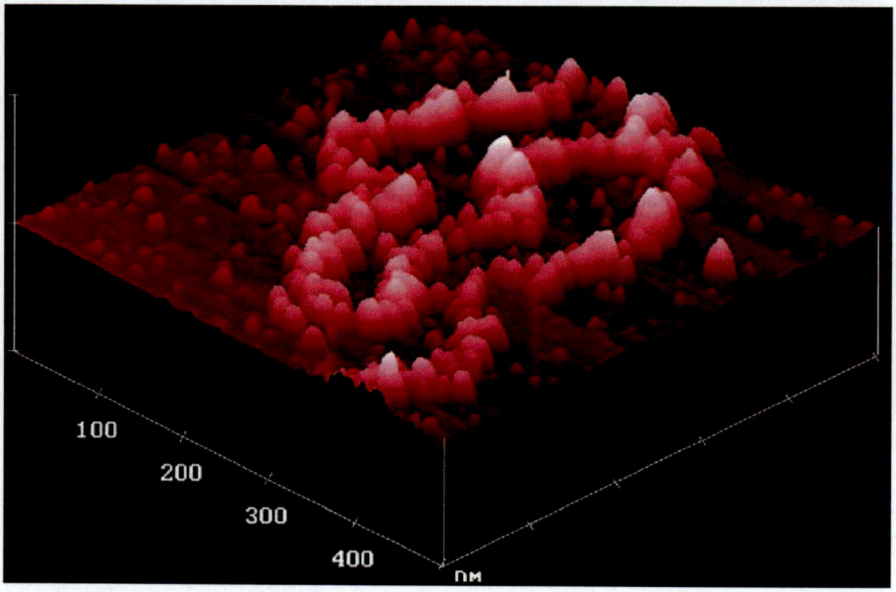

Fig. 13.1 500 nm × 500 nm AFM image of glutaraldehyde-fixed chromatin fiber that has been tilted 30° to emphasize the topographic height of the fiber above the surface. The chromatin fiber has an irregular three-dimensional arrangement of nucleosomes. Each vertical notch indicates a height scale of 25 nm. Adapted from Leuba et al. [70]

(also described in [8]). Ultimately, we adopted a mild glutaraldehyde treatment of the fibers before deposition, as had been used in electron microscopy [9] to fix the fibers prior to imaging. The resulting chromatin fibers exhibited an irregular three-dimensional arrangement of nucleosomes – these are the conditions used in the experiment depicted in Fig. 13.1.

13.3 Magnetic Tweezers

MT DNA/protein experiments typically have a DNA molecule tethered between a surface and a superparamagnetic bead such as a Dynal bead [10, 11]. An external magnetic field is used to apply a constant force and to keep the tethered superparamagnetic bead in the focus of a video camera. Changes in DNA length (manifested as vertical movement of the superparamagnetic bead) are measured over time by shifting the focus of the video camera. By using a tethered T7 RNA polymerase (RNAP) to bind DNA to a surface and the superparamagnetic bead to keep it elongated, the progression of the DNA through the RNAP (i.e., transcription) was manifested by vertical movement of the bead, followed by shifting the video camera focus. In order to discern whether passage of DNA through RNAP generated twist, a small biotinylated marker bead was bound to the large streptavidin-coated superparamagnetic bead. The marker

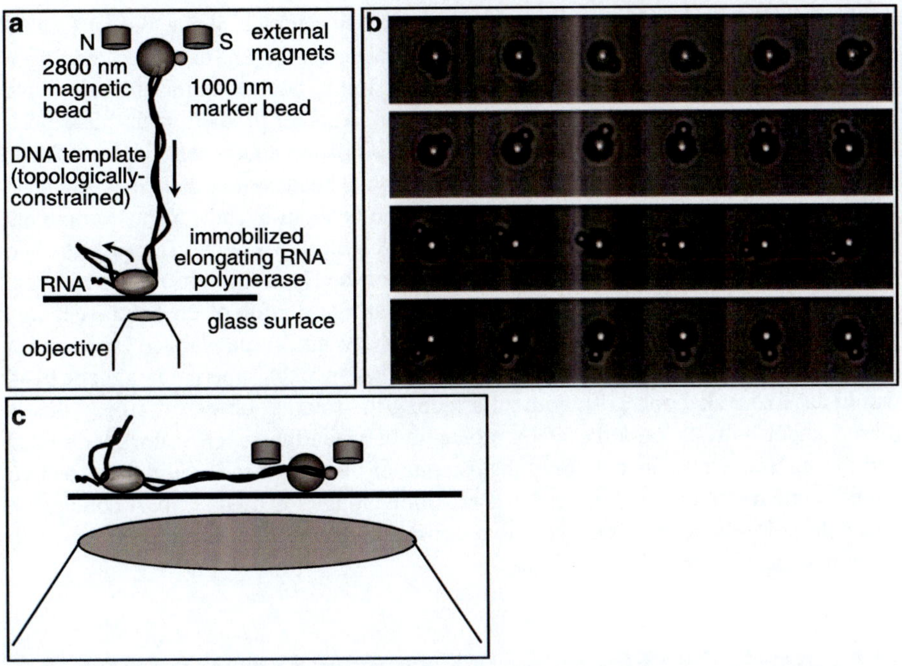

Fig. 13.2 Following transcription elongation occurring on a single RNA polymerase. (**a**) Schematic concept of the experiment, not to scale. (**b**) Individual frames following the rotation of a 1,000 nm biotinylated bead attached to a streptavidin-coated 2,800 nm superparamagnetic Dynal bead. The superparamagnetic bead is attached to a double-stranded DNA molecule, which is tethered to a T7 RNA polymerase molecule, adhered to a surface. Upon the addition of nucleotide triphosphates, the biotinylated bead was observed to rotate 30 times around the larger superparamagnetic bead suggesting that the passage of DNA through the RNA polymerase creates enough twist in the DNA to rotate the two-bead assembly. Adapted from Pomerantz et al. [12]. (**c**) Another potential geometry of the tethered superparamagnetic bead adapted from panel (**a**), not to scale. In this case both the RNA polymerase and the bead are on the surface. The attachment of the dsDNA template to the bead may inhibit rotation in some cases

bead uncovered rotation of the bead assembly during transcription (Fig. 13.2a, b). Using this arrangement, the magnetic tweezers facilitated capture of transcription occurring from a single RNAP (Fig. 13.2a, b) [12], movie is available at http://pubs.acs.org/doi/suppl/10.1021/nl0509125. We were able to observe the small bead rotate around the larger bead in 30 continuous rotations representing over 300 bp of transcription. A caveat was that this was the only movie of any sufficient length we were able to record. We had some instances of where the two-bead complex would make a 90° rotation only to be yanked back (unpublished data not shown); similar movies have been observed by Sakata-Sogawa and Shimamoto [13]. Moreover, when we calculated the torque that was generated [14, 15], it was in excess of 200 pN nm, which substantially exceeded the 35 pN nm maximum torque threshold that could be measured on naked DNA molecules in the elegant rotor bead experiments of Bryant et al. [16]. Steven B. Smith (personal communication) has suggested that the rotating superparamagnetic bead might not be suspended in solution above the surface as intended, but instead rotates/slides on the glass surface adjacent to the tethered transcribing RNAP (Fig. 13.2c). The superhelical tension in the DNA molecule may have generated spinning in the superparamagnetic bead. Another problem with this experiment is our inability to ascertain the exact location of the attachment of the DNA to the surface of the superparamagnetic bead. The Dynal bead is essentially a plastic bead soaked with ferric oxide particles that provide the bead an overall magnetic moment induced by the external magnetic field. The dsDNA molecule is not necessarily attached to the bottom of the Dynal bead, and the rotation of the Dynal bead in a magnetic field could be inhibited by DNA attachment to an equatorial portion of the bead (see, e.g., potential attachment in Fig. 13.2c). To avoid these caveats, future planned experiments would use rotor bead attachment [16, 17]. In that instance the superparamagnetic bead would act like a skyhook [17], without affecting the rotation of the rotor bead that is placed about halfway onto the DNA molecule between the superparamagnetic bead and the surface. Consequently, only the rotation of the rotor bead would be measured whereas one rotation would indicate transcription elongation of 10.2 bp. A conclusion about these kinds of experiments is that constructs have to be carefully prepared to avoid nonspecific surface interactions.

13.4 Single-Pair Fluorescence Resonance Energy Transfer

In a FRET experiment, two fluorophores with overlapping absorption and emission are used to precisely measure the distance between two locations. When the fluorophores are close together and depending on the orientation between the two dyes, energy transfer occurs and absorption at one fluorophore results in emission from the other fluorophore at a shifted wavelength. Structural modeling can enable us to predict the R_0 (Förster radius), the distance at which the 50% FRET occurs between a matched pair of fluorescence dyes. The extensive database of proteins and of protein/nucleic acid structures can facilitate the design of informative spFRET experiments in which the exact distances between fluorophores can be predicted based upon the locations of each fluorophore within a protein or a DNA molecule or in a model in which one fluorophore is in a protein and the other in bound DNA. For instance, programs such as PyMol (http://www.pymol.com) can facilitate placement of labels and modeling of the distances between labels. These models inform the design of experiments in which one configuration positions the distance between the two site-specifically located dyes at a nanometer or two below the R_0, while in the other configuration, the distance between the two site-specifically located dyes is 1–2 nm above the R_0. In this way, one can maximize the differences between two states of spFRET as a function of DNA, protein or DNA/protein configurations. It is convenient to make the low FRET state sufficiently above zero FRET such as at ~0.3 FRET. It is then easy to differentiate this low FRET signal from photoblinked or photobleached states that give zero FRET (in these states the fluorescent dyes enter temporary or permanent dark states, respectively, and become nonfluorescent and hence have no measurable fluorescence intensity).

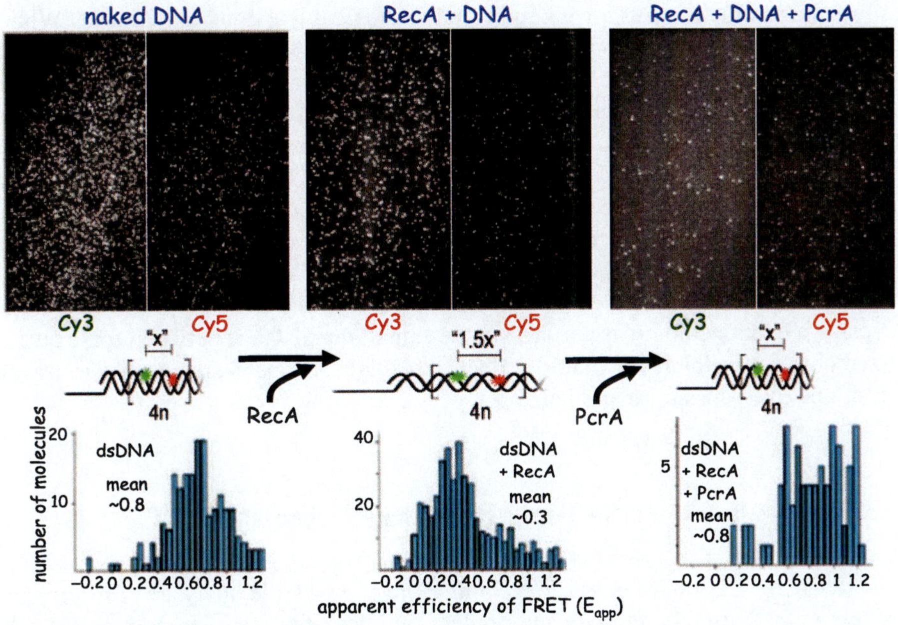

Fig. 13.3 PcrA reverses RecA induced double-stranded DNA lengthening. See text. Adapted from Anand et al. and Leuba et al. [23, 24]

Using electron microscopy, Stasiak and coworkers originally reported the 50% lengthening of DNA molecules by the addition of RecA [18]. Several studies with optical tweezers suggested RecA induced lengthening of individual DNA molecules [19–22] over a decade ago. We recently applied spFRET to study the RecA system [23, 24]. As demonstrated in Fig. 13.3, we prepared a dsDNA molecule containing a ssDNA tail such that a Cy3–Cy5 dye pair was separated by ~13 bp apart (a distance of ~4.4 nm) that was within the Förster radius for measurable spFRET. The top panels are video images using either filters for Cy3 (left halves) or Cy5 (right halves) of the same areas. Each individual light spot in the images represents the fluorescence intensity of an individual dye. In the naked DNA case, the high spFRET has a histogram peak with a mean of ~0.8 (shown in the graph below the raw data), indicating that the dyes are close together, as expected for naked DNA. Upon the addition of recombination protein RecA, the dsDNA is lengthened 1.5 times to a distance of about ~6.6 nm with a concomitant lowering of the histogram spFRET peak to ~0.3. Upon the addition of DNA helicase PcrA, we observe the spFRET in the histogram to rise again to ~0.8 units suggesting that PcrA has displaced the RecA molecules and the DNA molecule has returned to its naked dsDNA state. The rationale for performing these experiments was the observation that PcrA mutants are hyper-recombinagenic [25], suggesting a role of this DNA helicase with recombination proteins such as RecA. In addition to our measurements taken at various time points during the experiment, spFRET offers the advantage of following individual biological reactions in real time. While the final average results from the histograms in Fig. 13.3 could be obtained in a bulk experiment, these single-molecule experiments also allow one to obtain information on individual DNA molecules. Such individual information is critical for measuring the dynamics of RecA removal by PcrA (see below).

Since sites in DNA molecules can be specifically labeled, we believe that this robust system can be used to follow dynamics of processes moving along DNA molecules. Having two or more sets of dye pairs spaced along the DNA molecule so that there is no crosstalk from one pair of dyes to another, we hope to follow processes that sequentially interrogate one dye pair and then the next, each time causing a quantized increase or drop in the signal; the distance between these quantized changes in intensity should reflect the length of time a process has to travel from one dye pair site to another.

13.5 Single-Molecule Drug Discovery Using spFRET

An advantage of single-molecule techniques over ensemble analyses of molecular events is that discrete steps in intermolecular interactions can be captured. Such events would be masked during analysis of pooled samples. This could be beneficial in drug development because drugs that affect the same overall process could be shown to modulate distinct sequential effects. Ensemble analyses are less likely to distinguish between agents that target different steps in a clinically important mechanism. For example, inhibitors of reverse transcriptase can induce chain termination (as occurs with nucleoside analogues) or can inhibit the enzyme directly. Single-molecule spFRET analyses have uncovered an intricate choreography of movement of reverse transcriptase along nucleic acid duplexes that involved sliding between duplex DNA termini along with shifts from an RT into a polymerase role [26]. This model enabled a detailed analysis of HIV drug functionality. In another example of the power of single-molecule analyses, optical tweezers have been helpful in capturing the kinetics of movement of another polymerase, RNAP, and uncovered step motion along DNA as well as backtracking and pausing [27–29]. Polymerase kinetics can also be uncovered by FRET; in particular, three-color alternating laser excitation (3c-ALEX) can capture polymerase movement away from one fluorophore and toward another in a single experiment [30]. Such dissection of the kinetics of RNAP progression can augment the analysis and development of drugs aimed at inhibition of transcription initiation complexes or at inhibition of transcript elongation. By identifying additional steps in protein–DNA interaction, single-molecule studies can uncover transient states that represent new targets for drugs.

Most analyses to date have utilized naked DNA as a template for transcription studies; however, single-molecule techniques such as spFRET should facilitate the study of drugs and/or histone modifications on nucleosomal DNA that are linked to eukaryotic transcription or to DNA repair events. Insofar as experimental models incorporate nucleosomes and specific combinations of promoter elements, these experiments will be a more robust resource for identifying small molecules (or proteins) that can affect *specific* transcriptional sites within the genome, rather than indiscriminately modulating all transcripts. Ideally, single-molecule techniques will be directed at nucleosomal DNA that recapitulates the in situ response of corresponding promoter regions to drugs. These models could explain why only a fraction of the consensus sequence elements for a given transcription factor are active in the cell. It is believed that chromatin conformation and combinatorial transcription factor binding sequences contribute to this promoter specificity. Similarly, as histone deacetylase inhibitors (HDACs) and DNA methyltransferase inhibitors move into greater clinical use, it is evident that only subsets of genes are responsive to these agents. Single-molecule techniques could capture chromatin and sequence features that account for this selective responsiveness.

Single-molecule techniques may offer a sensitive readout for agents that can affect the initiation of transcription for select classes of promoters. In order to realize this potential, it would be advantageous to conduct the assays under conditions approximating physiology, i.e., using nucleosomal DNA. Challenges include the generation of informative and realistic nucleosomal constructs, the reproducibility of transcriptional modeling using such constructs, and the scaling of single-molecule assays to allow medium or high-throughput screening of candidate transcriptional modulators.

We have reported the generation of nucleosomes containing precisely positioned fluorophores on DNA that exhibited spFRET in the closed nucleosomal conformation, spontaneously fluctuated between open and closed conformations on a timescale of seconds, and lost FRET in salt-disrupted nucleosomes [31]. Widom's laboratory has reported short, spontaneous (10–50 ms) unwrapping of nucleosomes that rendered DNA accessible to protein binding to a target site that was positioned at one end *within* the nucleosome rather than in naked linker DNA between nucleosomes [32, 33]. Ensemble FRET was used to demonstrate durable trapping of nucleosomes in the open position by binding of the transcriptional repressor LexA [33]. Subsequent stopped flow, fluorescence anisotropy experiments exploited this protein trap to measure the kinetics of nucleosome unwrapping. These population-based experiments established that nucleosomes transiently unwrapped and that this feature facilitated transcription factor access. Single-molecule studies have allowed even more detailed analysis of these important unwrapping dynamics. Koopmans et al. recently conducted experiments that measured transient nucleosomal unwrapping by combining alternating laser excitation (ALEX) in conjunction with spFRET to enable determination of FRET efficiency and of fluorophore stoichiometry simultaneously [34]. The use of ALEX helped to resolve sample heterogeneity and allowed true conformational change in doubly labeled reconstituted nucleosomes to be distinguished from photobleaching and fluorophore blinking artifacts. In order to avoid immobilization artifacts, nucleosomes were imaged both in free solution and (to prevent diffusion-driven disassembly) in PAGE gels. The combination of ALEX-spFRET and fluorescence correlation spectroscopy (FCS) analysis of in-gel nucleosomes disclosed transient (~4.5 ms) unwrapping of DNA at nucleosomal ends ($K_{eq}=0.2$–0.6) and to a lesser degree 27 bp inside the nucleosome ($K_{eq}=0.1$).

Histone proteins and DNA remained associated, and nucleosomes did not dissociate during these unwrapping events. The above experiments indicate that transient fluctuations in nucleosomal wrapping (particularly at the ends of wrapped DNA) can facilitate access of nucleosomes to regulatory proteins binding at sites within the nucleosome.

The ability of single-molecule and ensemble techniques to precisely monitor the kinetics and positional features of chromatin remodeling make them suitable for the study of drugs that act to modulate transcription. This is significant because of the central role of transcriptional deregulation in many diseases such as cancer and developmental disorders.

Ligand-regulated transcription could be well suited to analysis using single-molecule techniques. Ligand addition rapidly modulates the formation of productive complexes on DNA or activates pre-formed complexes. The binding of agonist ligands to nuclear hormone receptors (NHR) triggers a conformational change in the ligand-binding domain that establishes a binding site for transcriptional coactivators. NHR act by binding single or repeated hexameric DNA motifs (generally 5′-PuGGTCA) as monomers or as homo- or hetero-dimers to activate genetic programs when they are bound by ligand. Different NHR bind to a recognition element depending on the spacing between the repeated motifs and the orientation of repeated motifs relative to each other. Because nuclear receptors orchestrate the signals of central cellular processes, it is not surprising that aberrant activation or downmodulation of nuclear receptor signaling contributes to a variety of diseases including diabetes and cancer. Truncated, mutated, or aberrantly activated nuclear receptors as well as overexpressed (AIB1) nuclear receptor coactivators have been associated with cancer, in some cases in a causal role [35]. Pharmacologic doses of nuclear receptor agonists or antagonists consequently have had therapeutic and/or preventive value in cancer. Ligands that have been used in this fashion include retinoids [36], synthetic estrogens [37], agonists of the PPARgamma nuclear receptor [38], and vitamin D [39].

Because of their substantial clinical impact, ligands that modulate NHRs would be a propitious focus for screening and analysis using single-molecule techniques. New ligands for NHRs (including "orphan" receptors for which ligands have not yet been identified) would comprise therapeutic drug candidates. The identification of NHR ligands has occurred primarily by screening libraries for NHR binding (e.g., [40, 41]) and sometimes guided by in silico analyses [42, 43]. However, the ability of candidate ligands to bind to NHRs does not establish the effect of that complex on transcription. Functional assessment of putative NHR ligands has generally used cell-based assays (e.g., [44–46]). In vitro models of ligand/receptor activity have been lacking. There are no extant high-throughput assays that screen concurrently for ligand binding and activity at nucleosome-bound NHRs. Given that only a portion of promoters harboring DNA-binding sites for NHRs (e.g., estrogen receptor) are activated by liganded estrogen receptor (ER) [47], it is important to study these interactions in the context of NHR-binding sites located within or adjacent to nucleosomes.

Many studies indicate that unliganded ER (as opposed to glucocorticoid receptor (GR) not bound to ligand) is predominantly nuclear [48–54]. Addition of estrogen leads to rapid recruitment of coactivators to bound nuclear estrogen receptor [54]. In addition to altering DNA-bound ER conformation to facilitate cofactor binding, estrogen induces subnuclear redistribution of ER and coactivators within the nucleus [48, 53, 55] and also decreases the k_{off} of DNA-bound ER [56]. Sequential chromatin immunoprecipitation has demonstrated that both unliganded ER and estrogen-bound ER both bound DNA cyclically with a period of roughly 20 min and recruited histone acetyl transferases to DNA. However, histone H3 dimethylation on R17 and H4 acetylation on K16 and ultimate recruitment of polymerase II only occurred following binding of estrogen to ER [57, 58]. These studies support a model in which the addition of estrogen to DNA-bound, unliganded estrogen receptor will trigger the formation of complexes that cause nucleosomal conformational changes and priming for transcription. Insofar as this process can be illuminated by bulk and single-molecule spFRET techniques, the conformational realignment of nucleosomes can be a marker for estrogen receptor agonists.

We are currently constructing a nucleosomal sensor to detect ER-mediated remodeling of mononucleosomes. The sensor is constructed to screen (at the ensemble level) candidate ligands that can activate (or prevent activation of) nucleosomal DNA containing estrogen response elements (EREs) (Fig. 13.4). At the single-molecule

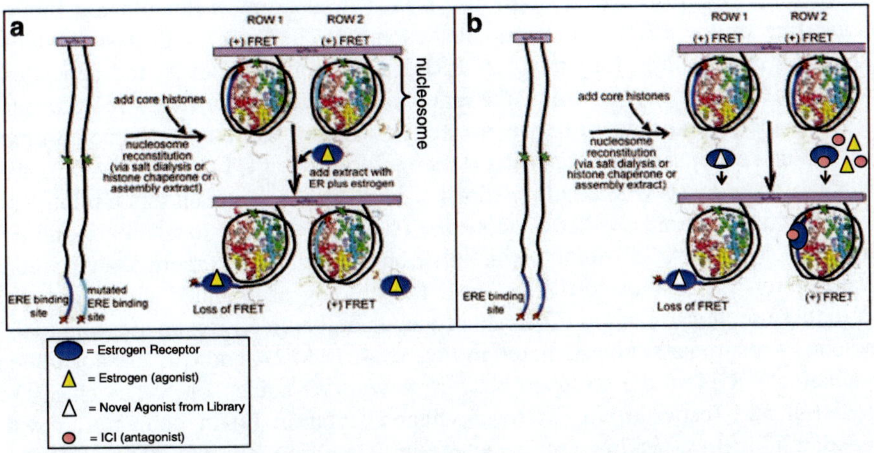

Fig. 13.4 Concept of nucleosomal biosensor use in drug discovery. Representative rows from 384-well plate are shown. On the *left side of each panel*, labeled DNA affixed to the well is shown. Two wells are shown in cross-section. Histones are added to package the DNA into nucleosomes, as shown, with FRET energy transfer from the donor to acceptor fluorophore in the closed nucleosomal conformation. In the schematic shown, the donor fluorophore is at the end of nucleosomal DNA and separated by 76 nucleotides from the acceptor fluorophore. An alternative construct positions both fluorophores at internal sites flanking the central dyad and spaced to permit FRET. The ERE at positions 11–27 may ultimately be positioned elsewhere in the nucleosome. (**a**) One row of wells contains wild-type estrogen response element in the DNA (*dark blue line*) while the other row of wells contains a mutated ERE that does not bind ER. Estrogen activates nucleosomes containing the wild-type ERE but not a mutated ERE that does not bind ER. (**b**) ERE-containing nucleosomes are shown. Novel agonists will bind to ER and open the nucleosome, causing FRET loss. In contrast, ER antagonists (shown as *pink circle*) or nonreactive compounds will not cause FRET loss. Antagonists are expected to prevent FRET loss associated with coadministered agonist

level, the kinetics of ligand-mediated remodeling will be studied. Ensemble and single-molecule analyses are complementary. An advantage of the single-molecule approach is that details about fast processes are obtained including kinetic information that may be lost in asynchronous populations. For instance, rate constants of nucleosomal opening following ligand addition can be calculated for individual nucleosomes. An advantage of the bulk approach is that the differences in agonist activity of different ligands can be quantified on a population basis that controls for variance between individual sensors.

Estrogen receptor readily accesses nucleosomes containing ER response elements. The receptor binds specifically and does not alter its binding position (as assayed by footprinting) as a function of ligand [59]. Metivier et al. [58] have conducted a minute-by-minute analysis of ER-mediated activation of a nucleosomal ERE using chromatin immunoprecipitation of MCF-7 cells to demonstrate the kinetics of estrogen-induced recruitment of proteins to the ER/nucleosome complex during transcription initiation. Interestingly, recruitment of coactivators and polymerase to the promoter was coordinated with cyclical methylation and demethylation of DNA bordering the transcription start site [60, 61]. Micrococcal nuclease accessibility assays showed that in the presence of estrogen, ER caused remodeling of the nucleosome, peaking at 60 min. In the absence of ligand, ER bound to the promoter and recruited a subset of cofactors to "prime" transcription but did not induce remodeling of the ERE-containing nucleosome [57]. These studies support the likelihood of detecting loss of DNA FRET upon initiation of ligand-dependent nucleosome remodeling and frame the timescale for initiation of FRET monitoring.

The nucleosomal sensor that we are studying uses a 164-bp "601" nucleosomal positioning sequence into which the ER recognition site agGTCAaagTGACCtg has been substituted at two internal positions. The positions at which this binding site was substituted within the "601" sequence [62] were chosen to maintain optimal alignment with AA/TT dinucleotide distribution [63] to comprise an optimal sequence for nucleosome formation [64]. The binding of liganded and unliganded ER to these sequences and to ER target sequence in the linker region adjacent to the nucleosome is under study. As noted in Fig. 13.4, TAMRA and Cy5 are positioned to generate FRET in the wrapped but not unwrapped nucleosome. The ability of agonist ligand (estradiol) or irrelevant ligand (vitamin D) in conjunction with ER-containing and -lacking cell extracts to initiate chromatin remodeling that partially unwraps the nucleosome leading to durable loss of FRET will be followed in ensemble assays. The ability of antagonist ligand (ICI182780) to disrupt estradiol-induced nucleosomal unwrapping serves as an additional control and sets the stage for screening of candidate ligands.

The timescale for single-molecule studies of the kinetics estrogen/ER chromatin remodeling should optimally capture events over seconds to minutes. The kinetics of E2 binding to ER ($K_a = 1.3 \times 10^6\ M^{-1} s^{-1}$ [65]) and co-activator assembly ($t^{-1/2}$ ~2 s [54]) suggest that changes in FRET should be readily discerned within the timeframe of the experiment prior to dye photobleaching. Photobleaching can be deterred with antioxidants such as Trolox [66] or improved oxygen scavenging systems [67].

We have also found the ALEX-spFRET system to be useful in discerning real from artifactual changes in FRET efficiency in these sorts of studies. Informative comparisons expected for the single-molecule studies include the impact of ER response element position on unwrapping kinetics, the kinetics of displacement of agonist with antagonist and vice versa, the impact of cellular origin of extracts (e.g., the ligand tamoxifen is an agonist in uterus and antagonist in breast) and the impact of specific cofactors determined through add-back to cofactor-depleted extract or through recombinant transcription reconstitution mixtures.

The advantage of an ALEX-spFRET-based system to study nucleosomal unwrapping as a marker of transcriptional initiation is that it provides a rapid readout that captures binding and remodeling. This is expected to be much more rapid and versatile than cell-based reporter gene assays and is more informative than assays that measure only ligand/receptor binding. Moreover, single-molecule systems are efficient, using much less input material than alternative assays. We foresee future adaptations of these nucleosome-based systems to incorporate three-color FRET [68] or ALEX-FRET [69] and to follow cofactor kinetics in conjunction with promoter unwinding and DNA-histone distancing.

Many caveats confront the successful accomplishment of nucleosomal single-molecule studies to dissect regulated transcription. These include the minimization and discernment of spontaneous nucleosome disassembly, optimizing protocols to minimize nonspecific protein binding artifacts from cell extracts, optimizing concentrations of components and positions of binding elements and fluorophores and ultimately adapting robust microfluidic screening platforms to incorporate viable nucleosomal sensors into high-throughput drug screens. The rapid advancement of sensing and analytical technologies as well as the expanding knowledge base and experience of investigators with these techniques will help propel the field beyond such obstacles.

13.6 Conclusions

While single-molecule techniques are still in an early phase, remarkable progress has been made using both home-built and standardized instrumentation. Not only has imaging fidelity become robust, enabling precise kinetic measurements, but also different technologies and modalities have been combined to yield synergistic benefits. Binding relationships uncovered in AFM can be further analyzed dynamically through ALEX-spFRET and/or optical tweezers. Precise characterization of the kinetics of protein/DNA or protein/protein interactions can follow because stepwise changes in interactions are not masked by averaging of pooled signals. Because of this degree of discrimination, it should be possible to distinguish between drugs that affect the same overall process, should they act on different targets or transition states. Given this potential, it will be important for investigators in the single-molecule field to forge scientific relationships with other biomedical scientists. Such partnerships can bring the benefits of these new analytical tools to bear on important physiologic problems (such as aberrant transcription in cancer) and on drugs to treat those conditions.

Acknowledgments We thank all our collaborators involved in the experiments described. This work was supported by GM077872 (S.H.L) and CA125514 (R.A.S.) from the National Institutes of Health.

References

1. Smith SB, Cui Y, Bustamante C (1996) Overstretching B-DNA: the elastic response of individual double-stranded and single-stranded DNA molecules. Science 271:795–799
2. Cluzel P, Lebrun A, Heller C, Lavery R, Viovy JL, Chatenay D, Caron F (1996) DNA: an extensible molecule. Science 271:792–794
3. Williams MC, Rouzina I, Bloomfield VA (2002) Thermodynamics of DNA interactions from single molecule stretching experiments. Acc Chem Res 35:159–166
4. McCauley MJ, Williams MC (2009) Optical tweezers experiments resolve distinct modes of DNA-protein binding. Biopolymers 91:265–282
5. van Mameren J, Gross P, Farge G, Hooijman P, Modesti M, Falkenberg M, Wuite GJ, Peterman EJ (2009) Unraveling the structure of DNA during overstretching by using multicolor, single-molecule fluorescence imaging. Proc Natl Acad Sci U S A 106:18231–18236
6. Williams MC, Rouzina I, McCauley MJ (2009) Peeling back the mystery of DNA overstretching. Proc Natl Acad Sci USA 106:18047–18048
7. Hansma HG, Vesenka J, Siegerist C, Kelderman G, Morrett H, Sinsheimer RL, Elings V, Bustamante C, Hansma PK (1992) Reproducible imaging and dissection of plasmid DNA under liquid with the atomic force microscope. Science 256:1180–1184
8. Allen MJ, Dong XF, O'Neill TE, Yau P, Kowalczykowski SC, Gatewood J, Balhorn R, Bradbury EM (1993) Atomic force microscope measurements of nucleosome cores assembled along defined DNA sequences. Biochemistry 32:8390–8396
9. Thoma F, Koller T, Klug A (1979) Involvement of histone H1 in the organization of the nucleosome and of the salt-dependent superstructures of chromatin. J Cell Biol 83:403–427
10. Leuba SH, Wheeler TB, Cheng CM, LeDuc PR, Fernandez-Sierra M, Quinones E (2009) Structure and dynamics of single DNA molecules manipulated by magnetic tweezers and or flow. Methods 47:214–222
11. Zlatanova J, Leuba SH (2003) Magnetic tweezers: a sensitive tool to study DNA and chromatin at the single-molecule level. Biochem Cell Biol 81:151–159
12. Pomerantz RT, Ramjit R, Gueroui Z, Place C, Anikin M, Leuba S, Zlatanova J, McAllister WT (2005) A tightly regulated molecular motor based upon T7 RNA polymerase. Nano Lett 5:1698–1703
13. Sakata-Sogawa K, Shimamoto N (2004) RNA polymerase can track a DNA groove during promoter search. Proc Natl Acad Sci USA 101:14731–14735
14. Harada Y, Ohara O, Takatsuki A, Itoh H, Shimamoto N, Kinosita K Jr (2001) Direct observation of DNA rotation during transcription by Escherichia coli RNA polymerase. Nature 409:113–115
15. Kinosita K Jr, Yasuda R, Noji H, Adachi K (2000) A rotary molecular motor that can work at near 100% efficiency. Philos Trans R Soc Lond B Biol Sci 355:473–89
16. Bryant Z, Stone MD, Gore J, Smith SB, Cozzarelli NR, Bustamante C (2003) Structural transitions and elasticity from torque measurements on DNA. Nature 424:338–341
17. Gore J, Bryant Z, Stone MD, Nollmann M, Cozzarelli NR, Bustamante C (2006) Mechanochemical analysis of DNA gyrase using rotor bead tracking. Nature 439:100–104
18. Stasiak A, Di Capua E, Koller T (1981) Elongation of duplex DNA by recA protein. J Mol Biol 151:557–564

19. Bennink ML, Scharer OD, Kanaar R, Sakata-Sogawa K, Schins JM, Kanger JS, de Grooth BG, Greve J (1999) Single-molecule manipulation of double-stranded DNA using optical tweezers: interaction studies of DNA with RecA and YOYO-1. Cytometry 36:200–208
20. Hegner M, Smith SB, Bustamante C (1999) Polymerization and mechanical properties of single RecA-DNA filaments. Proc Natl Acad Sci USA 96:10109–10114
21. Leger JF, Robert J, Bourdieu L, Chatenay D, Marko JF (1998) RecA binding to a single double-stranded DNA molecule: a possible role of DNA conformational fluctuations. Proc Natl Acad Sci USA 95:12295–12299
22. Shivashankar GV, Feingold M, Krichevsky O, Libchaber A (1999) RecA polymerization on double-stranded DNA by using single-molecule manipulation: the role of ATP hydrolysis. Proc Natl Acad Sci USA 96:7916–7921
23. Anand SP, Zheng H, Bianco PR, Leuba SH, Khan SA (2007) DNA helicase activity of PcrA is not required for the displacement of RecA protein from DNA or inhibition of RecA-mediated strand exchange. J Bacteriol 189:4502–4509
24. Leuba SH, Anand SP, Harp JM, Khan SA (2008) Expedient placement of two fluorescent dyes for investigating dynamic DNA protein interactions in real time. Chromosome Res 16:451–467
25. Petit MA, Dervyn E, Rose M, Entian KD, McGovern S, Ehrlich SD, Bruand C (1998) PcrA is an essential DNA helicase of Bacillus subtilis fulfilling functions both in repair and rolling-circle replication. Mol Microbiol 29:261–273
26. Liu S, Abbondanzieri EA, Rausch JW, Le Grice SF, Zhuang X (2008) Slide into action: dynamic shuttling of HIV reverse transcriptase on nucleic acid substrates. Science 322:1092–1097
27. Neuman KC, Abbondanzieri EA, Landick R, Gelles J, Block SM (2003) Ubiquitous transcriptional pausing is independent of RNA polymerase backtracking. Cell 115:437–447
28. Shaevitz JW, Abbondanzieri EA, Landick R, Block SM (2003) Backtracking by single RNA polymerase molecules observed at near-base-pair resolution. Nature 426:684–687
29. Vassylyev DG, Artsimovitch I (2005) Tracking RNA polymerase, one step at a time. Cell 123:977–979
30. Lee NK, Kapanidis AN, Koh HR, Korlann Y, Ho SO, Kim Y, Gassman N, Kim SK, Weiss S (2007) Three-color alternating-laser excitation of single molecules: monitoring multiple interactions and distances. Biophys J 92:303–312
31. Tomschik M, Zheng H, van Holde K, Zlatanova J, Leuba SH (2005) Fast, long-range, reversible conformational fluctuations in nucleosomes revealed by single-pair fluorescence resonance energy transfer. Proc Natl Acad Sci USA 102:3278–3283, Correction 2008 *105*, 10632
32. Li G, Levitus M, Bustamante C, Widom J (2005) Rapid spontaneous accessibility of nucleosomal DNA. Nat Struct Mol Biol 12:46–53
33. Li G, Widom J (2004) Nucleosomes facilitate their own invasion. Nat Struct Mol Biol 11:763–769
34. Koopmans WJ, Buning R, Schmidt T, van Noort J (2009) spFRET using alternating excitation and FCS reveals progressive DNA unwrapping in nucleosomes. Biophys J 97:195–204
35. Osborne CK, Bardou V, Hopp TA, Chamness GC, Hilsenbeck SG, Fuqua SA, Wong J, Allred DC, Clark GM, Schiff R (2003) Role of the estrogen receptor coactivator AIB1 (SRC-3) and HER-2/neu in tamoxifen resistance in breast cancer. J Natl Cancer Inst 95:353–361
36. Okuno M, Kojima S, Matsushima-Nishiwaki R, Tsurumi H, Muto Y, Friedman SL, Moriwaki H (2004) Retinoids in cancer chemoprevention. Curr Cancer Drug Targets 4:285–298
37. Jordan VC, Gapstur S, Morrow M (2001) Selective estrogen receptor modulation and reduction in risk of breast cancer, osteoporosis, and coronary heart disease. J Natl Cancer Inst 93:1449–1457
38. Wang T, Xu J, Yu X, Yang R, Han ZC (2006) Peroxisome proliferator-activated receptor gamma in malignant diseases. Crit Rev Oncol Hematol 58:1–14
39. Beer TM, Myrthue A (2004) Calcitriol in cancer treatment: from the lab to the clinic. Mol Cancer Ther 3:373–381
40. Cases M, Garcia-Serna R, Hettne K, Weeber M, van der Lei J, Boyer S, Mestres J (2005) Chemical and biological profiling of an annotated compound library directed to the nuclear receptor family. Curr Top Med Chem 5:763–772

41. Sun S, Almaden J, Carlson TJ, Barker J, Gehring MR (2005) Assay development and data analysis of receptor-ligand binding based on scintillation proximity assay. Metab Eng 7:38–44
42. Liao C, Liu B, Shi L, Zhou J, Lu XP (2005) Construction of a virtual combinatorial library using SMILES strings to discover potential structure-diverse PPAR modulators. Eur J Med Chem 40:632–640
43. Schapira M, Abagyan R, Totrov M (2003) Nuclear hormone receptor targeted virtual screening. J Med Chem 46:3045–3059
44. Beck V, Pfitscher A, Jungbauer A (2005) GFP-reporter for a high throughput assay to monitor estrogenic compounds. J Biochem Biophys Methods 64:19–37
45. Brown PJ, Smith-Oliver TA, Charifson PS, Tomkinson NC, Fivush AM, Sternbach DD, Wade LE, Orband-Miller L, Parks DJ, Blanchard SG, Kliewer SA, Lehmann JM, Willson TM (1997) Identification of peroxisome proliferator-activated receptor ligands from a biased chemical library. Chem Biol 4:909–918
46. Skretas G, Wood DW (2005) A bacterial biosensor of endocrine modulators. J Mol Biol 349: 464–474
47. Lin CY, Vega VB, Thomsen JS, Zhang T, Kong SL, Xie M, Chiu KP, Lipovich L, Barnett DH, Stossi F, Yeo A, George J, Kuznetsov VA, Lee YK, Charn TH, Palanisamy N, Miller LD, Cheung E, Katzenellenbogen BS, Ruan Y, Bourque G, Wei CL, Liu ET (2007) Whole-genome cartography of estrogen receptor alpha binding sites. PLoS Genet 3:e87
48. Htun H, Holth LT, Walker D, Davie JR, Hager GL (1999) Direct visualization of the human estrogen receptor alpha reveals a role for ligand in the nuclear distribution of the receptor. Mol Biol Cell 10:471–486
49. King WJ, Greene GL (1984) Monoclonal antibodies localize oestrogen receptor in the nuclei of target cells. Nature 307:745–747
50. Martinez ED, Rayasam GV, Dull AB, Walker DA, Hager GL (2005) An estrogen receptor chimera senses ligands by nuclear translocation. J Steroid Biochem Mol Biol 97:307–321
51. Picard D, Kumar V, Chambon P, Yamamoto KR (1990) Signal transduction by steroid hormones: nuclear localization is differentially regulated in estrogen and glucocorticoid receptors. Cell Regul 1:291–299
52. Press MF, Xu SH, Wang JD, Greene GL (1989) Subcellular distribution of estrogen receptor and progesterone receptor with and without specific ligand. Am J Pathol 135:857–864
53. Stenoien DL, Mancini MG, Patel K, Allegretto EA, Smith CL, Mancini MA (2000) Subnuclear trafficking of estrogen receptor-alpha and steroid receptor coactivator-1. Mol Endocrinol 14:518–534
54. Stenoien DL, Nye AC, Mancini MG, Patel K, Dutertre M, O'Malley BW, Smith CL, Belmont AS, Mancini MA (2001) Ligand-mediated assembly and real-time cellular dynamics of estrogen receptor alpha-coactivator complexes in living cells. Mol Cell Biol 21:4404–4412
55. Berno V, Amazit L, Hinojos C, Zhong J, Mancini MG, Sharp ZD, Mancini MA (2008) Activation of estrogen receptor-alpha by E2 or EGF induces temporally distinct patterns of large-scale chromatin modification and mRNA transcription. PLoS One 3:e2286
56. Sharp ZD, Mancini MG, Hinojos CA, Dai F, Berno V, Szafran AT, Smith KP, Lele TP, Ingber DE, Mancini MA (2006) Estrogen-receptor-alpha exchange and chromatin dynamics are ligand- and domain-dependent. J Cell Sci 119:4101–4116
57. Metivier R, Penot G, Carmouche RP, Hubner MR, Reid G, Denger S, Manu D, Brand H, Kos M, Benes V, Gannon F (2004) Transcriptional complexes engaged by apo-estrogen receptor-alpha isoforms have divergent outcomes. EMBO J 23:3653–3666
58. Metivier R, Penot G, Hubner MR, Reid G, Brand H, Kos M, Gannon F (2003) Estrogen receptor-alpha directs ordered, cyclical, and combinatorial recruitment of cofactors on a natural target promoter. Cell 115:751–763
59. Kraus WL, Kadonaga JT (1998) p300 and estrogen receptor cooperatively activate transcription via differential enhancement of initiation and reinitiation. Genes Dev 12:331–342
60. Kangaspeska S, Stride B, Metivier R, Polycarpou-Schwarz M, Ibberson D, Carmouche RP, Benes V, Gannon F, Reid G (2008) Transient cyclical methylation of promoter DNA. Nature 452:112–115

61. Metivier R, Gallais R, Tiffoche C, Le Peron C, Jurkowska RZ, Carmouche RP, Ibberson D, Barath P, Demay F, Reid G, Benes V, Jeltsch A, Gannon F, Salbert G (2008) Cyclical DNA methylation of a transcriptionally active promoter. Nature 452:45–50
62. Lowary PT, Widom J (1998) New DNA sequence rules for high affinity binding to histone octamer and sequence-directed nucleosome positioning. J Mol Biol 276:19–42
63. Satchwell SC, Drew HR, Travers AA (1986) Sequence periodicities in chicken nucleosome core DNA. J Mol Biol 191:659–675
64. Wang JP, Widom J (2005) Improved alignment of nucleosome DNA sequences using a mixture model. Nucleic Acids Res 33:6743–6755
65. Rich RL, Hoth LR, Geoghegan KF, Brown TA, LeMotte PK, Simons SP, Hensley P, Myszka DG (2002) Kinetic analysis of estrogen receptor/ligand interactions. Proc Natl Acad Sci USA 99:8562–8567
66. Rasnik I, McKinney SA, Ha T (2006) Nonblinking and long-lasting single-molecule fluorescence imaging. Nat Methods 3:891–893
67. Aitken CE, Marshall RA, Puglisi JD (2008) An oxygen scavenging system for improvement of dye stability in single-molecule fluorescence experiments. Biophys J 94:1826–1835
68. Hohng S, Joo C, Ha T (2004) Single-molecule three-color FRET. Biophys J 87:1328–1337
69. Kapanidis AN, Lee NK, Laurence TA, Doose S, Margeat E, Weiss S (2004) Fluorescence-aided molecule sorting: analysis of structure and interactions by alternating-laser excitation of single molecules. Proc Natl Acad Sci USA 101:8936–8941
70. Leuba SH, Yang G, Robert C, Samori B, van Holde K, Zlatanova J, Bustamante C (1994) Three-dimensional structure of extended chromatin fibers as revealed by tapping-mode scanning force microscopy. Proc Natl Acad Sci USA 91:11621–11625

Chapter 14
Long-Range Chromatin Interactions in Cells

Guo Ling Zhou, Li Xin, and De Pei Liu

14.1 Long-Range Regulatory Elements in Higher Eukaryotic Cells

To precisely control differentiation and development, eukaryotic organisms have evolved complicated genetic and epigenetic mechanisms to modulate gene expression. The definition of the gene as a functional unit in higher eukaryotic organisms goes far beyond the description of the promoter and coding regions. Regulatory elements, especially long-range regulatory elements, have profound influence. Aberrant gene expression resulting from improper positioning of long-range regulatory elements (e.g., due to chromosome translocation or deletion events) can result in loss of expression, ectopic expression, or enhanced expression that can disrupt normal tissue differentiation and development or promote the evolution of neoplasms.

Regulatory elements in genomic DNA have been studied intensively in multiple organisms. Elements are usually located upstream of the gene being regulated and are bound by transcription factors and other regulatory proteins that recruit coactivators or corepressors to control expression. Long-range regulatory elements are positioned far from the genes being regulated. Distances can be kilo base pair or mega base pair from genes controlled in *cis*, and elements can also be on different chromosomes, affecting expression in *trans*. Elements can function as enhancers to elevate gene expression, as silencers to repress expression, or as insulators (boundary elements) to block the influence of an enhancer or propagation of heterochromatin in *cis*. In addition, scaffold- or matrix-attachment regions (S/MAR) help sustain the partition of eukaryotic chromosomes into independent topological domains and thereby contribute to the regulation of gene expression.

To date, there has been no simple method to directly identify long-range regulatory elements. Because of their DNA sequence diversity and distant location from the

G.L. Zhou (✉)
Department of Genetics, Harvard Medical School, Center for Computational and Integrative Biology, Massachusetts General Hospital, 185 Cambridge Street, Boston, MA 02114-2790, USA
e-mail: gzhou@ccib.mgh.harvard.edu

M.C. Williams and L.J. Maher, III (eds.), *Biophysics of DNA-Protein Interactions*, Biological and Medical Physics, Biomedical Engineering,
DOI 10.1007/978-0-387-92808-1_14,

transcripts they regulate, it has been difficult to make reliable computational predictions. Although certain validated common features such as hypersensitivity to DNase I may be present these are not universal. Multiple unique properties including ATC-enriched sequence segments, base-unpairing regions (BURs) and stress-induced DNA duplex destabilization (SIDD) have been identified in certain S/MARs. However, there remain many discrepancies between the computational predictions of S/MARs within genomic sequences and the experimental observations [1–3]. That our present capacity for recognition of distant regulatory sequences is rudimentary at best is illustrated by the fact that many such sequences have been identified as a result of clues from highly penetrant human monogenic disorders. For example, some cases of β-thalassemia result from the translocation of the locus control region (LCR) of the β-globin gene cluster away from the adjacent genes [4]. Genetic investigations of aniridia, a congenital eye malformation, led to the discovery of one downstream regulatory region (DRR) of the *PAX6* gene located 130 kb from the transcript it controls [4]. The identification of regulatory sequences ~1 Mb from the sonic hedgehog (SHH) gene was similarly a consequence of the genetic study of limb developmental malformation in preaxial polydactyly (PPD) cases [4].

Experimental verification and fine structure dissection of distal regulatory sequences typically rely on reporter assays in cultured cells. More realistic animal models have been generated to reveal how structural alterations in such sequences modulate their target genes. Recent experimental advances have been made in the identification of chromosomal regulatory elements by detailed accounting of transcription factor-binding sites. In particular, ChIP on chip and chromosome conformation capture (3C) assays [5–7] provide powerful tools to elucidate the interaction between long-range regulatory elements and their regulated genes. Advances in 3C technology [8–11] have paved the way for explorations of novel interactions and chromatin organization.

14.2 Establishment and Regulation of Long-Range Chromatin Interactions

14.2.1 The Importance of DNA Looping in Establishment of Long-Range Intrachromosomal and Interchromosomal Interactions

The underlying mechanisms by which long-range regulatory elements modulate cellular gene expression have been sought for many years. Tracking, linking, and looping models have been proposed [12–14]. Because of technical limitations, little direct evidence had been presented to confirm or refute these proposals. Recently, RNA-Trap (tagging and recovery of associated proteins) methods [15], 3C, and related new technologies such as 4C, 5C, 6C, and QACT [5, 8–11, 16, 17] have been used to reveal higher order chromatin organization patterns in three-dimensions.

These data suggest that DNA looping plays a critical role in the establishment of long-range chromatin interactions.

DNA loops were first proposed many years ago [18]. Initially, it was believed that DNA loop formation was associated with efficient gene transcription. For decades, this relationship remained enigmatic. With increased understanding, the focus on chromatin structural influences on gene expression switched from linear control to three-dimensional control, as an accounting for both the physical and biological interactions between genes and regulatory elements. The notion of the "active chromatin hub (ACH)" has been proposed in connection with the regulation of the β-globin gene cluster to explain how the LCR interacts with different developmentally regulated globin genes through looping out of the interposed chromatin and gene activation in successive developmental stages [19, 20]. An α-ACH was identified in the α-globin gene cluster in erythroid cells but not in nonerythroid cells [21]. In addition, a cluster of housekeeping genes flanking the α-globin gene region was discovered to form a compartment-specific transcription factory to favor the expression of these genes in erythroid cells. In nonerythroid cells, the α-ACH is dissociated with this transcription factory and elevated expression is not observed [21]. At the insulin-like growth factor 2 (*Igf2*)/*H19* locus, parent-specific active and repressive chromatin compartments are apparently organized by linking different differentially methylated regions (DMRs) [22–24]. Another "prerepressive chromatin hub (pre-RCH)" formed by long-range interaction has been delineated at the *GATA-4* locus in undifferentiated Tera-2 cells. The pre-RCH dissociates when *GATA-4* is upregulated in retinoid-induced differentiating Tera-2 cells [25].

Modified 3C methods such as 4C, ACT, and QACT offer the prospect of identifying unknown interaction fragments [9, 10, 16, 26, 27]. Using these methods, interchromosomal (*trans*) interactions have been found to be important for controlling gene expression in some cases. For example, one allele of the *Igf2*/*H19* imprinting control region (ICR) on chromosome 7 has been found to be colocalized with one allele of *Wsb1*/*Nf1* on chromosome 11 [27]. The disruption of this association by deletion of ICR or knockdown of CTCF alters *Wsb1*/*Nf1* expression [27]. More recently, it has been shown that interchromosomal interaction is involved in olfactory receptor (OR) selection [28]. The results of 3C and fluorescent in situ hybridization (FISH) demonstrate that there is a specific association of an enhancer element, H, on chromosome 14 with multiple OR gene promoters on different chromosomes. The H enhancer interacts with only a single OR allele transcribed in a given sensory neuron [28]. In addition, using three-dimensional FISH, it was found that transient colocalization of two Xics (X-chromosome inactivation centers) occurs prior to X-chromosome inactivation. Interaction is controlled by a long-range element located at the 3′ end of the noncoding transcript Tsix [29]. Thus X-chromosome homologous pairing may preclude X-chromosome inactivation.

The preceding data suggest that spatially, temporally, and quantitatively regulated gene expressions not only require the integrity of local regulatory regions and protein-coding sequences, but also the correct chromatin organization and nuclear compartmentalization. This implies the need to further investigate higher order chromatin architecture including how it is organized and reorganized to modulate gene expression.

14.2.2 *Roles of DNA-Binding Proteins in Long-Range Chromatin Interactions*

Recent advances have made it possible to better understand the relationship between chromatin organization and gene regulation. The concepts of three-dimensional regulation have resulted from studies using the chromosome conformation capture (3C) assay, ChIP-loop assay, and other novel approaches. These results suggest that DNA loops bring distal DNA sequences into close proximity at several loci including globin genes [19–21], *Igf2/H19* [22–24, 30], *HoxB1* [31], Th2 cytokine loci [32–34], *Dlx5/Dlx6* [35], and MHC class-I loci [36, 37]. Studies are unraveling the chromatin "loopscape" and revealing mechanisms by which chromatin organization is dynamically modulated in response to differentiation and development stimuli. Two proteins, special AT-rich binding protein 1 (SATB1) and CCCTC-binding factor (CTCF) [38, 39], have been identified as global regulators in the processes of chromatin folding.

SATB1, a protein highly expressed in thymocytes but found broadly in most tissues, shows a "cage-like" nuclear distribution circumscribing the heterochromatin [32]. As a cell type-specific chromatin organizer, it recognizes a specialized MAR DNA sequence containing BURs characteristic of ATC-enriched sequence [32, 40]. SATB1 is thought to tether these DNA sequences in a folded conformation by anchoring them on the nuclear matrix [36]. To date, hundreds of SATB1-binding sites have been identified and nearby genes have been found to be dysregulated in SATB1-deficient thymocytes. SATB1 is ~800 amino acids in length and comprises a C-terminal homeodomain (HD) and two CUT domains that mediate high-affinity MAR DNA-binding activity in the MAR-binding domain (MBD) [41]. The N-terminal PDZ-like domain (PVPLS motif) may serve as an interface to interact with other partner proteins and/or aid in the homodimerization of SATB1 that is essential for DNA binding [42–44]. Additionally, posttranslational modifications afford SATB1 another level of complexity. It was demonstrated that phosphorylation and acetylation of SATB1 have opposite effects on transcription, leading to activation and repression, respectively [45, 46]. SATB1 regulates globin gene expression during erythroid differentiation [47], an effect that is mediated by remodeling of the local chromatin architecture. SATB1 binds to MARs in the LCR and the ε-globin promoter and influences the anchoring of the local chromatin [47, 48]. Its interactions with the poised β-ACH may be realized by association with CREB-binding protein, which possesses histone acetyltransferase activity, and by similarly influencing deacetylation during erythroid differentiation [47, 48]. Study of the role of SATB1 in regulating the coordinated expression of Th2 cytokine genes has revealed that it plays a central role in compacting the 200-kb cytokine locus into higher order transcriptionally active chromatin structure upon Th2 cell activation by working together with signaling proteins STAT4 and STAT6 and transcriptional factors GATA-3 and c-Maf [32]. In contrast, by using an IL2 or c-Myc promoter-derived luciferase reporter-based assay, it was found that SATB1 can mediate transcription repression through acetylation-dependent interaction with CtBP1 [46]. Further investigation will doubtlessly unravel the role of SATB1 in mediating the formation of repressive chromatin architecture. Previously, SATB1 was shown to act as a "docking site" for several chromatin-remolding enzymes including ACF,

ISWI, and HDAC1, which are implicated in suppressing gene expression through histone deacetylation and nucleosome remodeling at SATB1-bound MARs. Most recently, dual regulator roles of SATB1 were suggested by studies of the MHC class-I locus [39]. By cooperating with the promyelocytic leukemia (PML) oncoprotein, MAR-bound SATB1 can dynamically regulate the chromatin loopscape of the MHC class-I locus to alter gene expression within the locus [36, 37]. This suggests that SATB1 is a global regulator of long-range chromatin interactions in nuclei.

The universally expressed CTCF protein is an evolutionarily conserved 11-zinc finger transcriptional regulator that acts as a multifunctional regulator to control gene expression. CTCF binds DNA with different zinc finger subsets and interacts with different partner proteins [38]. Increasing evidence supports the view that CTCF is an important structural protein in the formation of chromatin loops and higher order chromatin organization. CTCF is associated with the nuclear matrix and the nuclear pore complex (NPC) [49, 50]. It recognizes insulator or boundary elements and may function together with other nuclear proteins such as nucleophosmin/B23, lamins, LAP2β, and Nu93 to tether chromatin to the nuclear periphery and position the promoter and nearby regulatory elements in a separate chromatin loop [49, 50]. Revealing insights have emerged from the study of CTCF-mediated parent-specific chromatin loop organization in the *Igf2/H19* locus. CTCF binds to the unmethylated maternal allele of the ICR in maternal cells and forms long-range intrachromosomal loops by interaction with differential methylated region 1 (DMR1) upstream of *Igf2*. This causes an enhancer blocking function by segregating the enhancers upstream of the *H19* and *Igf2* promoters into different chromatin compartments [22]. In addition to its enhancer blocking function, CTCF may prevent the spreading of DNA methylation on chromatin to maintain the boundaries between heterochromatin and euchromatin, thereby protecting gene expression [51]. These data suggest that CTCF is an important protein that facilitates the organization of genes and their regulatory elements into distinct compartments that govern their transcriptional status.

14.2.3 Potential Roles of Transcription Factors, Histone Modifications, and Noncoding RNAs

In addition to the two global bridge proteins described above, cell-specific, tissue-specific, or developmental stage-specific transcription factors, coactivators, and corepressors all play roles in long-range chromatin interactions. Often charged with unique physiological functions, these factors may be suitable targets for therapeutics to treat human diseases. However, many aspects of their roles in the establishment and maintenance of chromatin organization remain to be explored. Within the globin gene clusters, the erythroid-specific transacting factors, EKLF, GATA-1, FOG, and MafK/NF-E2, are indispensable factors that recruit active β-like globin genes to the ACH [52–55]. It is also known that chromatin remodeling and histone-modifying enzymes are recruited to transcriptionally active promoters [44, 56, 57], coupling the establishment of chromatin loop structure to promoter activation and pointing to an important functional link between *cis* regulatory sequences and the factors that bind them to the chromatin three-dimensional structure. One of the best-studied examples of cooperation between histone modification and large-scale chromatin architecture can be found in the CTCF-mediated formation of long-range intrachromosomal loops. Polycomb repressive complex 2 is recruited through the interaction of CTCF with Suz12, leading to allele-specific H3K27 methylation and to the suppression of maternal *Igf2* expression [56].

Although there are many examples of histone modifications associated with transcriptional regulation, it is difficult to distinguish whether they reflect causes or consequences in the orchestration of gene expression. Although chromatin looping at specific gene loci has been discovered, more effort is needed to explore the underlying mechanisms by which the global chromatin architecture is established, dynamically regulated, and maintained in different environments.

Increasing attention has also been paid to the multiple roles of noncoding RNAs. Noncoding transcripts may be associated with long-range chromatin interactions in cells. Noncoding transcripts have been predicted to cover ~60–70% of genome [58]. Approximately 72% of genome-mapped transcription units have antisense transcript counterparts [58–60]. In addition to functional noncoding RNAs like tRNA, rRNA, and snRNA, regulatory short noncoding RNAs such as siRNA, microRNA, and piwi-RNA are being intensively studied. Regulatory noncoding RNA distinct from the typically 21–31 nucleotide transcripts includes the larger noncoding RNAs ranging in length from 100 bp to several hundred kilo base pairs. Much of what we know about such regulatory RNAs comes from investigations of the noncoding Xist, Tsix and repA RNAs in X-chromosome inactivation [61, 62] and from *kcnq1ot1* transcripts in *KCNQ1* gene imprinting [63]. The mechanisms of noncoding transcripts in gene imprinting involve long-range chromatin reorganization in the nucleus. X-chromosome inactivation during female embryonic development involves the coating of one X-chromosome by Xist transcripts and heterochromatinization [64, 65]. The noncoding repA RNA within Xist can recruit PRC2 to the X through Ezh2, acting as the RNA-binding subunit. The antisense Tsix RNA inhibits this interaction. RepA depletion abolishes full-length Xist induction and trimethylation on lysine 27 of histone H3 of the X [62], suggesting its contribution to higher order chromatin organization at the X-inactivation center.

14.3 Biological Functions of Long-Range Chromatin Interactions in Cells

14.3.1 Cell Type-Specific, Differentiation-Specific, or Development-Specific Gene Expression

Development and differentiation are based on the regulated transcription of subsets of genes at appropriate times and locations. This complex regulation is mediated through *cis*-elements and their interaction with *trans*-factors. In contrast to the regulatory architecture of lower eukaryotic organisms such as yeast, the genomes of higher eukaryotic organisms contain huge numbers of regulatory elements dispersed over large genomic distance and only a small part of the genome encodes proteins. Understanding these long-range phenomena will be crucial for the appreciation of biological functions of genes under different physiological circumstances.

The globin gene loci are ideal models for the dissection of tissue- and development-specific expressions under the control of long-range chromatin interactions. Topological data have shown that tissue-specific globin gene expression is realized through the influence of the erythroid-specific ACH [20, 21, 66]. Switching expression from fetal- to adult-specific globin genes involves reconfiguration of the ACH leading to contact of the LCR with either fetal or adult globin gene promoters [19, 52]. Likewise, by changing higher order chromatin conformation, cytokine genes including Il4, Il5, and Il13 spanning a cluster of 120 kbp on chromosome 11 are activated during T-cell differentiation. The expression of these cytokine genes is under the control of a LCR, which is located 15 kbp upstream of Il13 and 60 kbp downstream of Il5 [67, 68]. The 3C analysis has revealed a complex looping structure in this region [33]. In contrast to the looping structure of the globin loci, the promoters of Il4, Il5, and Il13 all associate first in a "prepoised" hub in cells of the T-lineage, regardless of expression of the genes. During differentiation in T-cell and NK cell lineages, the LCR are recruited and the "prepoised" hub switches to a "poised" hub [33]. This structure is observed in all cells from the T-cell lineage, irrespective of the expression of the cytokine genes. LCR–promoter interactions become more significant with the expression of the cytokine genes in Th2 cells [33]. The unique looping structure is thought to allow a rapid induction of transcription upon Th2-cell activation.

In mammalian cells, the *GATA-4* locus has been investigated by using 3C and ChIP [25]. The authors suggest that this locus poses a pre-RCH in undifferentiated Tera-2 cells, which can be recognized by the presence of RNA PolII at the start site. *GATA-4* is expressed at a low to medium level in these cells [25]. However, the hub is also characterized by the PcG proteins and the H3K27me3 mark at loop intersection points and the embedded gene [25]. The PcG-associated hub is thought to provide a structure base for an efficient and flexible transcriptional repression in embryonic cells that can be reverted to an active conformation during differentiation.

Taken in combination, the results above provide compelling examples of the intricate network of interactions that can occur among genes and regulatory elements to drive cell type-specific and developmental stage-specific gene expressions. With the application of 3C technology, more examples of gene regulation by modulation of chromatin conformation are being added to the list. Recent cases include the MHC loci in lymphocytes [36], *Ifng* gene in T-helper cells [69], the *IgH* and *Igκ* loci in B cells [70], Ciz1-Lcn2 locus in glucocorticoid receptor activation [71], and *HoxB* [31] and some olfactory receptors [28].

14.3.2 Gene Imprinting and X-Chromosome Inactivation

It has been shown that long-range chromatin interactions are also involved in gene imprinting. At the imprinted *Igf2/H19* locus on chromosome 7, *Igf2* is expressed paternally and *H19* is expressed maternally. The loci are separated by approximately

100 kbp. DMRs in *Igf2* and *H19* regulate the reciprocal expression of the two genes by allowing them exclusive access to a shared set of enhancers [23]. On the maternal chromosome, two connected DMRs form an inactive chromatin loop. *Igf2* gene is placed within this negative chromatin loop preventing its promoter from interacting with the downstream enhancer and thereby keeping the maternal *Igf2* in a silent state [23]. This chromatin structure enables the maternal *H19* gene to make contact with the enhancer for active transcription. Deletion of either of two DMRs leads to activation of maternal *Igf2* transcription [23]. Moreover, a maternal allele-specific insulator is also found to participate in the long-range interaction. The ICR bears insulator activity and interacts directly with the maternal silenced *Igf2* promoter and enhancer blocking activation of the maternal *Igf2* gene. These studies indicate that long-range chromatin interactions can be involved in imprinting.

Mammalian dosage compensation involves silencing of one of two X chromosomes. Counting X chromosomes (counting) and selecting which X chromosome is to be inactivated in female cells (choice) are important events in mammalian cell differentiation. It has been demonstrated that the Xic controls these events and, in particular, the three noncoding RNAs Xist, Tsix, and RepA, conspire with long-range control elements in the Xic to differentially treat X chromosomes [62, 64, 65]. Recent studies suggest that both *cis* and *trans* long-range chromatin interactions are involved in X-inactivation [72]. A segment of the mouse Xic that lies several hundred kilo base pairs from Xist brings the two Xics together before the onset of X-inactivation [29]. This region can autonomously drive Xic *trans*-interactions even as an ectopic single-copy transgene [29]. Deleting Xite and Tsix perturbs pairing and counting/choice, whereas autosomal insertion induces de novo X-autosome pairing. Ectopic X-autosome interactions inhibit endogenous X–X pairing and block the initiation of X-chromosome inactivation. Thus, Tsix and Xite can function both in *cis* and in *trans* [29]. The 3C method has been used to map the higher order chromatin structure of Xic locus. The resulting data indicate that Xite makes looping interactions with Tsix and Xist associates with a novel noncoding gene *Jpx/Enox* [72]. These regions interact in a developmentally specific and sex-specific manner that is consistent with a regulatory role in XCI. Dynamic changes in three-dimensional chromatin architecture possibly lead to formation of separate chromatin hubs for Tsix and Xist that together regulate the initiation of X-chromosome inactivation.

14.4 Methods Used in Studying Long-Range Chromatin Interactions

14.4.1 3C

In eukaryotic nuclei, the transcription of genes is often controlled by regulatory elements that are located far from target promoters in *cis* or even on different chromosomes in *trans*.

Long-range *cis*- and *trans*-interactions can be studied at high resolution using chromosome conformation capture (3C) technology.

Basic 3C technology involves five steps (Fig. 14.1) [5]: (1) fixation – protein–protein and DNA–protein interactions are cross-linked with formaldehyde in living cells; (2) restriction digestion – the cross-linked chromatin is digested with restriction enzymes to release DNA–protein complex and create DNA fragments with cohesive termini; (3) selective ligation – digested chromatin is diluted and ligated and low DNA concentration favors intramolecular ligation; (4) cross-link reversal; and (5) DNA purification – the final purified DNA serves as the 3C template. The template is then analyzed by PCR, microarray hybridization, or sequencing.

Because a very low DNA concentration favors intramolecular ligation, the abundance of a particular ligation product of two DNA fragments reflects their spatial distance after cross-linking, rather than genomic linear distance. To analyze the 3C template by qPCR, a control template is generated. The basic steps of control template DNA preparation are (1) purification of genomic DNA or BAC clone DNA contains the genomic DNA region of interest; (2) complete restriction digestion with

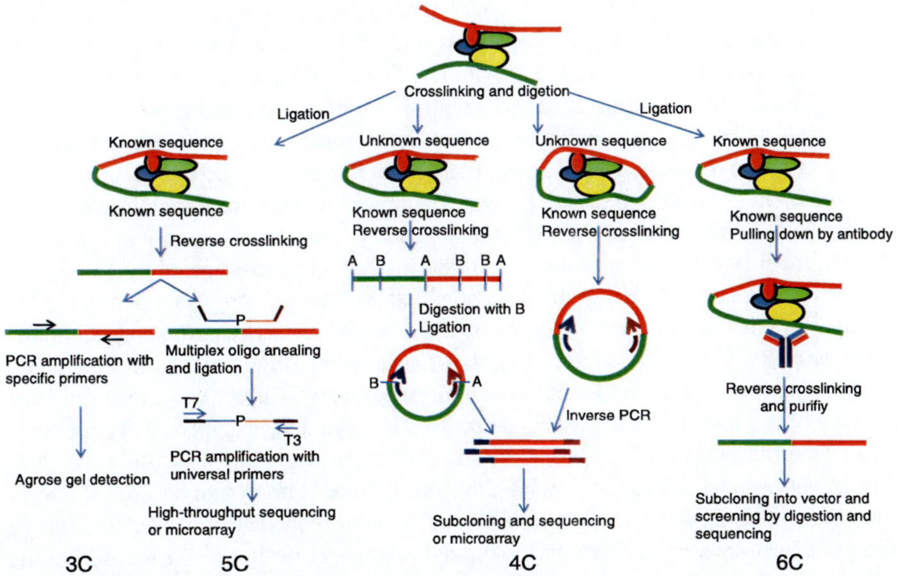

Fig. 14.1 Schematic flow chart of 3C, 5C, 4C, and 6C procedures. All assays start with formaldehyde cross-linking of living cells, digestion with the proper restriction enzyme, and intramolecular ligation. In the conventional 3C assay, ligated macromolecules undergo reversal of cross-links and are then amplified by specific primers to probe the conformation of the genomic region of interest. The 5C assay starts with 3C ligation library and adds a MLPA (multiplex ligation-dependent probe amplification) procedure to generate a 5C library for high-throughput sequencing or microarray hybridization analysis. In the 4C assay, circular ligation products are generated either after digestion with frequent cutting secondary restriction enzyme B (different from the enzyme A used in the 3C assay) and religation or by prolonged ligation time in the 3C assay. The resulting circular DNA molecules can be amplified by inverse PCR using divergent bait-specific primers. The PCR products are characterized by microarray hybridization or by quantitative sequencing to identify the unknown interacting DNA fragments. The 6C assay combines 3C with ChIP. Cross-linked DNA is immunoprecipitated using a specific antibody followed by cloning and sequence analysis

the same set of restriction enzymes used for 3C template preparation; (3) random ligation – the restriction digested DNA is ligated at a high DNA concentration; and (4) DNA purification – the 3C control DNA contains equal amounts of all possible ligation products, allowing for normalization of PCR amplification efficiency.

The amount of ligation product in the 3C and control DNA templates is determined by qPCR. The ratio of these two yields is a quantitative measure of the frequency with which the two corresponding DNA fragments interacted inside the nucleus in vivo. The spatial organization of the chromosomal region can be reconstructed by comparing the interaction frequencies among a number of restriction fragments throughout this region. It is important to note that 3C analysis reflects only a steady-state average measurement of a large population of cells.

14.4.2 5C

The 3C–carbon copy (5C) method [17] is a strategy based on 3C to allow large-scale, locus-wide analysis. In this method (Fig. 14.1), ligation junctions are selectively amplified by multiplex LMA (ligation mediated amplification), generating a quantitative carbon copy of a part of the initial 3C library. The amplification product can then be analyzed by microarray hybridization or high-throughput sequencing. LMA is widely used to detect and amplify specific target sequences using primer pairs that anneal next to each other on the same DNA strand. Only primers annealed next to each other can be ligated. The 5C primer pairs are designed to anneal across ligated junctions of head-to-head ligation products in the 3C library generated by the conventional approach. Once 5C primer pairs are annealed next to each other, nicks are ligated by Taq ligase. This step generates a 5C library that is a quantitative "carbon copy" of a part of the 3C library. Each 5C primer is designed to contain a unique sequence corresponding to the sense or antisense strand of the 3′-ends of restriction fragments. Since predicted forward and reverse primers for each restriction fragment are complementary, only one primer, either a forward or a reverse, can be used in a given 5C experiment. The primers also contain universal tails. The 5C library is further amplified with universal PCR primers that anneal to these universal tails. The final 5C library can be used for high-throughput analysis such as microarray hybridization or sequencing. The 5C technology allows analysis of a locus from multiple fixed points, permitting reconstruction of the intricate topology of the region being studied. Nevertherless, considering the big size of genome and the required huge number of 5C primers, it remains very difficult to perform genome-wide analysis by 5C.

14.4.3 4C and QACT

The 3C, 5C, and 6C technologies have been developed to analyze the interactions of known genomic elements. To identify unknown interactors with known baits, several related strategies have recently been developed. These methods create circular ligation templates by complete ligation of restricted chromatin DNA, so they are collectively termed circular 3C (or "4C") [10, 26]. Circular DNA templates then are amplified with two inverse primers designed to anneal to the known sequence (bait) within the circle.

In principle, there are two strategies to obtain circular DNA templates (Fig. 14.1). One strategy is to further decrease the DNA concentration and extend reaction time of the ligation step of 3C to enable both ends of restriction fragments to be ligated, forming a closed circular DNA template. During this step, the DNA is still cross-linked. After reversal of cross-links and purification, inverse amplification of the circular DNA from the known region is followed.

A second strategy generates circular DNA after reversal of cross-links and purification in 3C (Fig. 14.1). Ligation junctions are digested by a frequent cutting secondary restriction enzyme to create DNA fragments ~4 kbp in length. The resulting fragments are religated under conditions that favor the formation of self-ligated circles. A specific method termed quantitative-associated chromatin trap (QACT) [16] generates the QACT library through nested inverse PCR of 3C template DNA. The QACT DNA is used to create 19/20 bp tags of the unknown RCFs. Those tags are self-ligated to form a concatemer library. In both strategies, 4C products can be analyzed by either large-scale sequencing or microarray analysis.

14.4.4 6C

A novel 6C method involves the traditional 3C, ChIP, and a cloning procedure in a process termed combined 3C–ChIP–cloning (6C) [11]. This approach is used to identify the relationship between a particular protein factor and long-range chromatin interactions. The 6C assay (Fig. 14.1) begins with the usual 3C procedure. At the end of the 3C procedure, instead of de-cross-linking, the ligated chromatin is subjected to ChIP using an antibody against a protein of interest, typically the suspected "bridging protein." After immunoprecipitation, DNA cross-linking is reversed and DNA is purified. The purified DNA is ligated into a vector-containing cohesive DNA termini compatible with those generated in the 3C assay. This allows cloning of the 3C fragments immunoprecipitated by the antibody. Clones are screened by digestion with the same restriction enzyme that was used for the 3C assay and then sequenced from the both ends of the vector. Clones containing multiple fragments suggest the identity of the interacting partners. At the same time, each step in the procedure is monitored by controls run in parallel for 3C, ChIP, and cloning steps.

14.5 Perspective

Functional genomics allows exploration of the often elaborate gene expression regulatory patterns in cells. The concept of transcriptional regulation has been extended from linear DNA components to a three-dimensional nuclear organization. With the accumulation of knowledge about gene expression regulation, chromatin "loopscapes" and the chromatin "interactome" have been proposed to explain the complex interaction network in cell nuclei. With development of these technologies and advances in systems biology, further advances may reveal the entire chromatin "loopscape" in cells, deciphering the code of the chromatin "interactome" and annotating biological function during differentiation and development.

Acknowledgments We appreciate Prof. Brian Seed and Editors for critical reading and editing.

References

1. Evans K et al (2007) A comparative study of S/MAR prediction tools. BMC Bioinformatics 8:71
2. Girod PA et al (2007) Genome-wide prediction of matrix attachment regions that increase gene expression in mammalian cells. Nat Methods 4(9):747–753
3. Bode J et al (2006) Correlations between scaffold/matrix attachment region (S/MAR) binding activity and DNA duplex destabilization energy. J Mol Biol 358(2):597–613
4. Kleinjan DA, Lettice LA (2008) Long-range gene control and genetic disease. Adv Genet 61:339–388
5. Dekker J et al (2002) Capturing chromosome conformation. Science 295(5558):1306–1311
6. Lieb JD et al (2001) Promoter-specific binding of Rap1 revealed by genome-wide maps of protein-DNA association. Nat Genet 28(4):327–334
7. Ren B et al (2000) Genome-wide location and function of DNA binding proteins. Science 290(5500):2306–2309
8. Dostie J, Dekker J (2007) Mapping networks of physical interactions between genomic elements using 5C technology. Nat Protoc 2(4):988–1002
9. Hagege H et al (2007) Quantitative analysis of chromosome conformation capture assays (3C-qPCR). Nat Protoc 2(7):1722–1733
10. Simonis M et al (2006) Nuclear organization of active and inactive chromatin domains uncovered by chromosome conformation capture-on-chip (4C). Nat Genet 38(11):1348–1354
11. Tiwari VK et al (2008) A novel 6C assay uncovers Polycomb-mediated higher order chromatin conformations. Genome Res 18(7):1171–1179
12. Bulger M, Groudine M (1999) Looping versus linking: toward a model for long-distance gene activation. Genes Dev 13(19):2465–2477
13. Blackwood EM, Kadonaga JT (1998) Going the distance: a current view of enhancer action. Science 281(5373):60–63
14. Dean A (2006) On a chromosome far, far away: LCRs and gene expression. Trends Genet 22(1):38–45
15. Carter D et al (2002) Long-range chromatin regulatory interactions in vivo. Nat Genet 32(4):623–626
16. Di LJ et al (2008) Identification of long range regulatory elements of mouse alpha-globin gene cluster by quantitative associated chromatin trap (QACT). J Cell Biochem 105(1):301–312
17. Dostie J et al (2006) Chromosome conformation capture carbon copy (5C): a massively parallel solution for mapping interactions between genomic elements. Genome Res 16(10):1299–1309
18. Schleif R (1992) DNA looping. Annu Rev Biochem 61:199–223
19. Palstra RJ et al (2003) The beta-globin nuclear compartment in development and erythroid differentiation. Nat Genet 35(2):190–194
20. Tolhuis B et al (2002) Looping and interaction between hypersensitive sites in the active beta-globin locus. Mol Cell 10(6):1453–1465

21. Zhou GL et al (2006) Active chromatin hub of the mouse alpha-globin locus forms in a transcription factory of clustered housekeeping genes. Mol Cell Biol 26(13):5096–5105
22. Kato Y, Sasaki H (2005) Imprinting and looping: epigenetic marks control interactions between regulatory elements. Bioessays 27(1):1–4
23. Lopes S et al (2003) Epigenetic modifications in an imprinting cluster are controlled by a hierarchy of DMRs suggesting long-range chromatin interactions. Hum Mol Genet 12(3):295–305
24. Murrell A, Heeson S, Reik W (2004) Interaction between differentially methylated regions partitions the imprinted genes Igf2 and H19 into parent-specific chromatin loops. Nat Genet 36(8):889–893
25. Tiwari VK et al (2008) PcG proteins, DNA methylation, and gene repression by chromatin looping. PLoS Biol 6(12):2911–2927
26. Zhao Z et al (2006) Circular chromosome conformation capture (4C) uncovers extensive networks of epigenetically regulated intra- and interchromosomal interactions. Nat Genet 38(11):1341–1347
27. Ling JQ et al (2006) CTCF mediates interchromosomal colocalization between Igf2/H19 and Wsb1/Nf1. Science 312(5771):269–272
28. Lomvardas S et al (2006) Interchromosomal interactions and olfactory receptor choice. Cell 126(2):403–413
29. Bacher CP et al (2006) Transient colocalization of X-inactivation centres accompanies the initiation of X inactivation. Nat Cell Biol 8(3):293–299
30. Kurukuti S et al (2006) CTCF binding at the H19 imprinting control region mediates maternally inherited higher-order chromatin conformation to restrict enhancer access to Igf2. Proc Natl Acad Sci USA 103(28):10684–10689
31. Wurtele H, Chartrand P (2006) Genome-wide scanning of HoxB1-associated loci in mouse ES cells using an open-ended chromosome conformation capture methodology. Chromosome Res 14(5):477–495
32. Cai S, Lee CC, Kohwi-Shigematsu T (2006) SATB1 packages densely looped, transcriptionally active chromatin for coordinated expression of cytokine genes. Nat Genet 38(11):1278–1288
33. Spilianakis CG, Flavell RA (2004) Long-range intrachromosomal interactions in the T helper type 2 cytokine locus. Nat Immunol 5(10):1017–1027
34. Spilianakis CG et al (2005) Interchromosomal associations between alternatively expressed loci. Nature 435(7042):637–645
35. Horike S et al (2005) Loss of silent-chromatin looping and impaired imprinting of DLX5 in Rett syndrome. Nat Genet 37(1):31–40
36. Kumar PP et al (2007) Functional interaction between PML and SATB1 regulates chromatin-loop architecture and transcription of the MHC class I locus. Nat Cell Biol 9(1):45–56
37. Ottaviani D et al (2008) Reconfiguration of genomic anchors upon transcriptional activation of the human major histocompatibility complex. Genome Res 18(11):1778–1786
38. Filippova GN (2008) Genetics and epigenetics of the multifunctional protein CTCF. Curr Top Dev Biol 80:337–360
39. Galande S et al (2007) The third dimension of gene regulation: organization of dynamic chromatin loopscape by SATB1. Curr Opin Genet Dev 17(5):408–414
40. Cai S, Han HJ, Kohwi-Shigematsu T (2003) Tissue-specific nuclear architecture and gene expression regulated by SATB1. Nat Genet 34(1):42–51
41. Yamasaki K et al (2007) Structural basis for recognition of the matrix attachment region of DNA by transcription factor SATB1. Nucleic Acids Res 35(15):5073–5084
42. Galande S et al (2001) SATB1 cleavage by caspase 6 disrupts PDZ domain-mediated dimerization, causing detachment from chromatin early in T-cell apoptosis. Mol Cell Biol 21(16):5591–5604
43. Purbey PK et al (2008) PDZ domain-mediated dimerization and homeodomain-directed specificity are required for high-affinity DNA binding by SATB1. Nucleic Acids Res 36(7):2107–2122

44. Yasui D et al (2002) SATB1 targets chromatin remodelling to regulate genes over long distances. Nature 419(6907):641–645
45. Pavan Kumar P et al (2006) Phosphorylation of SATB1, a global gene regulator, acts as a molecular switch regulating its transcriptional activity in vivo. Mol Cell 22(2):231–243
46. Purbey PK et al (2009) Acetylation-dependent interaction of SATB1 and CtBP1 mediates transcriptional repression by SATB1. Mol Cell Biol 29(5):1321–1337
47. Wen J et al (2005) SATB1 family protein expressed during early erythroid differentiation modifies globin gene expression. Blood 105(8):3330–3339
48. Wang L et al (2009) Inter-MAR association contributes to transcriptionally active looping events in human beta-globin gene cluster. PLoS ONE 4(2):e4629
49. Brown CR et al (2008) Global histone acetylation induces functional genomic reorganization at mammalian nuclear pore complexes. Genes Dev 22(5):627–639
50. Yusufzai TM et al (2004) CTCF tethers an insulator to subnuclear sites, suggesting shared insulator mechanisms across species. Mol Cell 13(2):291–298
51. Lewis A, Murrell A (2004) Genomic imprinting: CTCF protects the boundaries. Curr Biol 14(7):R284–R286
52. Anguita E et al (2004) Globin gene activation during haemopoiesis is driven by protein complexes nucleated by GATA-1 and GATA-2. Embo J 23(14):2841–2852
53. Du MJ et al (2008) MafK/NF-E2 p18 is required for beta-globin genes activation by mediating the proximity of LCR and active beta-globin genes in MEL cell line. Int J Biochem Cell Biol 40(8):1481–1493
54. Kooren J et al (2007) Beta-globin active chromatin Hub formation in differentiating erythroid cells and in p45 NF-E2 knock-out mice. J Biol Chem 282(22):16544–16552
55. Vakoc CR et al (2005) Proximity among distant regulatory elements at the beta-globin locus requires GATA-1 and FOG-1. Mol Cell 17(3):453–462
56. Li T et al (2008) CTCF regulates allelic expression of Igf2 by orchestrating a promoter-polycomb repressive complex 2 intrachromosomal loop. Mol Cell Biol 28(20):6473–6482
57. Yasui DH et al (2007) Integrated epigenomic analyses of neuronal MeCP2 reveal a role for long-range interaction with active genes. Proc Natl Acad Sci USA 104(49):19416–19421
58. Mattick JS, Makunin IV (2006) Non-coding RNA. Hum Mol Genet 15(1):R17–R29
59. Mainguy G et al (2007) Extensive polycistronism and antisense transcription in the Mammalian Hox clusters. PLoS ONE 2(4):e356
60. Katayama S et al (2005) Antisense transcription in the mammalian transcriptome. Science 309(5740):1564–1566
61. Xu N et al (2007) Evidence that homologous X-chromosome pairing requires transcription and Ctcf protein. Nat Genet 39(11):1390–1396
62. Zhao J et al (2008) Polycomb proteins targeted by a short repeat RNA to the mouse X chromosome. Science 322(5902):750–756
63. Mohammad F et al (2008) Kcnq1ot1/Lit1 noncoding RNA mediates transcriptional silencing by targeting to the perinucleolar region. Mol Cell Biol 28(11):3713–3728
64. Boumil RM, Lee JT (2001) Forty years of decoding the silence in X-chromosome inactivation. Hum Mol Genet 10(20):2225–2232
65. Clerc P, Avner P (2003) Multiple elements within the Xic regulate random X inactivation in mice. Semin Cell Dev Biol 14(1):85–92
66. Vernimmen D et al (2007) Long-range chromosomal interactions regulate the timing of the transition between poised and active gene expression. Embo J 26(8):2041–2051
67. Fields PE et al (2004) Th2-specific chromatin remodeling and enhancer activity in the Th2 cytokine locus control region. Immunity 21(6):865–876
68. Lee GR et al (2003) Regulation of the Th2 cytokine locus by a locus control region. Immunity 19(1):145–153
69. Eivazova ER, Aune TM (2004) Dynamic alterations in the conformation of the Ifng gene region during T helper cell differentiation. Proc Natl Acad Sci USA 101(1):251–256

70. Liu Z, Garrard WT (2005) Long-range interactions between three transcriptional enhancers, active Vkappa gene promoters, and a 3′ boundary sequence spanning 46 kilobases. Mol Cell Biol 25(8):3220–3231
71. Hakim O et al (2009) Glucocorticoid receptor activation of the Ciz1-Lcn2 locus by long range interactions. J Biol Chem 284(10):6048–6052
72. Tsai CL et al (2008) Higher order chromatin structure at the X-inactivation center via looping DNA. Dev Biol 319(2):416–425

Index

M

N

O

P